Mahmoud Atef Abdulhamid (Ed.)
Polymer Membranes

Also of interest

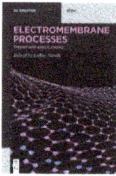

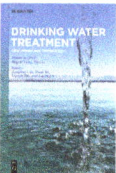

Polymer Membranes

Increasing Energy Efficiency

Edited by
Mahmoud Atef Abdulhamid

DE GRUYTER

Editor
Dr. Mahmoud Atef Abdulhamid
Center of Integrative Petroleum Research
College of Petroleum and Geosciences
King Fahd University of Petroleum and Minerals
Dhahran 31261
Saudi Arabia
Mahmoud.abdulhamid@kfupm.edu.sa

ISBN 978-3-11-079599-8
e-ISBN (PDF) 978-3-11-079603-2
e-ISBN (EPUB) 978-3-11-079613-1

Library of Congress Control Number: 2024930591

Bibliographic information published by the Deutsche Nationalbibliothek
The Deutsche Nationalbibliothek lists this publication in the Deutsche Nationalbibliografie;
detailed bibliographic data are available on the internet at http://dnb.dnb.de.

© 2024 Walter de Gruyter GmbH, Berlin/Boston
Cover image: Jian Fan/iStock/Getty Images Plus
Typesetting: Integra Software Services Pvt. Ltd.
Printing and binding: CPI books GmbH, Leck

www.degruyter.com

Preface

Polymer membranes are thin, selective barriers that can be used to separate different substances from a mixture. They are typically made from synthetic polymers such as polysulfone, polyethersulfone, and polyimides or polymers existing in nature like cellulose and chitosan. One of the key advantages of polymer membranes is their versatility. They can be designed to have a wide range of properties, such as pore size, permeability, and selectivity. This makes them suitable for a wide range of applications. Polymer membranes are considered relatively inexpensive, easy to fabricate, and energy-efficient which makes them a cost-effective solution for many separation problems. In recent years, there has been a growing interest in the development of new polymer membranes with improved performance and functionality. This is being driven by the need for more efficient and sustainable separation processes in a variety of industries. Up-to-date, polymer membranes have been employed in several applications including gas separation, water treatment, fuel cells, desalination, pervaporation, organic solvent nanofiltration, and catalysis. So far, there is no reference that cover various applications of polymer membranes. Therefore, this book will act as a reference document to researchers and students working or interested in the research area of polymer membranes.

The book is divided into the following 12 chapters:

1. In Chapter 1, the development of polyimide-based membranes from porous, nonporous, functionalized, and nonfunctionalized polyimides for gas separation applications is reviewed with particular emphasize method for polyimide synthesis and pure gas separation experiments.
2. In Chapter 2, natural gas sweetening and hydrogen sulfide removal using polymer membranes is reviewed. The chapter demonstrates the potential of utilizing polymer membranes to reduce the H_2S percentage in gas feeds before reaching the amine scrubbing system.
3. In Chapter 3, the different membrane fabrication methods, their advantages, and their disadvantages for water treatment application is described. A short description of the application of 3D printing technologies in membrane fabrication is also added. The comparison of properties and performance of 3D printing membranes with membranes prepared from conventional methods is also discussed.
4. In Chapter 4, the challenges that lie in enhancing the proton conductivity of natural polymers in both normal and high operating temperatures while conserving their mechanical strength and durability are outlined.
5. In Chapter 5, the different types of polymeric membranes that are used for catalysis applications, and the recent advances in the development of new polymeric membranes with improved catalytic performance, are reviewed.
6. In Chapter 6, a comprehensive overview of pervaporation, encompassing principles, polymer materials, membrane synthesis and modification, characterization techniques, transport mechanisms, performance evaluation, and recent advances is reviewed.

https://doi.org/10.1515/9783110796032-202

7. In Chapter 7, a comprehensive explanation of pervaporation, including the underlying principles and separation mechanisms of polymer membranes, as well as the various factors that impact their performance is reviewed.
8. In Chapter 8, the recent developments in electrospun nanofibrous membranes fabrication and their environmental applications including water treatment, catalysis, and air filtration are reviewed.
9. In Chapter 9, the key properties and advantages of polymer membranes in desalination and distillation, including their high selectivity, scalability, and potential for customization, are reviewed. Additionally, this chapter highlights recent advancements in membrane technology, such as the integration of novel materials, surface modifications, and innovative module designs.
10. In Chapter 10, the advancement in oil/water emulsion separation and fractionation using polymer membranes are reviewed. Various attempts for degumming and deacidification using membrane technology have also been covered.
11. In Chapter 11, the recent development of polymer membranes for organic solvent nanofiltration and the current challenges and limitations are reviewed. The chapter also addresses important key points that need to be considered for developing high-performance membranes for OSN.
12. In Chapter 12, the development of interpenetrating polymer networks (IPNs) and their applications in water treatment, gas separation, nonaqueous separation, and batteries are elaborated. The summary of the current state of IPNs-based membranes along with future outlooks is discussed.

As we grapple with increasingly complex global challenges, the role of advanced materials and technologies like polymer membranes becomes ever more crucial. This book has aimed to provide a comprehensive overview, from the molecular to the macroscopic, that scholars, industry professionals, and policymakers may find valuable. As the adage goes, "The best way to predict the future is to invent it." In the realm of polymer membranes for various applications, the tools for invention are more accessible than ever. It is upon the collective endeavors of scientists, engineers, and decision-makers to wield these tools effectively, steering us toward a future that is not just sustainable but also equitable and prosperous. The urgency of the challenges we face is matched only by the potential of the technologies we are developing. As we look forward, let's harness this potential to its fullest, ensuring that the next chapter in the story of polymer membranes is one of unequivocal success.

Acknowledgments

The editor would like to thank and express his appreciation to the authors of the chapters, who supported this effort in a professional and timely manner. The editor acknowledges College of Petroleum Engineering and Geoscience, King Fahd University of Petroleum and Minerals for the continuous support.

https://doi.org/10.1515/9783110796032-203

Contents

About the editor

Dr. Abdulhamid currently holds the position of Research Assistant Professor and serves as the Principal Investigator at the "Sustainable and Resilient Materials Lab" within the Center for Integrative Petroleum Research (CIPR), College of Petroleum Engineering and Geoscience (CPG) at King Fahd University of Petroleum and Minerals (KFUPM) in Saudi Arabia. With a background in organic and polymer chemistry, his primary focus is on the development of polymers and membranes tailored for separation applications. Dr. Abdulhamid earned his Ph.D. in Chemical Science from King Abdullah University of Science and Technology (KAUST) in 2019 and subsequently spent three years as a postdoc at KAUST, dedicated to advancing polymeric membranes for energy-efficient separations. His ongoing research endeavors encompass the development of polymers for diverse applications such as water treatment, gas separation, organic solvent nanofiltration, desalination, corrosion inhibition, and enhanced oil recovery. Dr. Abdulhamid's scholarly contributions include over 30 publications in reputable journals, and he has demonstrated innovation as an inventor and co-inventor on several patents.

https://doi.org/10.1515/9783110796032-205

List of contributors

Chapter 1
Zainah A. Aldhawi
Sustainable and Resilient Materials Lab
Center for Integrative Petroleum Research (CIPR)
College of Petroleum Engineering & Geosciences
(CPG)
King Fahd University of Petroleum and Minerals
Dhahran 31261
Saudi Arabia

Ibtisam I. BinSharfan
Sustainable and Resilient Materials Lab
Center for Integrative Petroleum Research (CIPR)
College of Petroleum Engineering & Geosciences
(CPG)
King Fahd University of Petroleum and Minerals
Dhahran 31261
Saudi Arabia

Mahmoud A. Abdulhamid
Sustainable and Resilient Materials Lab
Center for Integrative Petroleum Research (CIPR)
College of Petroleum Engineering & Geosciences
(CPG)
King Fahd University of Petroleum and Minerals
Dhahran 31261
Saudi Arabia

Chapter 2
Ali Hayek
Research and Development Center
Saudi Aramco
Saudi Arabia
email: ali.hayek.2@aramco.com

Chapter 3
Jisha Kuttiani Ali
Chemical Engineering Department
Khalifa University of Science and Technology
Abu Dhabi
United Arab Emirates
And
Center for Membranes and Advanced Water
Technology
Khalifa University of Science and Technology

Abu Dhabi
United Arab Emirates

Emad Alhseinat
Chemical Engineering Department
Khalifa University of Science and Technology
Abu Dhabi
United Arab Emirates
And
Center for Membranes and Advanced Water
Technology
Khalifa University of Science and Technology
Abu Dhabi
United Arab Emirates

Chapter 4
Abdullah Ali
Department of Chemical and Biological
Engineering
American University of Sharjah
P.O. Box 26666
Sharjah
United Arab Emirates

Amani Al-Othman
Department of Chemical and Biological
Engineering
American University of Sharjah
P.O. Box 26666
Sharjah
United Arab Emirates
email: aalothman@aus.edu

Muhammad Tawalbeh
Sustainable and Renewable Energy Engineering
Department
University of Sharjah
P.O. Box 27272
Sharjah
United Arab Emirates
And
Sustainable Energy and Power Systems Research
Centre
RISE
University of Sharjah

https://doi.org/10.1515/9783110796032-206

P.O. Box 27272
Sharjah
United Arab Emirates

Chapter 5
Wafa Suwaileh
Chemical Engineering Program
Texas A&M University at Qatar
Education City
Doha
Qatar
email: wafa.suwaileh@qatar.tamu.edu

Soheil Zarghami
Department of Chemical Engineering
Isfahan University of Technology
Iran

Saima Farooq
Department of Biological Sciences and
Chemistry
College of Arts and Science
University of Nizwa
Oman

Nipa Roy
Department of Physics
Yeungnam University
Gyeongsan 38541
Republic of Korea

Mohammad Boshir Ahmed
School of Engineering
Edith Cowan University
Joondalup
Australia

Chapter 6
Neelesh Ashok
Amrita School of Artificial Intelligence
Amrita Vishwa Vidyapeetham
Coimbatore, Tamil Nadu
India
email: neeleshashok@gmail.com

Sruthi M. S.
Amrita School of Artificial Intelligence
Amrita Vishwa Vidyapeetham

Coimbatore, Tamil Nadu
India

Taniya Rose Abraham
School of Energy Materials
Mahatma Gandhi University
Kottayam

Sabu Thomas
School of Energy Materials
Mahatma Gandhi University
Kottayam, Kerala
India

Chapter 7
Asmaa Selim
Renewable Energy Research Group
Institute of Materials and Environmental
Chemistry
Research Centre for Natural Science
1117 Budapest
Hungary
And
Chemical Engineering and Pilot Plat Department
Engineering and Renewable Energy Research
Institute National Research Centre
12622 Giza
Egypt

Chapter 8
Sadia Bano
Department of Chemistry
Lahore college for Women university
Lahore
Pakistan

Muhammad Altaf
Department of Chemistry
Chiang Mai University
Chiang Mai
Thailand

Sajda Bano
Department of Chemistry
University of Punjab
Lahore
Pakistan

Chapter 9
Neelesh Ashok
Amrita School of Artificial Intelligence
Amrita Vishwa Vidyapeetham
Coimbatore, Tamil Nadu
India
email: neeleshashok@gmail.com

Sruthi M. S.
Amrita School of Artificial Intelligence
Amrita Vishwa Vidyapeetham
Coimbatore, Tamil Nadu
India

Taniya Rose Abraham
School of Energy Materials
Mahatma Gandhi University
Kottayam, Kerala
India

Sabu Thomas
School of Energy Materials
Mahatma Gandhi University
Kottayam, Kerala
India

Chapter 10
Hind Yaacoubi
Sustainable and Resilient Materials Lab
Center for Integrative Petroleum Research (CIPR)
College of Petroleum Engineering & Geosciences (CPG)
King Fahd University of Petroleum and Minerals
Dhahran 31261
Saudi Arabia
email: Hindyaacoubi@outlook.fr

Mahmoud A. Abdulhamid
Sustainable and Resilient Materials Lab
Center for Integrative Petroleum Research (CIPR)
College of Petroleum Engineering & Geosciences (CPG)
King Fahd University of Petroleum and Minerals
Dhahran 31261
Saudi Arabia

Chapter 11
Mahmoud A. Abdulhamid
Sustainable and Resilient Materials Lab
Center for Integrative Petroleum Research (CIPR)
College of Petroleum Engineering & Geosciences (CPG)
King Fahd University of Petroleum and Minerals
Dhahran 31261
Saudi Arabia
email: Mahmoud.abdulhamid@kfupm.edu.sa

Chapter 12
Rifan Hardian
Advanced Membranes and Porous Materials Center
Physical Science and Engineering Division
King Abdullah University of Science and Technology (KAUST)
Thuwal 23955-6900
Saudi Arabia

Diana G. Oldal
Advanced Membranes and Porous Materials Center
Physical Science and Engineering Division
King Abdullah University of Science and Technology (KAUST)
Thuwal 23955-6900
Saudi Arabia

Zulfida Mohamad Hafis Mohd Shafie
Chemical Engineering and Biotechnology
School of Chemistry
Nanyang Technological University
62 Nanyang Drive
637459, Singapore
And
School of Chemical Engineering
Universiti Sains Malaysia
Penang
Malaysia
And
Laboratoire Réactions & Génie des Procédés
Université de Lorraine
Nancy
France

Mahmoud A. Abdulhamid
Sustainable and Resilient Materials Lab
Center for Integrative Petroleum Research (CIPR)
College of Petroleum Engineering & Geosciences
(CPG)
King Fahd University of Petroleum and Minerals
Dhahran 31261
Saudi Arabia

Gyorgy Szekely
Advanced Membranes and Porous Materials
Center
Physical Science and Engineering Division
King Abdullah University of Science and
Technology (KAUST)
Thuwal 23955-6900
Saudi Arabia

Zainah A. Aldhawi, Ibtisam I. BinSharfan, Mahmoud A. Abdulhamid*

Chapter 1
Polyimide-based membranes for gas separation applications

Abstract: Polymer membranes have emerged as a promising technology for gas separation in a variety of industrial and environmental applications. They offer a number of advantages over traditional methods, such as cryogenic distillation and absorption, including lower energy consumption, smaller footprint, and easier operation. Polymer membranes work by selectively permeating different gases at different rates. This is achieved by designing membranes with specific properties, such as pore size and surface chemistry. The most common types of polymer membranes for gas separation are dense and hollow fiber membranes. Dense membranes separate gases based on their solubility and diffusion in the polymer matrix, while porous membranes separate gases based on their molecular size and shape. Polymer membranes are used in a wide range of gas separation applications, including natural gas sweetening, air separation, hydrogen recovery, and carbon capture and storage. One of the key challenges in developing polymer membranes for gas separation is to achieve a balance between permeability and selectivity that overcomes the trade-off behavior. Researchers are developing new polymer membranes with improved permeability and selectivity through a variety of approaches, such as designing new polymer materials with tailored properties, developing composite membranes that combine different polymers or materials, and modifying the surface of polymer membranes to improve their separation performance. With continued research and development, polymer membranes are expected to become even more efficient and cost-effective, making them even more attractive for a wider range of applications. In this chapter, we review the involvement of polyimides in the preparation of dense membranes for gas separation. The chapter discusses the synthesis of polyimides and their gas separation performance.

Keywords: polyimides, membrane, gas separation, porous polymer

Acknowledgment: This work was supported by the College of Petroleum Engineering and Geoscience (CPG), King Fahd University of Petroleum and Minerals (KFUPM).

*Corresponding author: Mahmoud A. Abdulhamid**, Sustainable and Resilient Materials Lab, Center for Integrative Petroleum Research (CIPR), College of Petroleum Engineering and Geosciences (CPG), King Fahd University of Petroleum and Minerals, Dhahran 31261, Saudi Arabia, e-mail: mahmoud.abdulhamid@kfupm.edu.sa
Zainah A. Aldhawi, Ibtisam I. BinSharfan, Sustainable and Resilient Materials Lab, Center for Integrative Petroleum Research (CIPR), College of Petroleum Engineering and Geosciences (CPG), King Fahd University of Petroleum and Minerals, Dhahran, 31261, Saudi Arabia

https://doi.org/10.1515/9783110796032-001

1 Introduction

Membrane gas separation is a contemporary technology closely associated with numerous large-scale industrial applications. This technology is extensively utilized in various processes, such as natural gas purification, petrochemical and hydrogen recovery plants, the nitrogen separation industry, carbon capture and storage, and helium separation [1]. Monsanto Prism developed the first industrial membrane technique for gas separation in 1977 to modify the syngas (CO/H$_2$) ratio for a petrochemical process using polysulfone-based membranes. Later, they built the first membrane separation unit for hydrogen recovery from an ammonia purge gas stream in 1978. Despite the initial success of prism membranes, the expansion of membrane-based gas separation has been hindered by the lack of economically feasible methods for manufacturing high-performance membranes and designing effective modules [2, 3].

Recently, membrane-based separation technology as a promising separation technology has been widely used due to its high efficiency, low energy consumption, and eco-friendliness. The impetus for this development was the search for alternatives to traditional, energy-consuming gas separation processes, e.g., absorption and cryogenic distillation. These methods necessitate large, elaborate chemical plants and can be costly to operate and maintain. Table 1 compares traditional and membrane-based separation methods. Additionally, roughly 50% of industrial processes in the United States involve chemical separation techniques, with distillation accounting for approximately 50% of these processes. However, distillation is responsible for 8% of the total energy consumption in the industry. In contrast, membrane-based separation techniques can potentially consume up to 90% less energy than distillation (Figure 1) [4].

The synthetic membrane is defined as a selective thin layer with varied thicknesses and morphologies between two phases, allowing certain molecules to pass through while preventing others from doing so [6]. In the gas separation process, a gradient pressure is applied to a membrane, which lets the gases dissolve and diffuse through them, causing mass transfer across the film. Using membranes, different gas species can be separated from one another by pressure-driven variations in solubility and diffusivity. Later, the nonpermeating molecules are held on the feed stream side before being released from the membrane unit as the retentate stream (Figure 2). The performance of the membrane in gas separation is greatly influenced by the selection of materials used in the membrane fabrication [7].

Membranes can be classified based on their porosity. There are three different types of pore sizes based on the International Union of Pure and Applied Chemistry classification: microporous ($d_p < 2$ nm), mesoporous (2 nm $< d_p > 50$ nm), and macroporous ($d_p > 50$ nm) [8, 9]. However, membranes can also be classified based on the type of material utilized, with two broad categories being inorganic and polymeric membranes. Inorganic membranes typically include zeolites [10] (e.g., T [11], CHA [12], and MOR [13]), ceramics [14] (e.g., alumina–silica composite [15], zirconia–silica composite [16], titanium–silica composite [17], and perovskite oxide type [18]), Pd–metal-based

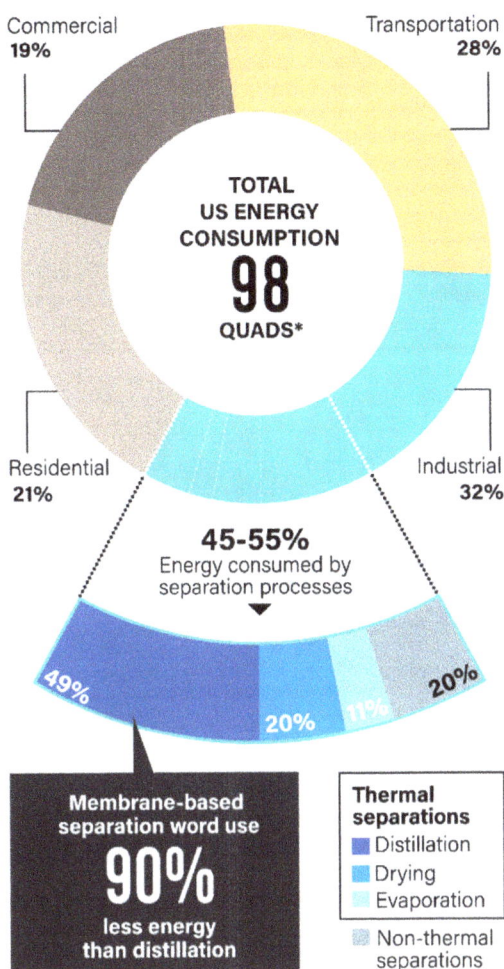

Figure 1: Total energy consumption by sector [4].

Table 1: Comparison between traditional methods and membrane-based separation methods [5].

Technology	Advantages	Disadvantages
Adsorption	Less energy is required for solvent regeneration, less corrosion, low cost, and easy to handle	Low adsorbent recovery, high operation coast, low capacity, temperature and pressure limitation, and low efficiency
Absorption	High product purity and high recovery	Less energy is required for solvent regeneration, complex step processes, equipment corrosion, and environmental hazards

Table 1 (continued)

Technology	Advantages	Disadvantages
Cryogenic distillation	High product purity	Highly energy-intensive process used only for high-concentration gases
Membrane-based gas separation	Suitable for bulk separation, low capital cost, easy and straightforward process, compatible with many techniques, easy to scale-up, less energy-intensive, and low toxicity	Moderate purity

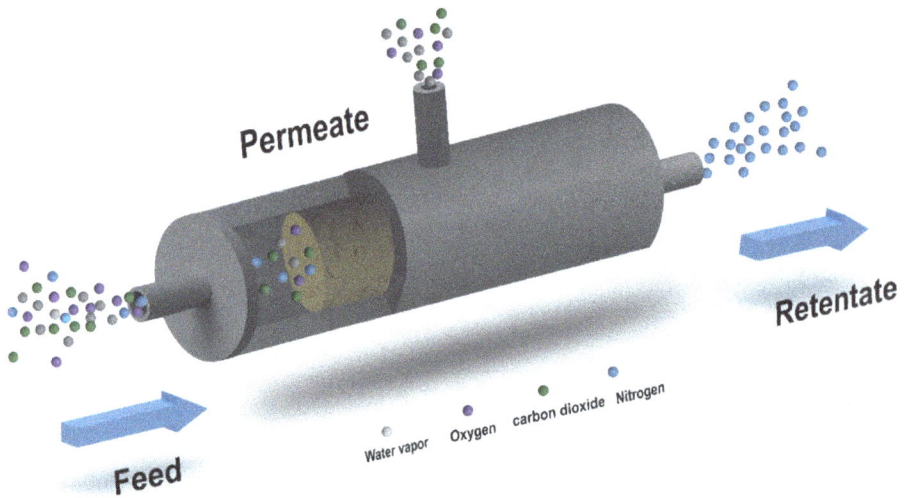

Figure 2: Gas diffusion principle in membranes.

membranes [19], and carbon membrane materials (e.g., graphene [20], graphene oxide [21], and carbon nanotube [22]). Inorganic membranes are well known for their high thermal stability, high permeability, and selectivity. On the other hand, the main drawbacks of these materials are the difficulty of producing defect-free thin films, high capital cost, manufacturing cost, poor resistance to poisoning, and low stability for the carbon-based membranes. During the last decade, substantial research efforts have been devoted to polymeric membranes to overcome the drawbacks of inorganic membranes. Table 2 lists a comparison between polymeric membranes and inorganic membranes. The two main critical parameters that are used to evaluate a given membrane's performance are permeability and selectivity. These parameters directly impact productivity and purity and are therefore crucial in determining the suitability of the membrane for specific applications. For the polymeric membranes, a trade-off between permeability and selectivity has been first investigated by Robeson's work in 1991 [23] and subsequent updates in 2008 [24]. He highlighted the progress made in polymer membranes for main gas pair separation processes such as O_2/N_2, CO_2/CH_4, CO_2/N_2, and H_2/CH_4. This study indicated, in short, that the more permeable membranes are usually less selective and vice versa (Figure 3). The trade-off limitations of polymeric membranes have motivated researchers in the membrane community to develop new and robust materials that can surpass this limitation by simultaneously achieving high permeability and selectivity.

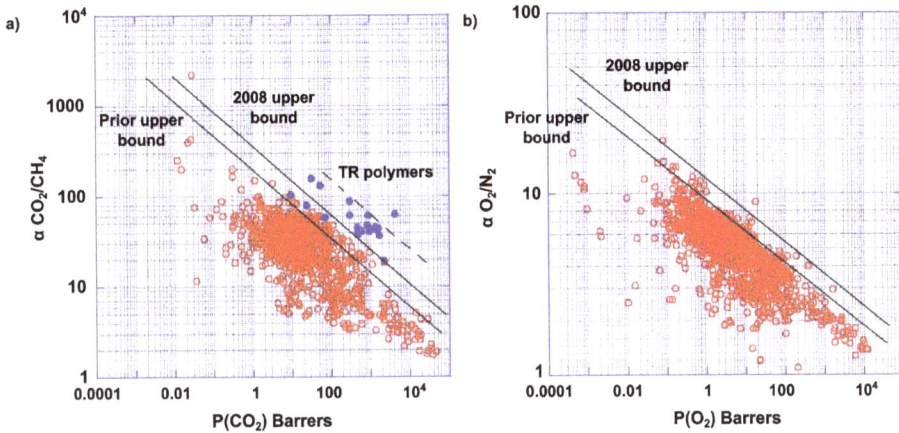

Figure 3: Upper bound correlation for different gas pair separation from Robeson's work in 2008.

Table 2: Polymeric membranes versus inorganic membranes [5, 25].

Membrane	Polymeric membrane	Inorganic membranes
Advantages	Low capital cost, handy fabrication processes and purification methods, and good selectivity	Long lifetime, excellent thermal and chemical stability, exhibit catalytic property, and suitable to use under aggressive conditions
Disadvantages	Pore thermal and thermal resistance, trade-off between permeability and selectivity	High capital cost, difficult to operate in a bulky organic environment, easily damaged, poor processability, and difficult to synthesize membrane with a large surface area

A suitable membrane candidate for gas separation should have the following properties: (1) medium to high permeability/flux; (2) high selectivity for the desired gas pairs; (3) environment stability and can tolerate some impurities in the feed gas; (4) mechanically robust and can be easily scaled up for mass production; (5) good thermal stability; (6) producing reproducibility, properties stable from batch to batch; (7) easy packing and can be easily embedded to the module; and (8) acceptable costs [26].

As a class of high-performance and promising membrane separation material, polyimide (PI) has been studied and applied to gas separation since the early 1960s [27].

During the last few decades, researchers have focused on designing new PIs and modifying existing materials to fit gas separation applications (Figure 4). A huge number of growing interests have been raised due to the use of PI-based membranes for gas separation due to their outstanding properties compared to other materials.

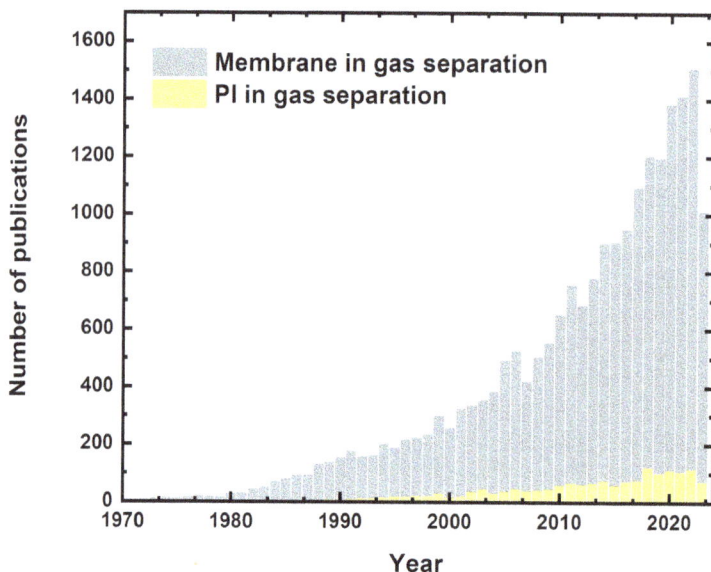

Figure 4: Number of publications per year on polyimide-based membrane for gas separation from 1980 to 2023 obtained from Scopus using the following keywords: gas separation, polyimides, and membrane.

2 Polyimide

In 1908, the first aromatic PI was synthesized and reported by Bogert and Renshawin [28, 29]. In the past, PIs had limited applications due to their lack of processability via melt polymerization [30]. However, DuPont's landmark discovery of Kapton in 1965, which was the first commercialized aromatic PI, marked a significant development in PI synthesis and processing [30]. On the other hand, the first aliphatic PI (API) was reported by Hirsch in 1971. This was followed by subsequent reports from Nakanishi in 1973, Dror in 1975, and Abajo in 1978 [29]. Since then, PIs have received great attention due to their unique and outstanding properties such as high thermal stability, superior chemical resistance, excellent electrical performance, and good mechanical properties which lead to their application in several robust fields, including gas separation, gasification, aviation, microsystems, aerospace, membranes, fuel cells, batteries, memory devices, sensors, optical devices, biomedical applications, hydrogels, and polymer matrices in composites/hybrid materials [27, 29, 31–35].

Aromatic PIs have received great attention in gas separation fields due to their unique and remarkable properties, such as thermal, mechanical, and chemical stability, and high gas permselectivity for different gases. The strong and rigid imide groups present on planar aromatic rings in the main chains of many PIs make them poorly soluble in common organic solvents due to the formation of strong intermolecular charge-transfer complexes (CTCs) between the aromatic rings. In addition, these imide groups can sometimes lead to poor gas permeability. As a result, the use of many PIs in the fabrication of high-performance membranes with superior separation properties is limited by these drawbacks [36–38]. There are only a few soluble aromatic PIs, such as Upilex® and P84®, and Matrimid®, which have been used for commercialization in the form of hollow fiber-type gas separation membranes by companies such as Air Liquide Co., Ube Industry, and Evonic Co. [39–41].

The solubility of aromatic PIs in organic solvents can be improved by introducing bulky or flexible groups, such as carbonyl, ether, and hexafluoroisopropylidene (6F) linkages, into the polymer backbone. This reduces the formation of intermolecular CTCs between the aromatic rings. Although PIs based on 4,4'-(hexafluoroisopropylidene)diphthalic anhydride (6FDA) monomers have been found to exhibit high selectivity and permeability for various gases, their high cost limits their commercial viability. Another potential approach to increase the solubility of aromatic PIs is to incorporate asymmetric nonpolar alicyclic structures into the polymer backbone, which can enhance solubility without significantly reducing the chemical resistance or mechanical properties of PI materials [37, 42, 43].

3 Synthesis of polyimides

The process of synthesizing PIs and using monomers is an essential factor in producing high-performance PIs that are suitable for various industrial applications. Typically, PIs are synthesized through a polycondensation reaction between organic diamines, which act as strong nucleophiles and tetracarboxylic acid dianhydride monomers, which are strong acceptor units. There are mainly two routes practiced in PI synthesis: one-step and two-step polymerization methods [44–47].

3.1 One-step polymerization method

A one-step synthetic procedure can be used to obtain PI without the need for a two-step synthesis method. The polycondensation reaction involves mixing equimolar amounts of dianhydride and diamine in a high boiling point aprotic solvent. The mixture is gradually heated to 120 °C, and a catalyst such as isoquinoline, tertiary amines, alkali metals, and zinc salts of carboxylic acids is added, followed by gradual heating up to 200 °C to ensure total imidization [48, 49]. Water produced during the reaction can be removed from the system by continuously flushing nitrogen gas through it or by using azeotropic agents. Common high boiling point solvents like nitrobenzene, *m*-cresol, *o*-dichlorobenzene, *p*-chlorophenol, *N*-methyl-2-pyrrolidone (NMP), dimethyla-cetamide, and α-chloronaphthalene are utilized [50] for this direct conversion. Unlike the two-stage procedure, isolating polyamic acid (PAA) is not necessary. Also, this method produces materials with a higher degree of crystallinity, which may be due to the increased solubility of the products in the solvent medium [29]. Nuclear magnetic resonance and Fourier transform infrared were used to confirm the complete imidization. The one-step polymerization method achieves 100% conversion without any detected defects. Furthermore, no isoimide was produced using this procedure [51]. Scheme 1 shows a generalized synthesis of PI from dianhydride and diamine.

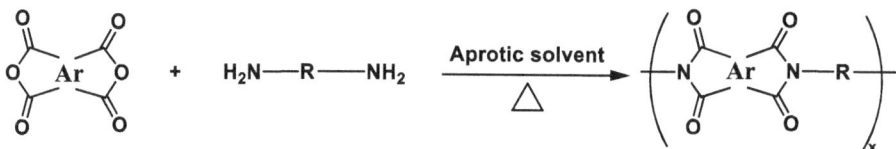

Scheme 1: One-step synthesis of polyimide from dianhydride and diamine.

Scheme 2 shows the common structures of aliphatic monomers that are utilized for the synthesis of APIs in previous studies [52–60]. The commercial monomers are dianhydrides A and B, and all diamines except 12 and 13, whereas dianhydrides C, D, E, and F are synthesized following the procedures described in the previous studies. Dianhydrides A and B have been found to impart good thermal and optical properties to PIs. However, the presence of a double bond in B can cause increased polarity, which reduces transparency so it is oxidized to C to obtain more utile results. Dianhydrides F and G are asymmetric, providing more free volume and decreasing interchain interactions. Mathews et al. investigated the effects of highly flexible and long-chain backbones in PI using diamines 1–4. Diamines 5 and 6 are used for the formation of *ortho*- and *para*-substitution of the imide group. Diamine 7 has a non-coplanar structure that may reduce the dielectric constant. Diamines 9, 10, and 11 are used to compare the properties with the presence of methyl groups. The structures of the diamines utilized to synthesize APIs with fluorine and siloxane groups are presented as 12–15. Those diamines can be used to study the effect of fluorine and siloxane moieties on API.

3.2 Two-step method for polyimide synthesis

The two-step polymerization process is the most common method that is used in the production of PIs [61]. This approach is suitable for a wide range of fully aromatic PIs and

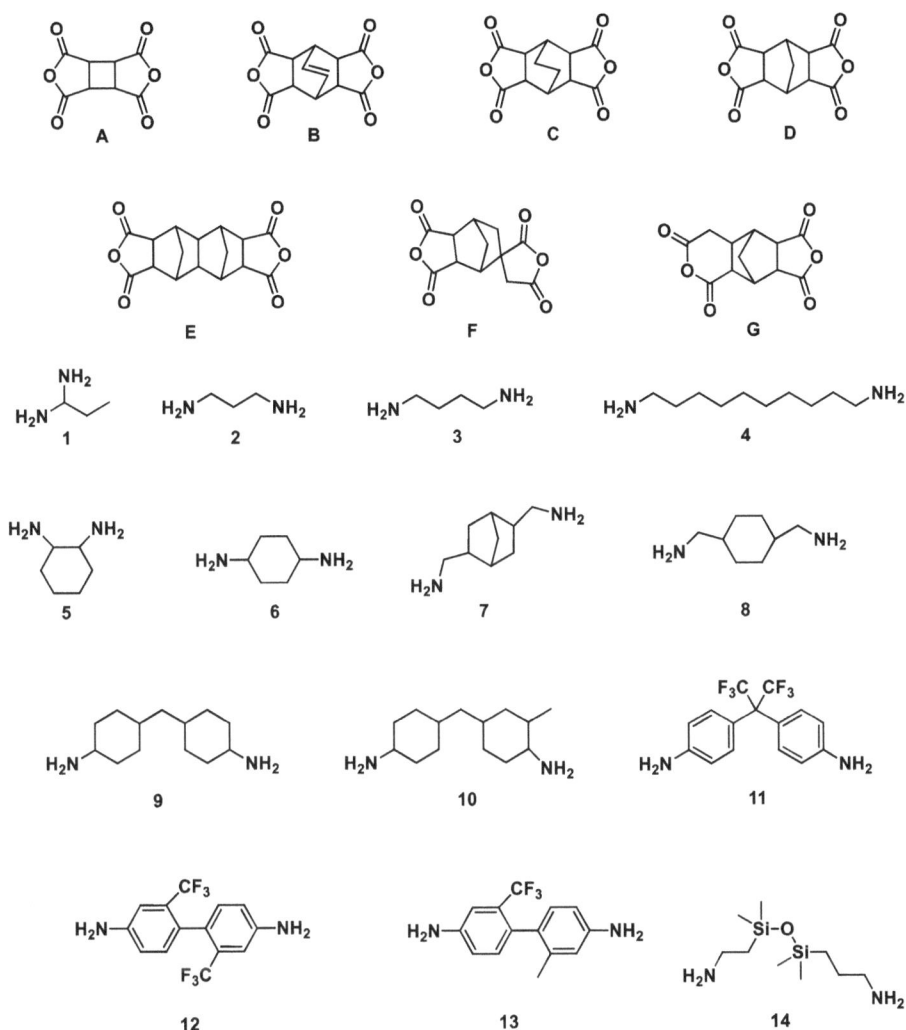

Scheme 2: Examples of diamine and dianhydride monomers used for polyimide synthesis: APIs.

some partial APIs [29]. The first step involves the polymerization of dianhydride and diamine monomers at $T < 50$ °C in aprotic solvent, such as N, N-dimethylacetamide, NMP, N,N-dimethylformamide, and dimethylsulfoxide (DMSO) to generate PAA solution [61, 62], as shown in Scheme 3.

In the second step, the PAA solution is transformed into the corresponding PI through either thermal imidization or a chemical imidization process at about 150 °C, and generally 300 °C is required to complete the process [61]. It is important to know that the obtained PAA solution is susceptible to degradation during storage under ambient conditions and it is sensitive to heat and moisture. Therefore, it is highly recommended to use newly synthesized PAA for the preparation of PI materials (e.g., films and fibers) to achieve higher quality [29]. This technique produces many commercially available PIs, including Kapton™ PI (pyromellitic dianhydride (PMDA) and 4,4'-oxydianiline (ODA)) and Matrimid® 5218 (3,3',4,4'-benzophenonetetracarboxylic dianhydride (BTDA) and diamino-phenylindane) [61].

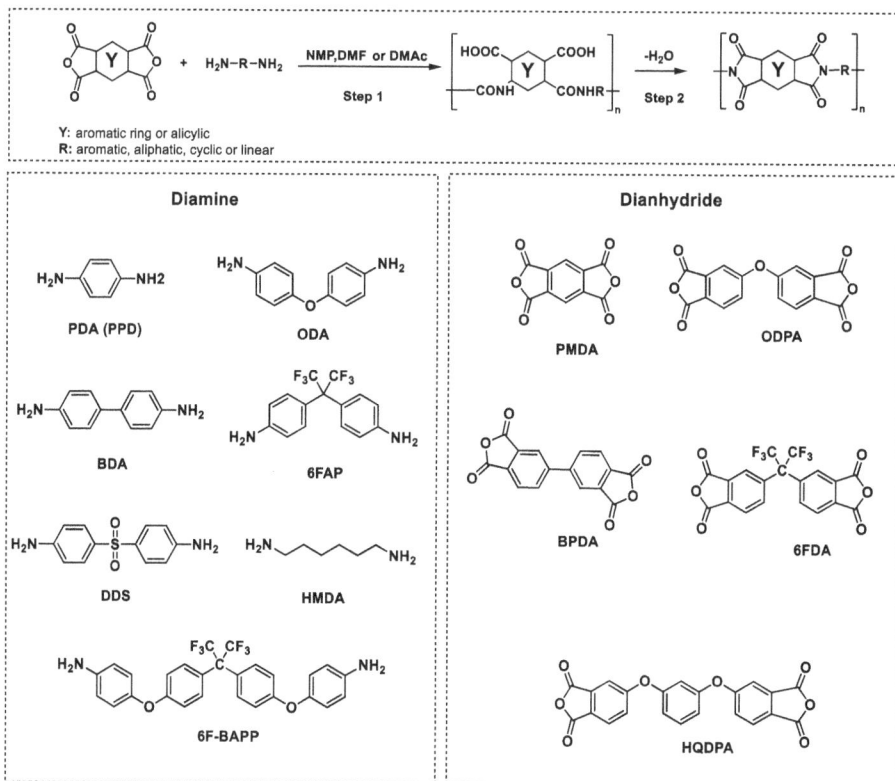

Scheme 3: Illustration of two-step method for synthesizing polyimides (PIs), and chemical structures of some commonly used diamines and dianhydrides.

3.3 Thermal imidization of poly(amic)acid

The thermal imidization route is commonly employed to convert intermediate poly (amic acid) to the final PI, especially when the desired end product is a film or coating. The film is first cast on proper support and then gradually heated up to the temperature range of 100–350 °C to remove all water molecules and create the five-membered imide ring. The literature has extensively discussed the optimal thermal cycle required to achieve close to 100% imidization of poly(amic acids) [63, 64]. Although there are numerous thermal cycles used for this purpose, they can be classified into two types:

– Gradual heating up to 250–350 °C, depending on T_g and stability of PI [63, 65, 66].
– Heating the poly(amic acid) mixture to 100 °C and holding for 1 h, then heating to 200 °C and holding for 1 h, and then heating to 300 °C and holding for 1 h before slowly cooling to the room temperature from 300 °C [67].

Regardless of the type of thermal cycle used, there are several factors that complicate the process of thermal imidization and determine the degree of imidization of the resulting PI. It is important to note that the imidization reactions occur in a highly concentrated and viscous solution, particularly during the initial and intermediate stages, and the residual solvent plays a crucial role for imidization in the later stages. Dipolar amide solvents are particularly effective in promoting imidization due to several reasons, which are:

1. Specific solvation allows the favorable conformation of the amic acid group to cyclize.
2. Plasticizing effect of the solvent to increase the mobility of the reacting functional groups.
3. The basicity of the amide solvent allows it to accept protons and may be responsible for the specific effect [68].

3.4 Chemical imidization of poly(amic) acids

The chemical imidization procedure is used to convert poly(amic) acids into PIs through cyclodehydration. This method involves the use of dehydrating agents to enhance the ring closure reaction, particularly when the PI product is insoluble [63, 69]. Various dehydrating agents, such as acetic anhydride, propionic anhydride, n-butyric anhydride, and benzoic anhydride, can be employed in this process. The presence of a tertiary amine catalyst is important to complete conversion. Examples of catalysts used include pyridine, trimethylamine, tetraalkylamines, methylpyridine, and lutidine [70, 71]. The outcome of the imidization reaction can be affected by several factors, such as dehydrating agent, catalyst, and reaction temperature. For instance, when pyridine is used as a catalyst, a mixture of imide and isoimide is formed. However, using trimethylamine can effectively eliminate isoimide formation [72]. In the presence of trialkylamines with high pK_a values (>10.65), high-molecular-weight PIs were produced. Despite the fact that these compounds have low basicity, high molecular weights were seen when using 2-methylpyridine and isoquinoline ($5.2 < pK_a < 5.7$) [71]. The reaction temperature also plays a crucial role in the imidization process, and it was found that increasing the reaction temperature from 20 to 100 °C significantly reduces the imidization time, allowing for complete imidization within 2 h instead of 15 h [51, 73, 74]. The chemical imidization mechanism has been proposed based on a kinetic study, as illustrated in Scheme 4.

The initial step of the mechanism involves the substitution of the hydroxyl group with a suitable leaving group like $-OOCCH_3$, to promote the ring closure. This occurs

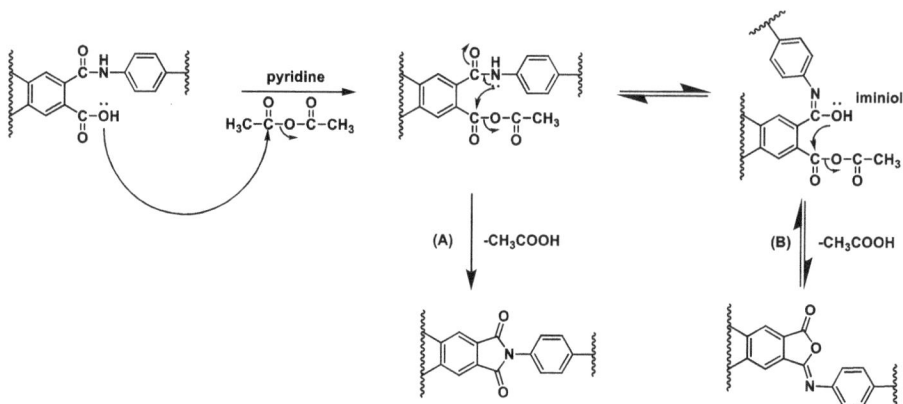

Scheme 4: Proposed mechanism for chemical dehydration of amic acid.

through a nucleophilic attack where the oxygen atom (nucleophile) reacts with the carbonyl group of acetic anhydrides (electrophile), which results in the release of acetic acid. The nitrogen atom in the amide group can have a dual role: it can either attack the carbonyl group, leading to ring closure and the formation of the desired imide (path A), or undergo tautomerization to form iminol, followed by ring closure and the formation of isoimide (path B). Isoimide can undergo thermal isomerization through a reverse reaction effect which begins with the attack of acetate ions, which act as a nucleophile, on the electrophilic isoimide [51, 75]. Unlike thermal treatment, which can impact the mechanical properties of PI and cause them to lose some of their molecular weight, chemical imidization can be utilized to produce high-molecular-weight polymers without affecting their mechanical properties. Nonetheless, its practical application in industries is limited due to the extensive usage of dehydrating agents and catalysts [76].

3.5 Other approaches to prepare polyimides

There are several methods for synthesizing PIs, including the traditional approach of using diamines and dianhydrides. However, other methods have also been developed, such as (1) reacting diisocyanates with dianhydrides, (2) combining diamines with dithioanhydrides, (3) employing bisdiene and bidienophiles in Diels–Alder reactions, (4) using bis(maleimide)s in Michael addition reactions with diamines, (5) utilizing silylated diamines and dianhydrides, and (6) utilizing diester acids and diamines.

3.5.1 From diisocyanates and dianhydrides

PIs can be synthesized using diisocyanates instead of diamines in the polymerization process. Previous studies have demonstrated that *N*-aryl and *N*-alkylphthalimides can be prepared by reacting phthalic anhydride with aromatic or aliphatic isocyanates. These reactions have proven to be effective in synthesizing homo- or co-PIs from diisocyanates and dianhydrides [77–80]. The process of polymerization using diisocyanates involves the hydrolysis of diisocyanate to form a diamine, which is converted into a PAA and then transformed into a PI using general imidization methods [81]. This approach is preferable because the solubility of diisocyanates in organic solvents is more than diamines and the reaction is less sensitive to moisture. However, one disadvantage of this method is the limited number of diisocyanates available when compared to the numerous diamines that can be obtained through synthetic or commercial sources. However, the high reaction rate between diisocyanates and tetracarboxylic acids makes this a potential route for PI preparation [81]. According to the study by Yeganeh et al., high-molecular-weight PIs were successfully produced by a direct reaction between diisocyanates and dianhydrides using microwave radiation. This reaction between isocyanate and anhydrides may produce a seven-membered ring intermediate, which removes carbon dioxide and forms a five-membered ring in the PI as shown in Scheme 5. The experimental results demonstrated that the PIs prepared using this approach had superior inherent viscosities and yields when compared to PIs synthesized through the traditional solution method [80].

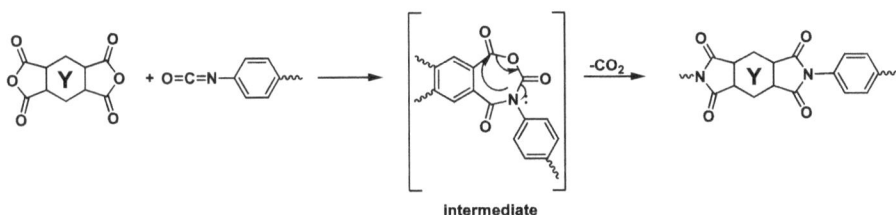

intermediate

Scheme 5: The proposed seven-membered ring intermediate and CO_2 release mechanism.

3.5.2 From diamines and dithioanhydrides

PIs can be synthesized using dithioanhydride monomers as starting materials. These monomers are not readily available and can be obtained by reacting sodium sulfide with aromatic tetracarboxylic acids. Imai et al. [82] employed this method to synthesize poly (amid thiocarboxylic acid)s, which were subsequently converted to PIs by removing hydrogen sulfide through heating, as illustrated in Scheme 6. Liou et al. [83] compared the production of PIs through the reaction between diamines and dithioanhydrides to those produced by the traditional two-step method and found that the inherent viscosities of

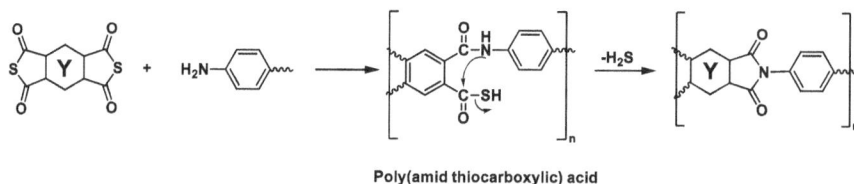

Scheme 6: Schematic representation of polyimide synthesis via dithioanhydrides.

PIs derived from dithioanhydrides were similar to those prepared by the traditional two-step method.

3.5.3 From bisdiene and bidienophiles in Diels–Alder reaction

The Diels–Alder reaction is a useful synthetic method in organic chemistry that involves a thermally driven [4 + 2] cycloaddition reaction between a conjugated 1,3-diene and a dienophile. This reaction provides a simple and efficient way to generate new bonds through inter- or intramolecular coupling as presented in Scheme 7. The reaction involves adding a dienophile to a conjugated diene to form a cyclic product known as an adduct. The furan rings are commonly used as heterocyclic dienes in Diels–Alder reactions [84]. Chi et al. synthesized PIs using an in situ Diels–Alder polymerization at 80 °C in the presence of NaI in DMSO. They used 1,4-bis[4-(methyloxy)phenyloxy]-2,3,6,7-tetrakis-(bromomethyl)benzene (MPBB) with four arylenebismaleimides to form PIs, as shown in Scheme 8. Differential scanning calorimetry and wide-angle X-ray diffractometry studies revealed that all aromatic PIs were completely amorphous. Also, UV–vis spectroscopy confirmed that APIs and DPAI were transparent at wavelengths longer than 375 nm [85].

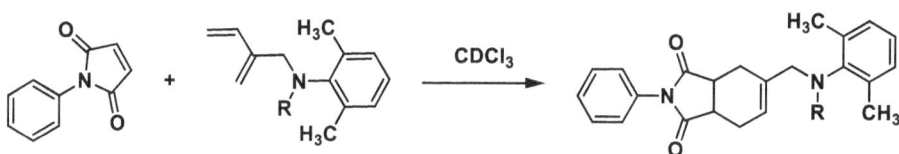

Scheme 7: Diels–Alder reaction for polyimide synthesis (model compound).

3.5.4 From bis(maleimides) and with diamines in Michael addition reactions

The Michael addition reaction is another method for preparing PIs, which involves the addition of diamines to bis(maleimides). The imide ring in PIs is not formed during the polymerization process in this method, but rather arises from the maleimide

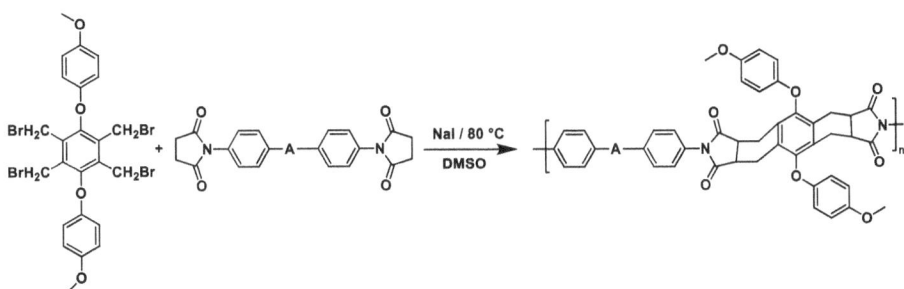

Scheme 8: Synthesis of polyimide using an in situ Diels–Alder.

structure. Bella et al. [86] have reported using the Michael addition reaction to synthesize PIs, as shown in Scheme 9. This technique is a general strategy for producing thin films from reactive or insoluble functional PIs [87].

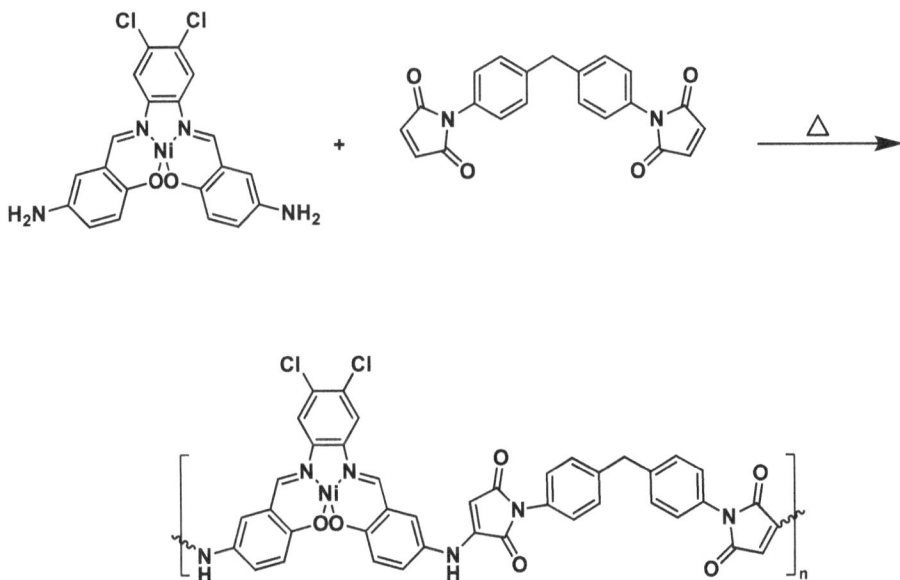

Scheme 9: Synthesis of polyimides from diamine and bis(maleimide).

3.5.5 From silylated diamines and dianhydrides

In 1967, Boldebuck and Klebe [88] reported the first synthesis of aromatic PIs using silylated diamines in a patent publication. However, using silylated amines has some limitations, such as the need to prepare and purify activated monomers, which are challenging to isolate due to their moisture sensitivity. In addition, silylated amines are more expensive than diamines [89]. To overcome these limitations, Kaneda et al. utilized in situ sily-lation of diamines that were produced by adding chloro(trimethyl)silane or other silylat-ing agents to the diamine solutions in the presence of a base, such as pyridine. This approach has been demonstrated to be a simple and convenient way to obtain high-molecular-weight PIs. Moreover, silylation can improve the low reactivity of diamines, particularly when using sterically hindered amines or amines with strong electron-withdrawing groups [90].

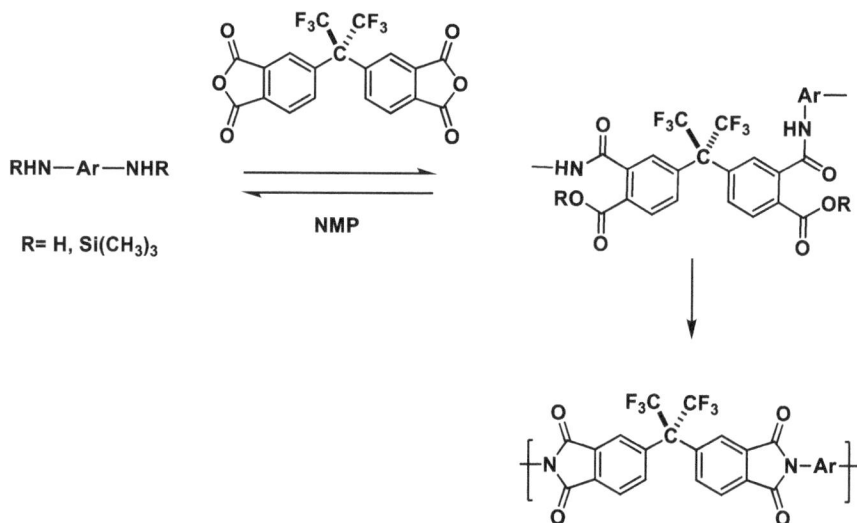

Scheme 10: Synthesis of polyimides from silylated diamines and dianhydrides.

3.5.6 From diester acids and diamines

PIs can be synthesized by using an alternative method, which involves the utilization of diester acids and diamines instead of the traditional dianhydrides. The process begins by converting tetracarboxylic acid into diester acid through an esterification reaction as shown in Scheme 11. In this step, the dianhydride is refluxed with excess alcohol in the presence of a tertiary amine acting as an acid acceptor. After removing the excess alcohol, diamines are added to produce the poly(amic) acids in a polar aprotic solvent. Then, the polymer can be obtained through solution or thermal imidization procedures. The production of PIs using this approach is suitable for commercial synthesis since it accepts the presence of some water in solvents and reactors [51, 91].

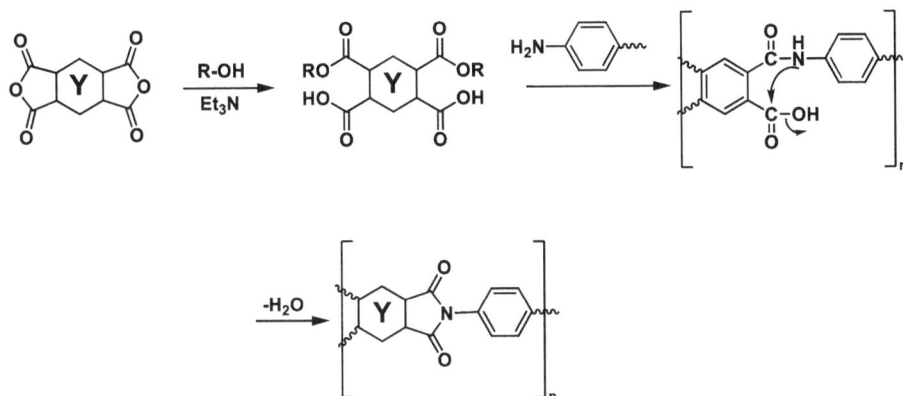

Scheme 11: Synthesis of polyimide via a diester acid intermediate.

4 Theory of gas transport through membranes

In principle, the gas separation by membranes is based on the transportation of gases across the membranes influenced by several factors such as membrane features including rigidity, molecular size, membrane polymer and porosity, and kinetic diameter of the gas molecules (Table 3), although the selectivity and permeability of the membrane determine the efficiency of the gas separation process [5]. Generally, the membranes can be classified into porous and nonporous. The porous membrane is known to be rigid and voided along their structure with interconnected pores randomly distributed. This is making the separation process using porous membrane very similar in structure and function to the conventional filter. In general, the dynamic diameter of the gases ranges from 0.26 to 0.45 nm, which is the range of the microporous materials; thus, only those molecules that differ considerably in size can be separated effectively by microporous membranes. Thus, in order to obtain an excellent performance membrane, the pore size of the polymers should be engineered to be in a proper range of microporosity. However, the used porous membranes in gas separation are well characterized by their high permeability and intrinsic low selectivity. Therefore, the gas diffusion through the pores of the microporous membranes is proportional to the molecule's velocity, that is, inversely proportional to the square root of the molecular mass (Knudsen diffusion). The flux through a porous membrane is much higher than through a nonporous one with about three- to fivefold.

In general, there are four types of diffusion mechanisms that can be utilized to affect the gas separation by porous membranes [93] (Figure 5):
1. Knudsen (or free molecule) diffusion
2. Surface diffusion

3. Capillary condensation
4. Molecular sieving

Table 3: Gases and vapor properties [92].

Gas	T_c (K)	V_c (cm³ mol⁻¹)	d_{L-J} (nm)	d_K (nm)	λ (nm)
He	5.2	57.5	0.255	0.26	280.9
H_2	33.2	64.9	0.283	0.289	177.5
N_2	126.2	89.3	0.380	0.364	94.7
O_2	154.6	73.5	0.347	0.346	103.9
CO_2	304	91.9	0.394	0.33	–
CH_4	191	98.6	0.376	0.38	–
C_2H_4	282.5	131.1	0.416	0.39	–
C_2H_6	305.3	147	0.444	–	–
C_3H_6	365.2	184.6	0.468	0.45	–
C_3H_8	369.9	200	0.511	0.43	–
H_2S	373	87.7	0.362	0.36	–

T_c, critical temperature of the gases; V_c, critical volume of the gases; d_{L-J}, Lennard–Jones diameter; d_k, kinetic diameter of the gases; and λ, mean free path of the gases (25 °C, 1 bar).

Knudsen diffusion may take place in a microporous membrane or through pinholes in dense polymeric membranes. It is a means of diffusion that occurs in a long pore with a narrow diameter (2–50 nm) because molecules frequently collide with the pore wall. This mode of transport is important when the mean free path of the gas molecules is greater than the pore size (Table 3). In such situations, the collisions of the molecules with the pore wall are more frequent than the collision between molecules. Separation selectivity with these mechanisms is proportional to the ratio of the inverse square root of the molecular weights. This mechanism is often prominent in macroporous and mesoporous membranes [94–96].

Surface diffusion can occur in parallel with the Knudsen diffusion. The gas molecules are adsorbed on the pore walls of the membrane and migrate along the surface. Surface diffusion increases the permeability of the components adsorbing more strongly to the membrane pores. At the same time, the effective pore diameter is reduced. Consequently, the transport of nonadsorbing components is reduced, and selectivity is increased. This positive contribution of surface diffusion only works for certain temperature ranges and pore diameters [95, 97].

Capillary condensation occurs if a condensed phase (partially) fills the membrane pores. If the pores are filled with condensed phase, only the species soluble in the condensed phase can permeate through the membrane. Fluxes and selectivities are generally high for capillary condensation. The appearance of capillary condensation, however, strongly depends on gas composition, pore size, and uniformity of pore sizes [94, 96].

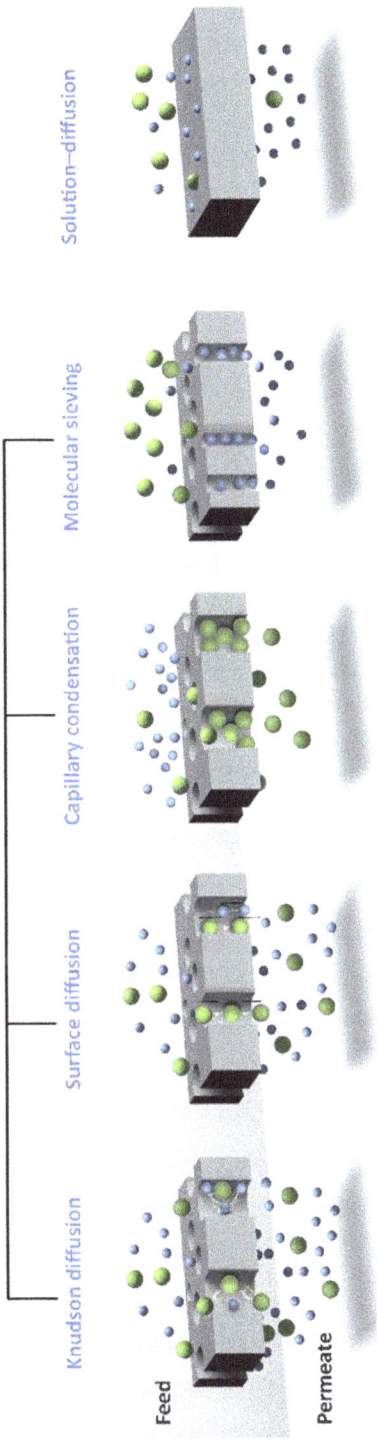

Figure 5: Schematic structure of the polymer-based membrane.

Molecular sieving occurs when pore sizes become sufficiently small (3.0–5.2 Å), leading to the separation of molecules that differ in kinetic diameter, which means when the pore size becomes so small, only the smaller gas molecules can permeate through the membrane [96].

In some cases, molecules can move through the membranes by more than one mechanism. Knudsen diffusion gives relatively low separation selectivity compared to surface diffusion and capillary condensation. Shape-selective separation or molecular sieving can yield high selectivity. The separation factor of these mechanisms depends strongly on the pore size distribution, temperature, pressure, and interaction between gases being separated, and the membrane surfaces [26].

Solution–diffusion is commonly acknowledged as the main transport mechanism of gases through nonporous polymeric membranes (dense membranes) [97]. Due to the studies of J. K. Mitchell, A. Fick, and T. Graham in the nineteenth century, it became evident that mass transfer in nonporous membranes is conceivable even with the lack of pores. However, during the measurements of the permeation rate of multiple gases through polymeric film, they noted that there is no monotonic correlation between permeation rate and gas diffusion coefficients. Later, among all gases tested, they observed a notable permeation rate for carbon dioxide distinguished by higher solubility in the membrane materials. This observation has led to the birth of solution–diffusion mechanism [98]. However, solution–diffusion mechanism is generally considered to be a three-step process. In the first step, the gas molecules are adsorbed by the membrane surface on the upstream end. Followed by the diffusion of the gas molecules through the polymer matrix and in the final step, the gas molecules evaporate on the down-end stream. According to this mechanism, the characteristic of the permeation rate, permeability coefficient P, can be presented as the following model:

$$P = DS$$

where D is the diffusion coefficient and S is the solubility coefficient.

5 Nonporous polyimides (dense polyimide)

Nonporous PI membranes are widely used for gas separation applications due to their exceptional properties and performance such as chemical resistance, excellent mechanical strength, and thermal stability. These properties make them ideal for a range of gas separation processes where high-performance membranes are required. Also, these membranes exhibit high selectivity but typically have a low gas transport rate. An important characteristic of PI dense membrane is that even gases of similar sizes can be separated if their solubility in the membrane differs significantly. Their nonporous structure is attributed to the presence of noncontinuous passages within the polymer chain matrix. The formation and disruption of these passages occur due to thermally induced motion of the polymer chains. Consequently, the transport of gas molecules relies on their movement through these passages. The gas transport rate and separation performance of the membrane depends on many factors such as the penetrant's activity (driving force) and operating conditions. The main mechanism of gas transport in dense PI membranes is solution–diffusion. This mechanism consists of three stages. In the initial stage, there is adsorption of gas molecules by the surface of the membrane on the upstream end. Subsequently, the gas molecules are diffused through the polymer matrix. In the last stage, there is evaporation of gas molecules at the downstream end [99, 100].

6FDA-based PI membranes have been investigated for gas separation using dense membranes containing different diamine moieties. Kim et al. used thermal imidization to prepare 6FDA-based PIs containing polar hydroxyl or carboxyl groups in diamine, such as 6FDA-BAPAF, 6FDA-DAP, and 6FDA-DABA (3,5-diaminobenzoic acid). These PIs were then utilized to produce composite membranes through the dip-coating technique, with poly(ether sulfone) membrane as a supporting layer. Scheme 12 illustrates the chemical structures of the 6FDA-based PIs synthesized by Kim et al. Among the various PIs studied, the 6FDA-DAP PI membrane demonstrated higher selectivity compared to others. On the other hand, the 6FDA-BAPAF PI membrane exhibited lower CO_2/CH_4 selectivity. The CO_2/N_2 selectivities of the membranes prepared using different PIs were similar to other dense 6FDA-based PI membranes in previous literature [100, 101], as presented in Table 4.

Table 4: Gas selectivity of the different 6FDA-based polyimides.

Polyimides	Selectivity	
	CO_2/N_2	CO_2/CH_4
6FDA-BAPAF	20.11	22.78
6FDA-DAP	29.26	78.82
6FDA-DABA	28.28	46.96
6FDA-3BDAF	27	48
6FDA-IPDA	23	43
6FDA-DAFO	22	60

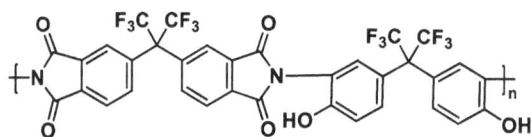

6FDA-BAPAF

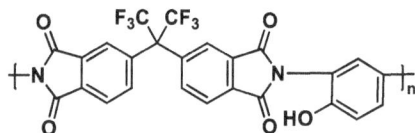

6FDA-DAP

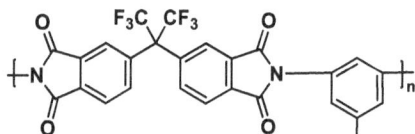

6FDA-DABA

Scheme 12: Chemical structure of 6FDA-based polyimides.

6 Porous polyimides

6.1 Functionalized polyimides

Aromatic PI membranes are known for their exceptional gas selectivity, which is attributed to their tightly packed molecular structure and strong intersegment interactions. Unfortunately, the difficulties associated with processing these materials have limited their use in high-end applications. Furthermore, the low solubility of PIs in most organic solvents, except for concentrated sulfuric acid, makes it challenging to produce thin films or use them for coating [102]. As a result, their large-scale application in gas separation is restricted by these limitations. Another unique characteristic of aromatic PIs is the production of intrachain charge complexes (CTCs), which has a direct impact on the solubility and permeability/selectivity. The CTC is the electronic interaction between the aromatic anhydride and diamine segments as an electron acceptor and electron donor, respectively (Scheme 13). However, the CTCs can limit the molecular mobility through membranes, narrowing the interchain space, which led to an increase in the selectivity.

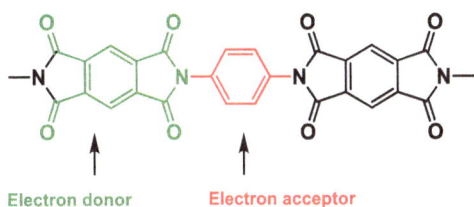

Scheme 13: Charge-transfer complex (CTC) segments in polyimides.

The strength of CTC interactions depends on the electron-withdrawing substituent, thermal property, and polymer chain packing of the PIs [34].

Regarding this issue, various techniques were established to enhance the solubility of aromatic PIs, including incorporation of flexible "kinks" attached to the diamines and dianhydrides monomers, such as ethers [103, 104], esters [105–108] (Scheme 14), or attaching pendant phenolic groups (Scheme 15) [102]. However, the kink structure is a crank and twisted noncoplanar structure. The introduction of such groups into polymer chains established a "kink" in the main chain that decreases the rigidity of the polymer backbone and inhibits close packing of the chains, which reduce the chains' alignment and disrupt the formation of efficient CTCs leading to enhanced solubility. Moreover, the introduction of kinked monomer units such as spiro and cardo diamines and dianhydride monomers (Scheme 16) to the PI backbone has a great influence to their solubility and permeability properties [27]. However, there are many commercial diamines and dianhydrides with kinked structures that are available in the market (Scheme 17).

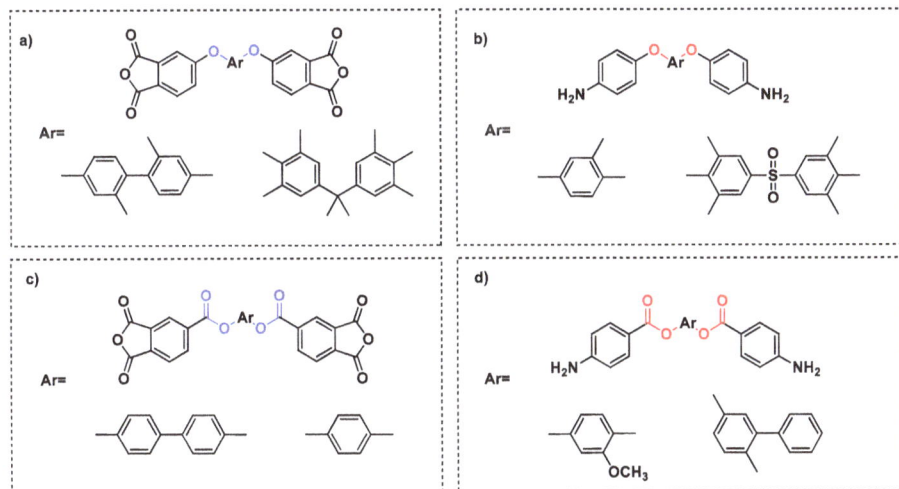

Scheme 14: Some ether linkages and ester linkages of (a, c) anhydride and (b, d) diamine monomer structures.

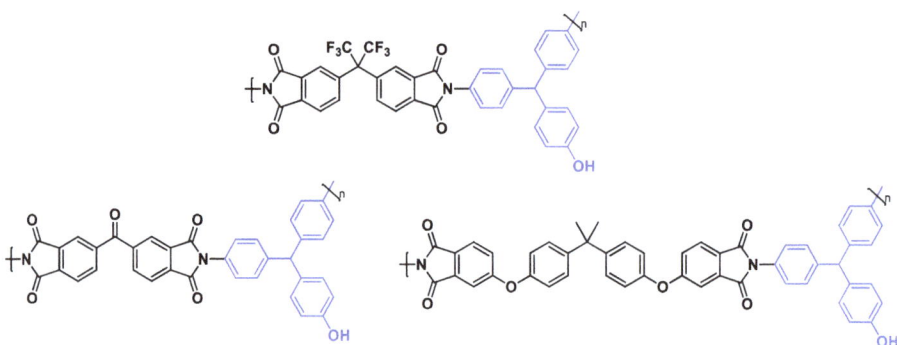

Scheme 15: Pendant phenolic groups in polyimides.

Furthermore, when a polymer membrane is exposed to a gas mixture in the gas separation processes, an interaction between the polymer chains and the gas molecule might occur within the polymer matrix. This interaction can cause the polymer to swell and increase increased segmentation mobility of polymer chains, resulting in higher gas permeability but lower selectivity. This phenomenon called plasticization of polymer. Additionally, plasticization can lead to deterioration of the polymer's structural and thermal stability, resulting in a decrease in the membrane's durability and life span. Consequently, it is necessary to address the impacts of plasticization to achieve optimal performance and stability of the membranes. By eliminating the effects of plasticization, gas separation membranes can preserve their selectivity and permeability, ultimately enhancing the efficiency and effectiveness of gas separation processes [109, 110]. The introduction of polar functional groups, such as carboxylic acid groups (–COOH) or hydroxyl groups (–OH), into the PI structure has been found to enhance the membrane's ability to selectively separate different sized molecules and resist plasticization caused by the penetration of gases. These improvements can be attributed to the tightening of the polymer chains through interchain or intrachain interactions facilitated by hydrogen bonding or π–π interactions [110, 111]. Furthermore, the incorporation of polar groups into the polymer structure can enhance the polymer's attraction toward polar gases like CO_2, H_2O, and NH_3, while decreasing the permeability of nonpolar gases such as N_2 and O_2. As a result, the selectivity of the PI membrane for specific gas separation processes can be improved [112]. Several hydroxyl-functionalized PIs were reported in the literature, as given in Table 5.

Moreover, few researchers reported novel PIs containing a carboxylic group. Koros group introduced carboxyl-functionalized PI 6FDA-DABA, which was prepared by reacting 6FDA with DABA and their copolyimides [115, 116]. These materials show great promise for natural gas separation, because of their ability for facile crosslinking (Scheme 18). Abdulhamid et al. have shown that the interdiction of –COOH group to the PI's backbone will tighten the structure which induced the size selectivity toward gases [117].

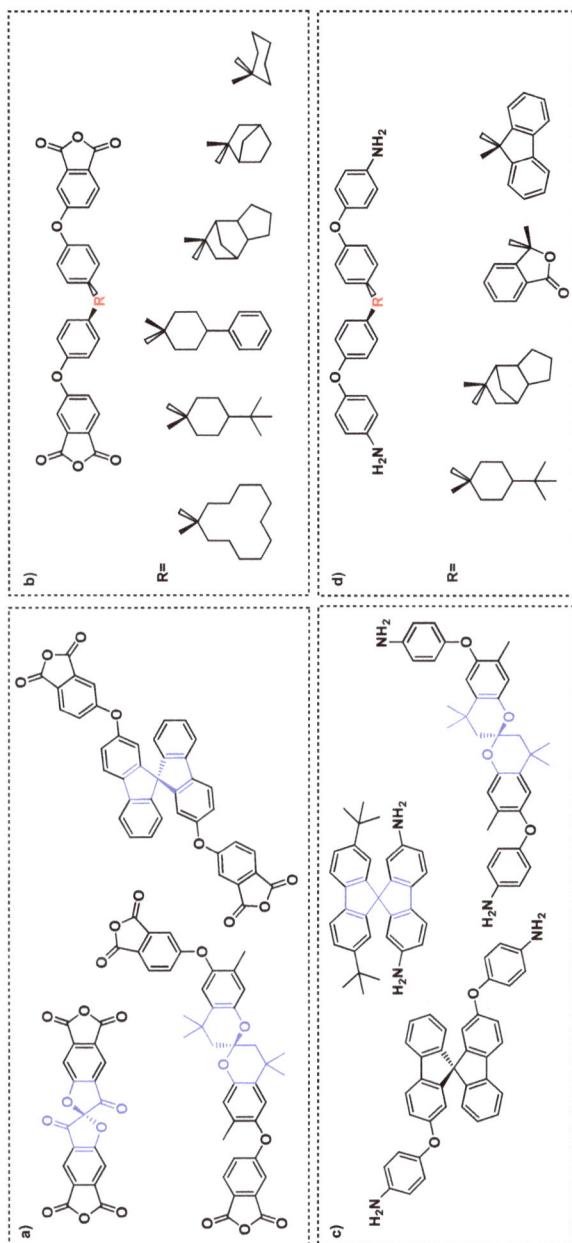

Scheme 16: Chemical structure of some kinked units such as (a) spiro dianhydrides, (b) cardo dianhydride, (c) spiro diamines, and (d) cardo diamine.

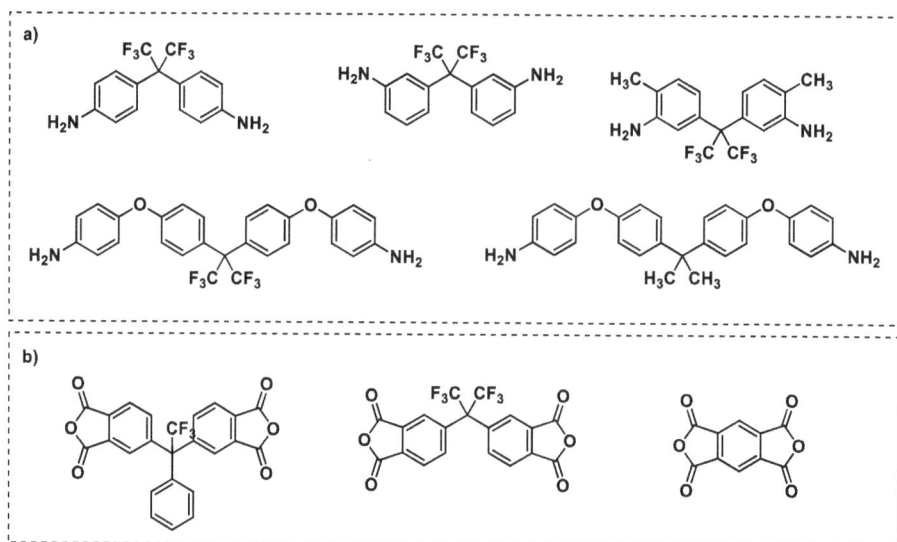

Scheme 17: Commercial kinked (a) diamines and (b) dianhydride monomers.

Table 5: Pure gas permeation properties of hydroxyl-functionalized low free volume polyimide.

Polymers	Permeability (barrer)					Selectivity			Reference
	H_2	N_2	O_2	CH_4	CO_2	H_2/N_2	O_2/N_2	CO_2/CH_4	
6FDA/DAP	38	0.37	2.5	0.12	11	103	6.8	92	[112]
6FDA/m-PD	46	0.62	3.6	0.2	14	–	5.9	70	[112]
6FDA/m-PD	–	0.87	5.4	0.35	20.3	–	6.2	58	[113]
6FDA/m-PD	–	0.29	2.1	0.1	7.3	–	7.3	76	[113]
6FDA/DAP	–	0.43	2.9	0.11	11	–	6.8	100	[113]
6FDA/DAR	34	0.28	1.9	0.085	8	121	6.8	94	[112]
6FDA/DAR	–	0.29	2.1	0.08	8.2	–	7.3	102	[113]
6FDA/APAFb	42.8	0.55	3.73	0.21	17.0	78	6.8	80	[114]
PMDA/APAFb	35.2	0.36	2.58	0.08	9.9	97	7.1	121	[114]
BPDA/APAFb	14.3	0.09	0.72	0.03	2.7	154	7.7	91	[114]
BTDA/APAFb	11.1	0.08	0.6	0.02	1.4	139	6.6	71	[114]
ODA/APAFb	13.4	0.06	0.54	0.02	1.8	232	9.3	108	[114]

Additionally, several methods can be used to enhance the permeability and selectivity of PIs like cross-linking, blending, incorporation of inorganic fillers, and surface modification. The cross-linking of PI chains can increase the rigidity of the polymer and reduce the free volume, which can increase the selectivity of the membrane without significantly decreasing the permeability. Besides that, cross-linking can also enhance the polymer resistance toward plasticization.

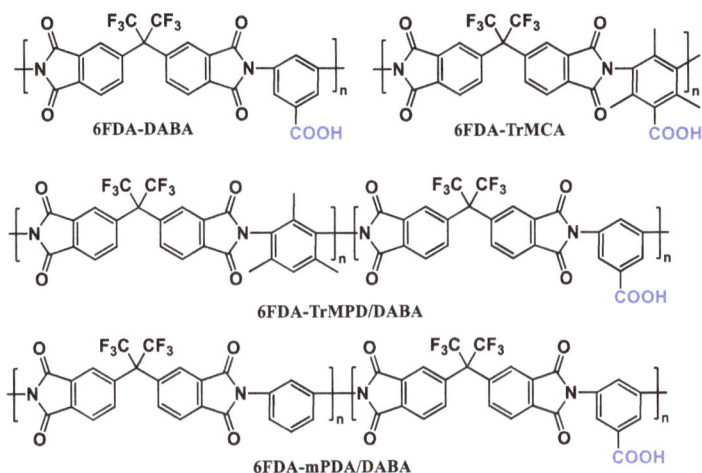

Scheme 18: Examples of polyimide and copolyimides contain carboxylic acid group.

Cross-linked PI as a derivative of 6FDA-based PI, 6FDA-diaminobenzoic acid 2,4,6-trimethyl-1,3-phenylenediamine (6FDA-DAM:DABA, Scheme 19), is a thermally cross-linkable polymer which has been broadly studied for gas separations [118, 119].

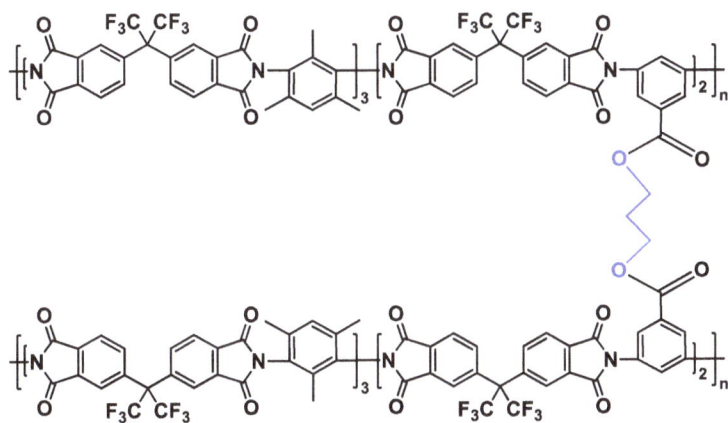

Scheme 19: Cross-linked 6FDA-DAM:DABA polyimide structure.

Blending PIs with other polymers, such as polyethylene oxide (PEO) [120], can improve the gas separation performance of the resulting membrane. The addition of PEO can increase the permeability of the membrane for certain gases. The incorporation of inorganic fillers, such as zeolites, metal-organic frameworks (MOFs), or silica nanoparticles, into the PI matrix can improve the gas separation performance of the membrane. These fillers can enhance the selectivity of the membrane by providing additional adsorption sites for specific gases, and can also increase the mechanical strength and thermal stability of the membrane [121].

Surface modification of the PI membrane can improve its gas separation properties by altering the surface chemistry or morphology of the membrane. This can be achieved through various methods, including plasma treatment, chemical grafting, or deposition of thin films. Surface modification can improve the selectivity and permeability of the membrane, as well as its stability and durability under harsh conditions [122].

However, recently, new generation of materials called polymer of intrinsic microporosity (PIM) has been developed. These materials have achieved tremendous success in air separation and hydrogen recovery applications. Due to their relatively weak performance for energy-intensive applications, condensable feeds such as CO_2/CH_4 and C_3H_6/C_3H_8 have encouraged interest in developing new-generation plasticization-resistant PIM materials. It has been widely studied how the PI structure engineering and the chain interactions correlated to their gas transport characteristics. The PI membrane permeability and gas selectivity could be improved by either disorganizing the polymer chain packing or encouraging interchain interactions by making the suitable choice of polymer backbone substituents [123].

6.2 Polyimides of intrinsic microporosity PIM-PIs

PIMs exhibit distinguished class of microporous organic materials that hold significant potential in gas separation membranes [124]. PIMs, belonging to the family of organic amorphous microporous materials, were initially reported in 1983 using poly[1-(trimethylsilyl)-1-propyne][(PTMSP)] was prepared by polymerizing 1-(trimethylsilyl)-1-propyne (MeC $\equiv$ CSiMe$_3$; TMSP) in the presence of a catalyst such as TaCl$_5$ or NbCl$_5$ [125]. However, the definition of intrinsic microporosity within a polymer is "a continuous network of interconnected intermolecular voids that forms as a direct consequence of the shape and rigidity of the component macromolecules" [126]. The structural rigidity and insufficient packing of the polymer chains is the main reason of the porosity of this class of polymers, where the good selection of monomers leads to such features. However, the high fractional free volume (FFV) caused by the insufficient packing of the polymeric chains creates voids within the microstructure and provides a new platform for this material to be used in gas separation and storage applications [127, 128]. In comparison to other microporous materials, PIMs have numerous benefits, including solubility, ease of processing, use of inexpensive monomers, flexible film creation, excellent temperature resistance, and exceptional mechanical capabilities. These characteristics make them particularly attractive for a wide range of applications. Recently, the concept of microporosity was extended to PIs by using kinked, contorted, and stable structures to obtain high gas performance combined with excellent solution processability, high thermal stability, and a unique platform for a wide range of possible modifications and tunability.

High gas selectivity is a characteristic of PI membranes that is typically attributed to their low free volume content and strong interchain interactions among their constituent polymer chains. However, compared to other polymeric membranes, the majority of PI membranes have very poor gas permeabilities, which presents a substantial challenge to their practical application in large-scale gas separation processes. From this point, intrinsically microporous PIs (PIM-PIs) were originated from PIM for the first time in 2007 by reacting spirobifluorene (SBF)-based diamine with PMDA [129]. The obtained PI was found to be easily soluble in conventional organic solvents, which also has surface area of about 550 m^2 g^{-1}. Moreover, Ghanem et al. [130, 131] reported the first PIM-PI for gas separation. Later, considerable studies have been dedicated toward synthesis of PIM-PIs with enhanced free volume and suitable microporosity dimensions to enhance their gas permeability while maintaining their high gas selectivity. Typically, to achieve high FFV in PI membranes, many methods were used such as the incorporation of bulky side groups or sterically hindered diamine and dianhydride building blocks into the polymer backbone. Thus, a variety of rigid or bulky diamine/dianhydride monomers containing triptycene (TRIP) [132–135], CANAL [48], pentiptycene [120, 136, 137], SBF/spirobisindane (SBI) [138, 139], large aliphatic group or pendent phenyl unit [140], and methyl groups [141, 142] were designed and synthesized, to prepare PI membranes with improved gas permeability. Also, a class of PIs with an intrinsic microporosity (PIM-PIs), which contained highly rigid and contorted semi-ladder bridged ring moieties like ethanoanthracene (EA) [139, 143–145], 1.5-methano-1,5-diazocine (Tröger's base, TB) [146–148], and TB-derived dianhydride (TBDA) [149], was reported.

Due to the variety of dianhydride and diamine moieties, the development of PIM-PI opens new possibilities for the design of PIM-based monomers. The diamine or dianhydride monomers could be given the contortion sites, which produce intrinsic microporosity. Importantly, in the PIM-PI system, varying the type of the side chains appended to diamines or dianhydrides impacts polymeric membrane properties, and in turn, gas separation performance. PIM-PIs could be further classified into two groups, depending on where the contortion sites are located in the polymer backbone: (1) PIM-PIs containing dianhydride contortion sites and (2) PIM-PIs containing diamine contortion sites (Figure 6).

6.2.1 PIM-PIs with contorted dianhydride sites

The McKeown's group [131] developed the first series of PIM-PIs with contorted dianhydride sites in 2008 by introducing SBI into dianhydride monomers, and various diamines were explored to synthesize PIM-PI-1 to PIM-PI-8 (Scheme 20). The as-prepared PIM-PI membranes showed significantly increased specific surface areas, FFV, and gas permeability as compared to others conventional low FFV PI membranes. The physical properties of PIs are summarized in Table 6. The CO$_2$ permeability of these PIM-PI mem-

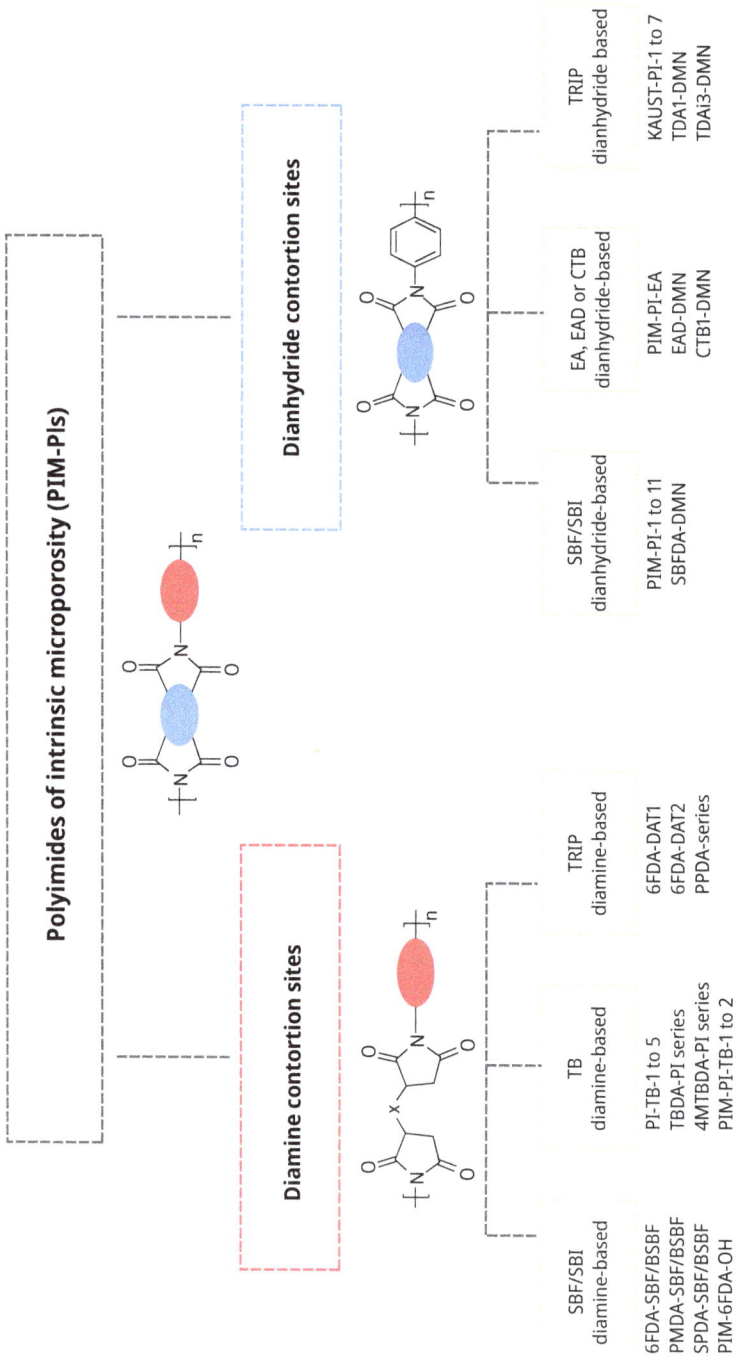

Figure 6: Classification of polyimides of intrinsic microporosity.

branes reached 3,700 barrier which is 770-fold higher than the conventional Matrimid. The gas transport properties of these PIs are mentioned in Table 7.

Table 6: The physical properties of the intrinsic microporous polyimides.

Polyimide	M_n (10^3 g mol^{-1})	Density (g cm^{-3})	S_{BET} (m^2 g^{-1})	FFV
PIM-PI-1	17	1.15	680	0.232
PIM-PI-2	30	–	500	–
PIM-PI-3	22	1.26	471	0.226
PIM-PI-4	11	1.26	486	0.228
PIM-PI-7	11	1.19	485	0.223
PIM-PI-8	54	1.14	683	0.231

M_n, the number-average molar mass of the polyimides; S_{BET}, the surface area determined from N_2 adsorption at 77 K; and FFV, the fractional free volume determined from the film density and van der Waals volume.

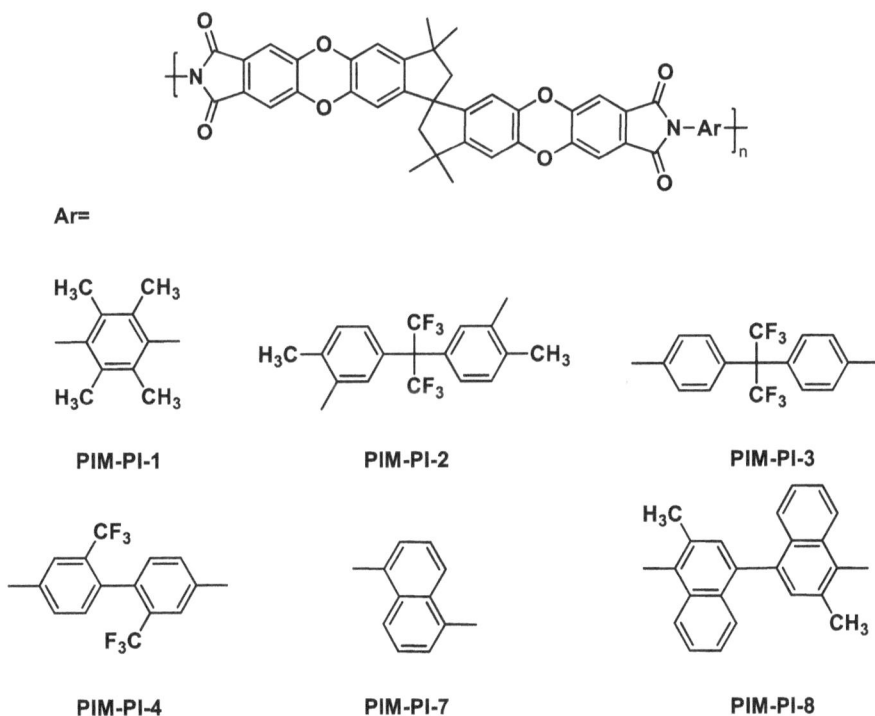

Ar=

PIM-PI-1　　　　　　PIM-PI-2　　　　　　　　PIM-PI-3

PIM-PI-4　　　　　　PIM-PI-7　　　　　　　　PIM-PI-8

Scheme 20: Examples of structures of PIM-PIs containing SBI.

Table 7: Gas permeability and selectivity of polyimides.

Polymer	Permeability (barrer)					Selectivity		
	H_2	N_2	O_2	CH_4	CO_2	H_2/N_2	O_2/N_2	CO_2/CH_4
PIM-PI-1	530	47	150	77	1,100	11.3	3.2	14.3
PIM-PI-2	220	9	39	9	210	24.4	4.3	23.3
PIM-PI-3	360	23	85	27	520	15.7	3.7	19.3
PIM-PI-4	300	16	64	20	420	18.8	4.0	21
PIM-PI-7	350	19	77	27	510	18.4	4.1	18.9
PIM-PI-8	1,600	160	545	260	3,700	10	3.4	14.2

Later, by replacing the flexible dibenzodioxane linkages in the SBI moieties in the dianhydrides, PIM-PI-9 to PIM-PI-11 have been developed, but they exhibited smaller gas permeability coefficients. Also, they developed PIM-PI-12 by replacing the SBI sites in PIM-PI-10 with EA, which exhibited O_2 permeability of 659 barrer [150].

Using more rigid and contorted bridging bicyclic structures like TB, EA, and TRIP units, high-performance PIM-PIs were developed with enhanced rigidity and internal free functional volume (FFV). Swaidan et al. [143] and Ghanem et al. [151] have developed a series of high-performance PIM-PIs, namely KAUST-PI-1 to KAUST-PI-7, by using a novel rigid contorted 3D dianhydride containing TRIP unit as the contortion sites (Scheme 21). Their physical properties were given in Table 8. The incorporation of 9,10-diakyltriptycene fused ring dianhydrides have demonstrated gains in permeability and selectivity over conventional low free volume 6FDA-based PIs. KAUST-PI-1, KAUST-PI-2, and KAUST-PI-3 exhibited excellent gas separation performance exceeding 2,008 upper bounds for most gas pairs, while KAUST-PI-7 showed high surface area of 837 $m^2\ g^{-1}$ with high permeability and lower gas selectivity due to the presence of bulky DMN which offers excessively large FFV. The gas separation properties of these PIM-PIs are illustrated in Table 9. More different PIM-PIs containing contorted dianhydride sites developed so far for gas separation are illustrated in Table 10. Recently, Abdulhamid [149] has prepared TBDA and synthesized two solution-processable PIs using 2,3,5,6-tetramethyl-p-phenylenediamine (TMPD) and 1,7-diamino-6H,12H-5,11-methanodibenzo[1,5]diazocine-2,8-diol (HTB). The as-prepared TBDA-TMPD polymer membrane exhibited CO_2 permeability of 1,457 barrer and CO_2/CH_4 selectivity of 17. Moreover, the aged TBDA-HTB exhibited notable performance that crossed 2,008 upper bound for H_2/CH_4 separation with H_2 permeability of 147 barrer and H_2/CH_4 selectivity of 129.

6.2.2 PIM-PIs with contorted diamines sites

With the variety of PIM-PI building blocks like SBF, SBI, TB, and TRIP units, the option of diamine type containing diverse sites of contortions is significantly more flexible than with bulky dianhydrides. Weber et al. [129] reported the first series of diamines

Ar=

| KAUST-PI-1 | KAUST-PI-2 | KAUST-PI-3 | KAUST-PI-4 |

| KAUST-PI-5 | KAUST-PI-6 | KAUST-PI-7 |

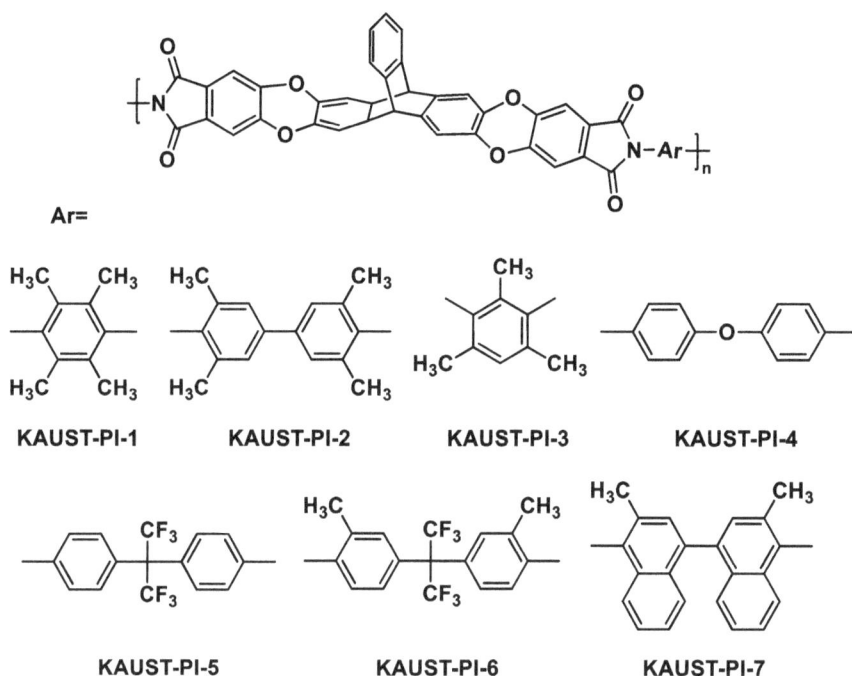

Scheme 21: Examples of triptycene-based PIM-PIs.

Table 8: Physical properties of the triptycene-based PIM-PI.

Polyimide	M_n (10^3 g mol^{-1})	PDI	S_{BET} (m^2 g^{-1})	Pore volume
KAUST-PI-1	158	2	750	0.531
KAUST-PI-2	120	1.8	740	0.49
KAUST-PI-3	82	1.9	760	0.57
KAUST-PI-4	73	2.1	420	0.282
KAUST-PI-5	83	2	650	0.53
KAUST-PI-6	153	1.6	500	0.61
KAUST-PI-7	140	1.7	840	0.59

M_n, average molecular weight of the polyimide; PDI, polydispersity of PIM-PIs; S_{BET}, BET surface area of PIM-PIs, obtained by N_2 adsorption–desorption measurement. Pore volume of the PIM-PIs is obtained from N_2 adsorption isotherm.

contorted with SBF sites in PIM-PIs. Later, two PIM-PI diamines containing SBF series were reported by Ma et al. [155] in 2013, as shown in Scheme 22. These polymers showed great gas transportation properties with notable dependence on the dianhydride (6FDA, PMDA, or SPDA) type. Moreover, the presence of Br substitution on the SBF structure has led to enhanced polymer chain rigidity which controlled the bond rotation, resulting

Table 9: Gas permeability and selectivity of the triptycene-based PIM-PIs.

Polymer	Permeability (barrer)					Selectivity		
	H_2	N_2	O_2	CH_4	CO_2	H_2/N_2	O_2/N_2	CO_2/CH_4
KAUST-PI-1	3,983	107	627	105	2,389	37	5.9	23
KAUST-PI-2	2,368	98	490	101	2,071	24	5.0	21
KAUST-PI-3	1,625	43	238	43	916	38	5.5	21
KAUST-PI-4	302	10.6	48	10.7	286	29	4.5	27
KAUST-PI-5	1,558	87	356	77	1,552	18	4.1	20
KAUST-PI-6	409	14.4	64.8	11	322	28	4.5	29
KAUST-PI-7	3,198	225	842	354	4,391	14	3.7	12

Table 10: Gas permeability and selectivity of PIM-PIs containing contorted dianhydride sites.

Polymer	Permeability					Selectivity			Reference
	H_2	N_2	O_2	CH_4	CO_2	H_2/N_2	O_2/N_2	CO_2/CH_4	
TDA1-DMN	3,050	182	783	216	3,700	16.7	4.3	17.1	[152]
TDAi3-DMN	2,233	160	594	211	3,154	14.0	3.7	14.9	[152]
SBFDA-DMN	2,966	226	850	326	4,700	13.1	3.8	14.4	[153]
PIM-PI-EA	4,230	369	1,380	457	7,340	11.5	3.7	16.1	[150]
EAD-DMN (172 mm)	4,703	480	1,586	707	8,070	9.8	3.3	11.4	[145]
EAD-DMN (70 mm)	2,856	171	655	225	3,500	16.7	3.8	15.5	[145]
EAD-DMN (23 mm)	1,289	43	213	55	979	29.9	4.9	17.8	[145]
EA-DMN (123 mm)	4,522	320	1,314	379	6,520	14.1	4.1	17.2	[145]
EA-DMN (60 mm)	3,291	165	783	154	3,321	20.0	4.8	21.6	[145]
EA-DMN (40 mm)	2,261	77.3	409	78.3	1,783	29.2	5.3	22.8	[145]
EA-DMN (15 mm)	1,844	42	263	44.8	1,100	43.9	5.6	24.6	[145]
CTB1-DMN	1,295	76.2	320	95.7	1,661	17.0	4.2	17.4	[154]
CTB2-DMN	1,150	39.9	206	40.4	948	28.8	5.2	23.5	[154]

in the gas permeability by threefold without losing the selectivity. In 2013, ladder-type polymers (PIM) containing highly rigid TB were reported. The membranes of these polymers show incredibly high permeability with good selectivity, which make their performance in gas separation located in a place higher than 2008 trade-off of Robeson's curve. From this point, a series of PIM-PI membranes were synthesized with TB as contortion sites in their building blocks. Earlier, a series of TB diamine-based PIM-PI (PI-TB-1 and PI-TB-2, Scheme 23) was prepared, displaying two different dianhydrides 6FDA and BTDA [146]. Both polymers PI-TB-1 and PI-TB-2 displayed surface areas of 540 m^2 g^{-1} and 270 m^2 g^{-1}, respectively. By comparing SBF diamine-based PIM-PIs such as 6FDA-SBF and PI-TB-1, the latter showed higher permeability with lower selectivity. Also, three additional TB diamine-based PIM-PIs were prepared using 6FDA, 3,3',4,4'-tetracarboxylicdiphenyl ether dianhydride (OPDA), and BTDA as dianhydrides [147] (Scheme 23). The resulting PI-PIM polymer (PI-TB-3) showed decreasing surface area and gas permeability but increased the selectivity compared to PI-TB-1. This is due to the presence of lower methyl groups in the TB sites in PI-TB-3 backbone which enhances the chain packing.

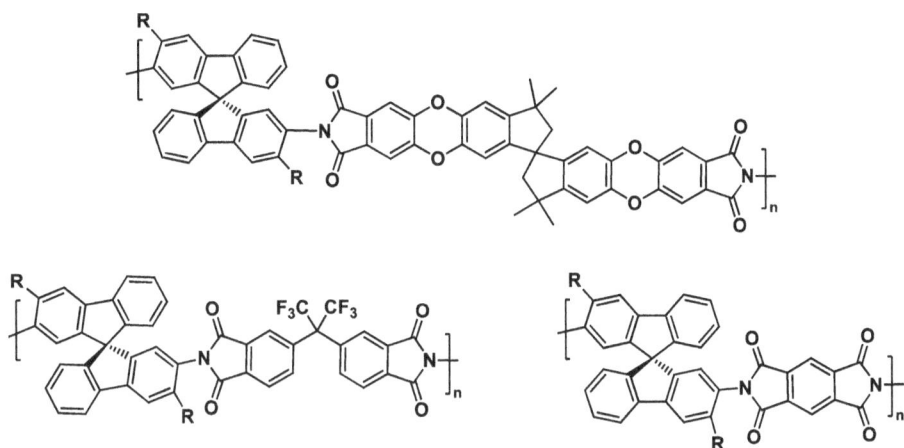

R= H or Br

Scheme 22: PIM-PIs with SBF as site of contortion in diamines.

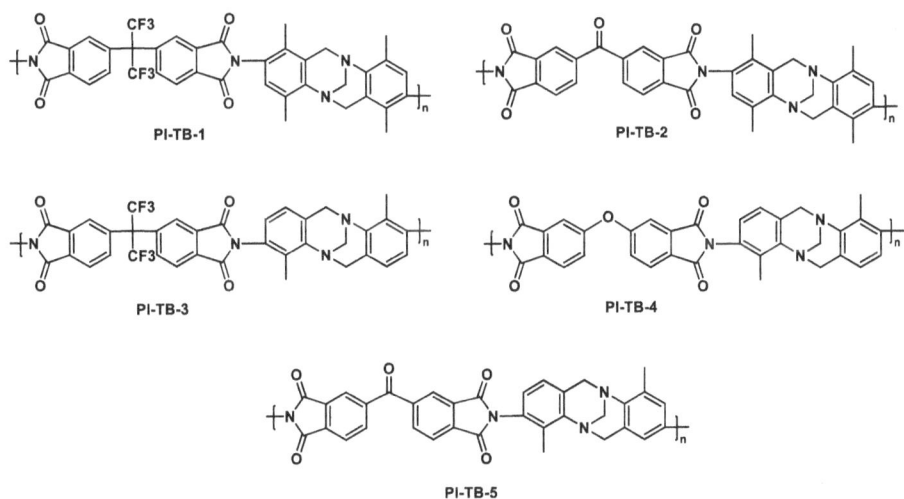

PI-TB-1

PI-TB-2

PI-TB-3

PI-TB-4

PI-TB-5

Scheme 23: Series of TB diamine-based PIM-PIs.

Moreover, the effect of methyl group positions in TB sites in two new diamines (*p*-methyl TBDA1 and *o*-methyl TBDA2) has been synthesized and studied with different dianhydrides 6FDA and 4,4-oxidiaphathalic anhydride (ODPA) [156]. The best performance was obtained by TBDA2-6FDA-PI with *o*-methyl with 285 barrer CO_2 permeability and CO_2/CH_4

selectivity of 36. Abdulhamid et al. [158] have investigated the effect of carbocyclic bridge (V-shaped) in the TB diamine in two PIM-PIs (6FDA-CTBDA and 6FDA-iCTBDA) on gas transport properties. Therefore, the carbocyclic PI displayed higher BET surface area and higher gas permeability with slightly lower gas pair selectivity compared to TB PI analogue 6FDA-TBDA. TRIP as contortion sites in diamines demonstrated highly rigid 3D paddle-like diamines as building blocks for PIM-PI, which afford high FFV. A highly FFV PIM-PI was prepared from 6FDA and 2,6-diaminotriptycene shows high gas permeability and selectivity with strong plasticization resistance for gas separation [158]. Wiegand et al. [159] have studied the effect of substitution of 1,4-triptycene diamines with substituted H, CH_3, and CF_3 groups on the gas permeability. All the three PIs prepared (6FDA-1,4-trip-para, 6FDA-1,4-trip-CH_3, and 6FDA-1,4-trip-CF_3) exhibited good selectivity but relatively low permeability compared with other PIM-PIs. They noticed that the gas increasing in the FFV as the substitutes changed from $CH_3 > H > CF_3$. Table 11 illustrates some PIM-PIs with contorted diamine and their gas transport properties.

Table 11: Gas separation performance of PIM-PIs with different contorted diamines.

Polymer	Permeability					Selectivity			Reference
	H_2	N_2	O_2	CH_4	CO_2	H_2/N_2	O_2/N_2	CO_2/CH_4	
6FDA-SBF	234	7.8	35.1	6.4	182	30	4.5	28.4	[155]
PMDA-SBF	230	8.5	35.5	9.1	197	27.1	4.2	21.6	[155]
SPDA-SBF	501	28.6	111	41.1	614	17.5	3.9	149	[155]
6FDA-BSBF	531	27	107	24.9	580	19.7	4	23.3	[155]
PMDA-BSBF	560	28.8	116	36.5	693	19.4	4	19	[155]
SPDA-BSBF	919	69	243	102	1340	13.3	3.5	13.1	[155]
PI-TB-1	607	31	119	27	457	19	3.8	17	[146]
PI-TB-2	134	2.5	14	2.1	55	53	5.5	26	[146]
PI-TB-3	299	9.5	42	6.7	218	31.4	4.5	32.7	[147]
PI-TB-4	40	1.3	3.5	1	13.5	32.1	2.8	13.5	[147]
PI-TB-5	53.8	1.9	4.9	1.7	19.6	28.7	2.6	11.5	[147]
PIM-PI-TB-1	612	42	133	44	662	14.6	3.2	15	[160]
PIM-PI-TB-2	582	34	123	31	595	17.1	3.6	19.2	[160]
4MTBDA-6FDA	1,446	133	408	116	1,672	10.4	3.1	14.4	[161]
4MTBDA-PMDA	3,300	290	1,080	390	4,460	11.4	3.7	11.4	[161]
4MTBDA-SBIDA	3,200	372	1,132	591	5,140	8.6	3.0	8.7	[161]
4MTBDA-SBFDA	2,901	264	941	371	4,476	11.0	3.6	12.1	[161]

7 Conclusion and future outlooks

PI-based membranes have emerged as a promising class of materials for gas separation applications due to their unique combination of properties, including high thermal and chemical stability, good mechanical strength, tunable permeability and selectivity, and ability to form thin and defect-free films. One of the key challenges in using PI-based membranes for gas separation is their inherent low permeability. However, researchers have developed a number of strategies to improve the permeability of PI-based membranes, such as incorporating nanofillers into the polymer matrix, developing composite membranes that combine PIs with other materials, such as inorganic membranes or porous polymers, and modifying the surface of PI membranes to improve their gas transport properties and develop PI of intrinsic microporosity which showed excellent permeability and moderate selectivity. As research and development in the field of PI-based membranes continues, we can expect to see significant advances in their performance and cost-effectiveness. This will make them even more attractive for a wider range of gas separation applications. The future of PI-based membranes for gas separation is very promising. Researchers are actively developing new strategies to improve the performance and cost-effectiveness of these membranes. One key area of research is the development of new PI materials with tailored properties, such as higher intrinsic permeability and selectivity. Another important area of research is the development of composite membranes that combine PIs with other materials to achieve synergistic effects. In addition to these material development efforts, researchers are also working to improve the manufacturing processes for PI-based membranes. This includes developing new methods for producing thin and defect-free films on a commercial scale. In the near future, we can expect to see PI-based membranes deployed in a variety of commercial and industrial settings.

References

[1] Yahya, E., Sean, G. P. and Ismail, A. F. 5 – Membrane Engineering in Gas Separation. In: Iulianelli, A., Cassano, A., Conidi, C. and Petrotos, K. (eds). Membrane Engineering in the Circular Economy, Elsevier, 2022, pp. 123–147. doi https://doi.org/10.1016/B978-0-323-85253-1.00010-1.

[2] Bernardo, P., Drioli, E. and Golemme, G. Membrane gas separation: A review/state of the art. Ind. Eng. Chem. Res., 2009, **48**: 4638–4663.

[3] Drioli, E., Giorno, L. and Macedonio, F. Membrane Engineering, Walter de Gruyter GmbH & Co KG, 2018.

[4] Sholl, D. S. and Lively, R. P. Seven chemical separations to change the world. Nature, *2016 5327600* 2016, **532**: 435–437.

[5] Sidhikku Kandath Valappil, R., Ghasem, N. and Al-Marzouqi, M. Current and future trends in polymer membrane-based gas separation technology: A comprehensive review. J. Ind. Eng. Chem., 2021, **98**: 103–129.

[6] Koros, W. J., Ma, Y. H. and Shimidzu, T. Terminology for membranes and membrane processes (IUPAC Recommendations 1996). Pure. Appl. Chem., 1996, **68**: 1479–1489.

[7] Ismail, A. F., Khulbe, K. C. and Matsuura, T. Gas separation membranes: Polymeric and inorganic. Gas Sep. Membr. Polym. Inorg., 2015, 1–331. doi: 10.1007/978-3-319-01095-3/COVER.

[8] Langmuir, I. The adsorption of gases on plane surfaces of glass, mica and platinum. J. Am. Chem. Soc., 1918, **40**: 1361–1403.

[9]   Rouquérol, J., *et al*. Recommendations for the characterisation of porous solids. Pure. Appl. Chem., 1994, **66**: 1739–1758.

[10]   Kosinov, N., Gascon, J., Kapteijn, F. and Hensen, E. J. M. Recent developments in zeolite membranes for gas separation. J. Memb. Sci., 2016, **499**: 65–79.

[11]   Deng, L. and Hägg, M. B. Techno-economic evaluation of biogas upgrading process using CO_2 facilitated transport membrane. Int. J. Greenh. Gas Control., 2010, **4**: 638–646.

[12]   Yu, L., Holmgren, A., Zhou, M. and Hedlund, J. Highly permeable CHA membranes prepared by fluoride synthesis for efficient CO_2/CH_4 separation. J. Mater. Chem. A., 2018, **6**: 6847–6853.

[13]   Miyamoto, M., *et al*. Effect of core-shell structuring of chabazite zeolite with a siliceous zeolite thin layer on the separation of acetone-butanol-ethanol vapor in humid vapor conditions. Chem. Eng. J., 2019, **363**: 292–299.

[14]   De Meis, D., Richetta, M. and Serra, E. Microporous inorganic membranes for gas separation and purification. InterCeram Int. Ceram. Rev., 2018, **67**: 16–21.

[15]   Gu, Y., Hacarlioglu, P. and Oyama, S. T. Hydrothermally stable silica–alumina composite membranes for hydrogen separation. J. Memb. Sci., 2008, **310**: 28–37.

[16]   Ahn, S. J., Takagaki, A., Sugawara, T., Kikuchi, R. and Oyama, S. T. Permeation properties of silica-zirconia composite membranes supported on porous alumina substrates. J. Memb. Sci., 2017, **526**: 409–416.

[17]   Gu, Y. and Oyama, S. T. Permeation properties and hydrothermal stability of silica–titania membranes supported on porous alumina substrates. J. Memb. Sci., 2009, **345**: 267–275.

[18]   Ito, W., Nagai, T. and Sakon, T. Oxygen separation from compressed air using a mixed conducting perovskite-type oxide membrane. Solid State Ion., 2007, **178**: 809–816.

[19]   Conde, J. J., Maroño, M. and Sánchez-Hervás, J. M. Pd-based membranes for hydrogen separation: Review of alloying elements and their influence on membrane properties. 2016, **46**: 152–177. http://dx.doi.org/10.1080/15422119.2016.1212379.

[20]   Li, H., *et al*. Ultrathin, molecular-sieving graphene oxide membranes for selective hydrogen separation. Science (80-.), 2013, **342**: 95–98.

[21]   Kim, H. W., *et al*. Selective gas transport through few-layered graphene and graphene oxide membranes. Science (80-.), 2013, **342**: 91–95.

[22]   Li, L., Song, C., Jiang, D. and Wang, T. Preparation and enhanced gas separation performance of Carbon/Carbon nanotubes (C/CNTs) hybrid membranes. Sep. Purif. Technol., 2017, **188**: 73–80.

[23]   Robeson, L. M. Correlation of separation factor versus permeability for polymeric membranes. J. Memb. Sci., 1991, **62**: 165–185.

[24]   Robeson, L. M. The upper bound revisited. J. Memb. Sci., 2008, **320**: 390–400.

[25]   Salleh, W. N. W., Ismail, A. F., Matsuura, T. and Abdullah, M. S. Precursor selection and process conditions in the preparation of carbon membrane for gas separation: A review. 2011, **40**: 261–311. http://dx.doi.org/10.1080/15422119.2011.555648.

[26]   Ma, X. H. and Yang, S. Y. Polyimide gas separation membranes. Adv. Polym. Mater. Synth. Charact. Appl., 2018, 257–322. doi: 10.1016/B978-0-12-812640-0.00006-8.

[27]   Liaw, D. J., *et al*. Advanced polyimide materials: Syntheses, physical properties and applications. Prog. Polym. Sci., 2012, **37**: 907–974.

[28]   Singh, L., Wadhwa, D. and Sedha, M. Review on structure, characteristics, and applications of cross-linked porous polyimides. J. Res. Med. Dent. Sci., 2022, **10**: 111–115.

[29]   Zhuang, Y., Seong, J. G. and Lee, Y. M. Polyimides containing aliphatic/alicyclic segments in the main chains. Prog. Polym. Sci., 2019, **92**: 35–88.

[30]   Meador, M. A. B. and Vivod, S. L. Encyclopedia of polymeric nanomaterials. Encycl. Polym. Nanomater., 2021, 1–11. doi: 10.1007/978-3-642-36199-9.

[31]   Ding, M. Isomeric polyimides. Prog. Polym. Sci., 2007, **32**: 623–668.

[32]   Fang, Q., *et al*. Designed synthesis of large-pore crystalline polyimide covalent organic frameworks. Nat. Commun., 2014, **5**: 4503.

[33] Hasegawa, M. Development of solution-processable, optically transparent polyimides with ultra-low linear coefficients of thermal expansion. Polymers (Basel), 2017, **9**: 520.

[34] Jiang, L. Y., Wang, Y., Chung, T. S., Qiao, X. Y. and Lai, J. Y. Polyimides membranes for pervaporation and biofuels separation. Prog. Polym. Sci., 2009, **34**: 1135–1160.

[35] Hasegawa, M. and Horie, K. Photophysics, photochemistry, and optical properties of polyimides. Progress in Polymer Science. 2001, **26**: 259–335.

[36] Morikawa, A., Furukawa, T. A. and Moriyama, Y. Synthesis and characterization of novel aromatic polyimides from bis(4-amino-2-biphenyl)ether and aromatic tetracarboxylic dianhydrides. Polym. J., 2005, **37**: 759–766.

[37] Park, C. Y., Kim, E. H., Kim, J. H., Lee, Y. M. and Kim, J. H. Novel semi-alicyclic polyimide membranes: Synthesis, characterization, and gas separation properties. Polymer (Guildf)., 2018, **151**: 325–333.

[38] Zhai, L., Yang, S. and Fan, L. Preparation and characterization of highly transparent and colorless semi-aromatic polyimide films derived from alicyclic dianhydride and aromatic diamines. Polymer (Guildf)., 2012, **53**: 3529–3539.

[39] Krol, J. J., Boerrigter, M. and Koops, G. H. Polyimide hollow fiber gas separation membranes: Preparation and the suppression of plasticization in propane/propylene environments. J. Memb. Sci., 2001, **184**: 275–286.

[40] Shalygin, M. G., Abramov, S. M., Netrusov, A. I. and Teplyakov, V. V. Membrane recovery of hydrogen from gaseous mixtures of biogenic and technogenic origin. Int. J. Hydrogen Energy, 2015, **40**: 3438–3451.

[41] Okamoto, K. -. I., Tanaka, K., Yokoshi, O. and Kita, H. The effect of morphology on sorption and transport of carbon dioxide in poly(4,4'-oxydiphenylene pyromellitimide). J. Polym. Sci, Part B: Polym Phys., 1989, **27**: 643–654.

[42] Sobana, S. and Panda, R. C. Review on modelling and control of desalination system using reverse osmosis. Rev. Environ. Sci. Biotechnol., 2011, **10**: 139–150.

[43] Sanders, D. F., *et al.* Energy-efficient polymeric gas separation membranes for a sustainable future: A review. Polymer (Guildf), 2013, **54**: 4729–4761.

[44] Ghosh, M. K. and Mittal, K. L. (Eds). Fundamentals, P. Applications. Taylor & francis, 1996.

[45] Sonnett, J. M., McCulloung, R. L., Beeler, A. J. and Gannett, T. P. The kinetics of one-step linear polyimide formation: Reaction of p-phenylenediamine and 4, 4-(hexafluoroisopropylidene) bis [phthalic acid]. Adv. Polyim. Sci. Technol. Lancaster Technomic. Publ. Co., 1993, 313–325.

[46] Kim, J., Lee, S. and Kim, S. Y. Incorporation effects of fluorinated side groups into polyimide membranes on their physical and gas permeation properties. J. Appl. Polym. Sci., 2000, **77**: 2756–2767.

[47] Feger, C., Khojasteh, M. M. and McGrath, J. E. *Polyimides: Materials, Chemistry, and Characterization: Proceedings of the Third International Conference on Polyimides, Ellenville, New York, November 2–4, 1988* Elsevier Publishing Company, 1989.

[48] Abdulhamid, M. A., Lai, H. W. H., Wang, Y., Jin, Z., Teo, Y. C., Ma, X., Pinnau, I. and Xia, Y. Microporous polyimides from ladder diamines synthesized by facile catalytic arene–norbornene annulation as high–performance membranes for gas separation. Chem. Mater., 2019, **31**(5): 1767–1774.

[49] Kreuz, J. A., Endrey, A. L., Gay, F. P. and Sroog, C. E. Studies of thermal cyclizations of polyamic acids and tertiary amine salts. J. Polym. Sci. Part A-1 Polym. Chem., 1966, **4**: 2607–2616.

[50] Hergenrother, P. M. Polyimides.

[51] Abdulhamid, M. A. Structure-property relationships of polyimides with intrinsic microporosity (PIM-PIs) and their gas transport properties. (2019). http://hdl.handle.net/10754/652449

[52] Seino, H., Mochizuki, A. and Ueda, M. Synthesis of aliphatic polyimides containing adamantyl units. J. Polym. Sci. Part A: Polym. Chem., 1999, **37**: 3584–3590.

[53] Mathews, A. S., Kim, I. and Ha, C. Fully aliphatic polyimides from adamantane-based diamines for enhanced thermal stability, solubility, transparency, and low dielectric constant. J. Appl. Polym. Sci., 2006, **102**: 3316–3326.

[54] Mathews, A. S., Kim, I. and Ha, C. Synthesis and characterization of novel fully aliphatic polyimidosiloxanes based on alicyclic or adamantyl diamines. J. Polym. Sci. Part A Polym. Chem., 2006, **44**: 5254–5270.

[55] Oishi, Y., *et al*. Synthesis of fluorine-containing wholly alicyclic polyimides by in situ silylation method. J. Photopolym. Sci. Technol., 2003, **16**: 263–266.

[56] Abdulhamid, M. A., Ma, X., Ghanem, B. S. and Pinnau, I. Synthesis and characterization of organo-soluble polyimides derived from alicyclic dianhydrides and a dihydroxyl–functionalized spirobisindane diamine. ACS Appl. Polym. Mater., 2019, **1**(1): 63–69.

[57] Seino, H., Sasaki, T., Mochizuki, A. and Ueda, M. Synthesis of fully aliphatic polyimides. High Perform. Polym., 1999, **11**: 255–262.

[58] Watanabe, Y., Sakai, Y., Ueda, M., Oishi, Y. and Mori, K. Synthesis of wholly alicyclic polyimides from N-silylated alicyclic diamines. Chem. Lett., 2000, **29**: 450–451.

[59] Watanabe, Y., Shibasaki, Y., Ando, S. and Ueda, M. Synthesis of semiaromatic polyimides from aromatic diamines containing adamantyl units and alicyclic dianhydrides. J. Polym. Sci. Part A Polym. Chem., 2004, **42**: 144–150.

[60] Watanabe, Y., *et al*. Synthesis of wholly alicyclic polyimides from N-silylated alicyclic diamines and alicyclic dianhydrides. Macromolecules, 2002, **35**: 2277–2281.

[61] Vanherck, K., Koeckelberghs, G. and Vankelecom, I. F. J. Crosslinking polyimides for membrane applications: A review. Prog. Polym. Sci., 2013, **38**: 874–896.

[62] Ding, Y., Hou, H., Zhao, Y., Zhu, Z. and Fong, H. Electrospun polyimide nanofibers and their applications. Prog. Polym. Sci., 2016, **61**: 67–103.

[63] Wilson, D., Stenzenberger, H. D. and Hergenrother, P. M. Polyimides, Springer Science & Business Media, 2013.

[64] Fukukawa, K. I. and Ueda, M. Recent progress of photosensitive polyimides. Polym. J., 2008, **40**: 281–296.

[65] Ghosh, M. K. and Mittal, K. L. Polyimides : Fundamentals and Applications, *TA – TT* – Marcel Dekker, New York, 1996, doi:LK – https://worldcat.org/title/34745932.

[66] Sroog, C. E. Polyimides. J. Polym. Sci. Macromol. Rev., 1976, **11**: 161–208.

[67] Harris, F. W. Synthesis of Aromatic Polyimides from Dianhydrides and Diamines BT – Polyimides, Wilson, D., Stenzenberger, H. D. and Hergenrother, P. M. (eds.). Springer, Netherlands, 1990, pp. 1–37. doi: 10.1007/978-94-010-9661-4_1.

[68] Ratta, V. Crystallization, Morphology, Thermal Stability and Adhesive Properties of Novel High Performance Semicrystalline Polyimides. *ProQuest Dissertations and Theses*, Virginia Polytechnic Institute and State University PP –, United States – Virginia, 1999.

[69] Wang, Z.-H., Chen, X., Yang, H.-X., Zhao, J. and Yang, S.-Y. The in-plane orientation and thermal mechanical properties of the chemically imidized polyimide films. Chinese J. Polym. Sci., 2019, **37**: 268–278.

[70] Meyer, G. W., *et al*. Phenylmaleimide encapped arylene ether imide oligomers. *Polymer (Guildf)*. Polym. Sci. USSR., 1996, **37**: 5077–5088.

[71] Vinogradova, S. V., *et al*. Chemical cyclization of poly (amido-acids) in solution. Polym. Sci. USSR., 1974, **16**: 584–589.

[72] Koton, M. M., *et al*. Investigation of the kinetics of chemical imidization. Polym. Sci. USSR., 1982, **24**: 791–800.

[73] Volksen, W. Condensation polyimides: Synthesis, solution behavior, and imidization characteristics. High Perform. Polym., 2005, **117**: 111–164.

[74] Koton, M. M., *et al*. Experimental and theoretical study of the effect of medium on chemical imidization. Polym. Sci. USSR., 1984, **26**: 2839–2848.

[75] Roderick, W. R. Dehydration of N-(p-chlorophenyl) phthalamic acid by acetic and trifluoroacetic anhydrides. J. Org. Chem., 1964, **29**: 745–747.

[76] Young, P. R., Davis, J. R. J., Chang, A. C. and Richardson, J. N. Characterization of a thermally imidized soluble polyimide film. J. Polym. Sci. Part A Polym. Chem., 1990, **28**: 3107–3122.

[77] Feldman, D. and Barbalata, A. Synthetic Polymers: Technology, Properties, Applications, Springer Science & Business Media, 1996.

[78] Ghosh, M. Polyimides: Fundamentals and Applications, CRC press, 2018.

[79] Verbicky, J. W. P., Jr. Wiley-Interscience, Encycl. Polym. Sci. Eng. Second Ed., 1988, **12**: 364–383.

[80] Takekoshi, T. Polyimides. In: New Polymer Materials, Springer, 2005, pp. 1–25.

[81] Pan, J., Lau, W. W. Y., Zhang, Z. F. and Hu, X. Z. Synthesis and properties of new copolymers containing hindered amine. J. Appl. Polym. Sci., 1996, **61**: 1405–1412.

[82] Imai, Y. and Kojima, K. Preparation of polyimides from pyromellitic dithioanhydride and aromatic diamines. J. Polym. Sci. Part A-1 Polym. Chem., 1972, **10**: 2091–2096.

[83] Liou, G., Hsiao, S., Ishida, M., Kakimoto, M. and Imai, Y. Synthesis and properties of new aromatic poly (amine-imide) s derived from N, N′-bis (4-aminophenyl)-N, N′-diphenyl-1, 4-phenylenediamine. J. Polym. Sci. Part A Polym. Chem., 2002, **40**: 3815–3822.

[84] Gheneim, R., Perez-Berumen, C. and Gandini, A. D. Alder reactions with novel polymeric dienes and dienophiles: Synthesis of reversibly cross-linked elastomers. Macromolecules, 2002, **35**: 7246–7253.

[85] Chi, J. H., Shin, G. J., Kim, Y. S. and Jung, J. C. Synthesis of new alicyclic polyimides by Diels-Alder polymerization. J. Appl. Polym. Sci., 2007, **106**: 3823–3832.

[86] Di Bella, S., et al. Film polymerization– a new route to the synthesis of insoluble polyimides containing functional nickel(II) Schiff base units in the main chain. Eur. J. Inorg. Chem., 2004, **2004**: 2701–2705.

[87] Liaw, D.-J., et al. Advanced polyimide materials: Syntheses, physical properties and applications. Prog. Polym. Sci., 2012, **37**: 907–974.

[88] Boldebuck, E. M. and Klebe, J. F. U.S. Patent No. 3,303,157. U.S. Patent and Trademark Office, 1967.

[89] Kaneda, T., et al. Wholly aromatic polyamides containing bridged biphenylylene groups, 1. Poly (3, 8-phenanthridinonediyl terephthalamide). Die Makromol. Chemie., 1982, **183**: 417–432.

[90] Muñoz, D. M., De la Campa, J. G., de Abajo, J. and Lozano, A. E. Experimental and theoretical study of an improved activated polycondensation method for aromatic polyimides. Macromolecules, 2007, **40**: 8225–8232.

[91] Moy, T. M., DePorter, C. D. and McGrath, J. E. Synthesis of soluble polyimides and functionalized imide oligomers via solution imidization of aromatic diester-diacids and aromatic diamines. Polymer (Guildf), 1993, **34**: 819–824.

[92] Matteucci, S., Yampolskii, Y., Freeman, B. D. and Pinnau, I. Transport of gases and vapors in glassy and rubbery polymers. Mater. Sci. Membr. Gas Vap. Sep., 2006, 1–47. doi: 10.1002/047002903X.CH1.

[93] Ismail, A. F., Khulbe, K. C. and Matsuura, T. Fundamentals of Gas Permeation Through Membranes. Gas Sep. Membr., 2015, 11–35. doi: 10.1007/978-3-319-01095-3_2.

[94] Koros, W. J. and Fleming, G. K. Membrane-based gas separation. J. Memb. Sci., 1993, **83**: 1–80.

[95] Ruthven, D. M. Diffusion in zeolites. Stud. Surf. Sci. Catal., 1995, **97**: 223–234.

[96] Sotirchos, S. V. and Burganos, V. N. Transport of gases in porous membranes. MRS Bull, 1999, **24**: 41–45.

[97] Javaid, A. Membranes for solubility-based gas separation applications. Chem. Eng. J., 2005, **112**: 219–226.

[98] Yampolskii, Y. Polymeric gas separation membranes. Macromolecules, 2012, **45**: 3298–3311.

[99] Bera, D., Chatterjee, R. and Banerjee, S. Aromatic polyamide nonporous membranes for gas separation application. E-Polymers, 2021, **21**: 108–130.

[100] Ismail, A. F., Khulbe, K. C. and Matsuura, T. Gas separation membranes. Switz. Springer, 2015, **10**: 973–978.

[101] Kim, K.-J., Park, S.-H., So, -W.-W., Ahn, D.-J. and Moon, S.-J. CO2 separation performances of composite membranes of 6FDA-based polyimides with a polar group. J. Memb. Sci., 2003, **211**: 41–49.

[102] Serbezeanu, D., Carja, I. D., Bruma, M. and Ronova, I. A. Correlation between physical properties and conformational rigidity of some aromatic polyimides having pendant phenolic groups. Struct. Chem., 2016, **27**: 973–981.

[103] Chen, Y. C., Su, Y. Y. and Lin, S. C. Organosoluble and colorless fluorinated poly(ether imide)s containing a bulky fluorene bis(ether anhydride) and various trifluoromethyl-substituted aromatic bis(ether amine)s: Synthesis and characterization. Iran. Polym. J. (English Ed., 2023, **32**: 499–511.

[104] Wang, Y. C., et al. Sorption and transport properties of gases in aromatic polyimide membranes. J. Memb. Sci., 2005, **248**: 15–25.

[105] Lee, J. S., Yan, Y. Z., Park, S. S., Ahn, S. K. and Ha, C. S. A novel diamine containing ester and diphenylethane groups for colorless polyimide with a low dielectric constant and low water absorption. Polym., 2022, **14**: 4504.

[106] Hasegawa, M., Saito, T. and Tsujimura, Y. Poly(ester imide)s possessing low coefficients of thermal expansion and low water absorption (IV): Effects of ester-linked tetracarboxylic dianhydrides with longitudinally extended structures. Polym. Adv. Technol., 2020, **31**: 389–406.

[107] Liaw, D. J., Hsu, C. Y., Hsu, P. N. and Lin, S. L. Synthesis and characterization of new highly organosoluble poly(etherimide)s derived from 1,1-bis{4-[4-(3,4-dicarboxyphenoxy)phenyl]-4-phenylcyclohexane} dianhydride. J. Polym. Sci. Part A Polym. Chem., 2002, **40**: 2066–2074.

[108] Hasegawa, M., Kasamatsu, K. and Koseki, K. Colorless poly(ester imide)s derived from hydrogenated trimellitic anhydride. Eur. Polym. J., 2012, **48**: 483–498.

[109] Zhang, M., et al. Approaches to suppress CO2-induced plasticization of polyimide membranes in gas separation applications. Process, 2019, **7**: 51.

[110] Wang, Y., et al. Polymers of intrinsic microporosity for energy-intensive membrane-based gas separations. Mater. Today Nano., 2018, **3**: 69–95.

[111] Staudt-Bickel, C. and Koros, W. J. Improvement of CO2/CH4 separation characteristics of polyimides by chemical crosslinking. J. Memb. Sci., 1999, **155**: 145–154.

[112] Ma, X., Abdulhamid, M., Miao, X. and Pinnau, I. Facile synthesis of a hydroxyl-functionalized Troger's base diamine: A new building block for high–performance polyimide gas separation membranes. Macromolecules, 2017, **50**(24): 9569–9576.

[113] Comesaña-Gándara, B., et al. Thermally rearranged polybenzoxazoles and poly(benzoxazole-co-imide)s from ortho-hydroxyamine monomers for high performance gas separation membranes. J. Memb. Sci., 2015, **493**: 329–339.

[114] Jung, C. H. and Lee, Y. M. Gas permeation properties of hydroxyl-group containing polyimide membranes. Macromol. Res., 2008, **16**: 555–560.

[115] Wind, J. D., Paul, D. R. and Koros, W. J. Natural gas permeation in polyimide membranes. J. Memb. Sci., 2004, **228**: 227–236.

[116] Qiu, W., Xu, L., Chen, C. C., Paul, D. R. and Koros, W. J. Gas separation performance of 6FDA-based polyimides with different chemical structures. Polymer (Guildf), 2013, **54**: 6226–6235.

[117] Abdulhamid, M. A., Genduso, G., Wang, Y., Ma, X. and Pinnau, I. Plasticization-resistant carboxyl-functionalized 6FDA-polyimide of intrinsic microporosity (PIM-PI) for membrane-based gas separation. Ind. Eng. Chem. Res., 2020, **59**: 5247–5256.

[118] Park, S. and Jeong, H. K. Cross-linked polyimide/ZIF-8 mixed-matrix membranes by in situ formation of ZIF-8: Effect of cross-linking on their propylene/propane separation. Membranes (Basel), 2022, **12**: 964.

[119] Hillock, A. M. W. and Koros, W. J. Cross-linkable polyimide membrane for natural gas purification and carbon dioxide plasticization reduction. Macromolecules, 2007, **40**: 583–587.

[120] Luo, S., et al. Highly CO2-selective gas separation membranes based on segmented copolymers of poly(Ethylene oxide) reinforced with pentiptycene-containing polyimide hard segments. ACS Appl. Mater. Interfaces, 2016, **8**: 2306–2317.

[121] Imtiaz, A., et al. ZIF-filler incorporated mixed matrix membranes (MMMs) for efficient gas separation: A review. J. Environ. Chem. Eng., 2022, **10**: 108541.

[122] Lee, T. H., *et al.* Surface modification of matrimid® 5218 polyimide membrane with fluorine-containing diamines for efficient gas separation. Membr., 2022, **12**: 256.

[123] Abdulhamid, M. A., Genduso, G., Ma, X. and Pinnau, I. Synthesis and characterization of 6FDA/3,5-diamino-2,4,6-trimethylbenzenesulfonic acid-derived polyimide for gas separation applications. Sep. Purif. Technol., 2021, **257**: 117910.

[124] Budd, P. M., *et al.* Solution-processed, organophilic membrane derived from a polymer of intrinsic microporosity. Adv. Mater., 2004, **16**: 456–459.

[125] Masuda, T., Isobe, E., Higashimura, T. and Takada, K. Poly[1-(trimethylsilyl)-1-propyne]: A new high polymer synthesized with transition-metal catalysts and characterized by extremely high gas permeability. J. Am. Chem. Soc., 1983, **105**: 7473–7474.

[126] McKeown, N. B. Polymers of intrinsic microporosity (PIMs). Polymer (Guildf), 2020, **202**: 122736.

[127] McKeown, N. B., *et al.* Polymers of intrinsic microporosity (PIMs): Bridging the void between microporous and polymeric materials. Chem. – A Eur. J., 2005, **11**: 2610–2620.

[128] Budd, P. M., McKeown, N. B. and Fritsch, D. Free volume and intrinsic microporosity in polymers. J. Mater. Chem., 2005, **15**: 1977–1986.

[129] Weber, J., Su, Q., Antonietti, M. and Thomas, A. Exploring polymers of intrinsic microporosity – microporous, soluble polyamide and polyimide. Macromol. Rapid Commun., 2007, **28**: 1871–1876.

[130] Ghanem, B. S., *et al.* Synthesis, characterization, and gas permeation properties of a novel group of polymers with intrinsic microporosity: PIM-polyimides. Macromolecules, 2009, **42**: 7881–7888.

[131] Ghanem, B. S., McKeown, N. B., Budd, P. M., Selbie, J. D. and Fritsch, D. High-performance membranes from polyimides with intrinsic microporosity. Adv. Mater., 2008, **20**: 2766–2771.

[132] Ghanem, B. S., Alghunaimi, F., Wang, Y., Genduso, G. and Pinnau, I. Synthesis of highly gas-permeable polyimides of intrinsic microporosity derived from 1,3,6,8-Tetramethyl-2,7-diaminotriptycene. ACS Omega, 2018, **3**: 11874–11882.

[133] Wiegand, J. R., *et al.* Synthesis and characterization of triptycene-based polyimides with tunable high fractional free volume for gas separation membranes. J. Mater. Chem. A, 2014, **2**: 13309–13320.

[134] Dai, S., Liao, R., Zhou, H. and Jin, W. Synthesis of triptycene-based linear polyamide membrane for molecular sieving of N2 from the VOC mixture. Sep. Purif. Technol., 2020, **252**: 117355.

[135] Zhou, H., *et al.* Microporous polyamide membranes for molecular sieving of nitrogen from volatile organic compounds. Angew. Chemie Int. Ed., 2017, **56**: 5755–5759.

[136] Luo, S., *et al.* Molecular origins of fast and selective gas transport in pentiptycene-containing polyimide membranes and their physical aging behavior. J. Memb. Sci., 2016, **518**: 100–109.

[137] Luo, S., *et al.* Pentiptycene-based polyimides with hierarchically controlled molecular cavity architecture for efficient membrane gas separation. J. Memb. Sci., 2015, **480**: 20–30.

[138] Shrestha, B. B., *et al.* A facile synthesis of contorted spirobisindane-diamine and its microporous polyimides for gas separation. RSC Adv., 2018, **8**: 6326–6330.

[139] Wang, Z., Wang, D. and Jin, J. Microporous polyimides with rationally designed chain structure achieving high performance for gas separation. Macromolecules, 2014, **47**: 7477–7483.

[140] Calle, M., *et al.* Local chain mobility dependence on molecular structure in polyimides with bulky side groups: Correlation with gas separation properties. J. Memb. Sci., 2013, **434**: 121–129.

[141] AlDawhi, Z. A., BinSharfan, I. I. and Abdulhamid, M. A. Carboxyl–functionalized polyimides for efficient bisphenol A removal: Influence of wettability and porosity on adsorption capacity. Chemosphere, 2023, **313**: 137347.

[142] Abdulhamid, M. A., Genduso, G., Wang, Y., Ma, X. and Pinnau, I. Plasticization-resistant carboxyl-functionalized 6FDA-polyimide of intrinsic microporosity (PIM–PI) for membrane-based gas separation. Ind. Eng. Chem. Res., 2019, **59**(12): 5247–5256.

[143] Swaidan, R., Al-Saeedi, M., Ghanem, B., Litwiller, E. and Pinnau, I. Rational design of intrinsically ultramicroporous polyimides containing bridgehead-substituted triptycene for highly selective and permeable gas separation membranes. Macromolecules, 2014, **47**: 5104–5114.

[144] Madkour, T. M. and Mark, J. E. Molecular modeling investigation of the fundamental structural parameters of polymers of intrinsic microporosity for the design of tailor-made ultra-permeable and highly selective gas separation membranes. J. Memb. Sci., 2013, **431**: 37–46.

[145] Ma, X. and Pinnau, I. Effect of film thickness and physical aging on intrinsic gas permeation properties of microporous ethanoanthracene-based polyimides. Macromolecules, 2018, **51**: 1069–1076.

[146] Zhuang, Y., *et al*. Intrinsically microporous soluble polyimides incorporating Tröger's base for membrane gas separation. Macromolecules, 2014, **47**: 3254–3262.

[147] Zhuang, Y., *et al*. High-strength, soluble polyimide membranes incorporating Tröger's Base for gas separation. J. Memb. Sci., 2016, **504**: 55–65.

[148] Zhuang, Y., *et al*. Soluble, microporous, Tröger's base copolyimides with tunable membrane performance for gas separation. Chem. Commun., 2016, **52**: 3817–3820.

[149] Abdulhamid, M. A. Tröger's base-derived dianhydride as a promising contorted building block for polyimide-based membranes for gas separation. Sep. Purif. Technol., 2023, **310**: 123208.

[150] Rogan, Y., *et al*. A highly permeable polyimide with enhanced selectivity for membrane gas separations. J. Mater. Chem. A, 2014, **2**: 4874–4877.

[151] Ghanem, B. S., *et al*. Ultra-microporous triptycene-based polyimide membranes for high-performance gas separation. Adv. Mater., 2014, **26**: 3688–3692.

[152] Ghanem, B., Alghunaimi, F., Ma, X., Alaslai, N. and Pinnau, I. Synthesis and characterization of novel triptycene dianhydrides and polyimides of intrinsic microporosity based on 3,3′-dimethylnaphthidine. Polymer (Guildf), 2016, **101**: 225–232.

[153] Ma, X., Ghanem, B., Salines, O., Litwiller, E. and Pinnau, I. Synthesis and effect of physical aging on gas transport properties of a microporous polyimide derived from a novel spirobifluorene-based dianhydride. ACS Macro. Lett., 2015, **4**: 231–235.

[154] Ma, X., Abdulhamid, M. A. and Pinnau, I. Design and synthesis of polyimides based on carbocyclic pseudo-Tröger's Base-derived dianhydrides for membrane gas separation applications. Macromolecules, 2017, **50**: 5850–5857.

[155] Ma, X., Salinas, O., Litwiller, E. and Pinnau, I. Novel spirobifluorene- and dibromospirobifluorene-based polyimides of intrinsic microporosity for gas separation applications. Macromolecules, 2013, **46**: 9618–9624.

[156] Wang, Z., Wang, D., Zhang, F. and Jin, J. Tröger's base-based Microporous polyimide membranes for high-performance gas separation. ACS Macro. Lett., 2014, **3**: 597–601.

[157] Abdulhamid, M. A., Ma, X., Miao, X. and Pinnau, I. Synthesis and characterization of a microporous 6FDA-polyimide made from a novel carbocyclic pseudo Tröger's base diamine: Effect of bicyclic bridge on gas transport properties. Polymer (Guildf), 2017, **130**: 182–190.

[158] Cho, Y. J. and Park, H. B. High performance polyimide with high internal free volume elements. Macromol. Rapid Commun., 2011, **32**: 579–586.

[159] Wiegand, J. R., *et al*. Synthesis and characterization of triptycene-based polyimides with tunable high fractional free volume for gas separation membranes. J. Mater. Chem. A, 2014, **2**: 13309–13320.

[160] Ghanem, B., Alaslai, N., Miao, X. and Pinnau, I. Novel 6FDA-based polyimides derived from sterically hindered Tröger's base diamines: Synthesis and gas permeation properties. Polymer (Guildf), 2016, **96**: 13–19.

[161] Lee, M., *et al*. Enhancing the gas permeability of Tröger's Base derived polyimides of intrinsic microporosity. Macromolecules, 2016, **49**: 4147–4154.

Ali Hayek

Chapter 2
Hydrogen sulfide removal from natural gas streams using polymeric membranes

Abstract: Natural gas is an important fossil fuel that is considered a relatively clean source of energy compared to other fossil products. A significant number of natural gas reserves are contaminated with several other gases, which from one side affect their calorific value, and from another side prohibit their distribution due to the health hazards consequent to the presence of toxic gases. A challenging contaminant in natural gas is the highly toxic gas hydrogen sulfide (H_2S). While several natural gas reserves worldwide contain high percentage of H_2S (i.e., up to 20 mol%), the allowed H_2S content of sales gas is 4 ppm only. Hence, the removal of H_2S is mandatory during gas processing. The current technology employed for H_2S removal from natural gas is based on liquid amines scrubbing process. This technology is very efficient with low H_2S content in gas feeds. However, it becomes economically challenging when dealing with high composition of H_2S in gas streams. In this case, to reduce the H_2S percentage in gas feeds before reaching the amine scrubbing system, a membrane module can be used in the upstream of the amine scrubber. The resulting technology is called "membrane-amine hybrid system." The performance of the membrane unit during this process becomes very important. Membrane's productivity (i.e., permeability) and efficiency (i.e., selectivity) are two important characteristics to be considered during the selection of the membrane. Equally important characteristics of membranes are the durability, physical aging, swelling, and plasticization. Therefore, researchers focused their work on solving questions pertaining to these characteristics. Some approaches were focused on the materials from which membranes are made of (i.e., glassy polymers or rubbery polymers), and some others focused on combining different materials to tailor the membrane properties as desired (i.e., mixed matrix membranes). In this chapter we will give an overview on how polymeric membranes are being studied in laboratories for their potential in H_2S removal and will go over all the works reported in literature to date.

Keywords: natural gas, hydrogen sulfide, polymeric membranes, glassy polymers, rubbery polymers, mixed matrix membranes

Ali Hayek, Research and Development Center, Saudi Aramco, Saudi Arabia,
e-mail: ali.hayek.2@aramco.com

https://doi.org/10.1515/9783110796032-002

1 Introduction

Natural gas (NG) is considered one of the cleanest energy sources produced from fossil fuels. The current use of natural gas is estimated to ~4.04 trillion cubic meters (TCM) worldwide by the year 2021. Its usage has increased by ~80% since 1998 [1]. The global demand of NG is forecasted to increase by ~27% by the year 2030 [2]. This is why oil and gas companies are looking to explore all possible NG reserves to increase their NG production. However, NG, with the main component methane (CH_4), is contaminated with other gases, such as nitrogen (N_2), heavy hydrocarbons (C_{2+}), helium (He), and acid gases (i.e., carbon dioxide CO_2, and hydrogen sulfide H_2S). In general, the presence of unharmful gas contaminants in NG lowers its high heating value. On the other hand, toxic gases, such as H_2S represent great danger to the human life. Therefore, for commercial use, all contaminants should be removed during NG processing to follow preset specifications depending on the sales market. For example, the US NG pipeline specifications are: $CO_2 < 2$ mol.%, $H_2S < 4$ ppm; N_2 and He < 4% total, and $H_2O < 120$ ppm [3, 4].

The removal of contaminants from NG streams is split into several processes; each process is designed to remove a specific type of contaminant. For example, the liquid amine scrubbing system is the conventional method used nowadays to remove acid gases (i.e., CO_2 and H_2S) from NG streams. While liquid amine scrubbing systems deliver natural gas with a quality that follows the US NG pipeline specifications, the system suffers from some drawbacks: (1) high capital and maintenance cost, (2) large foot print, (3) environmental impact, and (4) large energy demand. Therefore, the search for alternative methods for gas purification is of great interest. Membrane technology is potentially considered as one of the great substitutes to liquid amine scrubbing systems. The majority of membrane processes used nowadays is applied to carbon dioxide removal [5]. Substantial efforts are focused on producing membrane materials to be used for hydrogen sulfide removal. However, there is a great challenge that faces the membrane technology – to produce sales gas that follows the US natural gas pipeline specifications. Hence, several economic studies revealed the benefit of using membrane-amine hybrid system as illustrated in Figure 1 [6, 7].

Polymeric membranes, in particular, have been largely studied thanks to their good thermal stability, chemical resistance, and robust mechanical properties, which allow them to withstand the harsh operational conditions during gas processing. Several classes of polymers have been considered for this application, such as, cellulose acetate [8], polyimides [9], polysulfones [10], perfluoropolymers [11, 12], and so on.

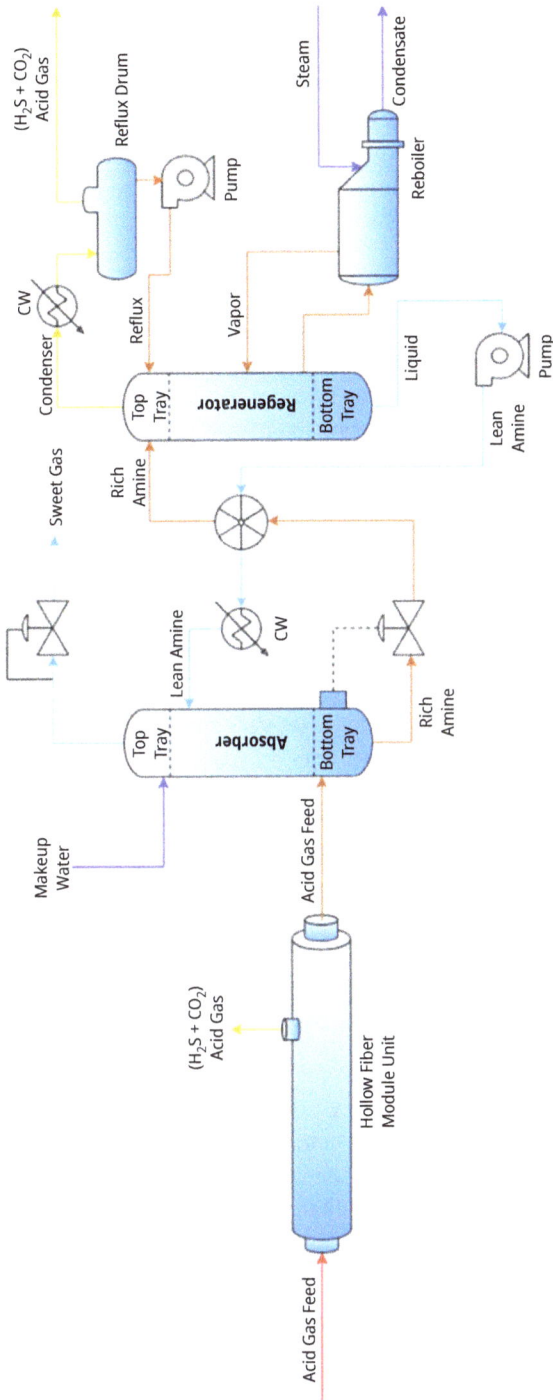

Figure 1: Schematic representation of a hybrid membrane/amine scrubbing system for sour natural gas processing.

2 Background and theory

The gas transport process through nonporous membranes occurs in three stages: the first stage involves the sorption of the gas penetrant into the membrane followed by the diffusion of the gas particle through the membrane matrix driven by concentration gradient, and the last stage of this process is represented by the desorption of the gas particle to the permeate side. Based on this, the transport process is called a "solution–diffusion" process [13, 14].

In general, a separation membrane is evaluated by its two intrinsic key characteristics: permeability (P) and selectivity (a). The permeability defines the productivity of the membrane, while the selectivity represents its efficiency. Experimentally, the permeability of a single gas stream through nonporous membranes can be determined using a constant-volume/variable-pressure system using the following expression:

$$P = 10^{10} \frac{V_d l}{p_f ART} \left[\left(\frac{dp_p}{dt} \right)_{ss} - \left(\frac{dp_p}{dt} \right)_{leak} \right] \tag{1}$$

where V_d is the permeate tube volume (cm^3), l is the membrane thickness (cm), p_f is the gas feed pressure (cmHg), A is the membrane effective surface area (cm^2), R is the universal gas constant ($R = 0.278$ cm^3.cmHg.cm^{-3}(STP).K^{-1}), T is the operational temperature (K), $\left(\frac{dp_p}{dt} \right)_{ss}$ is the steady-state (ss) pressure variation in the permeate side (cmHg), and $\left(\frac{dp_p}{dt} \right)_{leak}$ is the leak rate of the system, which is in most cases very small and thus could be neglected. The permeability unit is used as Barrer, where 1 Barrer = 10^{-10} cm^3(STP). cm.cm^{-2}.s^{-1}.cmHg^{-1}.

The ideal selectivity of the membrane for separating two distinguished gases A and B can be determined through their single gas permeability coefficients (P_A and P_B) using the following equation:

$$a_{A/B} = \frac{P_A}{P_B} \tag{2}$$

The permeability coefficient is governed by two main stages of the gas permeation process: sorption and diffusion. The sorption of a gas penetrant into the membrane, which is a thermodynamic process, depends mainly on the gas properties (condensability and gas-polymer affinity), however, the diffusion, which is a kinetic process, depends on the gas particle size (kinetic diameter). The smaller the kinetic diameter the higher the diffusion rate. Therefore, the permeability can be defined using the following expression:

$$P = D \times S \tag{3}$$

where D is the diffusivity coefficient (cm^2.s^{-1}), and S is the solubility coefficient (cm^3 (STP).cm^{-3}.cmHg^{-1}). Using this equation, the ideal selectivity expression could be modified using the solubility and diffusivity coefficients by:

$$\alpha_{A/B} = \frac{P_A}{P_B} = \left(\frac{D_A}{D_B}\right) \times \left(\frac{S_A}{S_B}\right) \tag{4}$$

The diffusivity coefficient can be experimentally determined using the time-lag method through the following expression:

$$D = \frac{l^2}{60} \tag{5}$$

where l is the membrane thickness (cm) and θ is the time lag (s). The solubility coefficient can thereafter be deduced from eq. (3) by:

$$S = \frac{P}{D} \tag{6}$$

For a gas mixture, however, the determination of the permeability coefficients of individual gases is determined using a constant-pressure/variable-volume permeation system. The system allows the determination of the permeate gas composition, a set of data needed to determine the permeability coefficient of a particular gas A, using the following expression:

$$P_A = P_{total} \frac{y_A (p_p - p_f)}{x_A p_p - y_A p_f} \tag{7}$$

where P_{total} is the permeability of total gas particles permeated through the membrane, x_A and y_A are the mole fractions of gas A in the feed and the permeate sides, respectively, and are determined experimentally using a gas chromatography analyzer connected to the system. The terms p_p and p_f are the partial pressures of gas A in the feed and the permeate sides, respectively.

P_{total} can be determined using the following expression:

$$P_{total} = \frac{Jl}{\Delta p} \tag{8}$$

where J is the penetrant flux ($cm^3(STP).cm^{-2}.s^{-1}$), l is the membrane thickness (cm) and Δp is the difference between the partial pressures of gas A at the feed and the permeate sides (cmHg).

The permeability coefficient could be affected by changing either the operational temperature or gas feed pressure or both. The dependence of the permeability, diffusivity, and solubility coefficients on operational temperature can be expressed using van't Hoff–Arrhenius types of equations:

$$P = P_0 Exp\left(-\frac{E_p}{RT}\right) \tag{9}$$

$$D = D_0 Exp\left(-\frac{E_D}{RT}\right) \tag{10}$$

$$S = S_0 Exp\left(-\frac{\Delta H_S}{RT}\right) \tag{11}$$

where P_0 (Barrer), D_0 (cm^2.s^{-1}), and S_0 (cm^3(STP).cm^{-3}.cmHg^{-1}) are pre-exponential factors, R is the universal gas constant ($R = 8.314$ kJ.mol^{-1}.K^{-1}), T is the operating temperature (K), E_P and E_D are the activation energies of permeation and diffusion (kJ.mol^{-1}), respectively, and ΔH_S is the enthalpy of sorption (kJ.mol^{-1}).

On the other hand, the dependence of permeability on feed pressure is governed by the dual sorption model, which takes into account Henry's sorption law (C_D) and Langmuir sorption law (C_H) as per the following expression of the gas concentration (C) within the membrane matrix:

$$C = C_D + C_H = k_D p + \frac{C'_H bp}{1 + bp} \tag{12}$$

where k_D is Henry's law coefficient (cm^3(STP).cm^{-3} polymer^{-1}.atm^{-1}), b is the Langmuir hole affinity parameter (atm^{-1}), C'_H is the capacity parameter (cm^3(STP).cm^{-3} polymer^{-1}), and p is the feed pressure (cmHg). Using Henry's law, which describes the relationship between the gas concentration (C) and gas solubility (S) and feed pressure (p):

$$C = Sp \tag{13}$$

Equation (12) could be modified to portray the dependence of permeability coefficient on feed pressure as follows:

$$P = k_D D_D + \frac{C'_H b D_H}{1 + bp} \tag{14}$$

where D_D and D_H are the diffusion of gas particles absorbed by each model (cm^2.s^{-1}).

It is noteworthy that in case of high concentration of gas particles within the membrane matrix and at high gas feed pressure, the mobility of polymer segments within the membrane will be affected, which will be reflected by an increase in the diffusion coefficient, and this is referred to as plasticization.

3 Physical properties of gas molecules

The distinct physical properties of gas molecules to separate within the mixture play an important role during the separation process through dense polymeric membranes (solution–diffusion mechanism). Two main keys determine the separation efficiency of polymeric membranes: (1) gas-polymer affinity (i.e., solubility), and (2) gas molecule size (i.e., diffusion). The electrostatic potentials mapping (ESPM) of some important gas molecules in natural gas streams are illustrated in Figure 2, and their kinetic diameter, dipole moments, and critical temperatures are listed in Table 1.

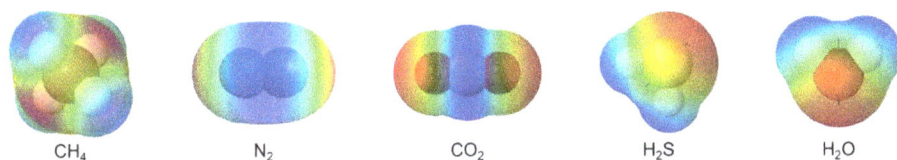

CH$_4$	N$_2$	CO$_2$	H$_2$S	H$_2$O

Figure 2: Electrostatic potentials mapping for some gas molecules from natural gas. Blue color: positive charge and red color: negative charge.

Table 1: Physical properties of some gas molecules within natural gas streams.

Gas molecule	CH$_4$	N$_2$	CO$_2$	H$_2$S	H$_2$O
Kinetic diameter (Å)	3.80	3.64	3.30	3.60	2.56
Dipole moment (Debye)	0.00	0.00	0.00	2.13	1.88
Critical temperature (K)	190	126	304	374	647

In general, nonpolar gas molecules have very low affinity to polymeric chains. Since CH$_4$ is nonpolar, its affinity to polymeric chains is very low, which makes its permeation through the membrane matrix very slow; in addition its kinetic diameter ($d_k = 3.80$ Å) is greater compared to other gas molecules in mixture. This aspect is advantageous for the use of polymeric membranes in NG processing. However, despite the nonpolar nature of CO$_2$ ($d_k = 3.30$ Å), its affinity to polymeric chains is greater than that of CH$_4$, due to the carbon-oxygen (C=O) polar bonds. In addition, its condensability ($T_c = 304$ K) is greater than that of other non-polar gas molecules, such as CH$_4$ and N$_2$.

The affinity of gas molecules to polymeric chains increases with their electrostatic nature: polar or nonpolar. In general, polar molecules have greater affinity to polymeric chains that are of polar nature. For example, H$_2$S molecule is of polar nature with high condensability ($T_c = 374$ K), and hence its solubility within membrane matrices is known to be higher than its nonpolar peers. Despite its greater kinetic diameter ($d_k = 3.60$ Å) its greater solubility enhances its permeation through the membrane matrix. On the other hand, this high affinity is not always advantageous, since the great H$_2$S-polymer interaction during the separation process may result into increasing the polymeric chains mobility, which results into membranes swelling and/or plasticizing. This is what makes the removal of H$_2$S from NG streams very challenging. Therefore, the molecular design of polymers to be used in H$_2$S removal is very crucial.

4 Gas permeation measuring systems

The gas permeation properties of dense membranes or thin film composites are measured in laboratories using mainly two different systems: (1) pure-gas permeation system and (2) mixed-gas permeation system.

4.1 Pure-gas permeation system

In general, pure-gas permeation system is denoted as constant-volume/variable-pressure permeation system. The notion of constant-volume/variable-pressure is derived from the fact that the permeate is collected in a known fixed volume vessel and that the permeate pressure changes over time. The permeability is measured using equation 1 depicted in the background and theory section, using the slope of the permeate pressure versus time steady-state curve (Figure 3c). A schematic representation of the pure-gas permeation system is illustrated in Figure 3b.

The system is mainly constructed from a membrane cell with three main openings (Figure 3a): (1) gas feed inlet, (2) gas reject outlet, and (3) gas permeate outlet. The gas feed inlet is connected through a high-pressure line to the gas cylinder. The line is equipped with a pressure regulator to set the testing feed pressure and a pressure transducer connected to a computer station (PC). The feed line is also connected to a vacuum pump to degas the feed side during the time lag diffusivity measurement as explained in the background and theory section (Figure 3d). The reject outlet is connected to a line equipped with a valve that serves to vent the gas after the measurement is completed. The permeate outlet is connected to a vessel with a known volume equipped with a pressure transducer connected to a computer station to measure the variation of the permeate pressure over time. The permeate side is connected to the vacuum pump, which serves to create a vacuum in the permeate side.

4.2 Mixed-gas permeation system

While some research groups that report the measurement of the mixed-gas permeation properties of membranes use also a modified constant-volume/variable-pressure permeation system to evaluate the performance of membranes during mixed-gas separation, herein, we discuss the constant-pressure/variable-volume mixed-gas permeation system, which is closer to the gas processing system used in gas plants. The mixed-gas permeation system is more complex that the pure-gas permeation system described previously, and it becomes more complex when dealing with gas mixtures containing H_2S due to the high risk associated to working with such a toxic gas.

Figure 4 illustrates a constant-pressure/variable-volume mixed-gas permeation system reported in literature for measuring the gas separation performance of mixed-gas

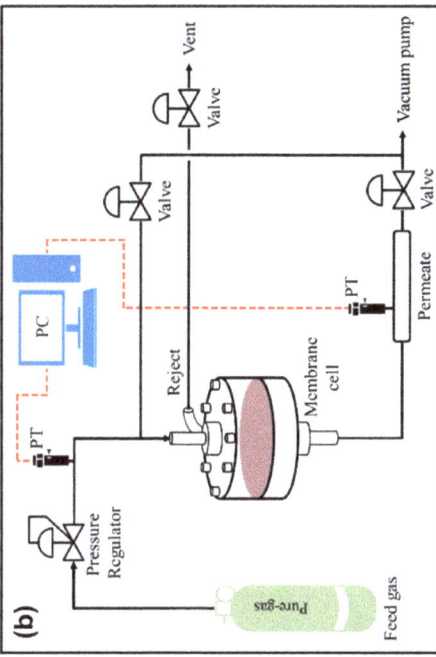

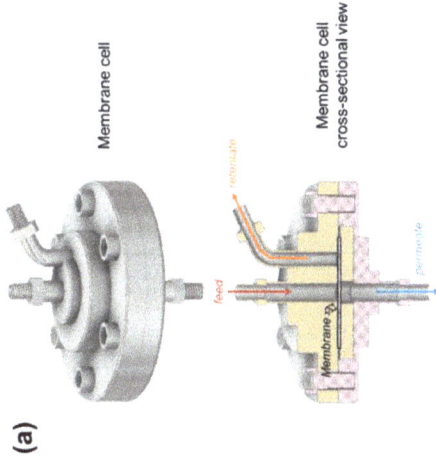

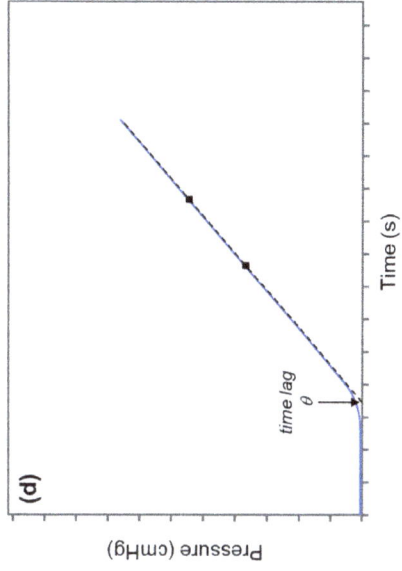

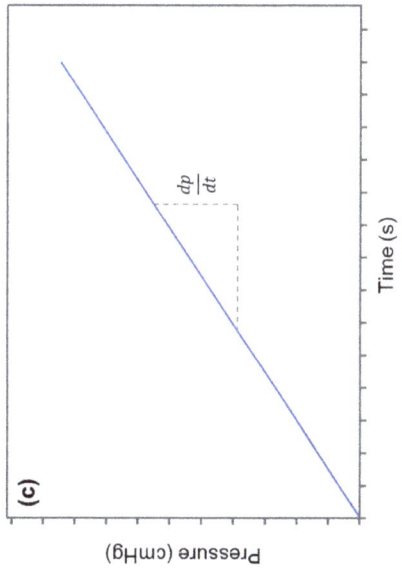

Figure 3: (a) High-pressure permeation membrane cell, (b) constant-volume/variable-pressure pure-gas permeation system, (c) pressure vs. time curve, and (d) pressure vs. time lag curve.

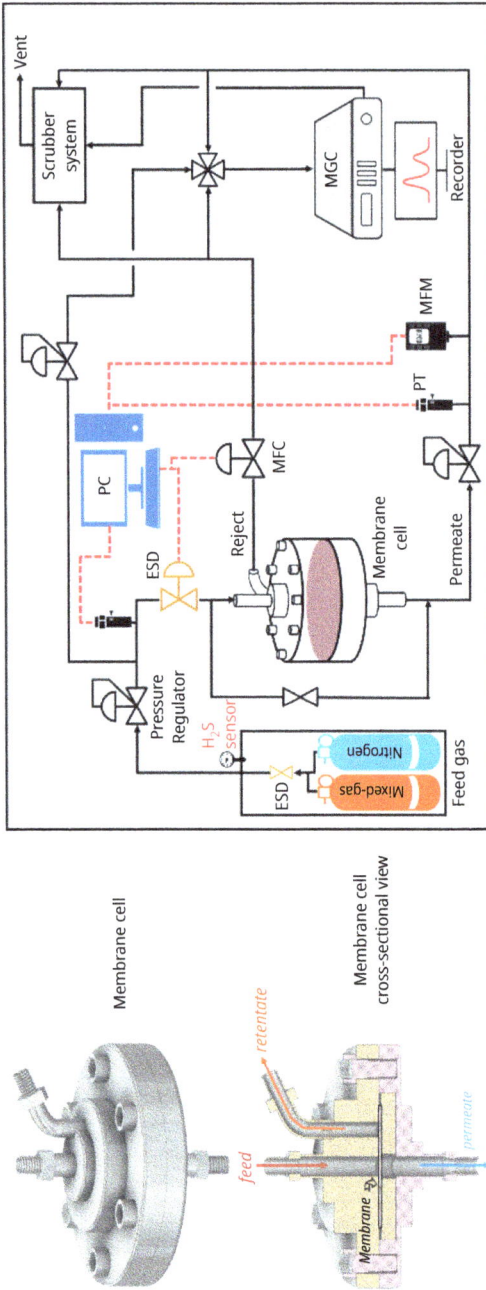

Figure 4: Constant-pressure /variable-volume mixed-gas permeation system.

permeation properties of dense membranes in the presence of H_2S. The system involves high safety measures to eliminate the risks associated with working with H_2S resulting from any leak, explosion, or measuring faults, such as the case of membrane breakage [15]. In this system, the feed cylinders are placed in a ventilated cabinet with an H_2S detector and an emergency shutdown valve (ESD). Moreover, the testing facility is usually equipped with built-in sensors and shutdown mechanism in case of leakage detection or membrane breakage. The reject stream is controlled by software to maintain the desired stage cut. Moreover, sample lines of feed, reject, and permeate streams are directed into a micro gas chromatography (MGC) to analyze the gas permeation across the membrane. After membrane conditioning, the flow of the permeate gas is monitored using mass flow meters (MFM) ranges from 0.1 to 1500 SCCM (air) for flow and gas composition stabilization. Finally, all streams are vented through a scrubber system (~40% w/w sodium hydroxide aqueous solution) to scrub out the H_2S before venting to the atmosphere. The operating feed pressure is controlled in the feed and the reject streams using an absolute pressure transmitter (PT).

4.3 Hollow fiber permeation system

When membranes are prepared into hollow fiber configuration, a slight change to the sample holder is applied to the prior described systems to test the membranes' permeation properties. Hollow fiber membrane (HFM) modules are usually prepared from a selected number of fibers potted in a bundle with a precise length in a stainless-steel tube to withstand high pressures (Figure 5a). From one side, the fibers are plugged using epoxy resin, and from the other side, the fibers are connected to the opening of the stainless-steel tube. The gas can be fed into the HFM module in two different methods: (1) shell-side feed or (2) bore-side feed [16]. The two gas feeding methods into HFM modules are illustrated in Figure 5b.

A simplified schematic representation for a constant-pressure/variable-volume permeation system to test hollow fiber membranes is illustrated in Figure 5c. In this example, the HFM module is connected to the gas cylinder through shell side. The gas feed pressure is adjusted through a pressure controller placed between the gas cylinder and the HFM module. Another line from the feed side is connected to a needle valve and a check valve to depressurize the system when needed. The retentate side is connected to a pressure transducer and a mass flow controller to adjust the stage cut. The permeate side of the HFM module is equipped with an MFM used to measure the permeate flow rate. Finally, the retentate and permeate lines are connected to a micro GC to analyze their gas compositions.

The membrane stage cut is defined as the fraction of feed gas that permeates through the membrane. Therefore, the stage cut is adjusted according to the following expression:

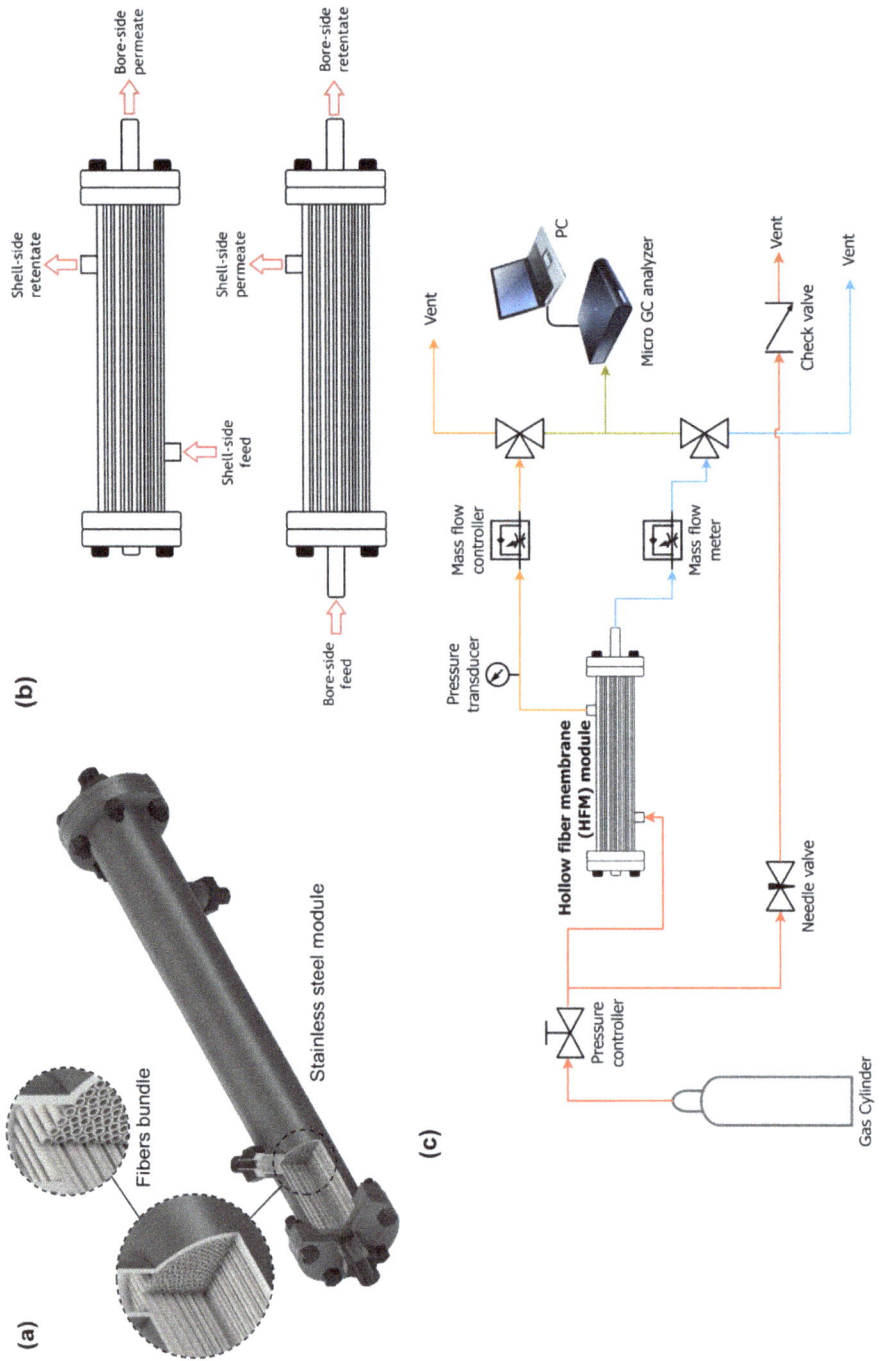

Figure 5: (a) High-pressure stainless-steel membrane module (b) Shell-side and bore-side feeds in HFM modules. (c) constant-pressure/variable-volume permeation system to test HFMs.

$$stage\ cut(\%) = \frac{permeate\ flow\ rate}{feed\ flow\ rate} \times 100$$

Experimentally, the feed flow rate is considered to be equal to the retentate flow rate plus the permeate flow rate and hence the retentate flow rate is measured accordingly. In industry, product recovery can be improved by increasing the stage cut, but it is relatively difficult to improve product purity, unless the pressure ratio is increased, or multi-stage membrane units are used [17].

5 Industrial membrane configurations

At the lab scale, most polymeric membranes are tested in the form of dense membranes, thin film composites (TFC), or hollow fibers. Dense membranes are usually prepared by solution casting method. A polymer solution with a known concentration is poured onto a flat surface, and the solvent is allowed to evaporate, forming a thin membrane with a predetermined thickness. On the other hand, thin film composites are prepared by coating a thin film of polymer over a porous support prepared from a cheap, commercially available polymer. The polymer thin film serves as the selective layer, and the porous support with no sieving effect helps withstand the high-pressure conditions. Finally, hollow fiber may be prepared in two forms: (1) asymmetric hollow fiber or (2) TFC hollow fiber. Asymmetric hollow fibers are spun from a single selected polymer, where the structure of the hollow fiber is engineered in a way that it is porous from the inside (bore side) and the size of the pores becomes smaller moving toward the outer wall (fiber shell), where the wall of the fiber becomes a dense selective layer.

TFC hollow fibers are prepared by coating a thin dense selective layer on the outer wall of a porous hollow fiber support prepared from a cheap, commercially available polymer. TFC hollow fibers are the cheapest to produce and therefore, they represent the desired industrial membrane configuration. The two hollow fiber types are illustrated in Figure 6.

For industrial application, membranes are fabricated mainly under two forms: (1) flat sheet or (2) tubular. In both cases, the membrane is constituted of two essential parts: a thin selective layer coated on a thick support. Two major membranes modules are produced commercially: (1) spiral wound and (2) hollow fiber. The two configurations are illustrated in Figure 7.

The spiral wound configuration consists of a number of membrane flat sheets rolled in a spiral shape around a central perforated tube, which represents the permeate path. The membrane layers are spaced by two spacers; one for the feed and another for the permeate. The gas stream is fed through the sides of the module, and the permeate travels through the permeate spacer in a spiral trajectory to reach the collecting central tube.

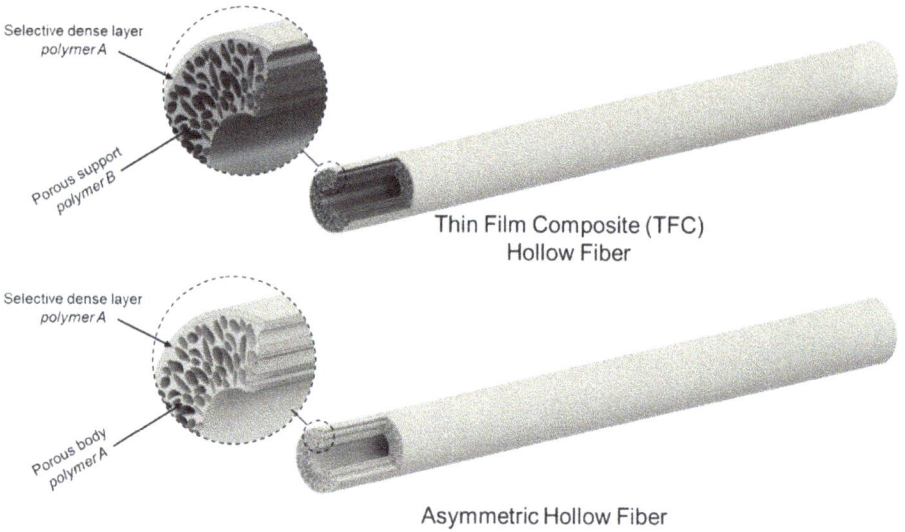

Figure 6: Asymmetric and thin film composite (TFC) hollow fibers.

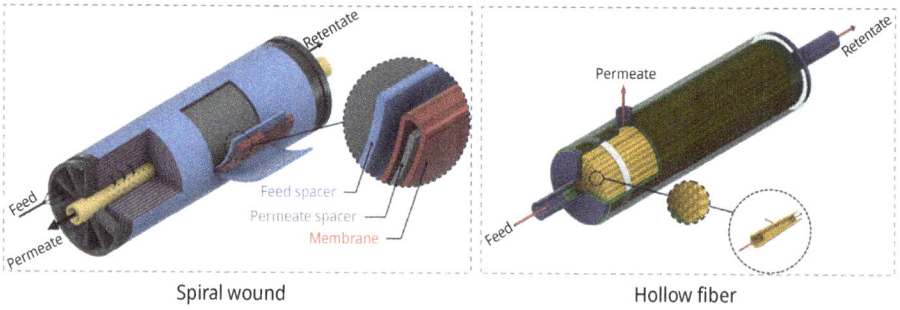

Figure 7: Commercially available industrial membranes: module configurations.

In general, hollow fiber modules are preferred over the spiral wound counterpart due to three main reasons: (1) more surface per volume translates into lower footprint, (2) lower capital expenditures (CAPEX), and (3) in some cases, simple module production. The lower footprint of hollow fiber modules derives from their high packing density of bundles of fibers, offering thereafter high contact surface area between gas molecules and the selective layer of membranes. This makes them easier to implement in existing gas plants where surface area availability is limited.

6 Polymeric membranes

A variety of polymeric materials have been prepared and studied for gas separation applications. The majority of these polymeric materials have been evaluated only using single gas streams or binary sweet mixed-gas streams. Due to the high toxicity of hydrogen sulfide, a limited number of these polymers were tested using sour mixed-gas under different format (i.e., dense membranes, TFC and hollow fiber). Two types of polymers were studied: (1) glassy polymers and (2) rubbery polymers.

6.1 Glassy polymers

Glassy polymers are amorphous polymers at temperatures below their glass transition temperatures (T_g). In other definition, a glassy polymer is any polymer with a T_g above the room temperature. In general, high values of T_g are associated with polymeric chains rigidity and stiffness. Therefore, the amorphous and stiff nature of glassy polymers allow them to form dense membranes with low free volume, which makes them size-selective membranes, and the separation occurs mainly through size discrimination of gas molecules. For instance, hydrogen (kinetic diameter = 2.89 Å) being a small molecule is 100 times more permeable than methane (kinetic diameter = 3.80 Å). Moreover, the excellent mechanical properties of glassy polymers allow their processability into dense films, TFCs, and hollow fibers.

Among the first glassy polymers studied for hydrogen sulfide removal from NG streams is cellulose acetate (CA). Cellulose acetate is considered as a low cost, industrial-standard membrane material. HFMs prepared from cellulose acetate are usually used for CO_2 removal from gas streams. The most studied cellulose acetate derivative is the triacetate derivative. The chemical structure of cellulose triacetate (CTA) is illustrated in Figure 8.

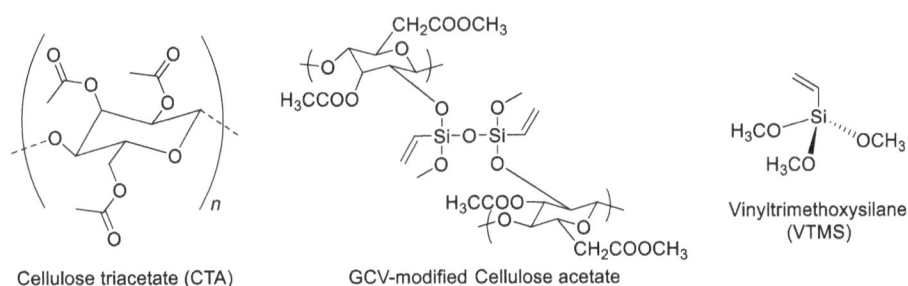

Cellulose triacetate (CTA)   GCV-modified Cellulose acetate

Figure 8: Chemical structures of cellulose triacetate (CTA) and its modified derivative GCV-modified CA [18–20].

The performance of dense membranes prepared from cellulose acetate separating a ternary mixture prepared from (65% CH_4/29% CO_2/6% H_2S) at a feed pressure of 10 atm and a temperature of 35 °C was studied. The H_2S permeability was recorded to be 2.13 Barrer and the H_2S/CH_4 selectivity was calculated to be 19 [18]. Another study performed on cellulose acetate dense membrane using a ternary mixture prepared from (60% CH_4/20% CO_2/20% H_2S) at a feed pressure of 48.3 bar and 35 °C. The H_2S permeability was recorded to be 39.7 Barrer and the H_2S/CH_4 selectivity was calculated to be 27.4 [19]. The obtained results of both studies are listed in Table 2.

Table 2: Sour mixed-gas transport properties of cellulose acetate at different testing conditions.

Polymer	$CH_4/CO_2/H_2S$	P (atm)/T (°C)	Permeability (Barrer)			Selectivity	
			CH_4	CO_2	H_2S	CO_2/CH_4	H_2S/CH_4
Cellulose	65/29/6	10/35	0.11	2.43	2.13	22	19
acetate (CA)	60/20/20	47.7/35	1.44	27.5	39.7	19.1	27.4

CTA HFMs were evaluated by Liu et al. for their performance in simultaneous removal of acid gases (H_2S and CO_2) in the presence of heavy hydrocarbons (i.e., C_2H_6, C_3H_8, and toluene). The CTA HFMs were tested at different gas feed pressures (up to 31.3 bar) and two different temperatures (35 °C and 50 °C) using two different gas mixtures with the following compositions: H_2S (20 mol%), CO_2 (5 mol%), C_2H_6 (3 mol%), C_3H_8 (3 mol%), toluene (100 ppm or 300 ppm), and CH_4 (balance gas). For example, at 35 °C and 31.3 bar, for a gas mixture with the aforementioned composition and in the presence of 100 ppm of toluene, the CTA HFMs exhibited H_2S and CO_2 permeances of ~140 GPU and ~115 GPU, respectively, with H_2S/CH_4 and CO_2/CH_4 selectivity coefficients of ~28 and ~22, respectively. Interestingly, the CTA HFM demonstrated plasticization benefits under the aggressive conditions employed during testing, indicating high separation potential of CTA HFMs for real-life NG sweetening industrial application [20].

Achoundong et al. reported the modification of cellulose acetate via grafting of vinyltrimethoxysilane (VTMS) to the hydroxyl groups of CA followed by condensation of silanols to form a polymer network. This modification is denoted as GCV-modification, and the modified CA is referred to as GCV-modified CA, since the modification includes a combination of grafting and cross-linking, hence the term "GCV" (Figure 8). Dense membranes prepared of GCV-modified CA were tested under aggressive conditions of feed pressures (up to 700 psia) using a ternary mixture (20/20/60 vol% of $H_2S/CO_2/CH_4$, respectively) at 35 °C [19]. For example, at 35 °C and 700 psia, the GCV-modified CA dense membrane exhibited H_2S and CO_2 permeability coefficients of 190 Barrer and 136 Barrer, respectively, with H_2S/CH_4 and CO_2/CH_4 selectivity coefficients of 27.5 and 20, respectively. The permeation properties of GCV-modified CA membrane at such aggressive conditions showed comparable selectivities to that of CA membrane and significantly higher H_2S and CO_2 permeability coefficients (more than 1 order of magnitude). This inexpensive polymer is the benchmark for membrane gas separations, and this approach may have an immediate application in the field.

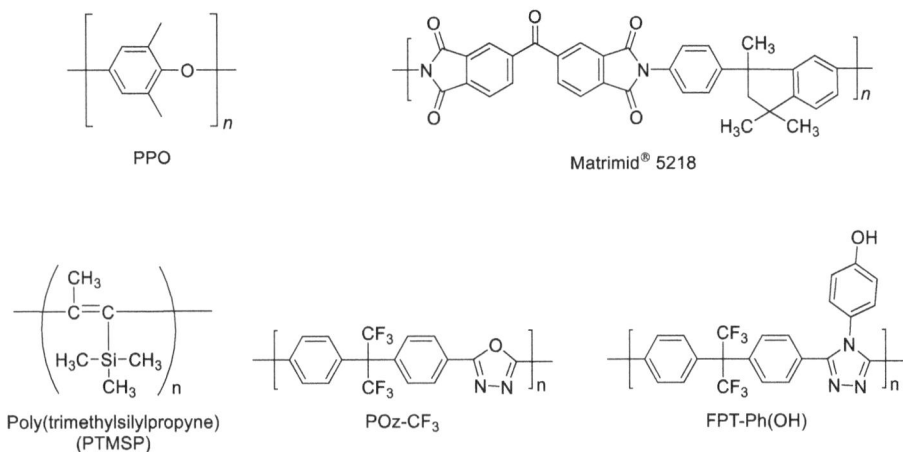

PPO

Matrimid® 5218

Poly(trimethylsilylpropyne)
(PTMSP)

POz-CF₃

FPT-Ph(OH)

Figure 9: Chemical structures of poly(phenylene oxide) (PPO), Matrimid® 5218, poly(trimethylsilylpropyne) (PTMSP), poly(1,3,4-oxadiazole) (POz-CF₃), and poly(1,2,3-triazole) (FPT–Ph(OH)) [21–30].

Other glassy polymers have been studied for their potential use in H_2S removal from NG. For example, poly(phenylene oxide) (PPO) is another example of low-cost produced polymer with high processability properties ($T_g \sim 490$ K). It can easily be prepared into HFMs that can withstand the harsh separation conditions during NG processing. It has been reported that PPO shows high permeabilities to gases, which is attributed to the absence of polar groups in the main chain of PPO. The gas permeation properties of PPO have been evaluated in the presence of H_2S in binary or real NG sample. The amount of H_2S in the studied mixtures varied between 101 ppm and 5,008 ppm [21, 22]. For example, Niknejad et al. studied the separation properties of PPO HFMs using a real-life NG sample that contained 3,360 ppm of H_2S in the presence of CO_2, N_2, H_2O, C_{2+}, and mercaptans. The membranes were tested at a feed pressure of 689 KPa and a temperature of 294.15 K. It was observed that the permeance of H_2S decreased as the feed pressure increased, which indicates that the competitive sorption dominates the plasticization effect. The H_2S/CH_4 separation factor of 3.2 obtained from the real-life NG sample was less than that obtained for a synthetic binary mixture with similar H_2S content, indicating the effect of competition on sorption sites between the various gas components in the mixture.

Another glassy polymer, poly(1-trimethylsilyl-1-propyne) (PTMSP), has been considered for H_2S removal from NG streams. PTMSP is considered as a super glassy polymer with a high excess fractional free volume (FFV = 0.343). The polymer is regarded as a nanoporous material due to the formation of nanoscale channels within the membrane matrix of PTMSP during chains packing, which is attributed to the restricted rotation around the double bond, and the bulkiness of the trimethylsilyl group. These intrinsic properties make of PTMSP a highly permeable polymer, which is even higher than its rubbery analogue poly(dimethylsiloxane) (PDMS). However, due to its large FFV, PTMSP suffers from rapid aging under the harsh conditions of gas separation, where its transport properties decay with a few weeks of usage [23, 24]. For example, Lokhandwala et al. reported the study of a PTMSP TFC membrane using a ternary mixture (balance CH_4/4% CO_2/800 ppm H_2S) at a feed pressure of 390 psig. The H_2S permeability was measured to be 101 GPU with a H_2S/CH_4 selectivity coefficient of 3.3 only [25]. These preliminary transport properties of PTMSP are not very attractive, especially with such a low H_2S content (800 ppm) gas mixture. In the same work, Matrimid® 5218 TFC membranes were studied using the same ternary mixture (balance CH_4/4% CO_2/800 ppm H_2S) at a feed pressure of 390 psig. The H_2S/CH_4 selectivity coefficient was found to be 9.5 [25]. These H_2S/CH_4 selectivity values are considered low for real-life industrial applications.

Poly(1,3,4-oxadiazole)s are a class of polymers that are known for their excellent thermal and mechanical properties, which makes it very easy to process them into thin films and hollow fibers [26]. Their pure-gas transport properties have been studied and demonstrated promising properties for gas purification application [26, 27]. Poly(1,3,4-oxadiazole)s can be converted through simple polycondensation reaction into their poly(1,2,4-triazole)s derivatives [28]. This modification allows the addition of new functional groups to the polymer backbone, leading to tailoring its properties accordingly. Despite their promising gas transport properties, their behavior in the presence of H_2S could not be predicted. Hayek et al. reported the gas separation properties of dense membranes prepared from the fluorinated poly(1,3,4-oxadiazole) POz-CF_3 and its hydroxyl functionalized poly(1,2,4-triazole) FPT-Ph(OH) using a quinary sour mixed-gas mixture containing 20 vol% H_2S (20/10/10/59/1 vol% of $H_2S/CO_2/N_2/CH_4/C_2H_6$, respectively) at various gas feed pressures (up to 700 psi) and 22 °C. The chemical structures of POz-CF_3 and FPT-Ph(OH) are illustrated in Figure 9. In contrast to the promising separation obtained during the sweet mixed-gas studies of POz-CF_3 membrane, the sour mixed-gas separation was mediocre. The H_2S permeability at 700 psi was 38.8 Barrer with a H_2S/CH_4 selectivity coefficient of 5.24 only. On the other hand, the hydroxyl-containing poly(1,2,4-triazole) FPT-Ph(OH) exhibited improved results compared to its parent POz-CF_3. At 700 psi, the H_2S permeability was 119 Barrer with a H_2S/CH_4 selectivity coefficient of 21.3. On the basis of this study, it could be deduced that the 1,2,4-triazole ring is more suitable for H_2S removal from NG than its parent 1,3,4-oxadiazole ring. Nevertheless, FPT-Ph(OH) could pave the way to develop new poly(1,2,4-triazole)s for NG sweetening.

Among the first glassy polymers evaluated for their potential use of high concentration H_2S removal from natural gas were those reported by Vaughn and Koros in 2014 [29]. This was a continuation to their preliminary results reported in 2012 [30]. In fact, they studied the effect of the chemical structure on the gas separation properties of polymers in the presence of H_2S in natural gas streams. Two main factors were evaluated: (1) effect of amide bonds and (2) effect of fluorine concentration in the polymer backbone. The chemical structures of the reported polymers are listed in Figure 10.

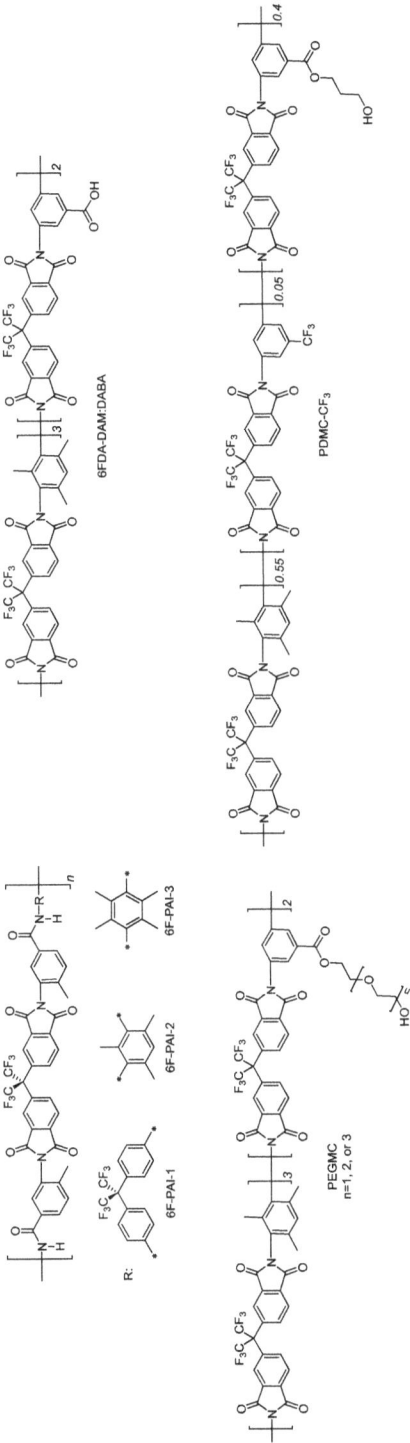

Figure 10: Chemical structures of some selected glassy polymers studied for H$_2$S removal from natural gas streams [31–37].

Regarding the amide vs. imide bonds, the study revealed that steric interactions around the amid bond are not as strong as around imide bonds. The rigid substituents into amide bonds resulted in lower free volume and polymer chain mobility. On the other hand, the fluorine content into the polymer backbone reduced the H_2S/polymer affinity, in an attempt to stabilize the membrane toward high condensable gases [31]. In fact, the ternary mixed-gas transport studies using 20/20/60 vol% of $H_2S/CO_2/CH_4$ at 40 psia H_2S partial pressure 6 F-PAI-2 plasticized and the H_2S/CH_4, dropped to below 10. Using the same ternary mixture feed, for the more fluorinated 6 F-PAI-1 the plasticization occurred at 80 psia H_2S partial pressure, while still showing attractive acid gas selectivity under such highly aggressive feed conditions.

In another work, Kraftschik et al. studied dense film membranes of the copolyimide 6FDA-DAM:DABA (3:2) for simultaneous removal of CO_2 and H_2S from sour NG streams (Figure 10) [32]. The pure-gas screening of freshly prepared membranes using pure H_2S exhibited significant plasticization at feed pressure of 1 bar. Therefore, sub-T_g annealing of membranes at two different temperatures (180 °C and 230 °C) was employed for more efficient chain packing and stronger interchain interactions to increase the plasticization resistance of membranes. It was observed that the higher the annealing temperature, the higher the chain packing within the membrane matrix. The annealed membranes were evaluated thereafter for their H_2S removal potential using two mixtures containing H_2S: (Mixture 1) 4.98/95.02 vol% of H_2S/CH_4, and (Mixture 2) 19.9/9.95/70.15 vol% of $CO_2/H_2S/CH_4$. The permeation tests were performed at 35 °C and feed pressures up to 62 bar. The permeation tests using Mixture 1 (4.98 vol% H_2S) demonstrated that the annealing method eliminated the plasticization at high feed pressures (up to 62 bar); however, both membranes exhibited signs of weak plasticization using Mixture 2 (9.95 vol% H_2S) while maintaining their separation efficiency. For example, the 6FDA-DAM:DABA (3:2) membrane annealed at 180 °C exhibited a H_2S permeability of ~32 Barrer with a H_2S/CH_4 of ~18 at 62 bar using Mixture 2 (19.9/9.95/70.15 vol% of $CO_2/H_2S/CH_4$). These results were considered very attractive compared to those obtained for the industrial-standard cellulose acetate.

Building on the attractive results obtained from the 6FDA-DAM:DABA (3:2) membranes, Kraftschik and Koros developed a series of cross-linkable membrane materials based on the 6FDA-DAM:DABA (3:2) polyimide backbone [33]. Short-chain poly (ethylene glycol) (PEG) molecules were used as cross-linking agents in an esterification-based cross-linking reaction (Figure 10). This synthetic modification is believed to solve some of the challenges related to H_2S-induced plasticization resistance and enhanced H_2S/CH_4 selectivity under aggressive sour gas feed conditions, such as the case during H_2S removal from NG. Therefore, sour mixed-gas permeation experiments were performed on dense films of these materials, using a 20/20/60 vol% of $H_2S/CO_2/CH_4$ mixture at feed pressures up to 62 bar and a temperature of 35 °C. Under these realistic conditions, for triethylene glycol (TEG) and tetraethylene glycol (TetraEG) cross-linking agents, attractive selectivities above 22 for H_2S/CH_4 were observed with no evident H_2S-induced plasticization.

Further studies were performed on the cross-linkable TEG 6FDA-DAM:DABA (3:2) polyimide; where Babu et al. prepared HFMs from this material to study their performance under the realistic membrane configuration that would be use industrially [34]. The HFM fabrication conditions were studied in detail, along with the effect of membrane's defects on its gas permeation properties. It was found that with posttreatment of the HFM outer layer with a polydimethylsiloxane (PDMS) layer, the high-pressure performance under mixed-gas feeds improved even further. The sour mixed-gas separation performance of these HFMs were conducted using two sour gas ternary mixtures (5/45/50 vol% of $H_2S/CO_2/CH_4$ and 20/20/60 vol% of $H_2S/CO_2/CH_4$) at a feed pressure of 500 psig and a temperature of 35 °C. It was observed that the H_2S/CH_4 selectivity of these HFMs exceeded that of their corresponding dense membranes, which was attributed to the polymer-chain alignment during the spinning process. The HFMs showed H_2S/CH_4 selectivity value as high as ~29 for a mixture with 5% H_2S and ~22 for a mixture with 20% H_2S, with the gas being fed shell-side of the hollow fiber at a pressure of 500 psig and a temperature of 35 °C. For both mixtures, and under the same conditions, the H_2S permeance was around 10 GPU. The HFMs have shown to retain their stability and efficiency under such aggressive sour gas feed conditions.

Despite the encouraging H_2S/CH_4 selectivity results obtained from the cross-linkable TEG 6FDA-DAM:DABA (3:2) HFMs, their H_2S permeance exhibited a significant loss after cross-linking due to the reduction in the excess free volume in the selective layer. To address this loss, Liu et al. evaluated the sour mixed-gas separation properties of HFMs prepared from semi-rigid cross-linkable 6FDA-based polyimides with bulky CF_3 groups (PDMC – CF_3) (Figure 10) [35]. The HFMs' performance was evaluated using three different ternary sour gas mixtures (0.5/20/79.5 vol% of $H_2S/CO_2/CH_4$; 25/5/70 vol% of $H_2S/CO_2/CH_4$; 20/20/60 vol% of $H_2S/CO_2/CH_4$). At 450 psi and 35 °C, and for the mixture with 0.5% H_2S content, the H_2S permeance was measured ~32 GPU with an H_2S/CH_4 selectivity ~30. For higher content sour natural gas (20% and 25% H_2S) the performance of the HFMs slightly decreased; however, plasticization was not observed for the cross-linked HFMs, and no permeance decline was observed due to cross-linking. Such robust high-performing HFMs are needed for sour natural gas purification.

To further investigate this type of polyimides, Liu et al. studied 6FDA-DAM:DABA dense membranes with different DAM:DABA molar ratios (3:2, 1:1, and 1:2) [36]. The pure-gas screening results showed that tuning the DAM:DABA ratio in polymer backbone show different effects on the H_2S/CH_4 separation performance. If fact, it was found that for a higher DAM:DABA ratio, higher H_2S permeability coefficients were attained with a little decrement in H_2S/CH_4 selectivity; however, the plasticization tendency was increased as well. The dense membranes were tested using a ternary natural gas mixture consisting of 25/5/70 vol% of $H_2S/CO_2/CH_4$ at 35 °C with a total pressure up to 700 psi. It was demonstrated that polymers with more DAM exhibit more attractive performance. For example, 6FDA-DAM:DABA (3:2) membrane exhibits H_2S permeability of 106.7 Barrer with H_2S/CH_4 of 22.4.

The incorporation of the DAM moiety became very popular after the study reported by Liu et al. on the plasticization benefit of 6FDA-DAM dense membranes tested under a ternary gas mixture (20/20/60 vol% of $H_2S/CO_2/CH_4$) at a feed pressure of 46 bar and a temperature of 35 °C [37]. They reported that the plasticization can be regarded as a tool for performance optimization. In fact, 6FDA-DAM represented a case where plasticization can provide performance benefit for H_2S/CH_4 separation. The 6FDA-DAM membrane showed H_2S permeability of 495 Barrer with H_2S/CH_4 selectivity of ~31 under the aforementioned testing conditions. Such H_2S/CH_4 separation performance for aggressive high-pressure feeds exceeds that of rubbery polymers, making the glassy materials ideal for processing NG feeds containing H_2S.

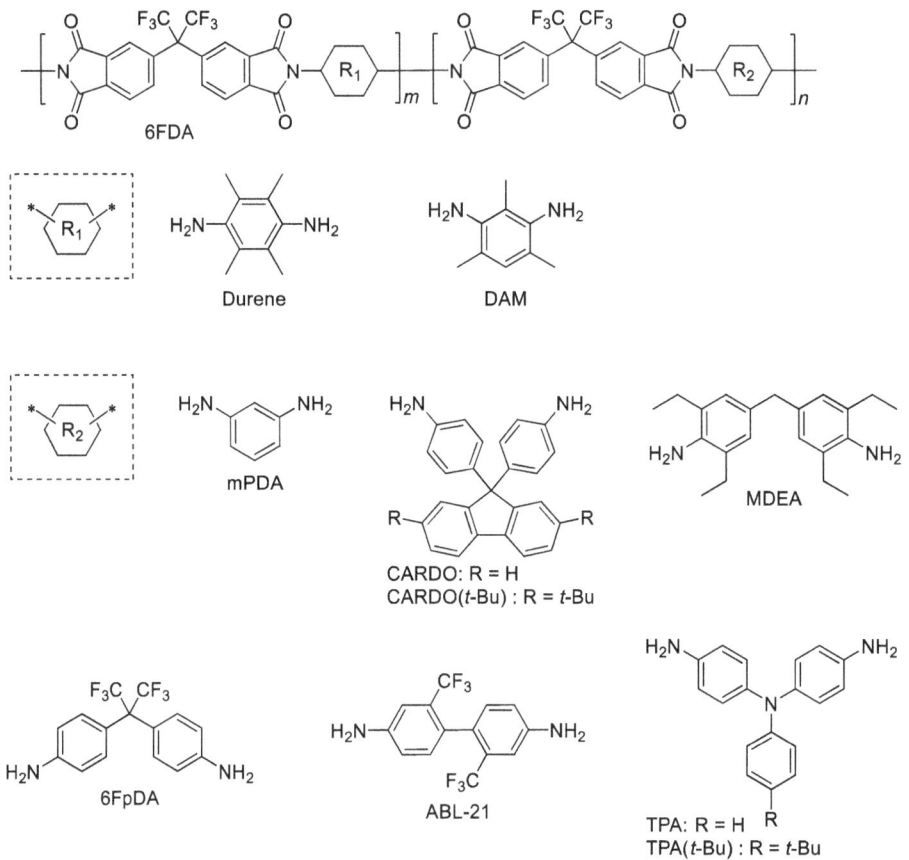

Figure 11: Chemical structures of some selected polyimides studied for H_2S removal from natural gas streams [38–46].

Yahaya et al. evaluated the sour mixed-gas separation properties of dense membranes prepared from different block lengths of 6FDA-*m*PDA/6FDA-Durene copolyimide (Figure 11), using three quaternary sour gas mixtures (1/10/30/59 vol%, 10/10/20/ 60 vol%, and 20/10/10/60 vol% of $H_2S/CO_2/N_2/CH_4$, respectively) at feed pressures up to 500 psi and a temperature of 22 °C [38]. It was found that the H_2S permeability and H_2S/CH_4 selectivity increase with the increase of H_2S content (or partial pressure) in the mixture. The H_2S/CH_4 selectivity was measured as 23 for a mixture with 20 vol% H_2S at a feed pressure of 500 psi.

Following up on the promising results obtained from their previous studies, the research group synthesized CARDO-based 6FDA-Durene/CARDO random copolyimides [CARDO = 9,9-bis(4-aminophenyl)fluorene, Figure 11] with different Durene:CARDO molar ratios (3:1, 1:1, and 1:3) [39]. Their dense membranes were evaluated for H_2S removal from NG using two quinary sour gas mixtures (10/10/20/3/57 vol%, and 20/10/ 10/3/57 vol% of $H_2S/CO_2/N_2/C_2H_6/CH_4$, respectively) at feed pressures up to 34 bar and 22 °C. The twisted and somewhat bulky CARDO moiety introduced to the polymer backbone disrupts the chain packing within the membrane matrix and hence leads to increasing the excess free volume, and therefore enhances the H_2S permeation thorough the membrane. For example, for the mixture containing 20 vol% H_2S and at feed pressure of 21 bar, the H_2S permeability was measured as 38.1 Barrer with a H_2S/CH_4 selectivity of 19.1.

In a similar fashion, Alghannam et al. studied the sour mixed-gas permeation properties of dense membranes prepared from block 6FDA-Durene/6FDA-CARDO co-polyimides with different block lengths, using three different quinary sour gas mixtures (10/10/20/1/59 vol%, 20/10/10/3/57 vol%, and 36/10/4.4/1/48.6 vol% of $H_2S/CO_2/N_2/$ C_2H_6/CH_4, respectively) at feed pressures up to 41 bar and a temperature of 22 °C [40]. For example, in the presence of 36% of H_2S and a feed pressure of 27.6 bar, the H_2S permeability and H_2S/CH_4 selectivity were measured as 275.8 Barrer and 20.08, respectively. Later, TFC prepared from the block 6FDA-Durene/6FDA-CARDO copolyimide and the transport properties were tested using a quinary sour gas mixture containing 10/10/20/1/59 vol% of $H_2S/CO_2/N_2/C_2H_6/CH_4$, respectively, at feed pressures up to 700 psi and 22 °C [41]. For instance, the H_2S permeance at 700 psi was measured as 220 GPU with a H_2S/CH_4 selectivity of 15.2. This study demonstrated that the preliminary results obtained on dense membranes can be reproduced for TFCs, which is a closer configuration to that used in an industrial setting.

Based on aforementioned studies, Yahaya et al. prepared a series of random and block DAM-based copolyimides with a variety of diamine monomers known for their high CO_2/CH_4 permselectivity: CARDO, 6FpDA, and ABL-21 (Figure 11) [42]. While the developed membranes are expected to exhibit high CO_2/CH_4 selectivity in a non-sour mixed-gas, their unconventional behavior in the presence of H_2S was difficult to predict.

Since polymers tend to behave in an unconventional manner in the presence of H_2S molecules, their H_2S removal potential may not be predicted from their pure-gas or sweet mixed-gas separation properties. Hence, selected monomers with high CO_2 permeability coefficient and/or CO_2/CH_4 selectivity in a non-sour mixed-gas were used to prepare copolyimides and then tested their sour mixed-gas separation properties.

Hayek et al. synthesized random and block 6FDA-based copolyimides using durene as a permeability-enhancing monomer combined with CO_2/CH_4 selectivity-enhancing monomers: MDEA [43], and 6FpDA [44]. The H_2S removal performance of their dense membranes was evaluated using a quinary sour gas mixture 20/10/10/1/59 vol% of $H_2S/CO_2/N_2/C_2H_6/CH_4$, respectively, at elevated feed pressures. For example, for the block 6FDA-6FpDA/6FDA-Durene (4:1), and in the presence of 20 vol% H_2S at 500 pi, the CO_2/CH_4 selectivity was measured as 20.8, which represents ~53% drop from the CO_2/CH_4 selectivity of 44.5 in the absence of H_2S. The H_2S/CH_4 selectivity was measured as 13.1. This is usually explained by the competition on sorption sites within the membrane matrix between the acid gas molecules, H_2S and CO_2.

To improve the permeation properties of polyimides, Hayek et al. introduced bulky *tert*-butyl groups into the polymer backbone. Bulky groups tend to disrupt the chain packing within the membrane matrix, and hence increase the excess fractional free volume (FFV). Since gas diffusivity increases with the increment in FFV, the H_2S permeation is expected to increase following this methodology. Therefore, the performance of dense membranes prepared from 6FDA-TPA (TPA = triphenylamine, Figure 11) [15] and 6FDA-TPA(*t*-Bu) and 6FDA-CARDO and 6FDA-CARDO(*t*-Bu) [CARDO(*t*-Bu) = 9,9-bis(4-aminophenyl)-2,7-ditert-butylfluorene, Figure 11] [45] homopolyimides was tested using a quinary sour mixed-gas (20/10/10/1/59 vol% of $H_2S/CO_2/N_2/C_2H_6/CH_4$, respectively) at elevated feed pressures. While the non-sour mixed-gas permeation results were as expected from the molecular design of all the polymers, the sour mixed-gas results were case-dependent. For example, the sour mixed-gas permeation results for 6FDA-TPA and 6FDA-TPA(*t*-Bu) were very similar. The effect of the *tert*-butyl group was not prominent. This was attributed to some swelling/plasticization due to the H_2S/polymer affinity, which eliminated the effect of the bulky side group in TPA(*t*-Bu). However, this was not observed in the case of membranes prepared from the homopolyimides 6FDA-CARDO and 6FDA-CAROD(*t*-Bu). For example, when testing both membranes using a quinary sour mixed-gas (20/10/10/1/59 vol% of $H_2S/CO_2/N_2/C_2H_6/CH_4$, respectively) at a feed pressure of 500 psi and 22 °C, it was observed that the *tert*-butyl groups introduced to the CARDO moiety of 6FDA-CAROD(*t*-Bu) increased the H_2S permeability 3.8 fold (148 Barrer) compared to that of 6FDA-CARDO (38.8 Barrer), with a H_2S/CH_4 selectivity of 16.1.

Following the enhancement obtained from the structural modification in the previous work, Hayek et al. prepared block 6FDA-Durene/6FDA-CARDO(*t*-Bu) copolyimide and evaluated the performance of its dense membrane using the same quinary mixture and conditions of pressure and temperature as their previous work [46]. For instance, at a feed pressure of 500 psi, the H_2S permeability of the modified 6FDA-Durene/6FDA-CARDO(*t*-Bu) membrane was measured as 456 Barrer, which represents ~27% increase compared to the non-modified 6FDA-Durene/6FDA-CARDO copolyimide (358 Barrer). The H_2S/CH_4 selectivity, however, dropped by ~26% due to the permeability-selectivity trade-off relationship previously known for glassy polymers.

Yi et al. reported for the first time the H_2S removal performance of membranes prepared from a hydroxyl-functionalized polymer of intrinsic microporosity (PIM-6FDA-OH,

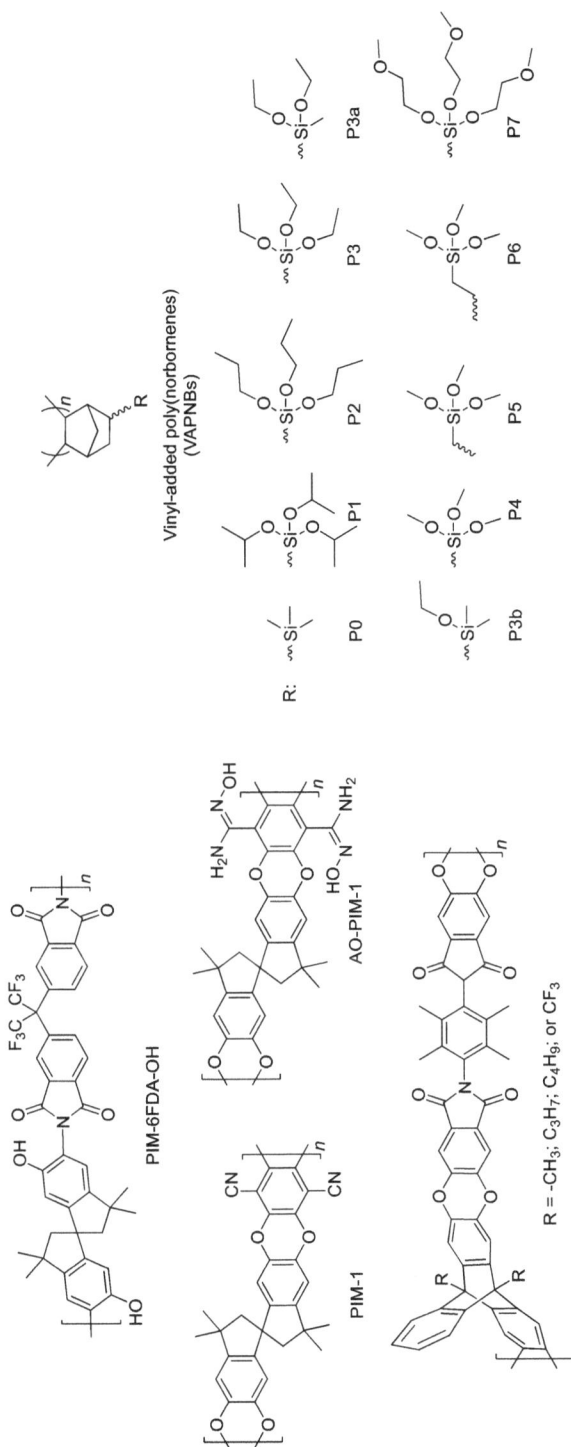

Figure 12: Chemical structures of polymers with intrinsic microporosity (PIM)s and poly(norbornene)s studied for H$_2$S removal from natural gas streams [47–50].

Figure 12) [47]. The sour mixed-gas permeation properties were measured using a ternary gas mixture containing 15/15/70 vol% of $H_2S/CO_2/CH_4$, respectively, at feed pressures up to 48 bar and a temperature of 35 °C. Due to the high affinity of H_2S molecules to the polymeric chains, the membrane suffered from evident H_2S-induced plasticization starting from a feed pressure ~4.5 bar. The mixed-gas H_2S permeability of the membrane kept increasing up to highest feed pressure tested (48 bar), which was not the case of CO_2 and CH_4, due to the dominance of H_2S molecules (i.e., higher condensability and polarity) over the sorption sites within the membrane matrix. Despite the evident plasticization, the membrane did not lose its efficiency, where the mixed-gas H_2S/CH_4 selectivity at 48 bar was measured as 30.0, with a H_2S permeability of 63.0 Barrer. These results are considered attractive under such aggressive conditions of H_2S content and feed pressure.

Based on the results of the study obtained with PIM-6FDA-OH, Yi et al. studied the performance of the prototype ladder polymer PIM-1 (Figure 12) for H_2S removal from natural gas [48]. In a similar way to sweet mixed-gas performance, the PIM-1 membrane demonstrated high H_2S permeability (~19,000 Barrer) with a H_2S/CH_4 selectivity ~30 when tested using a ternary sour gas mixture (20/20/60 vol% of $H_2S/CO_2/CH_4$, respectively) at a feed pressure of 77 bar and 35 °C. However, PIM-1 is very well known to suffer from severe physical aging conditions due to the weak interactions between the polymeric chains within the membrane matrix. To mitigate this issue, Yi et al. prepared amidoxime-functionalized polymer of intrinsic microporosity (AO-PIM1, Figure 12) through modification of the nitrile groups in PIM-1. Under the same testing conditions listed for PIM-1, the AO-PIM1 membrane demonstrated unprecedented separation performance under such aggressive separation conditions. At 77 bar, and in the presence of 20 vol% H_2S, the AO-PIM1 membrane H_2S-permeability was measured to be greater than 4,000 Barrer with a high H_2S/CH_4 selectivity of 75. Moreover, continuous aging studies under harsh testing conditions demonstrated that the AO-PIM1 membrane maintained its performance for acid gas treatment without detrimental aging effect.

Ghasemnejad-Afshar et al. studied the theoretical separation potential of membranes prepared from branched triptycene-based polymers of intrinsic microporosity (PIM)s using ab initio calculations, molecular dynamics (MD) and Monte Carlo (MC) simulations (Figure 12) [49]. The triptycene moiety is substituted with different groups: methyl (CH_3), iso-propyl (C_3H_7), sec-butyl (C_4H_9), and trifluoromethyl (CF_3). The substituents serve as bulky groups to disrupt the polymer's chain packing within the membrane matrix, leading to a change in the available free volume, which affects the diffusion of gas molecules. In fact, it was found that the calculated FFV increased due to presence of side groups in the following order: $C_4H_9 > C_3H_7 > CF_3 > CH_3$. Moreover, taking into account the sorption factor, it was found that the polymer membrane with the side group CF_3 has the strongest tendency among the other side groups to adsorb H_2S, due to the polarity of the CF_3 group. Nevertheless, based on both separation factors (i.e., solubility and diffusion) the overall performance of the membrane of the polymer with the sec-butyl (C_4H_9) side chain is the best among all other side groups studied. These simulation techniques demonstrate a new method to design new polymeric materials with the desired gas separation performance, especially in the case of unconventional behavior of H_2S.

Lawrence et al. reported the performance of dense membranes prepared from various alkoxysilyl-substituted vinyl-added poly (norbornene)s (VAPNBs) for the removal of H_2S from NG streams [50]. The VAPNBs vary in the pendent alkoxysilyl side substitution (Figure 12). The sour mixed-gas testing was performed using two gas mixtures with different compositions: (ternary mixture) 5/3/92 vol% of $H_2S/CO_2/CH_4$, and (quinary mixture) 20/10/10/3/57 vol% of $H_2S/CO_2/N_2/C_2H_6/CH_4$, respectively, at a feed pressure of 800 psi and a temperature of 25 °C. The comparison between the gas permeation properties of the 10 polymers revealed that shortened alkyl chain length with increased polarity is beneficial to fine-tune sour gas separation performance. For instance, the reduced chain length promotes the polymer/penetrant interactions and the polar groups containing oxygen enhance the gas solubility parameters. For example, P7, which possesses relatively long 2-methoxyethoxy side chains and short terminal methoxy groups demonstrated the highest performance when tested using the quinary mixture. The H_2S permeability was measured as 6,715 Barrer with a H_2S/CH_4 selectivity of 40.5. These sour gas permeation properties are among the highest ever reported for glassy polymers in literature.

Based on the aforementioned various sour gas separation performance of membranes prepared from glassy polymers, it was observed that these membranes suffer from permeability-selectivity trade-off relationship. When the H_2S permeability increases, the H_2S/CH_4 selectivity decreases, and vice versa. Pure-gas upper bounds have been reported in literature by Robeson for a variety of gas pairs, except for H_2S/CH_4 due to lack of significant data [51, 52]. Furthermore, theoretical (2015, [53]) and experimental (2018, [54] and 2021, [55]) CO_2/CH_4 sweet mixed-gas upper bounds have been proposed. However, sour mixed-gas separation is even more complex, due to the presence of H_2S in gas streams. While several H_2S/CH_4 trade-off lines were reported in literature based on the performance of specific materials prepared, in each case, the empirical parameters were drastically different. In 2021, Hayek et al. proposed a new H_2S/CH_4 upper bound derived from sour gas separation data obtained from glassy membranes as illustrated in Figure 13 [55].

While pure-gas permeation data are largely available in literature for a wide variety of polymeric membranes, there is a shortage in sour mixed-gas separation data, due to strict regulations to operate with H_2S due to its high toxicity. This is what makes the generation of an empirical pure-gas upper bound somewhat straightforward, which is the opposite of the generation of an empirical sour mixed-gas upper bound. Another complexity of this task resides in the multitude of sour gas separation testing conditions performed in literature. For instance, Hayek et al. identified 15 different sour gas mixtures (binary, ternary, or quinary), with several H_2S contents, along with different compositions of other gases in the mixture. Furthermore, the variety of total feed pressures and testing temperatures provides an additional challenge to compare data obtained from different studies.

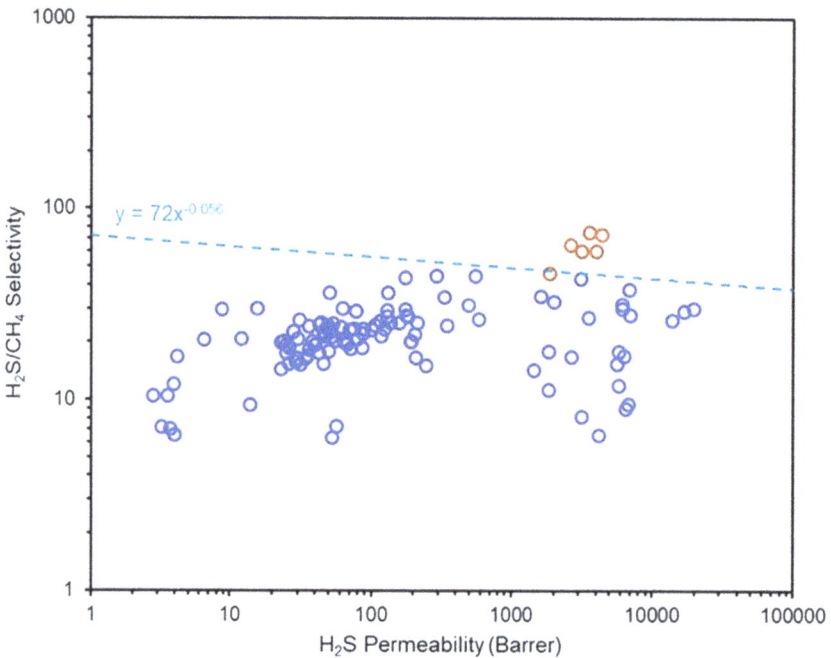

Figure 13: 2021 H_2S/CH_4 sour mixed-gas upper bound [55].

6.2 Rubbery polymers

Rubbery membranes are known to have ultra-high free volume, which makes them solubility-selective membranes. The diffusivity-selectivity becomes irrelevant due to the high free volume within the membrane matrix. They can also be defined as polymers with glass transition temperature (T_g) lower than room temperature or gas processing temperature. The main discriminating factor during the gas molecules separation is their condensability; which determines their relative affinity to rubbery polymeric materials. The higher the condensability, the higher the permeation coefficient; and therefore, hydrogen sulfide (H_2S) is much more favored over methane (CH_4). Due to the high free volume of rubbery membranes matrices, they are prone to plasticization at high feed pressures, which negatively impacts their permeability and selectivity coefficients [23]. Several rubbery polymeric membranes have been evaluated for their potential use in H_2S removal from NG as illustrated in Figure 14.

Poly(ether urethane)s (PU) and poly(ether urethane urea)s (PUU) have been used to study separation performance in two ternary sour gas mixtures (70.8% CH_4/27.9% CO_2/1.30% H_2S and 69.4% CH_4/18.1% CO_2/12.5% H_2S) at feed pressures varied between 4 atm and 13.6 atm and temperatures varied between 20 °C and 35 °C. For example, for the PUU membranes, the obtained H_2S permeability coefficients were recorded be-

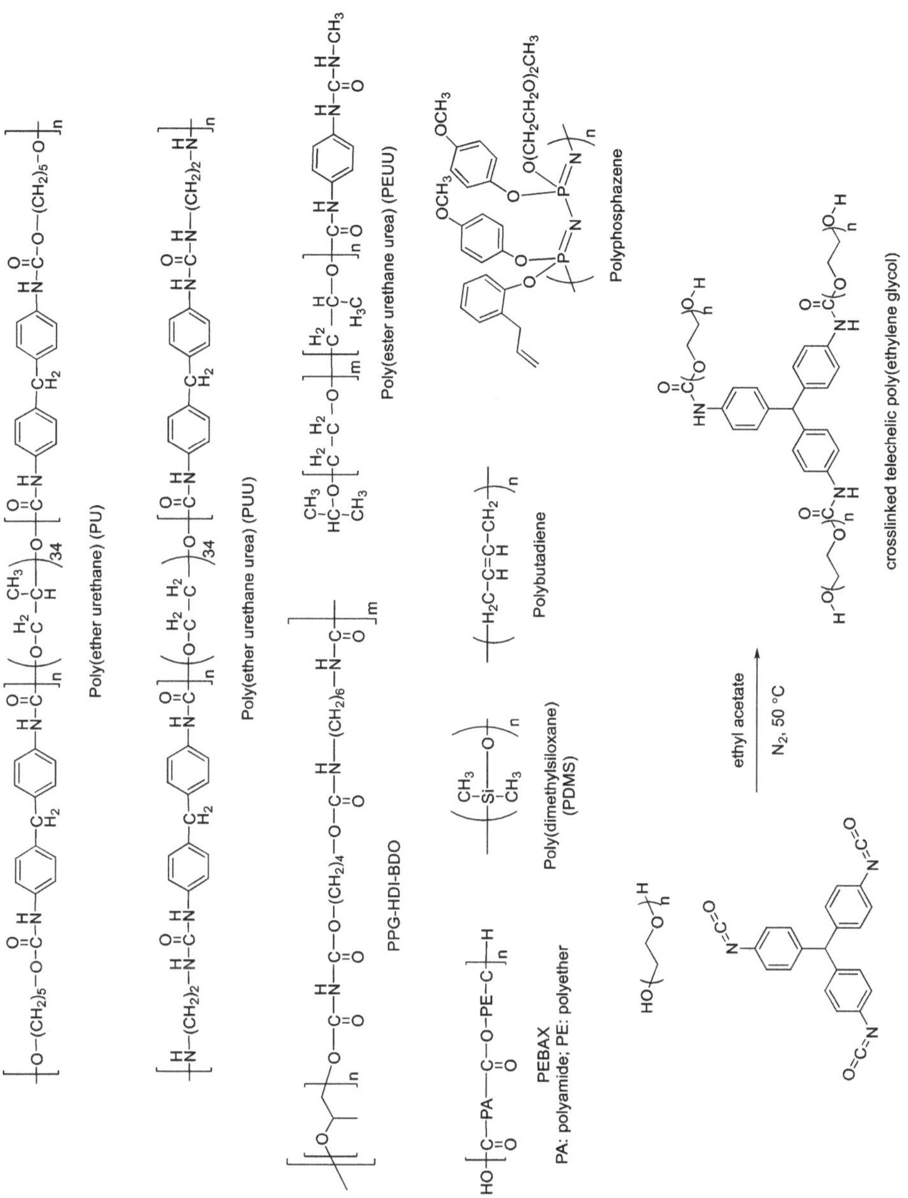

Figure 14: General chemical structures of rubbery polymers used for H$_2$S removal from natural gas streams [18, 25, 56–61].

tween 95 Barrer and 130 Barrer and the H_2S/CH_4 selectivity was as high as 100 at 13.6 atm and 20 °C. The results were compared to dense membranes prepared from PEBAX^{TM} [poly(amide-co-ether)] using the same ternary mixtures at a feed pressure of 10 atm and a temperature of 35 °C. The obtained H_2S permeability coefficients were recorded between 7.6 Barrer and 888 Barrer and the H_2S/CH_4 selectivity between 15 and 54 [18].

Poly(ester urethane urea) TFC membranes were studied using two ternary mixtures (93.2% CH_4/6.2% CO_2/0.6% H_2S, and 91.6% CH_4/5.4% CO_2/3% H_2S) at feed pressures varied between 10 bar and 30 bar and temperatures varied between 35 °C and 55 °C. The studies showed an H_2S permeance of 60 GPU and a maximum H_2S/CH_4 selectivity of 43, which tends to decrease to a minimum of 12 with increasing feed pressures up to 30 bar [56].

Polybutadiene TFC membranes were studied using a ternary mixture (balance CH_4/4% CO_2/800 ppm H_2S) at a feed pressure of 390 psig. The H_2S/CH_4 selectivity coefficients were found to be 3.3 and 8.4, respectively. Moreover, Silicon rubber (PDMS) composite membranes were studied using a ternary mixture (balance CH_4/4% CO_2/650 ppm H_2S) at a feed pressure of 95 psig. The H_2S/CH_4 selectivity coefficients were found to be 6.9. These H_2S/CH_4 selectivity coefficients are considered very low, especially for such low content of H_2S gas in the mixture [25].

Another polyurethane prepared from polypropylene glycol (PPG), hexamethylene diisocyanate (HDI), and 1, 4-butane diol (BDO) moieties, is denoted by BDO-PPG-HDI. The performance of its membranes was evaluated using H_2S-containing binary (99.9814% CH_4/0.0186% H_2S and 99.925% CH_4/0.075% H_2S) and ternary (97.5% CH_4/2.1% CO_2/0.4% H_2S and 97.24% CH_4/2.1% CO_2/0.66% H_2S) gas mixtures at different operating conditions of temperatures and pressures [57]. Despite the low concentration of H_2S in the gas mixtures compositions, the slight change on concentration revealed a clear effect on the performance of the membranes. Moreover, the presence of CO_2 in the mixture led to a drop in the H_2S permeability and H_2S/CH_4 selectivity. The best performance of the BDO-PPG-HDI membrane was measured at 30 bar and 25 °C in the presence of 0.066 mol% of H_2S. The H_2S permeability was measured to be 790 Barrer and the H_2S/CH_4 selectivity was calculated to be 27.2. On the other hand, when the feed pressure increased from 20 bar to 30 bar, the membrane showed signs of plasticization, where the permeation of all gas molecules increased. This was attributed to the change on the dynamic free volume of the rubbery membrane. However, since the separation is solubility-selective, the sorption of CO_2 and H_2S within the membrane matrix increased more than that of CH_4, which led to an increase in H_2S/CH_4 selectivity.

Orme et al. studied a new class of rubbery dense membranes prepared from three polyphosphazenes containing three pendant groups of the polymer backbone: 2-(2-methoxyethoxy)ethanol (MEE), 4-methoxyphenol, and 2-allylphenol (Figure 14) [58, 59]. The molar ratios between the three pendant groups were varied to tailor the physical properties of the polymer: 6/75/19 mol% (polymer 1), 48/48/4 mol% (polymer 2), and 74/24/2 mol% (polymer 3) of MEE/4-methoxyphenol/2-allylphenol, respectively. For instance, the MEE group is regarded as a hydrophilic group; however, the 4-methoxyphenol and 2-allylphenol are considered as hydrophobic groups. The sour mixed-gas separation performance of their dense membranes was evaluated using a sour gas mixture containing 3% each of Ar, N_2, O_2, CH_4, CO_2, and H_2S in a balance of helium at a total feed pressure of 206 kpa (30 psi) and a temperature of 30 °C. The H_2S permeability and H_2S/CH_4 selectivity coefficients of the membranes increased with the increase of the MEE mol% in the polymer backbone, which also decreased the polymer's glass transition temperature (T_g) according to the following order: polymer 3 (74 mol% MEE) > polymer 2 (48 mol% MEE) > polymer 1 (6 mol% MEE). For example, for polymer 3, the H_2S permeability was measured as 1,103 Barrer with a H_2S/CH_4 selectivity of 78.8. This work represents an example of a class of polymers where a variety of pendant groups can be used to tailor the permeation properties of polymeric membranes according to the impurity content of the natural gas reserve.

Harrigan et al. evaluated the sour gas separation performance of five grades of Pebax® (MV 1074, MH 1657, 2533, MV 3000, and 5513) dense membranes using two gas mixtures with different compositions: (ternary mixture) 5/3/92 vol% of $H_2S/CO_2/CH_4$, and (quinary mixture) 20/10/10/3/57 vol% of $H_2S/CO_2/N_2/C_2H_6/CH_4$, respectively, at three feed pressures of 200, 500, and 800 psi and a temperature of 25 °C [60]. It was found that the overall performance of membranes depends largely on block type and relative block content. For instance, when subjected to a 20% H_2S-containing feed, Pebax® 1074 and Pebax® 1657, which contain 50% PE and 60% PE by weight, respectively, exhibited the highest H_2S/CH_4 selectivity of 79.9–80.6 at 800 psi. From the structure-property relationship study, it was concluded that while the separation is solubility-driven, diffusivity selectivity plays an important role during the process. This is why, a balanced combination of highly permeable PE blocks and structurally supportive PA blocks led to a high H_2S/CH_4 selectivity in Pebax® 1074 and Pebax® 1657. However, all the studied membranes suffered from severe plasticization in the presence of 20% H_2S in the mixture. Nevertheless, the increasing hydrogen bonding in Pebax® 1074 and Pebax® 1657 reduced performance loss associated with plasticization.

In another work, Harrigan et al. evaluated the sour gas separation performance of TFCs with a selective layer prepared from cross-linked poly(ethylene glycol) (PEG, Figure 14) [61]. A series of telechelic PEG oligomers with average molecular weights between 200 g.mol^{-1} and 2050 g.mol^{-1} were used to fabricate the membranes and their sour gas separation performance was measured using a ternary sour gas mixture (5/3/92 vol% of $H_2S/CO_2/CH_4$, respectively). It was found that the membrane performance is tunable. For low molecular weight PEG derivatives, the H_2S/CH_4 selectivity was greater than 60, and for higher molecular weight derivatives the H_2S/CH_4 selectivity reached values higher than 110. This was attributed to the cross-linking being effective in eliminating PEG crystallinity for molecular weights (Mws) less than 1,000 g.mol^{-1}; this is why, the cross-linked PEG membranes reside in both the rubbery (Mw > 1,000 g.mol^{-1}) and glassy (Mw < 1,000 g.mol^{-1}) regimes. Another factor that affected the gas permeation of these membranes is the cross-linking density. Therefore, a balance between molecular weight and cross-linking density can achieve optimal sour gas separation performance.

6.3 Mixed-Matrix Membranes

It is very well known at this stage that polymeric membranes suffer from permeability-selectivity tradeoff relationship; when the permeability increases, the selectivity decreases, and vice versa. Based on the separation mechanism through polymeric matrices, the two main keys controlling the separation are solubility and diffusivity. Tailoring the membrane's gas separation productivity and efficiency through controlling the diffusivity is challenging, and cannot overcome the permeability-selectivity tradeoff relationship, due to the slight difference in gas molecules kinetic diameters. On the other hand, tailoring the membrane's permeation properties through the solubility factor (i.e., gas/polymer affinity) is a more promising approach, due to the difference in polarity between the gas molecules. Hence, researchers developed hybrid membranes, also called mixed-matrix membranes (MMMs), constituted from polymeric matrix and a selective adsorbent [62]. These adsorbents could be of different nature, such as metal organic frameworks (MOFs) [63], silica nanoparticles [64, 65], carbon nanotubes [66, 67], zeolites [68, 69], and other organic and inorganic porous materials [62].

The most important factor during the preparation of MMMs is the chemical affinity between the adsorbent and the polymer matrix in order to avoid the formation of defects resulting from phase separation. Gas molecules travel through all these defects equally and the membrane loses its sieving functionality and hence its efficiency. In general, standalone adsorbents may possess better gas separation properties than MMMs, however, their processability into industrial scale membrane units is very challenging. MMMs combine the distinct molecular separation properties of selective adsorbents and take advantage of the processability and mechanical stability of polymers.

Several MMMs were developed to study their gas transport properties, however, very few were evaluated for H_2S removal from natural gas. This is mainly due to the chemical stability of adsorbents toward H_2S molecules. Recently, several MMMs using MOFs have been prepared and their H_2S separation from gas mixtures has been reported. Moreover, in addition to stability towards H_2S, MOF-based membranes face other challenges, such as the excessively large and/or flexible apertures in most MOFs, their scale up cost, and compatibility to polymeric matrices to avoid agglomeration and sedimentation. Examples of MOF-based MMMs studied in literature for H_2S removal from NG are illustrated in Figure 15.

Ahmad et al. reported the preparation of MMMs using three different types of Zr-MOFs (UiO-66, UiO-66-NH_2, and UiO-66-NH-$COCH_3$) mixed with the polyimide 6FDA-DAM [70]. The gas separation performance of 6FDA-DAM and its Zr-MOF MMMs was studied using 30:70 vol% CO_2:CH_4 feed mixture at 20 bar and 35 °C, before adding 5 vol% of H_2S, making the feed composition to 30/5/65 vol% CO_2/H_2S/CH_4. The separation performance after H_2S exposure was also investigated at 20 bar and 35 °C. Upon the addition of 5 vol% H_2S in the mixed gas, the CO_2 permeability in all samples decreased. 6FDA-DAM MMMs showed a higher CO_2 permeability reduction in the presence of H_2S, compared to the neat 6FDA-DAM membrane. The observation exhibited the influence of Zr-MOFs in the

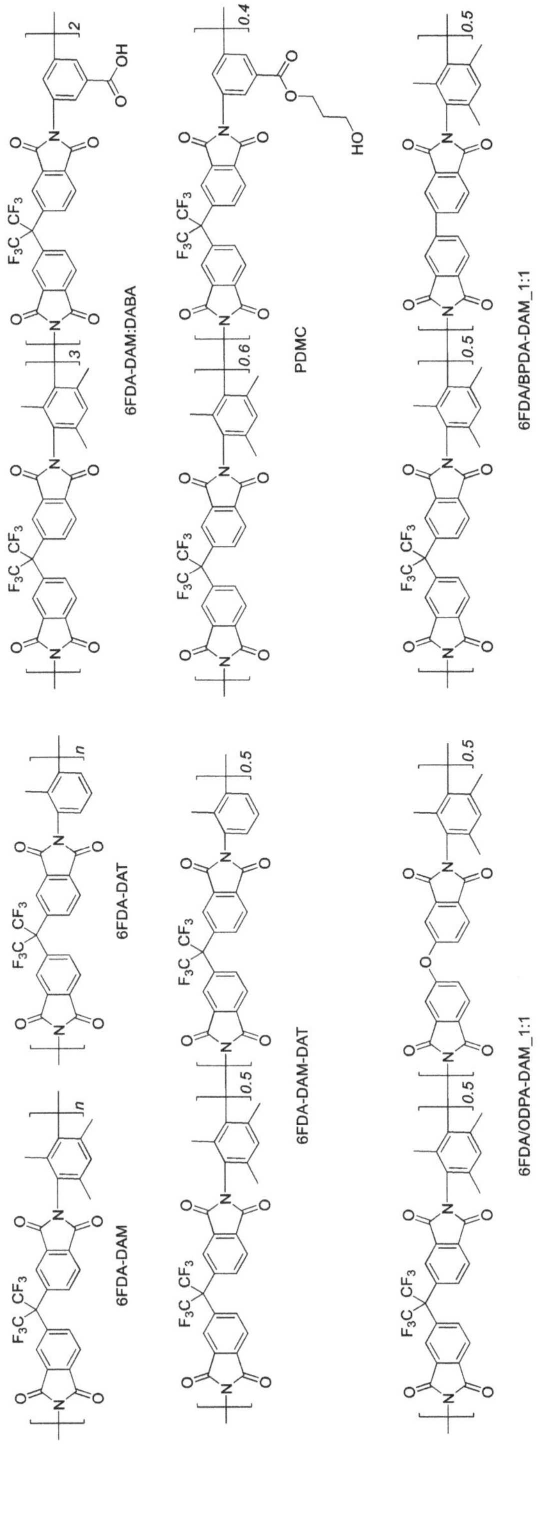

6FDA-DAM:DABA

PDMC

6FDA/BPDA-DAM_1:1

6FDA-DAT

6FDA-DAM

6FDA-DAM-DAT

6FDA/ODPA-DAM_1:1

Figure 15: Chemical structures of polymers used to prepare mixed-matrix membranes for H_2S removal from natural gas [70–74].

MMMs, of which their active metal sites also preferentially adsorb H_2S and thus reduce their CO_2 adsorption capacity. The best performance was obtained with the 16 wt.% UiO-66-NH-COCH$_3$/6FDA-DAM membrane. The H_2S permeability and H_2S/CH_4 selectivity were measured to be 193 Barrer and 16.2, respectively.

Liu et al. reported detailed studies on MMMs prepared from four 6FDA-based polyimides (6FDA-DAM, 6FDA-DAM/DABA_3:2 random copolymer, 6FDA/BPDA-DAM_1:1, 6FDA/ODPA-DAM_1:1) and three different MOFs (Y-*fum*-fcu-MOF, Zr-*fum*-fcu-MOF and Zr-*mfum*-fcu-MOF). The different MOFs were chosen to tailor the pore aperture size to control the sieving properties of the crystalline materials – different combinations of MOF/polymer with 10 wt.% and 20 wt.% loadings. The mixed-gas permeation tests were performed using three $H_2S/CO_2/CH_4$ gas mixtures in different $H_2S/CO_2/CH_4$ compositions (20/20/60, 25/5/70, and 0.5/20/79.5, respectively), at pressures up to ~56 bar and a temperature of 35 °C [71]. The developed hybrid membranes show extraordinary H_2S permeability and H_2S/CH_4 separation efficiencies at 56 bar, surpassing the empirical tradeoff relationships for polymer membranes under high pressures (>30 bar). Moreover, the hybrid membrane performances are highly tunable in terms of permeability vs. selectivity by choosing different polymer matrices while preserving the attractive H_2S and CO_2 simultaneous removal feature. For example, for the 10 wt.% Zr-*fum*-fcu-MOF/6FDA-DAM-ODPA membrane at 56.2 bar and 35 °C, the H_2S permeability and H_2S/CH_4 selectivity coefficients were measured to be 359 Barrer and 62.7, respectively.

Liu et al. reported the gas permeation properties of hybrid membranes prepared from two different rare earth (RE) MOF with a face-centered cubic (fcu) topology (RE-fcu-MOF). The pore size was controlled using two different types of linkers [1,4-naphthalenedicarboxylate (*naph*) or fumarate (*fum*)], conferring the pore system entrance with the desired molecular sieving capabilities appropriate for NG purification. The fine-tuned Y-*fum*-fcu-MOF and Eu-*naph*-fcu-MOF adsorbents offer molecular-sieving-based separations. Subsequent introduction of different loading of sub micrometer-sized crystals into polymers (6FDA-DAM or 6FDA-DAM:DABA) yields highly enhanced membrane molecular sieving properties for the challenging $H_2S/CO_2/CH_4$ (20/20/60 vol%, respectively) separation [72]. The separation performances of hybrid membranes with different MOF loads were evaluated under high feed pressures up to 55 bar at a temperature of 35 °C. For example, for a 30 wt.% Y-fum-fcu-MOF/6FDA-DAM hybrid membrane, the H_2S permeability and H_2S/CH_4 selectivity were measured at 6.9 bar and 35 °C to be 470 Barrer and 23.4, respectively. Moreover, preliminary results of 13 wt.% Y-*fum*-fcu-MOF/6FDA-DAM hybrid HFM was applied for the separation of 20/20/60 $H_2S/CO_2/CH_4$ mixtures at 6.7 bar and 35 °C. The H_2S/CH_4 selectivity of the HFM was measured to be similar to that obtained for dense membranes. These results are considered promising, not only in terms of gas separation, but even more, for the demonstration of the preparation of hybrid HFM, which is the configuration used in industrial applications.

In another work, Liu et al. reported the preparation of MMMs using three different MOFs (NbOFFIVE-1-Ni, AlFFIVE-1-Ni, and SIFSIX-3-Cu) and two 6FDA-based polyimides (6FDA-DAM, and PDMC). The pore structure of the MOFs was finely tuned by varying the metal and the pillar while using the same ligand. Through the fine-tuning of MOF structures and interfaces with the selected polymer matrices, the MOFs-based membranes demonstrated simultaneous removal of H_2S and CO_2 from NG with performance far beyond the state-of-the-art polymer membrane [73]. Therefore, two MMMs (20 wt.% NbOFFIVE-1-Ni/6FDA-DAM and 20 wt.% AlFFIVE-1-Ni/6FDA-DAM) were evaluated for their potential in simultaneously removing H_2S and CO_2 from CH4 stream using a highly aggressive model natural gas feed (20% H_2S, 20% CO_2, and 60% CH_4) at a feed pressure of 6.9 bar and a temperature of 35 °C. The H_2S permeability coefficients were measured to be 401 Barrer and 458 Barrer, and the H_2S/CH_4 selectivity coefficients were calculated to be 20.7 and 18.6 for NbOFFIVE-1-Ni/6FDA-DAM and AlFFIVE-1-Ni/6FDA-DAM, respectively. The differences in gas affinity and aperture size between NbOFFIVE-1-Ni and AlF-FIVE-1-Ni led to the variation of transport properties in their corresponding MMMs. More importantly, these MOF apertures exhibit molecular sieving properties that increase the diffusion selectivity for both H_2S and CO_2 over CH_4.

Most recently, Datta et al. reported the preparation of three MMMs using three different polyimides (6FDA-DAM, 6FDA-DAM-DAT and 6FDA-DAT) and a nickel-based fluorinated MOF, AlFFIVE-1-Ni. The MOF was prepared into nanosheets configuration with a completely exposed (001) facet. The load of MOF into the polymer matrix was adjusted to around 60 wt.%. The three MMMOFs are denoted by: (001)-AlFFIVE(58.9)/6FDA-DAM, (001)-AlFFIVE(59.6)/6FDA-DAM-DAT, and (001)-AlFFIVE(60.3)/6FDA-DAT. The performances of the membranes were evaluated using three ternary gas mixtures ($H_2S/CO_2/CH_4$: 1/9/90, 2/18/80, and 5/5/90) at a feed pressure of 10 bar and a temperature of 35 °C. For example, for the (001)-AlFFIVE(58.9)/6FDA-DAM membrane, and using the ternary mixture $H_2S/CO_2/CH_4$: 5/5/90, the H_2S permeability was measured to be 573 Barrer and the H_2S/CH_4 selectivity was calculated to be 39.0 [74]. This promising performance was attributed to the in-plane alignment and extremely high loading of (001) nanosheets. The nanosheets enhanced the sieving of gas molecules based on their kinetic diameter during the diffusion through the membranes' matrices. The results confirm the benefit of preparing MOF configuration into oriented nanosheets, allowing the desired orientation of the 1D channels parallel to the gas diffusion direction, which improves the overall performance of the membranes.

7 Conclusion and outlook

Membrane-based sour natural gas (NG) sweetening is a promising technology to replace or assist the currently used liquid amine scrubbing process. In fact, amine scrubbing technology is feasible to treat NG streams with high acid gas content (>20 vol% H_2S), however, the treatment becomes very costly. Placing a membrane module upstream to the amine scrubbing unit helps reduce the acid gas content of the gas stream before reaching the scrubbing column. This will actually reduce the use of lean amine solvent, and therefore, reduces its circulation rate and amine regeneration, which reduces the power consumption and improves the economics of the entire process.

However, membrane-based technology needs to provide answers to several challenges before being implemented in the field. One important aspect is the membrane productivity (i.e., permeability) and efficiency (i.e., selectivity). High membrane productivity is needed to allow the gas processing at the needed daily rate, and the high membrane selectivity is needed to reduce the hydrocarbon loss due to methane slippage to the permeate side, which affects the economics of the entire process. Another equally important aspect is the membrane stability in the presence of hydrogen sulfide under the aggressive operational conditions of pressure and temperature. The strong affinity between the polar gas molecule hydrogen sulfide and polymeric chains may result in membrane swelling, plasticization, or speeding of its physical aging.

Three main polymeric membrane types have been discussed in this chapter: (1) glassy polymers, (2) rubbery polymers, and (3) mixed-matrix membranes (MMMs).

Glassy polymers represent a class of membranes where the separation is diffusivity-selective. The separation occurs through the discrimination between the gas molecules kinetic diameters. The main advantage of glassy polymers is their processability into thin films or hollow fibers. However, while many examples have been shown to improve the membrane's permeability and selectivity, the problem of physical aging and plasticization at high feed pressure remains an important challenge to overcome.

On the other hand, the sour gas separation performance of membranes prepared from rubbery polymers is very attractive in terms of both permeability and selectivity. In fact, rubbery polymers exhibit high H_2S permeability and high H_2S/CH_4 selectivity which is rarely achieved by glassy polymers. However, the poor mechanical properties of rubbery polymers render their processability very challenging, and their membranes are prone to physical aging and plasticization.

Moreover, MMMs have demonstrated unequal potential to tailor the membranes performance as needed. The design of the inorganic filler (i.e., organic frameworks, MOFs) allow the fine-tuning of the acid gas permeation through the membrane matrix via targeted gas/MOF interactions. Despite the attractive performance of MMMs during H_2S removal from natural gas streams, their processability remains a challenging task. Moreover, the polymer/MOF affinity is important to maintain structural integrity of membranes, to avoid non-sieving channels, which are considered as defects that alter the membrane performance. However, recently several MMM hollow fibers have been reported in literature, mainly for CO_2 removal, which demonstrates the possibility of solving the challenge of MMMs' processability.

Finally, carbon molecular sieves (CMS) membranes have shown great potential in CO_2 removal from gas streams. Their mechanical stability has been improved over time, and high-pressure tests have been performed on CMS hollow fiber membranes. This type of membranes has great potential in H_2S removal from NG streams.

References

[1]   Global natural gas consumption 2021 | Statista. (n.d.). https://www.statista.com/statistics/282717/global-natural-gas-consumption/ (accessed November 15, 2022).

[2]   BP Energy Outlook 2030 shows increasing impact of unconventional oil and gas on global energy markets | News and insights | Home, Bp Global. (n.d.). https://www.bp.com/en/global/corporate/news-and-insights/press-releases/bp-energy-outlook-2030-shows-increasing-impact-of-unconventional-oil-and-gas-on-global-energy-markets.html (accessed November 15, 2022).

[3]   Baker, R. W. and Lokhandwala, K. Natural gas processing with membranes: An overview. Ind. Eng. Chem. Res., 2008, **47**: 2109–2121. https://doi.org/10.1021/ie071083w.

[4]   Duval, S. Chapter 2 – Natural gas sweetening. In: Wang, Q. (Ed.) Surface Process, Transportation, and Storage, Gulf Professional Publishing, 2023, pp. 37–78. https://doi.org/10.1016/B978-0-12-823891-2.00007-7.

[5]   Castel, C., Bounaceur, R. and Favre, E. Membrane processes for direct carbon dioxide capture from air: Possibilities and limitations. Front. Chem. Eng., 2021, **3**: https://www.frontiersin.org/articles/10.3389/fceng.2021.668867. (accessed November 15, 2022).

[6]   Hamad, F., Qahtani, M., Ameen, A., Vaidya, M., Duval, S., Bahamdan, A. and Otaibi, F. Treatment of highly sour natural gas stream by hybrid membrane-amine process: Techno-economic study. Sep. Purif. Technol., 2020, **237**: 116348. https://doi.org/10.1016/j.seppur.2019.116348.

[7]   Ameen, A. W., Budd, P. M. and Gorgojo, P. Superglassy polymers to treat natural gas by hybrid membrane/Amine processes: Can fillers help? Membranes, 2020, **10**: 413. https://doi.org/10.3390/membranes10120413.

[8]   Vatanpour, V., Pasaoglu, M. E., Barzegar, H., Teber, O. O., Kaya, R., Bastug, M., Khataee, A. and Koyuncu, I. Cellulose acetate in fabrication of polymeric membranes: A review. Chemosphere, 2022, **295**: 133914. https://doi.org/10.1016/j.chemosphere.2022.133914.

[9]   Ma, X.-H. and Yang, S.-Y. Chapter 6 – Polyimide gas separation membranes. In: Yang, S.-Y. (Ed.) Advanced Polyimide Materials, Elsevier, 2018, pp. 257–322. https://doi.org/10.1016/B978-0-12-812640-0.00006-8.

[10] Shokri, A. Synthesize and characterization of polysulfone membrane for the separation of hydrogen sulfide from natural gas. Surf. Interfaces., 2021, **25**: 101233. https://doi.org/10.1016/j.surfin.2021.101233.

[11] Arcella, V., Ghielmi, A. and Tommasi, G. High performance perfluoropolymer films and membranes. Ann. N. Y. Acad. Sci., 2003, **984**: 226–244. https://doi.org/10.1111/j.1749-6632.2003.tb06002.x.

[12] Okamoto, Y., Chiang, H.-C. and Merkel, T. Chapter 5 – Perfluoropolymers for gas separation membrane applications. In: Ameduri, B. and Fomin, S. (Eds.) Fascinating Fluoropolymers and Their Applications, Elsevier, 2020, pp. 143–155. https://doi.org/10.1016/B978-0-12-821873-0.00005-9.

[13] Graham, T. On the law of the diffusion of gases. J. Membr. Sci., 1995, **100**: 17.

[14] Wijmans, J. G. and Baker, R. W. The solution-diffusion model: A review. J. Membr. Sci., 1995, **107**: 1.

[15] Hayek, A., Alsamah, A., Qasem, E. A., Alaslai, N., Alhajry, R. H. and Yahaya, G. O. Effect of pendent bulky groups on pure- and sour mixed-gas permeation properties of triphenylamine-based polyimides. Sep. Purif. Technol., 2019, **227**: 115713. https://doi.org/10.1016/j.seppur.2019.115713.

[16] Rackley, S. A. 8 – Membrane separation systems. In: Rackley, S. A. (Ed.) Carbon Capture and Storage, (2nd ed.), Butterworth-Heinemann, Boston, 2017, pp. 187–225. https://doi.org/10.1016/B978-0-12-812041-5.00008-8.

[17] Xiao, W., Gao, P., Dai, Y., Ruan, X., Jiang, X., Wu, X., Fang, Y. and He, G. Efficiency separation process of H2/CO2/CH4 mixtures by a hollow fiber dual membrane separator. Processes, 2020, **8**: 560. https://doi.org/10.3390/pr8050560.

[18] Chatterjee, G., Houde, A. A. and Stern, S. A. Poly(ether urethane) and poly(ether urethane urea) membranes with high H2S/CH4 selectivity. J. Membr. Sci., 1997, **135**: 99–106. https://doi.org/10.1016/S0376-7388(97)00134-8.

[19] Achoundong, C. S. K., Bhuwania, N., Burgess, S. K., Karvan, O., Johnson, J. R. and Koros, W. J. Silane modification of cellulose acetate dense films as materials for acid gas removal. Macromolecules, 2013, **46**: 5584–5594. https://doi.org/10.1021/ma4010583.

[20] Liu, Y., Liu, Z., Morisato, A., Bhuwania, N., Chinn, D. and Koros, W. J. Natural gas sweetening using a cellulose triacetate hollow fiber membrane illustrating controlled plasticization benefits. J. Membr. Sci., 2020, **601**: 117910. https://doi.org/10.1016/j.memsci.2020.117910.

[21] Niknejad, S. M. S., Savoji, H., Pourafshari Chenar, M. and Soltanieh, M. Separation of H2S from CH4 by polymeric membranes at different H2S concentrations. Int. J. Environ. Sci. Technol., 2017, **14**: 375–384. https://doi.org/10.1007/s13762-016-1156-3.

[22] Chenar, M. P., Savoji, H., Soltanieh, M., Matsuura, T. and Tabe, S. Removal of hydrogen sulfide from methane using commercial polyphenylene oxide and Cardo-type polyimide hollow fiber membranes. Korean. J. Chem. Eng., 2011, **28**: 902–913. https://doi.org/10.1007/s11814-010-0437-7.

[23] Gugliuzza, A., Iulianelli, A. and Basile, A. 10 – Membranes for hydrocarbon fuel processing and separation. In: Basile, A. and Nunes, S. P. (Eds.) Advanced Membrane Science and Technology for Sustainable Energy and Environmental Applications, Woodhead Publishing, 2011, pp. 295–338. https://doi.org/10.1533/9780857093790.3.295.

[24] Kocherlakota, L. S., Knorr, D. B., Foster, L. and Overney, R. M. Enhanced gas transport properties and molecular mobilities in nano-constrained poly[1-(trimethylsilyl)-1-propyne] membranes. Polymer, 2012, **53**: 2394–2401. https://doi.org/10.1016/j.polymer.2012.03.067.

[25] Lokhandwala, K. A. and Baker, R. W. Sour gas treatment process including membrane and non-membrane treatment steps. US5407466A, 1995, https://patents.google.com/patent/US5407466A/en (accessed November 17, 2022).

[26] Chisca, S., Bettahalli, N. M. S., Musteata, V. E., Vasylevskyi, S., Hedhili, M. N., Abou-Hamad, E., Karunakaran, M., Genduso, G. and Nunes, S. P. Thermal treatment of hydroxyl functionalized polytriazole and its effect on gas transport: From crosslinking to carbon molecular sieve. J. Membr. Sci., 2022, **642**: 119963. https://doi.org/10.1016/j.memsci.2021.119963.

[27] Hensema, E. R., Sena, M. E. R., Mulder, M. H. V. and Smolders, C. A. Gas separation properties of new polyoxadiazole and polytriazole membranes. Gas. Sep. Purif., 1994, **8**: 149–160. https://doi.org/10.1016/0950-4214(94)80025-1.

[28] Chisca, S., Marchesi, T., Falca, G., Musteata, V.-E., Huang, T., Abou-Hamad, E. and Nunes, S. P. Organic solvent and thermal resistant polytriazole membranes with enhanced mechanical properties cast from solutions in non-toxic solvents. J. Membr. Sci., 2020, **597**: 117634. https://doi.org/10.1016/j.memsci.2019.117634.

[29] Vaughn, J. T. and Koros, W. J. Analysis of feed stream acid gas concentration effects on the transport properties and separation performance of polymeric membranes for natural gas sweetening: A comparison between a glassy and rubbery polymer. J. Membr. Sci., 2014, **465**: 107–116. https://doi.org/10.1016/j.memsci.2014.03.029.

[30] Vaughn, J. and Koros, W. J. Effect of the amide bond diamine structure on the CO2, H2S, and CH4 transport properties of a series of novel 6FDA-based polyamide–imides for natural gas purification. Macromolecules, 2012, **45**: 7036–7049. https://doi.org/10.1021/ma301249x.

[31] Merkel, T. C. and Toy, L. G. Comparison of hydrogen sulfide transport properties in fluorinated and nonfluorinated polymers. Macromolecules, 2006, **39**: 7591–7600. https://doi.org/10.1021/ma061072z.

[32] Kraftschik, B., Koros, W. J., Johnson, J. R. and Karvan, O. Dense film polyimide membranes for aggressive sour gas feed separations. J. Membr. Sci., 2013, **428**: 608–619. https://doi.org/10.1016/j.memsci.2012.10.025.

[33] Kraftschik, B. and Koros, W. J. Cross-linkable polyimide membranes for improved plasticization resistance and permselectivity in sour gas separations. Macromolecules, 2013, **46**: 6908–6921. https://doi.org/10.1021/ma401542j.

[34] Babu, V. P., Kraftschik, B. E. and Koros, W. J. Crosslinkable TEGMC asymmetric hollow fiber membranes for aggressive sour gas separations. J. Membr. Sci., 2018, **558**: 94–105. https://doi.org/10.1016/j.memsci.2018.04.028.

[35] Liu, Z., Liu, Y., Liu, G., Qiu, W. and Koros, W. J. Cross-linkable semi-rigid 6FDA-based polyimide hollow fiber membranes for sour natural gas purification. Ind. Eng. Chem. Res., 2020, **59**: 5333–5339. https://doi.org/10.1021/acs.iecr.9b04821.

[36] Liu, Z., Liu, Y., Qiu, W. and Koros, W. J. Molecularly engineered 6FDA-based polyimide membranes for sour natural gas separation. Angew. Chem. Int. Ed., 2020, **59**: 14877–14883. https://doi.org/10.1002/anie.202003910.

[37] Liu, Y., Liu, Z., Liu, G., Qiu, W., Bhuwania, N., Chinn, D. and Koros, W. J. Surprising plasticization benefits in natural gas upgrading using polyimide membranes. J. Membr. Sci., 2020, **593**: 117430. https://doi.org/10.1016/j.memsci.2019.117430.

[38] Yahaya, G. O., Qahtani, M. S., Ammar, A. Y., Bahamdan, A. A., Ameen, A. W., Alhajry, R. H., Sultan, M. M. B. and Hamad, F. Aromatic block co-polyimide membranes for sour gas feed separations. Chem. Eng. J., 2016, **304**: 1020–1030. https://doi.org/10.1016/j.cej.2016.06.076.

[39] Yahaya, G. O., Mokhtari, I., Alghannam, A. A., Choi, S.-H., Maab, H. and Bahamdan, A. A. Cardo-type random co-polyimide membranes for high pressure pure and mixed sour gas feed separations. J. Membr. Sci., 2018, **550**: 526–535. https://doi.org/10.1016/j.memsci.2017.10.063.

[40] Alghannam, A. A., Yahaya, G. O., Hayek, A., Mokhtari, I., Saleem, Q., Sewdan, D. A. and Bahamdan, A. A. High pressure pure- and mixed sour gas transport properties of Cardo-type block co-polyimide membranes. J. Membr. Sci., 2018, **553**: 32–42. https://doi.org/10.1016/j.memsci.2018.02.042.

[41] Yahaya, G. O., Choi, S.-H., Sultan, M. M. B. and Hayek, A. Development of thin-film composite membranes from aromatic cardo-type co-polyimide for mixed and sour gas separations from natural gas. Global. Chall., 2020, **4**: 1900107. https://doi.org/10.1002/gch2.201900107.

[42] Yahaya, G. O., Hayek, A., Alsamah, A., Shalabi, Y. A., Ben Sultan, M. M. and Alhajry, R. H. Copolyimide membranes with improved H2S/CH4 selectivity for high-pressure sour mixed-gas separation. Sep. Purif. Technol., 2021, **272**: 118897. https://doi.org/10.1016/j.seppur.2021.118897.

[43] Hayek, A., Yahaya, G. O., Alsamah, A., Alghannam, A. A., Jutaily, S. A. and Mokhtari, I. Pure – and sour mixed-gas transport properties of 4,4′-methylenebis(2,6-diethylaniline)-based copolyimide membranes. Polymer, 2019, **166**: 184–195. https://doi.org/10.1016/j.polymer.2019.01.056.

[44] Hayek, A., Yahaya, G. O., Alsamah, A. and Panda, S. K. Fluorinated copolyimide membranes for sour mixed-gas upgrading. J. Appl. Polym. Sci., 2020, **137**: 48336. https://doi.org/10.1002/app.48336.

[45] Hayek, A., Alsamah, A., Yahaya, G. O., Qasem, E. A. and Alhajry, R. H. Post-synthetic modification of CARDO-based materials: Application in sour natural gas separation. J. Mater. Chem. A., 2020, **8**: 23354–23367. https://doi.org/10.1039/D0TA06967A.

[46] Hayek, A., Alsamah, A., Saleem, Q., Alhajry, R. H., Alsuwailem, A. A. and Jassim, F. I. Modified CARDO-based copolyimides with improved sour mixed-gas permeation properties. ACS Appl. Polym. Mater., 2022, https://doi.org/10.1021/acsapm.2c01527.

[47] Yi, S., Ma, X., Pinnau, I. and Koros, W. J. A high-performance hydroxyl-functionalized polymer of intrinsic microporosity for an environmentally attractive membrane-based approach to decontamination of sour natural gas. J. Mater. Chem. A., 2015, **3**: 22794–22806. https://doi.org/10.1039/C5TA05928C.

[48] Yi, S., Ghanem, B., Liu, Y., Pinnau, I. and Koros, W. J. Ultraselective glassy polymer membranes with unprecedented performance for energy-efficient sour gas separation. Sci. Adv., 2019, **5**: eaaw5459. https://doi.org/10.1126/sciadv.aaw5459.

[49] Ghasemnejad-Afshar, E., Amjad-Iranagh, S., Zarif, M. and Modarress, H. Effect of side branch on gas separation performance of triptycene based PIM membrane: A molecular simulation study. Polym. Test., 2020, **83**: 106339. https://doi.org/10.1016/j.polymertesting.2020.106339.

[50] Lawrence, J. A., Harrigan, D. J., Maroon, C. R., Sharber, S. A., Long, B. K. and Sundell, B. J. Promoting acid gas separations via strategic alkoxysilyl substitution of vinyl-added poly(norbornene)s. J. Membr. Sci., 2020, **616**: 118569. https://doi.org/10.1016/j.memsci.2020.118569.

[51] Robeson, L. M. Correlation of separation factor versus permeability for polymeric membranes. J. Membr. Sci., 1991, **62**: 165–185. https://doi.org/10.1016/0376-7388(91)80060-J.

[52] Robeson, L. M. The upper bound revisited. J. Membr. Sci., 2008, **320**: 390.

[53] Lin, H. and Yavari, M. Upper bound of polymeric membranes for mixed-gas CO2/CH4 separations. J. Membr. Sci., 2015, **475**: 101–109. https://doi.org/10.1016/j.memsci.2014.10.007.

[54] Wang, Y., Ma, X., Ghanem, B. S., Alghunaimi, F., Pinnau, I. and Han, Y. Polymers of intrinsic microporosity for energy-intensive membrane-based gas separations. Mater. Today. Nano., 2018, **3**: 69–95. https://doi.org/10.1016/j.mtnano.2018.11.003.

[55] Hayek, A., Shalabi, Y. A. and Alsamah, A. Sour mixed-gas upper bounds of glassy polymeric membranes. Sep. Purif. Technol., 2021, **277**: 119535. https://doi.org/10.1016/j.seppur.2021.119535.

[56] Mohammadi, T., Moghadam, M. T., Saeidi, M. and Mahdyarfar, M. Acid gas permeation behavior through poly(ester urethane urea) membrane. Ind. Eng. Chem. Res., 2008, **47**: 7361–7367. https://doi.org/10.1021/ie071493k.

[57] Sadeghi, M., Talakesh, M. M., Arabi Shamsabadi, A. and Soroush, M. Novel application of a polyurethane membrane for efficient separation of hydrogen sulfide from binary and ternary gas mixtures. ChemistrySelect, 2018, **3**: 3302–3308. https://doi.org/10.1002/slct.201703170.

[58] Orme, C. J., Harrup, M. K., Luther, T. A., Lash, R. P., Houston, K. S., Weinkauf, D. H. and Stewart, F. F. Characterization of gas transport in selected rubbery amorphous polyphosphazene membranes. J. Membr. Sci., 2001, **186**: 249–256. https://doi.org/10.1016/S0376-7388(00)00690-6.

[59] Orme, C. J. and Stewart, F. F. Mixed gas hydrogen sulfide permeability and separation using supported polyphosphazene membranes. J. Membr. Sci., 2005, **253**: 243.

[60] Harrigan, D. J., Yang, J., Sundell, B. J., Lawrence, J. A., O'Brien, J. T. and Ostraat, M. L. Sour gas transport in poly(ether-b-amide) membranes for natural gas separations. J. Membr. Sci., 2020, **595**: 117497. https://doi.org/10.1016/j.memsci.2019.117497.

[61] Harrigan, D. J., Lawrence, J. A., Reid, H. W., Rivers, J. B., O'Brien, J. T., Sharber, S. A. and Sundell, B. J. Tunable sour gas separations: Simultaneous H2S and CO2 removal from natural gas via crosslinked telechelic poly(ethylene glycol) membranes. J. Membr. Sci., 2020, **602**: 117947. https://doi.org/10. 1016/j.memsci.2020.117947.

[62] Dong, G., Li, H. and Chen, V. Challenges and opportunities for mixed-matrix membranes for gas separation. J. Mater. Chem. A., 2013, **1**: 4610–4630. https://doi.org/10.1039/C3TA00927K.

[63] Seoane, B., Coronas, J., Gascon, I., Benavides, M. E., Karvan, O., Caro, J., Kapteijn, F. and Gascon, J. Metal–organic framework based mixed matrix membranes: A solution for highly efficient CO2 capture?. Chem. Soc. Rev., 2015, **44**: 2421–2454. https://doi.org/10.1039/C4CS00437J.

[64] Hassanajili, S., Khademi, M. and Keshavarz, P. Influence of various types of silica nanoparticles on permeation properties of polyurethane/silica mixed matrix membranes. J. Membr. Sci., 2014, **453**: 369–383. https://doi.org/10.1016/j.memsci.2013.10.057.

[65] Ahn, J., Chung, W.-J., Pinnau, I. and Guiver, M. D. Polysulfone/silica nanoparticle mixed-matrix membranes for gas separation. J. Membr. Sci., 2008, **314**: 123–133. https://doi.org/10.1016/j.memsci. 2008.01.031.

[66] Singh, S., Varghese, A. M., Reddy, K. S. K., Romanos, G. E. and Karanikolos, G. N. Polysulfone mixed-matrix membranes comprising poly(ethylene glycol)-grafted carbon nanotubes: Mechanical properties and CO2 separation performance. Ind. Eng. Chem. Res., 2021, **60**: 11289–11308. https://doi.org/10.1021/acs.iecr.1c02040.

[67] Zhang, H., Guo, R., Hou, J., Wei, Z. and Li, X. Mixed-matrix membranes containing carbon nanotubes composite with hydrogel for efficient CO2 separation. ACS Appl. Mater. Interfaces., 2016, **8**: 29044–29051. https://doi.org/10.1021/acsami.6b09786.

[68] Bastani, D., Esmaeili, N. and Asadollahi, M. Polymeric mixed matrix membranes containing zeolites as a filler for gas separation applications: A review. J. Ind. Eng. Chem., 2013, **19**: 375–393. https://doi.org/10.1016/j.jiec.2012.09.019.

[69] Klaysom, C. and Shahid, S. Chapter 6 – Zeolite-based mixed matrix membranes for hazardous gas removal. In: Lau, W.-J., Ismail, A. F., Isloor, A. and Al-Ahmed, A. (Eds.) Advanced Nanomaterials for Membrane Synthesis and Its Applications, Elsevier, 2019, pp. 127–157. https://doi.org/10.1016/B978-0-12-814503-6.00006-9.

[70] Ahmad, M. Z., Peters, T. A., Konnertz, N. M., Visser, T., Téllez, C., Coronas, J., Fila, V., de Vos, W. M. and Benes, N. E. High-pressure CO2/CH4 separation of Zr-MOFs based mixed matrix membranes. Sep. Purif. Technol., 2020, **230**: 115858. https://doi.org/10.1016/j.seppur.2019.115858.

[71] Liu, Y., Chen, Z., Qiu, W., Liu, G., Eddaoudi, M. and Koros, W. J. Penetrant competition and plasticization in membranes: How negatives can be positives in natural gas sweetening. J. Membr. Sci., 2021, **627**: 119201. https://doi.org/10.1016/j.memsci.2021.119201.

[72] Liu, G., Chernikova, V., Liu, Y., Zhang, K., Belmabkhout, Y., Shekhah, O., Zhang, C., Yi, S., Eddaoudi, M. and Koros, W. J. Mixed matrix formulations with MOF molecular sieving for key energy-intensive separations. Nat. Mater., 2018, **17**: 283–289. https://doi.org/10.1038/s41563-017-0013-1.

[73] Liu, G., Cadiau, A., Liu, Y., Adil, K., Chernikova, V., Carja, I.-D., Belmabkhout, Y., Karunakaran, M., Shekhah, O., Zhang, C., Itta, A. K., Yi, S., Eddaoudi, M. and Koros, W. J. Enabling fluorinated MOF-based membranes for simultaneous removal of H2S and CO2 from natural gas. Angew. Chem. Int. Ed., 2018, **57**: 14811–14816. https://doi.org/10.1002/anie.201808991.

[74] Datta, S. J., Mayoral, A., Murthy Srivatsa Bettahalli, N., Bhatt, P. M., Karunakaran, M., Carja, I. D., Fan, D., Graziane, P., Mileo, M., Semino, R., Maurin, G., Terasaki, O. and Eddaoudi, M. Rational design of mixed-matrix metal-organic framework membranes for molecular separations. Science, 2022, **376**: 1080–1087. https://doi.org/10.1126/science.abe0192.

Jisha Kuttiani Ali, Emad Alhseinat*

Chapter 3
Recent progress in modification methods of polymeric membranes for water treatment

Abstract: Polymer membranes are the fundamental base of the membrane-based wastewater treatment process. The production of membranes with perfect properties and high separation performance with antifouling is a challenging problem in the water treatment process. The creation of polymer membranes relies on various membrane fabrication processes. However, the membrane fabrication process is still facing the problem in the creation of perfect membranes that are suitable for different processes. To achieve better membranes, it is very important to compare currently used fabrication methods, and clarify their advantages and disadvantages in terms of the properties and performance of membranes. This chapter briefly describes the different membrane fabrication methods, their advantages, and their disadvantages. A short description of the application of 3D printing technologies in membrane fabrication is also added. The comparison of properties and performance of 3D printing membranes with membranes prepared from conventional methods is also discussed.

Keywords: polymer membranes, membrane modification, phase inversion, surface grafting, surface coating, interfacial polymerization, 3D printing

1 Introduction

Since membrane-based water treatment processes contribute nearly greater than 50% of the total world's water treatment volume, application of this method mainly includes the removal of dyes, dissolved organics, viruses, and pigments, heavy metal removal, desalination, oily wastewater treatment, and photodegradation of organic hazardous materials, [1, 2]. Membrane-based water treatment processes can be categorized based on the membrane material type and its structure in addition to the driving strength, the size range of contaminants, and the separation mechanism [2, 3]. Polymeric membrane for membrane-based water treatment is a cutting-edge guide

---
*Corresponding author: Emad Alhseinat, Chemical Engineering Department, Khalifa University of Science and Technology, Abu Dhabi, United Arab Emirates; Center for Membranes and Advanced Water Technology, Khalifa University of Science and Technology, Abu Dhabi, United Arab Emirates, e-mail: emad.alhseinat@ku.ac.ae
Jisha Kuttiani Ali, Chemical Engineering Department, Khalifa University of Science and Technology, Abu Dhabi, United Arab Emirates; Center for Membranes and Advanced Water Technology, Khalifa University of Science and Technology, Abu Dhabi, United Arab Emirates

https://doi.org/10.1515/9783110796032-003

that mainly focuses on different categories of treatment methods such as microfiltration (MF), ultrafiltration (UF), nanofiltration (NF), reverse osmosis (RO), and photocatalytic membrane treatments [4, 5]. In recent years, the development of membrane-based technology has great interest in academia and industry due to the very broad range of polymeric membranes with countless advantages [5, 6]. The main advantages of polymeric membranes include (i) a wide variety of polymer materials and are easily marketable; (ii) generation of various types of selective barriers such as pore, concentration and affinity of loaded polymeric materials via a flexible and robust system; (iii) large-scale production at reasonable cost; (iv) stable industrial-scale performance in different contaminated water chemistry; and (iv) easy production of flat sheet, hollow fiber, capillary, or tubular membranes in accordance with industrial needs [3]. Typically, membranes hold an asymmetric structure consisting of a dense layer, also known as a separation layer or active layer, and a loose layer or support layer or substrate (Figure 1(a)) [7]. Generally, the separation layer is very thin and has smaller pores, which helps to decrease the mass transfer resistance and promises high selectivity during the membrane separation processes [7]. The support layer is highly porous and has good mechanical strength, which acts just as support for the separation layer and, hence, reduces the resistance of transport of water. The transport of water through the active layer to the support layer is shown in Figure 1(b). Since the feed solutions directly contact the separation layer, the physical and chemical properties of the separation layer have significant impacts on the performance of membranes [8]. Figure 2 depicts a representative classification of polymeric membranes based on the nature, material, macroscopic properties, microscopic properties, pore size, and surface charge of the membrane [9, 10].

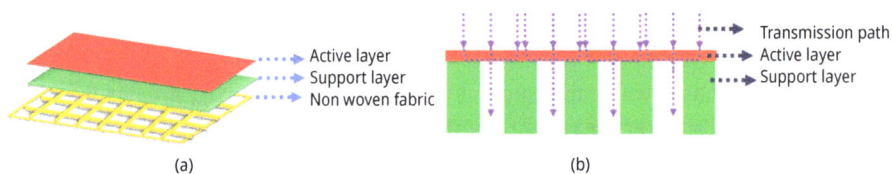

Figure 1: (a) Structure of membrane and (b) transport of water through different layers of membrane.

The commonly used polymers for the construction of polymeric membranes include polysulfone (PSF), polyethersulfone (PES), polyamide (PA), polyacrylonitrile (PAN), cellulose acetate (CA), chitosan (C), and polyvinylidene fluoride (PVDF). CA is a common polymer used to make MF, UF, and RO membranes, while PSF, PAN, PES, PA, C, and PVDF are usually applied to create UF and MF membranes; PA and PAN have been used to develop different membranes for MF, UF, NF, and RO [1] (PM22). Table I summarizes the different polymeric membranes with their chemical structures and their advantages, and membrane process [7, 9, 11]. As discussed in Table 1, the polymeric membranes have high thermal stability, chemical stability, and mechanical strength.

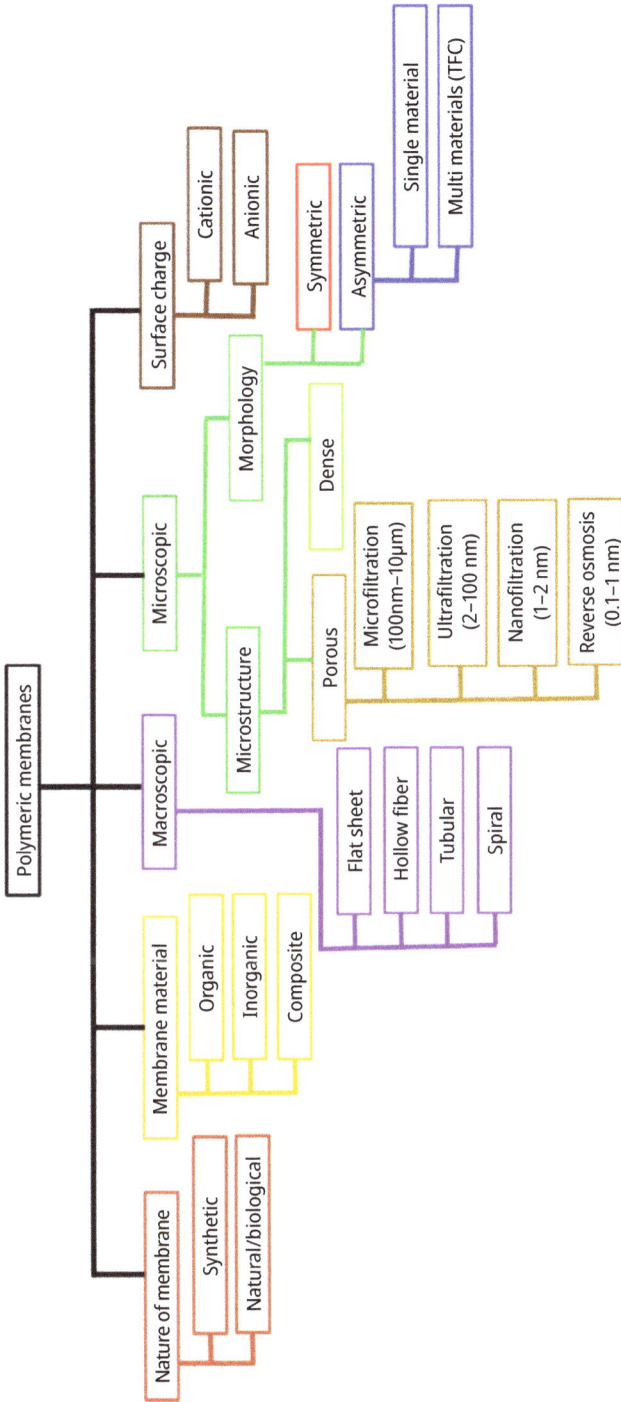

Figure 2: Schematic diagram representing the different categories of membranes used in wastewater treatment. The classification is based on the nature, material, macroscopic, microscopic, and surface charge of the membrane.

However, the choice of polymeric membranes for wastewater treatment mainly depends on various factors, including pore size distribution, wetting susceptibility, porosity, cost, polymer flexibility, fouling resistance, low tortuosity, regeneration/fouling recovery, resistance to cleaning agents, and high rejection performance [11]. These properties basically depend on the properties of polymer material that is used for the creation of membranes and mechanisms for the formation of membranes [11].

Table 1: Different polymeric membranes with their chemical structures and their advantages, and membrane processes.

Membrane materials	Monomer units	Properties	Membrane processes
Polyether sulfone		High thermal stability, mechanical strength, relatively low porosity, and weak solvent resistance	MF/UF
Polysulfone		High mechanical strength, high thermal stability, relatively low porosity, and weak solvent resistance	MF/UF
Polyamide		Heat stability, high chemical stability, and high elongation	RO/NF
Aromatic polyamide		Heat stability, high chemical stability, high tensile strength at low weight, low thermal shrinkage, and excellent dimensional stability	RO
Polyimide		High mechanical strength, high thermal stability, and relatively poor oxidation resistance	MF/UF
Polysulfonamide		High chemical stability and high thermal stability	MF/UF//RO
Polybenzimidazole		High-temperature stability, unusual resistance to organic solvents, excellent mechanical properties, and interesting electrical properties	NF
Polyacrylonitrile		High thermal stability, solvent resistance to multiple chemicals, and strong hydrophilicity	MF/UF/RO

Table 1 (continued)

Membrane materials	Monomer units	Properties	Membrane processes
Polyvinylchloride		Oil/chemical stability, poor temperature stability, high mechanical properties, flame-retardant properties, high mechanical stability, chemical durability, low price, and strong hydrophobicity	RO, MF/UF
Polycaprolactum		High tensile strength, high resistance to acids and alkalis, and high resistance to abrasion	MF/UF
Polypropylene		High mechanical strength, high chemical durability, strong hydrophobicity, low surface porosity, and unevenly distributed surface pores	MF/UF
Polyethylene		High mechanical stability, high chemical durability, uniform pore structure, high surface porosity, and strong hydrophobicity	MF/UF
Polyvinylidene fluoride		Thermal resistance, good chemical stability, mechanical strength, and strong hydrophobicity	MF/UF
Polytetrafluoroethylene		Good chemical stability, high mechanical strength, thermal resistance, and strong hydrophobicity	MF/UF
Cellulose acetate		Heat stability, a wide range of solvent tolerance, and easy film formation	Polymer inclusion membranes
Chitosan		High tensile strength and high hydrophilicity	RO

As discussed, the separation performance of polymeric membranes mainly depends on the high porosity, uniform pore size, appropriate wettability, surface charge, and so on. However, the active layer of many polymeric membranes possesses only limited properties. In addition, the chemical environment of contaminated water such as pH, nature of the contaminant, and temperature of the feed solution also affected the membrane performance [12]. Moreover, currently used polymeric membranes faced several serious drawbacks for long-term usage on an industrial scale. The critical challenge is high fouling, which leads to low permeability and inferior separation performance. Moreover, the cost of the process was also enhanced because intensive physical and chemical washing might be required, or the membrane replacement needed based on the extent of fouling. Another challenge is its low resistance to chlorine, which is a common disinfectant in water treatment process as chlorine can react with electron-rich functional groups in polymeric membranes that damage the membrane surface [12]. Hence, it is difficult to improve one without compromising the other for the current polymeric membrane. Thus, membranes with high permeability and selectivity or high rejection performance, as well as strong antifouling properties and high chlorine resistance are needed to develop. One of the possible solutions to overcome these limitations creates the next-generation membranes via modification of currently used polymer membranes using different techniques.

2 Membrane modification methods

Low hydrophilicity, fouling and anticompaction properties, low separation, and non-recyclable ability are the most important challenges in membrane technologies which lead to poor membrane performance and shorten the life span of the membrane [13]. Regarding the aim of the membrane process applications, polymer membrane modification in the surface of the membrane or into membrane pores depends on the desired morphology, porous nature, and application [2]. The most common methods for the modification of polymeric membranes include phase inversion, blending technique, surface grafting, surface coating, and interfacial polymerization (IP) [1, 14]. The selection of membrane fabrication methods relies on polymeric material properties, the structure of the membrane, the nature of the feed solution, and its application [14, 15]. It is believed that different membrane fabrication methods help to improve the properties of membranes, including tuning pore size, synthesizing selective layers, improving fouling resistance, regulating wettability, enhancing mechanical robustness, enabling (photo)catalytic properties, improving chemical stability, and building transitional layers [8]. Figure 3 summarizes the different membrane modification techniques. In the following sections, different modification methods for improving the properties and separation performance have been discussed.

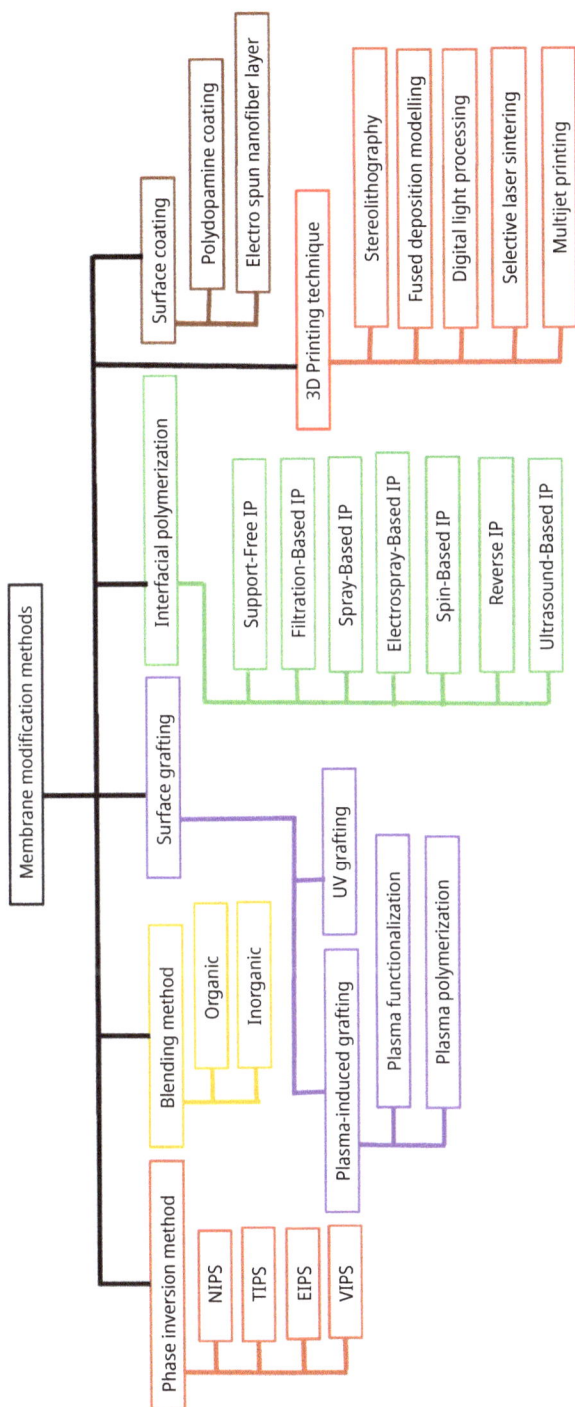

Figure 3: Scheme for the classification of membrane modification techniques.

2.1 Phase inversion method

Among different techniques, phase inversion is the most commonly used method to prepare both asymmetric and symmetric polymeric membranes as low-cost, simple processing, and flexible production scale [14, 16]. In phase inversion, the homogeneous polymer solution is transformed from liquid to solid state in a controlled manner by the demixing process. In this method, two main solvents determine the formation mechanism of membranes. The first solvent is for creating a homogeneous polymer solution, and the second solvent is known as nonsolvent for the precipitation of membranes during the immersion process in the coagulation bath. The properties of modified membranes mainly depend on the exchange rate between solvent and nonsolvent during precipitation via demixing. The main steps included in the phase inversion process are represented in Figure 4. During the immersion process, membrane formation depends on the polymer precipitation rate in the nonsolvent bath. The instantaneous demixing created membranes with a thin skin layer and finger-like morphology in the sublayer (Figure 5(a)) while delayed demixing results in a dense top layer and a characteristic sponge-like structure (Figure 5(b)) [10]. Moreover, to improve the membrane properties, additives such as polymers (polyvinylpyrrolidone (PVP) or polyethylene glycol (PEG)), low-molecular-weight chemicals including salts (LiCl), weak cosolvents (ethanol, propanol, and acetone), weak nonsolvents (glycerol or ethylene glycol), organic acids (propionic acid), and inorganic acids (acetic acid and phosphoric acid) have been added [2, 17, 18]. Additives can enhance the properties of membranes via coagulation exchange, including increasing pore formation, enhancing hydrophilicity, and improving pore interconnectivity. Moreover, membrane morphology from a porous surface to a denser layer is also affected by air humidity, temperature, and exposure times. In addition, different membrane morphologies are also achieved depending on the rate of solvent–nonsolvent exchange [2].

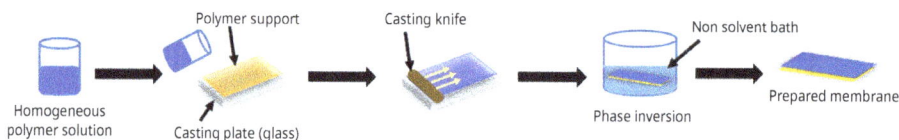

Figure 4: Schematic of membrane preparation by the phase inversion process.

Due to the formation of polymer membranes from a liquid state to a solid state by demixing, the phase inversion process can be done in different ways, including nonsolvent-induced phase separation (NIPS), thermally induced phase separation (TIPS), evaporation-induced phase separation (EIPS), and vapor-induced phase separation (VIPS) [2]. Figure 6 summarizes the four experimental approaches in the phase inversion process. One of the most widely used methods in research and industry is immersion precipitation or NIPS [19, 20]. In this method, a homogeneous solution of polymer

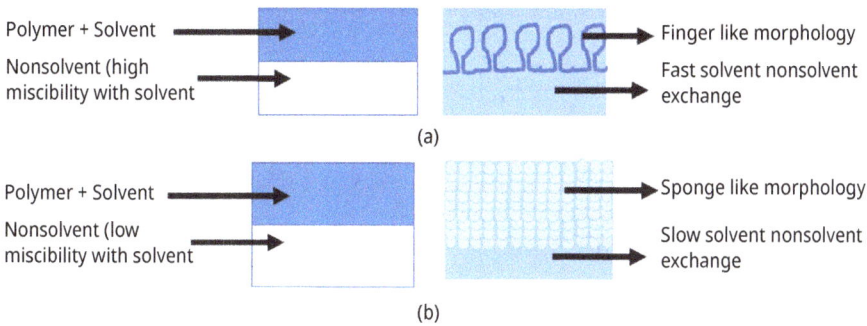

Figure 5: (a) Instantaneous mixing and (b) delayed mixing during the phase inversion process.

in a suitable solvent is prepared and later cast on a suitable support material followed by immersion in a nonsolvent coagulation bath, resulting in a polymeric membrane with an asymmetric structure (Figure 6(a)) [10, 21]. The factors affecting the properties of membranes prepared via this method are casting solution composition, coagulation bath composition, choice of solvent and nonsolvent, the temperature of the coagulation bath, evaporation time, air humidity, and temperature. Whereas in the TIPS method, the hot homogeneous polymer solution in an appropriate solvent with a high boiling point is prepared at an elevated temperature and cast into the desired shape (Figure 6(b)) [21–23]. Then, it is cooled to induce phase separation for the membrane formation and finally the remaining solvent is removed by solvent extraction resulting in high-strength porous polymeric membrane. The porosity and strength of the membrane are highly dependent on the cooling rate, use of additives, and temperature gradient. The third method is EIPS. In this process, a polymer solution is prepared in volatile solvents and a less volatile nonsolvent, and then cast on the suitable support followed by the evaporation of solvents producing dense anisotropic membrane film (Figure 6(c)) [19]. However, in this method, polymer membrane morphology on the final film depends on polymer and nonsolvent concentration, relative humidity, casting solution, thickness, rate of evaporation, and temperature. In VIPS, the casting solution containing polymer in a specific solvent is exposed to nonsolvent vapor in a vapor chamber (Figure 6(d)) [23]. The formation of membrane structure occurs through vapor absorption. Main advantages of this method are the membrane without macrovoids. Moreover, air humidity, temperature, and exposure times result in membranes with different morphologies [24].

2.2 Blending methods

The blending methods are the simplest route to improve the hydrophilicity and antifouling properties of polymeric membranes. In this method, a casting solution is created using two or more compounds that are known as fillers with polymer in a

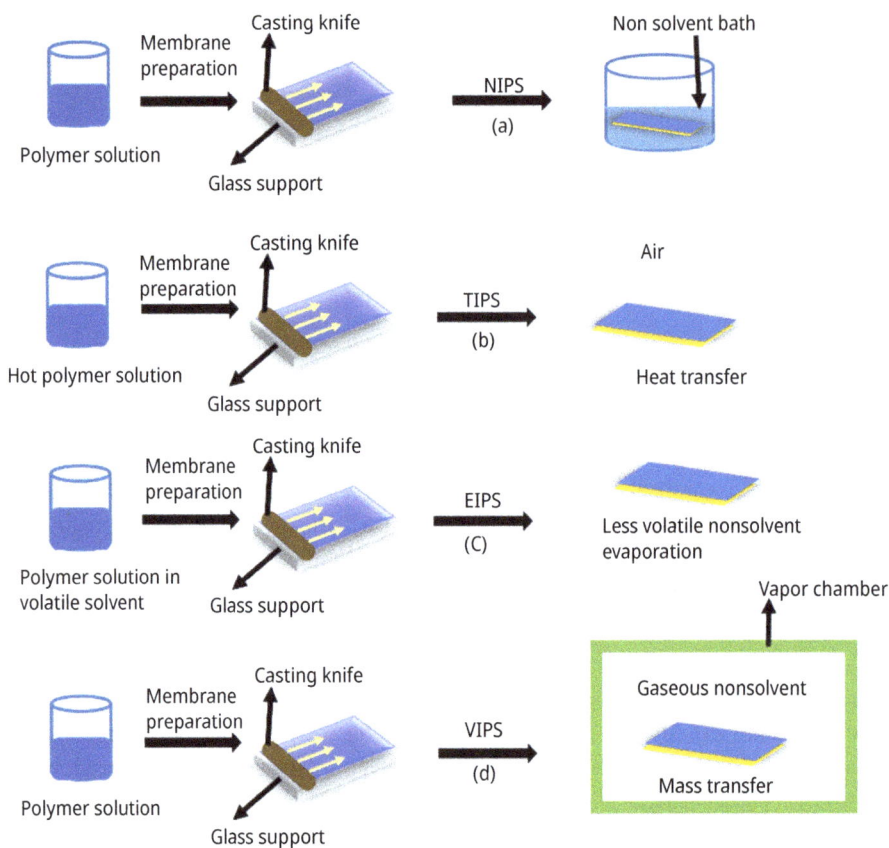

Figure 6: Different steps in (a) NIPS, (b) TIPS, (c) EIPS, and (d) VIPS [1].

suitable solvent via physical mixing. Here, phase separation or inversion methods are usually applied to create the membranes. The disadvantages of this method are limited compatibility between fillers and polymers due to differences in the hydrophilic nature of polymers and fillers [25, 26]. Moreover, this method also faces another issue such as the leaching of blended compounds (fillers) after long-term use of polymeric membranes. The generally used fillers are inorganic compounds, polymers, and organic compounds. Polymeric membranes with inorganic fillers are known as mixed matrix membranes (MMM). Most researchers are interested to create MMM because of its high thermal stability, mechanical stability, enhanced permeability, good separation performance, and porous structure due to the uniform dispersion of inorganic fillers in the polymeric membrane surface [27]. Several inorganic materials have been used as fillers to MMMs, including titanium dioxide, alumina, graphene oxide, Mxene, and different metal nanoparticles (silver, copper, nickel, etc.) [7]. However, the use of nanoparticles as fillers for creating MMM has faced two problems such as aggregation of nanoparticles during cast solution preparation and also during membrane fabrication as well as leaching of nanoparticles. To overcome these problems, porous fillers including mesoporous silica, clays, zeolite, halloysite, sepiolite, covalent organic framework, and metal organic framework have been used to control the stability of nanoparticles in MMM [2]. In the case of organic fillers and polymers, organic compounds or polymers (e.g., PEG and PVP) with hydrophilic properties enhance the hydrophilicity of membranes and antifouling properties [3, 28]. Here, due to the almost same physical and chemical properties of the polymer and organic compound, there is perfect compatibility between polymer matrix and fillers. As a result, a highly efficient membrane surface with enhanced antifouling properties and separation can be developed. In addition, organic fillers, especially polymers, also acted as pore-forming agents or porosity enhancers during the phase inversion process.

2.3 Surface grafting

To improve antifouling properties, surface grafting is an efficient method. In this method, hydrophilic chains, functional groups, or electrostatically charged groups are introduced into the membrane surface through covalent bonding [29, 30]. In this method, plasma-induced grafting and UV grafting are the main methods to activate the surface of the membrane for grafting approaches. In plasma-induced grafting, plasma (an electrically quasi-neutral gas) was used for generating the functional groups, radicals, or electrons using microwaves or radio frequency waves [30]. In this method, the nature of functional groups for grafting mainly determines the plasma gas used. The plasma activation using oxygen gas introduces functional groups such as hydroxyl, carboxylic acid, or peroxide. On the other hand, the plasma initiation using carbon dioxide or carbon monoxide helps to introduce the hydroxyl, ketone, aldehyde, ester, and carboxylic acid groups, whereas the plasma from nitrogen or ammonia produces amides and amines. An inert gas such as helium or argon is also used with other gases and monomers to generate homogeneous plasma discharge. Further creation of polar groups in the polymer matrix after treatment mainly occurred in the presence of air.

Based on the functional group's generation and polymerization, plasma-induced grafting can be classified into two types such as plasma functionalization or plasma activation and plasma polymerization or plasma deposition [31]. In plasma functionalization or plasma activation, firstly, functional groups are created in the polymeric material when it is exposed to plasma, and further grafting takes place when exposed to atmospheric air (Figure 7(a)). On the other hand, in plasma polymerization or plasma deposition, fragmentation of monomer and functional groups/radical site generation on polymeric membrane surface occurs in the presence of plasma discharge (Figure 7(b)). Later, reactive fragments of polymer initiate the graft polymerization resulting in plasma-deposited polymer coating on the substrate in the gas phase [32].

UV grafting photochemical-initiated graft polymerization or UV grafting method is a very popular method among researchers due to its low cost, simplicity, and versatility [30, 33]. Moreover, this method enables the antifouling properties of the polymer after grafting. Here, with or without a photoinitiator, active sites or free radicals are generated due to the homolytic cleavage of the bond upon UV light irradiation, and these radical sites induce the grafting process in the polymer backbone. Table 2 lists a summary of recent studies using different surface grafting and its results.

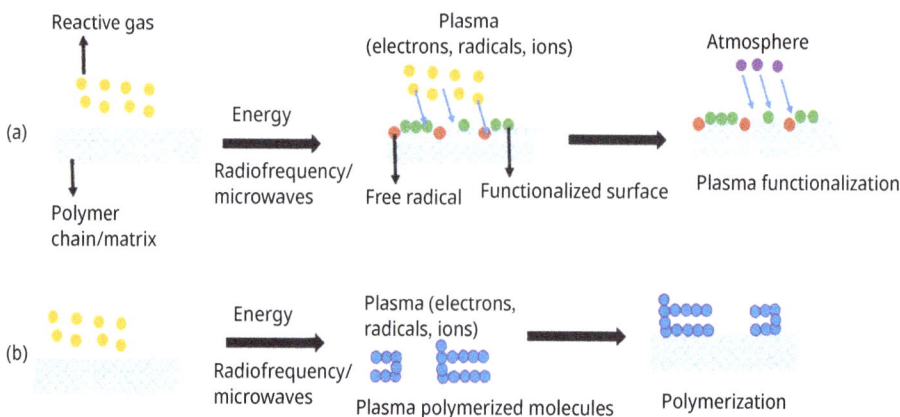

Figure 7: (a) Plasma functionalization or plasma activation and (b) plasma polymerization or plasma deposition.

2.4 Surface coating

The surface coating mainly consists of the deposition of an active layer on the surface of membranes causing minimal structural effects [38, 39]. Here, a strong covalent bond generates at the substrate and coating interface resulting in enhanced performance and long-term stability. This method is very simple, cost-effective, and environmentally friendly. Polydopamine coating and electrospun nanofiber layer are the two main methods to develop a coating on the surface of the membrane [40]. In the polydopamine coating method, polydopamine has generally used the platform to generate covalently grafted functional layers on the surface of the membrane [41]. Because it has strong adhesion properties and can easily undergo self-polymerization in alkaline conditions between pH 7.5 and 8.5 in the presence of oxygen. This will result in the formation of a very thin layer on the support layer. Moreover, the coating of polydopamine will enhance hydrophilicity and fouling resistance. On the other hand, the electrospun nanofiber layer is a method that produces fibers with very small diameters on the surface of membranes [40]. In this method, a high-voltage source injects charge with certain polarity into a polymer solution; later, it is accelerated toward a collector

Table 2: Different surface grafting methods studied in literature.

Membrane	Surface grafting method	Additives	Results	Reference
Polysulfone (PS)	Argon–oxygen plasma treatment	Natural seed basil gum NPs	Enhanced water permeability, high fouling resistance, higher BSA rejection rate	[29]
Polypropylene	Oxygen plasma treatment	Deposition of TiO$_2$ nanoparticles	Enhanced membrane hydrophilicity	[34]
Polypropylene	UV grafting polyacrylic acid and benzophenone as photoinitiator	–	Higher water permeate flux and rejection rate	[30]
Polyethersulfone	Grafted allylamine monomers in the presence of UV light	–	Higher hydrophilicity, enhanced water permeate flux, and good fouling resistance	[33]
Polyethersulfone	Grafting of acrylic acid monomers in the presence of UV light	–	Enhanced water permeability and less salt diffusion	[35]
Polyethylene terephthalate	Plasma-induced graft polymerization	–	Higher hydrophilicity	[36]
Polyvinylidene fluoride	Plasma-induced self-polymerization of polyacrylic acid	Self-assembly of ZnO nanoparticles	Higher hydrophilicity, enhanced water flux antifouling properties	[31]
Polyvinylidene fluoride	Benzophenone-initiated UV grafting	Graphene oxide	Antibacterial activity solute rejection properties, no change in membrane permeability	[37]
Polyvinylidene fluoride	Argon plasma treatment	Polystyrene sulfonate deposition	Selective to divalent anions, high stability, and decreased average pore radius	[32]

of opposite polarity resulting in the layer of nanofiber on the surface of the membrane. Due to the formation of nanofibers, membranes exhibit tunable porosity, high interconnectivity, high surface area to volume ratio, and effective functionalization. Table 3 discusses the recent studies of surface coating with results.

Table 3: Recent studies of surface coating.

Membrane	Surface coating method	Additives	Results	Reference
Poly(vinylidene fluoride)	Electrospun PVDF nanofiber coating	SiO$_2$ nanoparticles	Superhydrophobicity stable flux	[38]
Poly(vinylidene fluoride)	Electrospun nanofiber coating	TiO$_2$ nanoparticles	Enhanced hydrophilicity and higher antifouling behavior	[40]
Poly(vinylidene fluoride)	Dopamine coating	In situ immobilized copper nanoparticles	Enhanced surface hydrophilicity and antibacterial activity	[39]
Poly(vinylidene fluoride) and polysulfone	Polydopamine layer		High water permeate flux and higher retention	[41]
Polysulfone	Polydopamine layer	Silver nanoparticles	Increased pure water flux, fouling resistance, and good antibacterial activity	[42]
	Poly(acrylic acid) and -poly(vinyl alcohol) electrospun layer	–	Enhanced hydrophilicity, high fouling resistance, and antimicrobial activity	[17]
Polysulfone	Polydopamine layer		Increased hydrophilicity and decreased membrane permeability due to excessive deposition	[43]
Poly(ether imide)	Polydopamine layer	Immobilized silver nanoparticles	High permeation and separation performance and antibiofouling properties	[44]

2.5 Interfacial polymerization (IP)

IP is one of the powerful fabrication methods in membrane modification because it creates a defect-free uniform membrane layer [45, 46]. In this method, a continued uniform polymer film growth through copolymerization between two immiscible reactive monomers in different solvent phases (the most used solvent phases are the aqueous phase and organic phase) [14]. During the IP process, the support layer is dipped in the first monomer in the aqueous phase and then the second monomer solution in the organic phase is added to start the copolymerization. The generally used monomers for the IP modification of NF and RO membranes comprise amine monomers (*m*-phenyl diamine, piperazine, triethyl amine, etc.) and second monomers having the acyl chloride (trimesoyl chloride, camphorsulfonic acid, 2,2′,5,5′-biphenyl tetraacyl chloride,

2,2′,4,4′-biphenyl tetraacyl chloride, etc.) [46, 47]. In this method, properties, morphology, and performance of modified membranes depend on the type of monomer and its concentration, properties of solvent phases, polymerization reaction time, posttreatment conditions, and temperature. Generally, IP has been applied to produce thin active layer on the NF and RO membranes [46]. Figure 8 illustrates the typical procedure of thin-film composite (TFC) membrane fabrication via the IP technique.

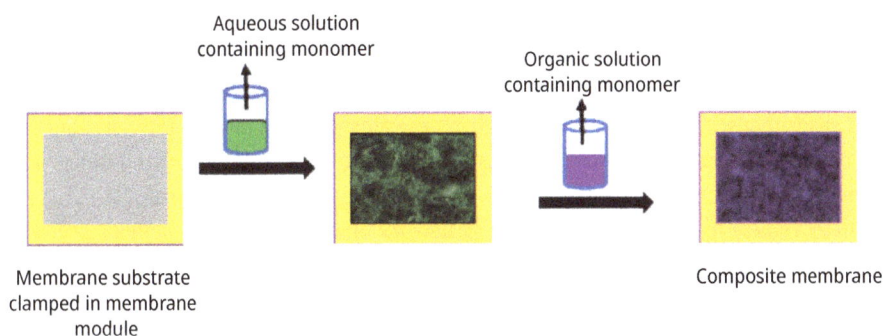

Aqueous solution
containing monomer

Organic solution
containing monomer

Membrane substrate
clamped in membrane
module

Composite membrane

Figure 8: Different steps in interfacial polymerization.

However, the IP process has faced significant drawbacks, including requiring a huge volume of organic solvents and monomers needed to complete polymerization. Thus, the IP process is economically unfeasible and non-eco-friendly at the industrial scale. Another drawback of this method is the application of either rubber roller or air gun to remove excess solution from the substrate surface which will negatively affect the membrane surface [46]. Hence, advancement in novel/modified IP techniques developed since 2013 can significantly improve the characteristic properties of polymer membranes [46]. Compared to normal IP processes, novel/modified IP techniques are more sustainable and environmentally friendly. The following section briefly describes the advanced IP methods.

2.5.1 Support-free IP

Support-free IP, also known as free-standing IP, is a technique in which a polymer membrane layer is created without any support [48]. In this technique, the polymer film is formed on the interface of the solution, and it will float in the excess aqueous phase. Later generated polymer film is transferred onto a suitable substrate or support membrane by draining the solvent phases. Mostly, the free-standing polymer layer or film is manually raised, followed by attachment on the support surface [48]. The main advantage of this method is that there is no interference from any substrate or support membrane during polymerization. The membranes produced from this method have a smooth uniform surface with enhanced properties as a higher degree of cross-linking compared to the normal IP method [49]. The support-free IP technique is applied for the modification of NF and RO membranes, as well as the FO and organic solvent NF process [49]. The advanced support-free methods are the dual-layer slot coating (DSC) technique, aqueous template IP (ATIP), agar gel IP technique, and in situ free IP (IFIP) [46]. In the DSC method, the simultaneous and nonstop spreading of two reactive monomers generates an unsupported polymer layer through in situ polymerization [50]. Later, it adhered to a support membrane. In the agar gel IP technique, generate high-performance free-standing polymer film using the hydrogel temperature at the organic phase–hydrogel interface [51]. In ATIP method, the pressure controlled rolling left a nanoscale aqueous layer on the surface of the substrate [46, 52]. Due to the presence of an aqueous template, the morphology of the membrane surface has dense rigid nanostructures and shows high pure water permeability. In IFIP, an aqueous layer is left on the substrate before being exposed to the organic solution microdroplets permitting the formation of a polymer layer with a thickness of nanometer scale [53]. This results in extremely high flux, especially for NF and RO membranes. These methods are very suitable to fabricate the thin polymer film with excellent performance without any support.

2.5.2 Filtration-based IP

The vacuum filtration-based IP technique is newly found to be a promising fabrication method for TFN membranes [27]. In this method, a thin layer of nanomaterials or monomers is deposited on the substrate surface by vacuum filtration method prior to the formation of the polymer layer. The main advantages of this method are uniform distribution of monomers/nanomaterials on the support layer, avoiding the wastage of chemicals, especially monomers/nanomaterials. The main advantages of this method are easy removal of excess solution from the surface of substrate without disturbing the coating, improved hydrophilicity, high water permeate flux, and uniform distribution of particles on the surface of coating [54].

2.5.3 Spray-based IP

In this method, a sprayer is used to introduce the monomer in either aqueous or organic phase onto the surface of substrate [55]. The pressurized gas containing air or nitrogen is ejected using airbrush or air gun to produce microscale dispersion of the monomers in solvent phases. Later, stepwise diffusion takes place, resulting in the formation of loose polymer layer. The membrane obtained from this method shows improved water permeability and high rejection performance compared to normal IP process [56]. However, this approach is not feasible for the simultaneous application of both phases as less wettability of the substrate results in a decrease in the hydrophilicity of membranes. In addition, longer spraying time adversely affected the surface area of the membrane and will create lower water permeability.

2.5.4 Electrospray-based IP

The electrohydrodynamic process (electrospraying and electrospinning) is widely employed to generate nanoparticle coating or nanofibers on the support [57]. In this method, the solution is ejected from the needle as coulombic repulsion force from the strong electric field. This repulsion is helpful to overcome the surface tension resulting in the formation of fine droplets from liquid jets. Meanwhile, continuous jets are produced due to the stronger intermolecular forces in the solution. Finally, a transition occurs from a polymeric solution to a solid-state nonwoven mat of nanofibers due to the bending and stretching. This method is very often applied in the modification of microporous membranes with high porosity [58, 59]. The main advantages of this method are: minimum chemicals are needed and can control the thickness while it is a highly energy-required process and cannot generate membranes in large sizes.

2.5.5 Spin-based IP

Spin-based IP is generally applied for the creation of dense polymer layer with controllable thickness [60, 61]. So, rotational time and its speed, and solution viscosity are the major factors that determine the thickness, properties, and morphology of the membrane. In this method, the centrifugal shearing force is raised on the porous support due to the spinning motion, and this will cause the spreading of the reactant from the center toward the outer edge of the substrate resulting in the formation of a thin film with uniform thickness [61]. Compared to the normal IP process, spin-based IP displays enhanced hydrophilicity, high water permeate flux, uniform distribution of monomers on thin layers, and improved separation process [61, 46].

2.5.6 Reverse IP

The reverse IP technique is very similar to the normal IP method except sequences of the aqueous phase and organic phase are reversed during the reverse IP process [9]. This technique is applied to highly hydrophobic substrates to improve the affinity of water [7]. In this technique, the porous substrate is exposed to the organic phase instead of the aqueous solution compared to the normal IP method since substrates are highly hydrophobic. Thus, the monomer from the aqueous phase is moved to the organic interface and hence partial in the organic phase. However, since direct contact between the organic phase and substrate, cross-linked polymer film can be generated from the top aqueous phase toward the bottom of the substrate. As a result of this, pores are developed on the top surface of polymer film. Due to the primary bulk diffusion, the top denser structure is achieved while the loose structure is generated from the secondary stepwise diffusion to the surface of substrate [46]. However, the polymer film obtained from this method has multiple defects because of rapid evaporation of organic solvent from organic phase. To overcome this problem, the gelatin interlayer is applied on the substrate surface prior to the reverse IP process. Studies revealed that this extra interlayer can enhance the improved water permeate flux due to the synergistic interaction between gelatin and monomers [62, 46]. However, reverse IP method may increase the cost of process because additional pretreatment method is needed to hydrophilic membranes to employ this method. In addition, the uncontrolled polymer growth due to the limiting fabrication methods that self-terminate may create difficulties to control the polymer film thickness, porosity, surface chemistry, and roughness [46].

2.5.7 Ultrasound-based IP

The ultrasound-based technique is a novel IP method used to develop TFC membrane for NF and FO processes [63]. In this method, cavitation is created using ultrasound waves, and this will enhance the efficiency of mixing at the interface phase resulting in loose polymer layer (Figure 9). The membrane created from this method displays high water permeate flux and separation process due to the formation of a thicker but looser polymer layer [63]. The high ultrasonic frequency and sonication time will affect the thickness and roughness of membranes as well as the degree of cross-linking and antifouling properties [45]. Table 4 summarizes the advantages and disadvantages of different IP processes.

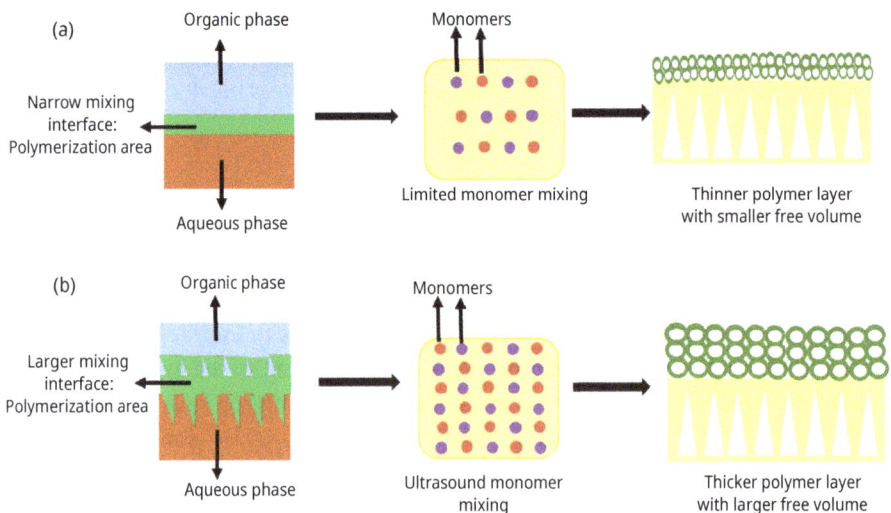

Figure 9: Schematic diagram: (a) normal IP and (b) ultrasonic-assisted IP for membrane fabrication.

Table 4: Advantages and disadvantages of different interfacial processes.

IP process	Advantages	Disadvantages
Support-free IP	High scalability and precision, uniform porosity, and able to form polymer file even at very low monomer concentration	Very hard to transfer or attach polymer film onto the substrate
Filtration-based IP	applicable to generate 2D nanosheets on the substrate, good stability of nanoparticles/monomers on the substrate, and negligible leaching of nanomaterials/monomers during filtration	Not applicable for 3D nanomaterials as particle size of 3D materials much smaller than substrate's pore size, low scalability, and not able to control the thickness of polymer layer
Spin-based IP	Can create highly uniform polymer layer and fast process	Chemical wastage including nanomaterials or monomer cannot avoid during spinning, low scalability, and cannot control shearing force
Spray-based IP	Least use of chemicals/nanomaterials, good scalability, can control thickness of polymer layer, and fast process	No stability evaluation and economic analysis
Electrospray-based IP	Minimum chemicals are used, moderate scalability, and can control the thickness of polymer layer in nanometer scale	Comparatively slow process (>1 h), cannot produce large sheet of membrane, and high energy requirement
Reverse IP	Applicable for hydrophobic substrate	Difficult to generate defect-free TFC membrane
Ultrasound-based IP	High water permeate flux due to the presence of nanovoids	Limited studies

2.6 3D printing technology in membrane-based water treatment

Recently, additive manufacturing or 3D-printed polymer technology adds an extra dimension to the membrane-based wastewater treatments, especially oil/water separation applications [64, 65]. This is considered as a promising technological advancement due to the development of modern 3D printing techniques with high printing resolution and high energy efficiency [66]. This method can provide better control on the pores and its structure, and porosity of membranes. Thus, this will enable development of structures with high water permeate flux without the use of external energy and pressure [66, 67]. Basically, in this technology, add one layer at a time (layer-by-layer manufacturing process) to produce the 3D structure by printing based on a 3D computer-aided design (CAD) model. Different kinds of 3D printing techniques are available, including stereolithography (STL), fused deposition modeling (FDM), digital light processing (DLP), selective laser sintering (SLS), and multijet printing (MJP) [66]. These techniques work on the same basic concept to create the final object. In this method, firstly, CAD model is created, which is then converted into any of the 3D printing techniques (STL, FDM, DLP, SLS, MJP) to get 3D file [66]. The generated 3D file is preprocessed by specific software, where process parameters such as 3D part orientation into the build volume and slicing parameters are defined. Finally, the information is sent to the 3D printer that executes layer-by-layer manufacturing resulting in membranes with desired structure [66, 68].

In recent years, 3D printing techniques have achieved remarkable developments in membrane module design, fabrication of composite membrane, and development of oil–water separation and wastewater treatment materials [66]. But major challenges in 3D printing methods are cost, printing resolution, speed, and material selection compared to conventional membrane fabrication methods [66]. Figure 10 illustrates the materials used in conventional and 3D printing membranes. As shown in Figure 10, only 12% of materials are available for 3D membrane printing compared to materials available for conventional membrane fabrication methods [66]. More precisely, 3D printing systems are limited to polymeric materials as this method is not compatible with all polymers. Moreover, different 3D printing techniques are specific for different available polymeric materials to print the membranes. For example, FDM can apply to polymers, including polylactic acid, polylactide-co-glycolide, and polyethylene terephthalate. While the SLS technique is applied to print PA 12 and PSF membranes, SLA is used for Tangoplus membranes and diurethane dimethacrylate-co-polyethylene glycol diacrylate (DUDA-co-PEGDA) polymer [66, 69].

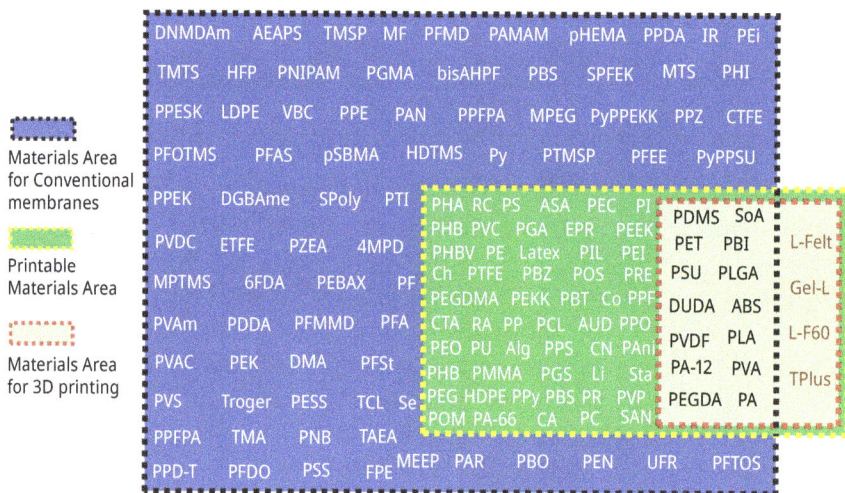

Figure 10: Polymers used for the manufacture of membranes: conventional membranes versus 3D-printed membranes.

In conventional polymeric membranes, the porosity, structure, interconnectivity, and pore structure cannot be controlled due to the difficulties in regulating the preparation parameters, whereas the 3D printing membranes obtained from a CAD object can enable to control membrane parameters to attain the desired structure [69]. Moreover, the 3D printing membranes with grooved or embossed structures can easily be generated with larger surfaces than those of flat membranes from conventional methods. Thus, patterned membranes can exhibit enhanced transport performance and alleviating fouling properties. In addition, this method can offer great production flexibility via easier fabrication of complex structures than conventional membrane fabrication methods [67]. More precisely, any complex geometric shape can be planned and created using 3D printing methods (Figure 11). However, major drawback of this method is the resolution limits. But currently used 3D printing methods are able to get high resolution in the z-dimension but the same precision cannot be achieved in the x- and y-axes [67].

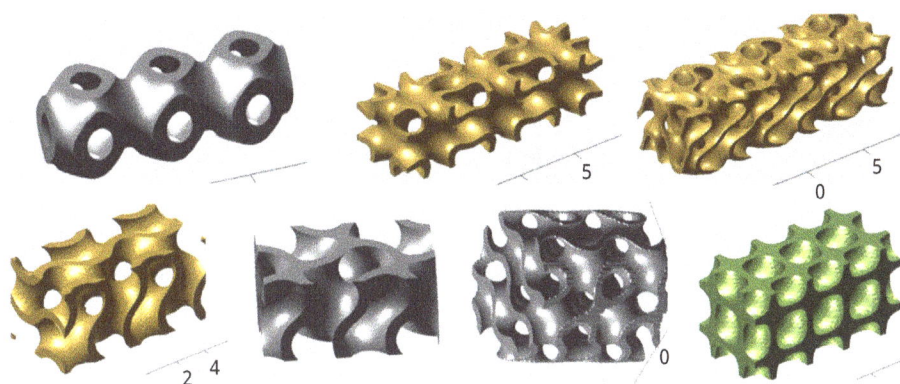

Figure 11: Different shapes can be created by 3D printing technology.

2.6.1 Properties and performance of 3D printing polymer membranes

Thickness, pores, surface roughness, hydrophilicity or hydrophobicity, and mechanical properties are the main properties of the membrane which determine the separation performance [67]. Generally, membranes fabricated using 3D printing methods are thicker than conventional membranes. For oil and water application, 3D-printed membranes exhibit a thickness of range 800–500 μm, while conventional membranes show a range of 150–250 μm [69]. Furthermore, the cation exchange membranes created using the FDM method display a membrane thickness of 2,000 μm. This is attributed that layer-by-layer addition in 3D printing and single-layer height is not shown below 25 μm [67, 69]. Moreover, to achieve structural integrity, a multilayer is needed, which leads to thicker membranes. On the other hand, the pores of 3D-printed membranes are larger than those of conventional membranes because the pore size varies according to the structure of 3D printing membranes and depends on the resolution of the 3D printing technology [67, 70]. The most available 3D printers are not able to print below submicron resolution and also the actual product resolution is lower than the nominal 3D printer resolution. Another important property is surface roughness. The surface roughness of 3D-printed membranes depends on the 3D printing technology used [70]. Generally, it is found to be lower surface roughness compared to conventional membranes. The next important factors are the hydrophobicity or hydrophilicity, and mechanical properties of 3D printing membranes. 3D printing membranes are superhydrophobic membranes and exhibit high mechanical properties due to surface structure and higher thickness compared to conventional membranes for the same polymer and the same method of application. Moreover, 3D printing membranes exhibit higher water permeate flux due to higher porosity, and display quite stable performance with high rejection [70, 71]. Table 5 summarizes the materials, 3D printing techniques, and applications of 3D printing membranes.

Table 5: Materials, 3D printing techniques, and applications of 3D printing membranes.

Materials	3D-printed section	3D printing techniques	Applications	Reference
Polyamide 12	Membrane	SLS	Oil–water separation	[64]
Metaphenylene diamine (MPD) and trimesoyl chloride (TMC)	Thin-film composite membrane	Electrospraying (for the active layer)	RO	[72]
Polylactic acid and polybenzimidazole (PBI)	Membrane	Solvent cast printing (SCP) and FFF	–	[68]
Polysulfone	Membrane	SLS	Oil–water separation	[65]
PVDF	Membrane	3D printing near-field electrospinning (NFES)	Filtration	[73]
Kankara clay powder + maltodextrin powder	Ceramic membrane	Inkjet	Filtration	[74]
Fluorinated diamine incorporated into m-phenylenediamine-based polyamide	Thin-film composite membrane	Inkjet printing + interfacial polymerization	RO and NF	[75]

3 Conclusion and outlook

In this chapter, the main attention is focused on different polymer membrane fabrication methods including conventional methods (phase inversion, blend type, surface coating, surface grafting, and IP) as well as the application of 3D printing technology in the fabrication of membranes. Based on the discussion, these methods are helpful to improve the different properties and performance of membranes for various water treatments. Hence, it is concluded that membrane modification methods are material-specific, and also dependent on the chemistry of the polymeric membranes to be modified. More precisely, one specific modification technique is only appropriate to certain polymeric membranes with selective chemical properties. Thus, the same method cannot be applied to different types of polymeric membranes. Moreover, conventional modification strategies usually may not completely modify fine pores in the surface of membranes resulting in homogeneous modification. In addition, excessive coating or deficient coating may happen in the membrane pores leading to blockage of coating or the presence of unmodified pores. Consequently, conventional modification methods usually faced challenges, including nonuniformity, pore blocking tedious processing, and insufficient efficiency. Whereas the application of 3D printing technology in the membrane fabrication methods can resolve problems in the currently used conventional modification methods. The 3D printing technology can provide better control of the pores and their structure, and porosity of membranes and able to create different geometry for membranes. As a result, 3D printing membranes with stable and high permeability, enhanced antifouling properties, high regeneration capacity, and high performance can be achieved. But this technology is still facing challenges such as limited applicability of polymer materials, large-scale production of membranes for industrial application, low resolution, the need for a unique printing system incorporating both membrane and other components, and cost compared to conventional fabrication methods. Hence, further investigations are needed to demonstrate the application of 3D printing technology in membrane applications.

References

[1] Dong, X., et al. Polymers and solvents used in membrane fabrication: A review focusing on sustainable membrane development. Membranes (Basel), 2021, **11**(5): 309.
[2] Díez, B. and Rosal, R. A critical review of membrane modification techniques for fouling and biofouling control in pressure-driven membrane processes. Nanotechnol. Environ. Eng., 2020, **5**(2): 15.
[3] Nagandran, S., et al. The recent progress in modification of polymeric membranes using organic macromolecules for water treatment. Symmetry, 2020, **12**(2): 239.
[4] Jiang, S., et al. A comprehensive review on the synthesis and applications of ion exchange membranes. Chemosphere, 2021, **282**: 130817.
[5] Charcosset, C. Classical and recent developments of membrane processes for desalination and natural water treatment. Membranes (Basel), 2022, **12**(3): 267.
[6] Ahmad, A., et al. Recent trends and challenges with the synthesis of membranes: Industrial opportunities towards environmental remediation. Chemosphere, 2022, **306**: 135634.
[7] Li, X., et al. Regulating the interfacial polymerization process toward high-performance polyamide thin-film composite reverse osmosis and nanofiltration membranes: A review. J. Membr. Sci., 2021, **640**: 119765.

[8]   Xiong, S., et al. Atomic layer deposition for membrane modification, functionalization and preparation: A review. J. Membr. Sci., 2022, **658**: 120740.

[9]   Hosseini, S. S., et al. Recent progress in development of high performance polymeric membranes and materials for metal plating wastewater treatment: A review. J. Water Process. Eng., 2016, **9**: 78–110.

[10] Bera, S. P., Godhaniya, M. and Kothari, C. Emerging and advanced membrane technology for wastewater treatment: A review. J. Basic. Microbiol., 2022, **62**(3–4): 245–259.

[11] Warsinger, D. M., et al. A review of polymeric membranes and processes for potable water reuse. Prog. Polym. Sci., 2016, **81**: 209–237.

[12] Wen, Y., et al. Polymeric nanocomposite membranes for water treatment: A review. Environ. Chem. Lett., 2019, **17**(4): 1539–1551.

[13] Huyan Shi, L. X., Gao, A. and Zhou, Q. Dual layer hollow fiber PVDF ultra-filtration membranes containing Ag nano-particle loaded zeolite with longer term anti-bacterial capacity in salt water. Water Sci. Technol., 2016, **73**(9): 2159–2167.

[14] Amiri, S., et al. Recent trends in application of nanoscale zero-valent metals and metal single atoms in membrane processes. J. Environ. Chem. Eng., 2022, **10**(3): 107457.

[15] Vinh-Thang, H. and Kaliaguine, S. Predictive models for mixed-matrix membrane performance: A review. Chem. Rev., 2013, **113**(7): 4980–5028.

[16] Liu, F., et al. Progress in the production and modification of PVDF membranes. J. Membr. Sci., 2011, **375**(1–2): 1–27.

[17] Díez, B., Amariei, G. and Rosal, R. Electrospun composite membranes for fouling and biofouling control. Ind. Eng. Chem. Res., 2018, **57**(43): 14561–14570.

[18] Aroon, M. A., et al. Morphology and permeation properties of polysulfone membranes for gas separation: Effects of non-solvent additives and co-solvent. Sep. Purif. Technol., 2010, **72**(2): 194–202.

[19] Lalia, B. S., et al. A review on membrane fabrication: Structure, properties and performance relationship. Desalination, 2013, **326**: 77–95.

[20] Kahrs, C. and Schwellenbach, J. Membrane formation via non-solvent induced phase separation using sustainable solvents: A comparative study. Polymer, 2020, **186**: 122071.

[21] Li, D., et al. Membrane formation via thermally induced phase separation (TIPS): Model development and validation. J. Membr. Sci., 2006, **279**(1–2): 50–60.

[22] Sabir, A., et al. Conjugation of silica nanoparticles with cellulose acetate/polyethylene glycol 300 membrane for reverse osmosis using MgSO4 solution. Carbohydr. Polym., 2016, **136**: 551–559.

[23] Ahmad, N. N. R., et al. Current advances in membrane technologies for saline wastewater treatment: A comprehensive review. Desalination, 2021, **517**: 115170.

[24] Pervin, R., Ghosh, P. and Basavaraj, M. G. Tailoring pore distribution in polymer films via evaporation induced phase separation. RSC Adv., 2019, **9**(27): 15593–15605.

[25] Van der Bruggen, B. Chemical modification of polyethersulfone nanofiltration membranes: A review. J. Appl. Polym. Sci., 2009, **114**(1): 630–642.

[26] El-Aswar, E. I., et al. A comprehensive review on preparation, functionalization and recent applications of nanofiber membranes in wastewater treatment. J. Environ. Manage., 2022, **301**: 113908.

[27] Wu, M.-B., et al. Thin film composite membranes combining carbon nanotube intermediate layer and microfiltration support for high nanofiltration performances. J. Membr. Sci., 2016, **515**: 238–244.

[28] Chakrabarty, B., Ghoshal, A. K. and Purkait, M. K. Preparation, characterization and performance studies of polysulfone membranes using PVP as an additive. J. Membr. Sci., 2008, **315**(1–2): 36–47.

[29] Tafreshi, J. and Fashandi, H. Environmentally Friendly modification of polysulfone ultrafiltration membrane using organic plant-derived nanoparticles prepared from basil seed gum (BSG) and Ar/O2 low-pressure plasma. J. Environ. Chem. Eng., 2019, **7**(4): 103245.

[30]   Yang, S.-J., et al. Preparation of hydrophilic polypropylene hollow fiber membranes by UV modification. Integr. Ferroelectr., 2016, **169**(1): 83–89.

[31]   Laohaprapanon, S., et al. Self-cleaning and antifouling properties of plasma-grafted poly(vinylidene fluoride) membrane coated with ZnO for water treatment. J. Taiwan. Inst. Chem. Eng., 2017, **70**: 15–22.

[32]   Sandoval-Olvera, I. G., et al. Ultrafiltration membranes modified by PSS deposition and plasma treatment for Cr(VI) removal. Sep. Purif. Technol., 2019, **210**: 371–381.

[33]   Igbinigun, E., et al. Graphene oxide functionalized polyethersulfone membrane to reduce organic fouling. J. Membr. Sci., 2016, **514**: 518–526.

[34]   Jaleh, B., et al. Improving wettability: Deposition of TiO(2) nanoparticles on the O(2) plasma activated polypropylene membrane. Int. J. Mol. Sci., 2019, **20**(13): 3309.

[35]   Abdul Rahman, A. F. H. B. and Abu Seman, M. N. B. Polyacrylic-polyethersulfone membrane modified via UV photografting for forward osmosis application. J. Environ. Chem. Eng., 2018, **6**(4): 4368–4379.

[36]   Trachevskyi, V., et al. Surface polymerization of monomers on the polyethylene terephthalate membrane in low temperature plasma for water treatment. Chem. Chem. Technol., 2018, **12**(1): 64–68.

[37]   Kaneda, M., et al. Photografting graphene oxide to inert membrane materials to impart antibacterial activity. Environ. Sci. Technol. Lett., 2019, **6**(3): 141–147.

[38]   Efome, J. E., et al. Enhanced performance of PVDF nanocomposite membrane by nanofiber coating: A membrane for sustainable desalination through MD. Water Res., 2016, **89**: 39–49.

[39]   Li, R., et al. A novel strategy to develop antifouling and antibacterial conductive Cu/polydopamine/polyvinylidene fluoride membranes for water treatment. J. Colloid. Interface. Sci., 2018, **531**: 493–501.

[40]   Kumar, M. and Jaafar, J. Preparation and characterization of TiO2 nanofiber coated PVDF membrane for softdrink wastewater treatment. Environment & Ecosystem Science, 2018, **2**(2): 35–38.

[41]   Gao, F., et al. Aged PVDF and PSF ultrafiltration membranes restored by functional polydopamine for adjustable pore sizes and fouling control. J. Membr. Sci., 2019, **570–571**: 156–167.

[42]   Wu, H., et al. Preparation and characterization of antifouling and antibacterial polysulfone ultrafiltration membranes incorporated with a silver-polydopamine nanohybrid. J. Appl. Polym. Sci., 2018, **135**(27): 46430.

[43]   Kasemset, S., et al. Effect of polydopamine deposition conditions on polysulfone ultrafiltration membrane properties and threshold flux during oil/water emulsion filtration. Polymer, 2016, **97**: 247–257.

[44]   Saraswathi, M. S. S. A., et al. Polydopamine layered poly (ether imide) ultrafiltration membranes tailored with silver nanoparticles designed for better permeability, selectivity and antifouling. J. Ind. Eng. Chem., 2019, **76**: 141–149.

[45]   Yang, L., et al. Recent advances in graphene oxide membranes for nanofiltration. ACS Appl. Nano Mater., 2022, **5**(3): 3121–3145.

[46]   Seah, M. Q., et al. Progress of interfacial polymerization techniques for polyamide thin film (nano) composite membrane fabrication: A comprehensive review. Polymers (Basel), 2020, **12**(12): 2817.

[47]   Lee, T. H., et al. ZIF-8 particle size effects on reverse osmosis performance of polyamide thin-film nanocomposite membranes: Importance of particle deposition. J. Membr. Sci., 2019, **570–571**: 23–33.

[48]   Trivedi, J. S., et al. Multifunctional amines enable the formation of polyamide nanofilm composite ultrafiltration and nanofiltration membranes with modulated charge and performance. J. Mater. Chem. A., 2018, **6**(41): 20242–20253.

[49]   Song, X., et al. Confined nanobubbles shape the surface roughness structures of thin film composite polyamide desalination membranes. J. Membr. Sci., 2019, **582**: 342–349.

[50] Sung-Joon, P., et al., Fabrication of high-performance reverse osmosis membranes via dual-layer slot coating with tailoring interfacial adhesion, J. Memb. Sci., 2020, **614**: 118449.

[51] Ma, Z. Y., et al. Polyamide nanofilms synthesized via controlled interfacial polymerization on a "jelly" surface. Chem. Commun. (Camb), 2020, **56**(53): 7249–7252.

[52] Chi, J., et al., Thin-film composite membranes with aqueous template-induced surface nanostructures for enhanced nanofiltration, J. Memb. Sci., 2019, **589**: 117244.

[53] Jiang, C., et al. Ultrathin film composite membranes fabricated by novel in situ free interfacial polymerization for desalination. ACS Appl. Mater. Interfaces., 2020, **12**(22): 25304–25315.

[54] Wang, Z., et al. Nanoparticle-templated nanofiltration membranes for ultrahigh performance desalination. Nat. Commun., 2018, **9**(1): 2004.

[55] Tsuru, T., et al. Multilayered polyamide membranes by spray-assisted 2-step interfacial polymerization for increased performance of trimesoyl chloride (TMC)/m-phenylenediamine (MPD)-derived polyamide membranes. J. Membr. Sci., 2013, **446**: 504–512.

[56] Ang, M., et al. Assessing the performance of thin-film nanofiltration membranes with embedded montmorillonites. Membranes (Basel), 2020, **10**(5): 79.

[57] Wang, Y., et al. Physicochemical properties of gelatin films containing tea polyphenol-loaded chitosan nanoparticles generated by electrospray. Mater. Des., 2020, **185**: 108277.

[58] He, K., et al. Biodegradation of pharmaceuticals and personal care products in the sequential combination of activated sludge treatment and soil aquifer treatment. Environ. Technol., 2020, **41**(3): 378–388.

[59] Su, C., et al. Robust superhydrophobic membrane for membrane distillation with excellent scaling resistance. Environ. Sci. Technol., 2019, **53**(20): 11801–11809.

[60] Cao, L., et al. Alumina film deposited by spin-coating method for silicon wafer surface passivation. J. Mater. Sci. Mater. Electron., 2020, **31**(3): 2686–2690.

[61] Zheng, Q., et al. Annealing temperature impact on Sb2S3 solar cells prepared by spin-coating method. Mater. Lett., 2019, **243**: 104–107.

[62] Qanati, O., et al. Thin-film nanofiltration membrane with monomers of 1,2,4,5-benzene tetracarbonyl chloride and ethylene diamine on electrospun support: Preparation, morphology and chlorine resistance properties. Polym. Bull., 2017, **75**(8): 3407–3425.

[63] Shen, L., et al. High-performance thin-film composite polyamide membranes developed with green ultrasound-assisted interfacial polymerization. J. Membr. Sci., 2019, **570–571**: 112–119.

[64] Yuan, S., et al. Super-hydrophobic 3D printed polysulfone membranes with a switchable wettability by self-assembled candle soot for efficient gravity-driven oil/water separation. J. Mater. Chem. A., 2017, **5**(48): 25401–25409.

[65] Yuan, S., et al. Structure architecture of micro/nanoscale ZIF-L on a 3D printed membrane for a superhydrophobic and underwater superoleophobic surface. J. Mater. Chem. A., 2019, **7**(6): 2723–2729.

[66] Tijing, L. D., et al. 3D printing for membrane separation, desalination and water treatment. Appl. Mater. Today., 2020, **18**: 100486.

[67] Thiam, B. G., et al. 3D printed and conventional membranes-a review. Polymers (Basel), 2022, **14** (5): 1023.

[68] Singh, M., et al. Additive manufacturing of mechanically isotropic thin films and membranes via microextrusion 3D printing of polymer solutions. ACS Appl. Mater. Interfaces., 2019, **11**(6): 6652–6661.

[69] Fijoł, N., Aguilar-Sánchez, A. and Mathew, A. P. 3D-printable biopolymer-based materials for water treatment: A review. J. Chem. Eng., 2022, **430**: 132964.

[70] Mohd Yusoff, N. H., et al. Recent advances in polymer-based 3D printing for wastewater treatment application: An overview. J. Chem. Eng., 2022, **429**: 132311.

[71] Issac, M. N. and Kandasubramanian, B. Review of manufacturing three-dimensional-printed membranes for water treatment. Environ. Sci. Pollut. Res. Int., 2020, **27**(29): 36091–36108.

[72] Chowdhury, M. R., et al. 3D printed polyamide membranes for desalination. Science, 2018, **361** (6403): 682–686.

[73] Yang, Y., et al. 3D-printed biomimetic super-hydrophobic structure for microdroplet manipulation and oil/water separation. Adv. Mater., 2018, **30**(9): 1704912.

[74] Hwa, L. C., et al. Integration and fabrication of the cheap ceramic membrane through 3D printing technology. Mater. Today Commun., 2018, **15**: 134–142.

[75] Badalov, S., Oren, Y. and Arnusch, C. J. Ink-jet printing assisted fabrication of patterned thin film composite membranes. J. Membr. Sci., 2015, **493**: 508–514.

Abdullah Ali, Amani Al-Othman*, Muhammad Tawalbeh

Chapter 4
Polymer membranes: general principles and applications in fuel cells

Abstract: The membrane electrode assembly is the core of a fuel cell that contains the embedded ionic conducting membrane. Ions produced at the anode, when the fuel is electro-oxidized, are transported via the polymeric conducting membrane to the cathode to complete the circuit. The effective functioning of ions is crucial to maintain fuel cell performance. For proton exchange membrane (PEM) fuel cells, in particular, protons are transported through the membrane. A good PEM must demonstrate high proton conductivity, resistance to fuel crossover, high mechanical strength, durability, and superior water retention capability. Although Nafion membranes are widely employed in fuel cells, by virtue of their high conductivity and electrochemical stability, they suffer from drawbacks such as the high cost and performance degradation at higher temperatures above the boiling point of water. High-temperature operations are preferred because of the accelerated fuel cell kinetics, possible recovery of beneficial heat, and enhanced tolerance of catalysts for contaminants. Therefore, modified Nafion membranes were investigated. The modification incorporated inorganic fillers. Examples are metal organic frameworks and ionic liquids in the Nafion polymer matrix. Nafion-free membranes were also investigated using other polymers. Furthermore, eco-friendly and cost-effective natural polymers, such as cellulose and chitosan, have also been successfully doped to enhance their proton conductivity. However, their mechanical stability was compromised. This chapter outlines the challenges that lie in enhancing the proton conductivity of natural polymers in both normal and high operating temperatures while conserving their mechanical strength and durability.

Keywords: fuel cells, polymeric membranes, proton exchange membranes, nafion, proton conductivity

*Corresponding author: Amani Al-Othman, Department of Chemical and Biological Engineering, American University of Sharjah, P.O. Box, Sharjah 26666, United Arab Emirates, e-mail: aalothman@aus.edu
Abdullah Ali, Department of Chemical and Biological Engineering, American University of Sharjah, P.O. Box, Sharjah 26666, United Arab Emirates
Muhammad Tawalbeh, Sustainable and Renewable Energy Engineering Department, University of Sharjah, PO. Box, Sharjah 27272, United Arab Emirates; Sustainable Energy and Power Systems Research Centre, RISE, University of Sharjah, P.O. Box, Sharjah, 27272, United Arab Emirates

https://doi.org/10.1515/9783110796032-004

Nomenclature

CP	Calcium phosphate
[dema][TfO]	Diethylmethylammonium trifluoromethanesulfonate
[HMIM][C_4N^{3-}]	1-Hexyl-3-methylimidazolium tricyanomethanide
[EMIM][$CH_3O_3S^-$]	1-Ethyl-3-methylimidazolium methanesulfonate
FC	Fuel cell
FeSPP	Ferric sulfophenylphosphate
HT-PEMFC	High-temperature proton exchange membrane fuel cell
ILs	Ionic liquids
PEM	Polymer electrolyte membrane
MOFs	Metal organic frameworks
NS	Nanosheets
PANI	Polyaniline
PBI	Polybenzimidazole
PTFE	Polytetrafluoroethylene
PWA	Phosphotungstic acid
RH	Relative humidity
SEM	Scanning electron microscope
SPAEK	Sulfonated poly(arylene ether ketones)
SPEEK	Sulfonated polyether ether ketone
TEM	Transmission electron microscope
TIC	Titanium carbide
ZrP	Zirconium phosphate

1 Introduction

Polymer electrolyte membrane fuel cells (PEMFCs) that employ hydrogen as a fuel are clean energy systems that exhibit emission-free operation. They are characterized by rapid start-up, high efficiency, and high power densities [1, 2]. The PEMFCs have attracted increasing interest in portable electric devices and electric vehicles over the last decade. In the case of portable electric devices, they display merits such as silent operation, long operation duration, and immediate refilling and self-discharge [3], and for electric vehicles application, they demonstrate the complete solid structure and high energy density by weight of hydrogen gas [4]. The operation of PEMFC is highly affected by the stoichiometry of the reactants, humidity, temperature, and pressure, wherein alterations of any of these parameters impact the stability, water, and heat management of the cell, overall cell conductivity, and gas diffusivity [5, 6]. PEMFC is composed of a negative electrode called the anode, a positive electrode called the cathode, and a conductive electrolyte material embedded between the porous electrodes. In PEMFC, the polymer electrolyte membrane functions as the proton conducting material, transporting them from the anode to the cathode. Fuel, such as hydrogen, is supplied at the anode, while the oxidant is supplied at the cathode. Redox reactions occur as illustrated in Eqs. (1), (2), and (3) that produce protons and electrons. Electrons flow through the external wired circuit to produce a DC current while the proton ions diffuse across the membrane to the cathode, meet with reduced oxygen, and form water [7]. The following half reactions occur on an acidic electrolyte-based PEMFC:

At the anode:

$$H_2 \rightarrow 2H^+ + 2e^- \tag{1}$$

At the cathode:

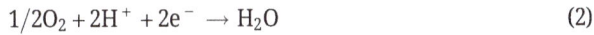

$$1/2O_2 + 2H^+ + 2e^- \rightarrow H_2O \tag{2}$$

Overall reaction:

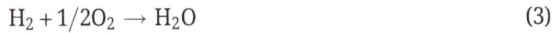

$$H_2 + 1/2O_2 \rightarrow H_2O \tag{3}$$

As observed, water is produced as the main product of the reaction, which contributes to the reason for its swelling attention among researchers. Protons pass through the membrane electrode assembly, as illustrated in Figure 1 [8]. The membrane is considered the core of the fuel cell (FC) system consisting of the gas diffusion layer and a scattered catalyst layer. Current FC research is mostly engaged in developing membranes and polymers to achieve and enhance the PEMFC performance, durability, and stability, particularly for high-temperature applications, while also ensuring that the membranes remain cost-effective.

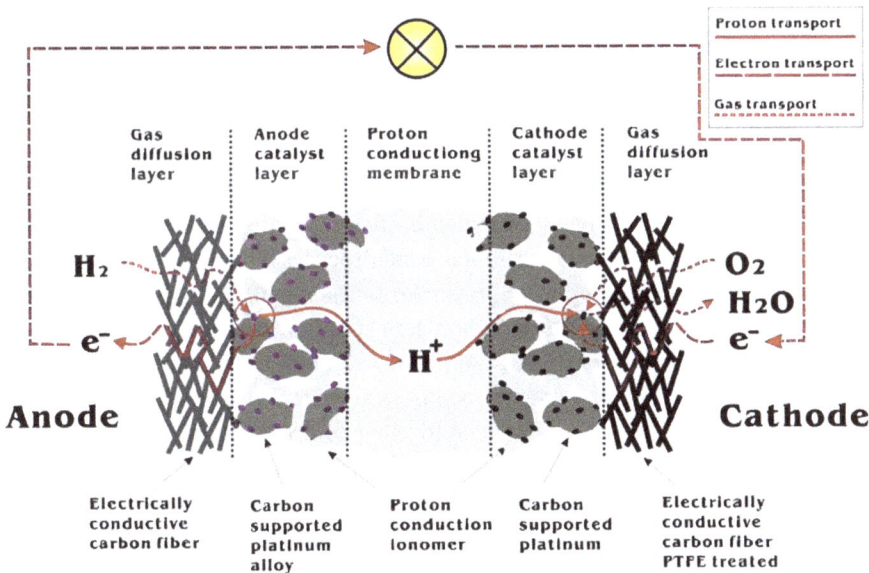

Figure 1: Membrane electrode assembly of a typical PEMFC [8].

Theoretically, proton conduction occurs via two proposed mechanisms in the literature: Grotthuss hopping and the vehicular mechanism [9]. Grotthuss mechanism accounts for the proton conduction across a hydrogen-bonded structure of water molecules [10], while the vehicular mechanism explains the conductivity based on the diffusion of protons via a carrier [11]. Regardless, a good PEM should demonstrate the following characteristics [12]: (i) exceptional proton conductivity, (ii) superior water retention characteristics, (iii) increased stability in electrochemical environments, (iv) resistance to fuel crossover, (v) low-priced material, (vi) easy fabrication, (vii) thermal stability, (viii) good chemical stability, (ix) high durability, and (x) effective performance. This chapter aims at exploring the conventional Nafion membranes, modified Nafion membranes with emphasis on selected modifying materials (inorganic compounds, ionic liquids (ILs), and metal organic frameworks (MOFs)), and non-Nafion membranes with a focus on sulfonated polyether ether ketone (SPEEK) membranes as a cost-effective polymeric substrate, and concludes with challenges and future directions involving high-temperature operations and incorporation of natural polymers, such as cellulose and chitosan, as lucrative and eco-friendly alternatives. Selected membrane modification and doping experiments have also been highlighted with their overall impact on membrane property enhancement.

2 Conventional perfluorinated-sulfonated membranes

Grubb and Niedrach first proposed a solid ion-exchange membrane FC based on cross-linked polystyrene with sulfonic acid group-based membrane in 1960 [13]. Perfluorosulfonic acid-based ionomer membranes superseded them due to their enhanced electrochemical stabilities and outstanding proton conductivities [14]. The replaced membranes are commonly referred to as Nafion and were originated by Dupont in 1966 [15]. Structurally, Nafion membranes consist of sulfonated polytetrafluoroethylene (PTFE), in which the sulfonic acid dispenses the hydrophilic properties while PTFE imparts the hydrophobic properties to the membrane [16]. Sulfonic acid clusters are connected to the fluorocarbon side chain of the fluorinated backbone [17], as illustrated in Figure 2 [11]. A 1 nm narrow channels and 4 nm sulfonated ion clusters are connected, thereby creating a cluster network in the membrane [18]. The hydrophilic sulfonic acid groups assemble together to produce ionic clusters, which allow the protons and water to pass [19]. Nafion membranes demand a fully hydrated atmosphere exhibiting proton conductivities of 0.1 S/cm in such conditions [20]. Additionally, Nafion membranes possess high chemical stability with 60,000 h of FC operation; the performance, albeit, deteriorates at temperatures greater than 80 °C and reduced relative humidity (RH) [21, 22].

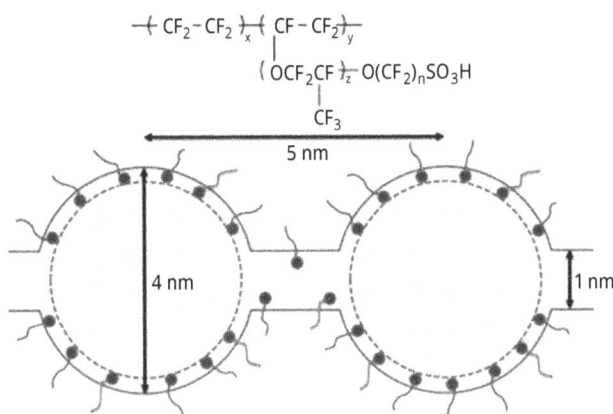

Figure 2: Chemical structure of Nafion and cluster network illustration [11].

Some researchers propose that the overall performance of FC could be determined by the Nafion membrane thickness and its hydration [23]. A thin membrane provides less mass transfer resistance for the diffusing protons [24]. Although the membrane demonstrates excellent performance within its operating temperature of about 80 °C, it undergoes dehydration at higher temperatures leading to performance deterioration by the reduced proton conductivities [25]. Additionally, the high cost of Nafion membrane, around \$711 for a 61×50 cm^2, greatly inhibits its extensive manufacturing and large-scale commercial rollout [26, 27]. Furthermore, degradation of the membrane occurs, that is, sparked by the generation of certain chemicals such as hydrogen peroxide (H_2O_2) [28, 29]. H_2O_2 is formed by the presence of oxygen (O_2) on the catalyst layer that, eventually, is reduced by the diffused protons on the Pt/C catalyst [30]. The degradation of H_2O_2 leads to the formation of hydroxyl and hydroperoxyl radicals, OH$^{\bullet}$ and OOH$^{\bullet}$, respectively. The OH radicals could also be produced by the reaction of metal ion impurities, such as Fe^{2+}, with H_2O_2 and also by the reaction of acidic PEM with H_2O_2 [31]. These OH radicals attack the Nafion membrane causing its degradation. Therefore, to ensure uncompromised membrane performance, especially at high temperatures, researchers have not only proposed modifications to Nafion membrane but also suggested the utilization of Nafion-free membranes.

3 Modified Nafion and Nafion-free membranes

The performance of PEM fuel can be advanced by either doping appropriate chemicals to the Nafion membrane itself or by deploying nonfluorinated membranes that are less costly. The primary aim is to advance the performance of the membrane at high temperatures and anhydrous conditions, resist fuel crossover, and increase water retention.

3.1 Modified Nafion membranes

Membrane dehydration at higher temperatures could be reduced by introducing hygroscopic fillers into the Nafion membrane [32, 33]. The increased tortuosity and crystallinity of the fillers could lead to significantly reduced water loss, and consequently, enhanced proton conduction of the membrane [34]. Besides this, a dehydrated Nafion membrane renders the ionic channels becoming prone to a collapse, thereby affecting proton conduction [35]. Sorbents could be used to inflate the ionic channels in a manner such that they have no impact on the mechanical properties of Nafion [11]. The objective of modifying a Nafion membrane is to improve its conductivity and upgrade its water retention abilities at higher temperatures. Materials analyzed to advance the properties of Nafion membranes in this section are inorganic materials, MOFs, and ILs.

3.1.1 Inorganic materials

Nafion membranes could be reinforced by doping inorganic fillers into the membrane. Inorganic materials could be added to the polymer solution that is used to fabricate/or cast the membrane. In general, the inorganic materials can be broadly classified into three categories: proton conducting (e.g., zirconium phosphate (ZrP)), inert hygroscopic (e.g., TiO_2 and Al_2O_3), and hydrophilic coupled with proton conductivity (e.g., functionalized zeolite) [36]. Addition of hygroscopic oxide accommodates proton transfer when FC operates at high temperatures by assisting in maintaining the RH of the membrane [37]. Water retention is enhanced by virtue of their high porosity and their operation as effective Lewis acid sites that assist in increased water absorption [38]. Methanol could be prevented from passing (if direct hydrocarbon FCs are used) since the inorganic materials fill the pores [39]. Homogeneous dispersion across the polymer matrix is a common challenge faced in utilizing such inorganic fillers; hence, nanosized, high surface area hydrophilic fillers have been analyzed, which were not only observed to produce compact membranes and prevent methanol crossover but also increase water retention, by providing suitable hydrogen bonding locations, and preserve the mechanical strength of the membrane [40]. However, it is essential to utilize an optimum amount of this hydrophilic filler since an excess may result in deterioration of the mechanical properties, increased water uptake, and high dimensional swelling [41].

Figure 3(a) and (b) displays the TEM and SEM images of the monodisperse spherical calcined silica nanoparticle fillers prepared by Jin et al. [42], respectively. Figure 3(c) and (d) shows the cross-sectional SEM images for the composite Nafion membrane containing 1 and 3 wt.% silica fillers, respectively. The two composite membranes were prepared using brittle breakage with liquid nitrogen. It is evident from Figure 3(c) and (d) that the excessive amount of the filler caused the formation of silica agglomerates, while at lower amount of the filler leads to a more homogeneous dispersion of the fillers.

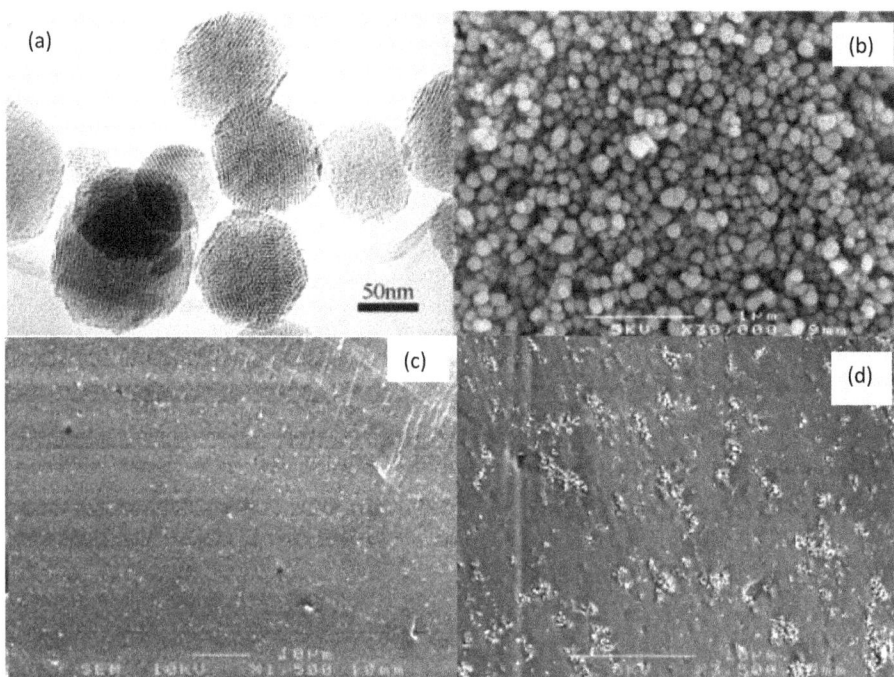

Figure 3: (a) TEM image of hydrophilic calcined silica nanoparticle fillers, (b) SEM image of hydrophilic calcined silica nanoparticle fillers, (c) SEM image of the cross section for the composite Nafion membrane containing 1 wt.% silica fillers, and (d) SEM image of the cross section for the composite Nafion membrane containing 3 wt.% silica fillers (adapted from [42]).

3.1.2 Metal organic frameworks (MOFs)

The MOFs are composed of metal-containing nodes and organic binders and are highly porous crystalline solids of coordination polymers [43]. They are deployed in numerous applications, such as water splitting, carbon dioxide (CO_2) reduction, solar cells, lithium ion batteries, supercapacitors, and FCs as illustrated in Figure 4 [44]. They possess large specific surface areas that accommodate facile migration of protons and aid in water retention, thus, enabling the proton conductivity [45]. In addition, fuel and oxidant crossover are prevented, given the small pore size of MOFs; they selectively allow certain particles to pass [46]. Modifications of membranes by MOFs are categorized into two methods: (a) the pores of MOFs could be immersed with proton carriers such as inositol phytic acid (myo-inositol hexaphosphonic acid) (phytic@MIL), poly(ionic liquid)@MIL (PIL@MIL), and acids@MIL [47–49]; and (b) the organic MOF ligands could be modified with functional groups, such as $-NH_2$, $-SO_3H$, and $-2COOH$, to increase the hydrophilicity and acidity [45]. The MOFs are classified into various "series," such as the MIL series, University of Oslo (UiO) series, and zeolitic imidazolate framework (ZIF) series, under which, after appropriate modifications and doping, various MOFs are obtained.

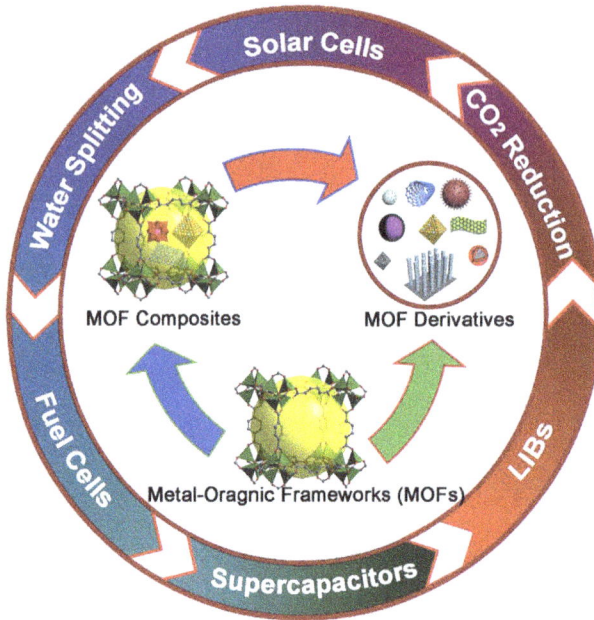

Figure 4: Applications of MOFs [44].

3.1.3 Ionic liquids (ILs)

The ILs exhibit a melting temperature of less than 100 °C at atmospheric pressure, and they are molten organic salts at room temperature [50]. Various combinations of organic cations and an organic or an inorganic anion are made to tune the IL structure, consequently, providing the benefit of flexible solvation features [51]. Figure 5 illustrates the commonly utilized cations and anions to create different ILs [52]. Nafion membranes incorporating ILs demonstrate high proton conductivities. For example, imidazole-based ILs possess nitrogen sites that can create hydrogen bonds with the sulfonic group in Nafion, thereby permitting easy proton hopping in reduced water content systems [53]. Additionally, ILs are not prone to evaporation at room temperature, given their low vapor pressure making them ideal for utilization in FCs exposed to high temperatures and low humidity [54]. Although ILs demonstrate promising features, studies on Nafion membranes modified with ILs are limited, with most of the work concentrated on hydrogen FCs [55]. Table 1 lists some studies employing the mentioned materials along with the addition of alcohols. Examples of the materials deployed and their overall impact have also been summarized.

Common cations

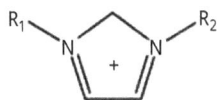

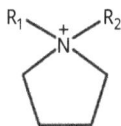

imidazolium   pyrrolidinium   pyridinium

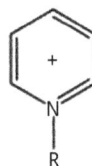

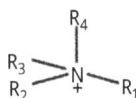

tetraalkylammonium   tetraalkylphosphonium

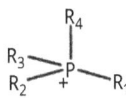

Common anions

Cl^- BF_4^- PF_6^-

$CF_3CO_2^-$ $CF_3SO_3^-$ $N(CF_3SO_2)_2^-$

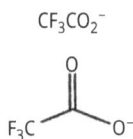

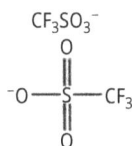

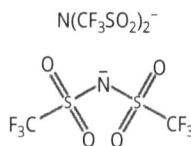

Figure 5: Commonly employed cations and anions in ILs [52].

Table 1: Summary of selected studies on modified Nafion membranes.

Material	Examples	Impact	Reference
Inorganic Materials	Hygroscopic oxides such as Al_2O_3	Porosity enhanced, fuel cross-over reduced, and water retention enhanced.	[56]
	Zirconia and silica	Silica-doped membrane exhibited increased water uptake and zirconia-doped membrane demonstrated increased conductivity at 90% RH and 80 °C. Mechanical properties and water handling upgraded.	[11]
	CeO_2	Durability of PEM improved by the elimination of free radicals. Addition of TIC boosted thermal stability and tensile strength.	[57]

Table 1 (continued)

Material	Examples	Impact	Reference
MOFs	ZIF-8 grown on graphene oxide	Enhanced proton conductivity of 280 mS/cm at 40 RH and 120 °C.	[58]
	MIL-53 (Al) and CPO-27 (Mg)	CPO-27 (Mg) composite membrane demonstrated increased water uptake and boosted proton conductivity. Enhanced power densities at all temperatures.	[59]
	Phytic@MIL101	Eleven times higher proton conductivity at 80 °C than recast Nafion, amelioration in mechanical properties, and refined swelling resistance observed.	[47]
	GO@UiO-66-NH$_2$	Increased proton conductivity in high RH and anhydrous conditions.	[60]
ILs	IL containing imidazolium cation	The doping level of phosphoric acid was enhanced, which upgraded proton conductivity when temperature increased without humidification and under anhydrous conditions.	[61]
	Pyridinium	One hundred times higher proton conductivity observed at elevated temperature of 120 °C for the composite membrane compared to the dry Nafion 117.	[62]
Alcohols	Methanol and ethanol	Nafion ionic clusters expanded. Proton conductivity depended on the molar volume of the solute with the conductivity decreasing at higher molar volumes.	[63]

3.2 Nafion-free membranes

The goal of developing non-Nafion membranes is to overcome the drawbacks of Nafion membranes, reduce the PEM FC capital cost, and simultaneously ensure uncompromised membrane performance at high temperature which is the current industrial demand [64]. These membranes are manufactured by introducing proton conductors such as glycerol, ZrP, zirconium silicate, ILs, silica, and calcium phosphate (CP) into polymer substrates [65–68]. As observed, some materials used for modifying Nafion membranes can be utilized to prepare Nafion-free membranes as well. Furthermore, many modified membranes emerged as possible alternatives for Nafion. Examples are the thermoplastic polymer polyether ether ketone (PEEK), SPEEK, and polybenzimidazole (PBI). Table 2 summarizes some of the Nafion-free membranes studied in literature with their corresponding results.

Table 2: Examples of studied Nafion-free composite membranes.

Type of composite membrane	Synthesis method	Results	Reference
Glycerol/ZrP/PTFE	The spin coating technique was used. Suspensions are composed of $ZrOCl_2 \cdot 8H_2O$, ethanol, isopropanol, water, and glycerol.	Unaltered conductivity was observed for a membrane operated at high inlet operating temperatures (200 °C and $y_{H2O} = 0.86$) of a direct propane fuel cell.	[69]
PANI/IL/PTFE (the IL used was [HMIM] $[C_4N^{3-}]$)	A slurry paste like mixture was prepared by dissolving the prepared PANI protonated salt in the IL along with the ZrP powder. The prepared slurry paste was mixed with isopropanol after which the final paste was painted over a PTFE sheet placed between two Teflon hoops.	The composite membrane demonstrated a high proton conductivity of 0.02 S/cm.	[64]
CP/IL/glycerol/PTFE (ILs used were [HMIM] $[C_4N^{3-}]$ and [EMIM] $[CH_3O_3S^-]$)	The spin coating technique was used. Suspensions were composed of glycerol, ILs, isopropanol, water, and $CaCl_2$. The membrane was soaked in H_3PO_4 after complete addition of solution.	The proton conductivity of CP/PTFE membrane was enhanced by the addition of ILs. The CP/PTFE/[HMIM]$[C_4N^{3-}]$ demonstrated a reasonable proton conductivity of 3.14×10^{-3} S/cm at 200°C and completely anhydrous conditions.	[65]
PBI/FeSPP-PWA	The hot-pressed method was employed to fabricate the PTFE membrane with the hot solution containing the proton conductor FeSPP–PWA.	Enhanced mechanical and dimensional stability, reduced methanol permeability, and increased proton conductivity at high temperatures (170–180 °C) were observed.	[70, 71]
SPEEK/(ZrP-NS)	SPEEK/ZrP-NS suspension was prepared that was cast onto Teflon plate left at 60 °C overnight. Obtained membranes were peeled off and appropriately dried.	Hydrogen bonds formed enhanced the chemical stability of the membranes and improved proton conductivities at 150 °C.	[72]

3.2.1 Merits of SPEEK membranes

The SPEEK membranes have demonstrated increasing interest for utilization as PEM in FCs. They come under the category of nonfluorinated membranes that exhibit advanced chemical stability and cost-effectiveness [73–76]. The SPEEK membranes also demonstrate enhanced thermal stabilities, increased proton conductivities, high mechanical strength, reduced fuel crossover, and facile operation [74]. The SPEEK membranes are formed by the incorporation of sulfonic acid functional groups to the PEEK polymer backbone. Sulfonic acid functional groups are responsible for the enhanced hydrophilicity that ultimately results in increased proton conductivity. The aromatic backbone of the PEEK provides the mechanical and thermal strength despite chemical modifications [75]. This explains the reason as to why SPEEK membranes have long been subject to chemical modifications by doping the matrix with fillers such as TiO_2 [76], ZrP [72], and heteropoly acids such as tungstophosphoric acid [33, 77] and silicotungstic acid [78]. The objective of doping is to further enhance the proton conductivity and improve SPEEK's durability and mechanical strength [79, 80]. Additionally, organic fillers such as carbon nanotubes have been utilized to enhance the performance and durability, given their outstanding physical, thermal, electrical, and mechanical properties [81]. Importantly, the commercial availability of PEEK polymers and their capability of acquiring properties of other compounds make them more attractive and hence are more extensively utilized compared to SPEEK, which is also another widely studied sulfonated aromatic main chain polymer [82]. Figure 6 distinguishes between the different aromatic chains and illustrates the evolution of SPEEK and sulfonated polyaryl ether ketone (SPAEK) from polyether ketone ketone (PEKK).

4 Challenges and future directions

High-temperature operation and the successful incorporation of natural polymers into PEM are pressing areas of research. The solutions of this area will catalyze the FC technology to its near full potential. High-temperature operation is advantageous because of accelerated FC kinetics, possible recovery of useful heat, and upgraded tolerance of catalysts for contaminants [64]. High-temperature PEMFCs (HT-PEMFCs) are particularly advantageous in stationary power generation applications, such as the combined heat and power generation, given their improved water management and superior plate designs [83, 84]. However, their extensive commercialization is inhibited because of their reduced durability engendered by unsatisfactory membrane performance at elevated temperatures, even though the efficiency and power production of this FC is high [85]. The working life of vital FC components such as the gasket, electrode plate, and the catalyst relies on the working life of the membranes; hence, dete-

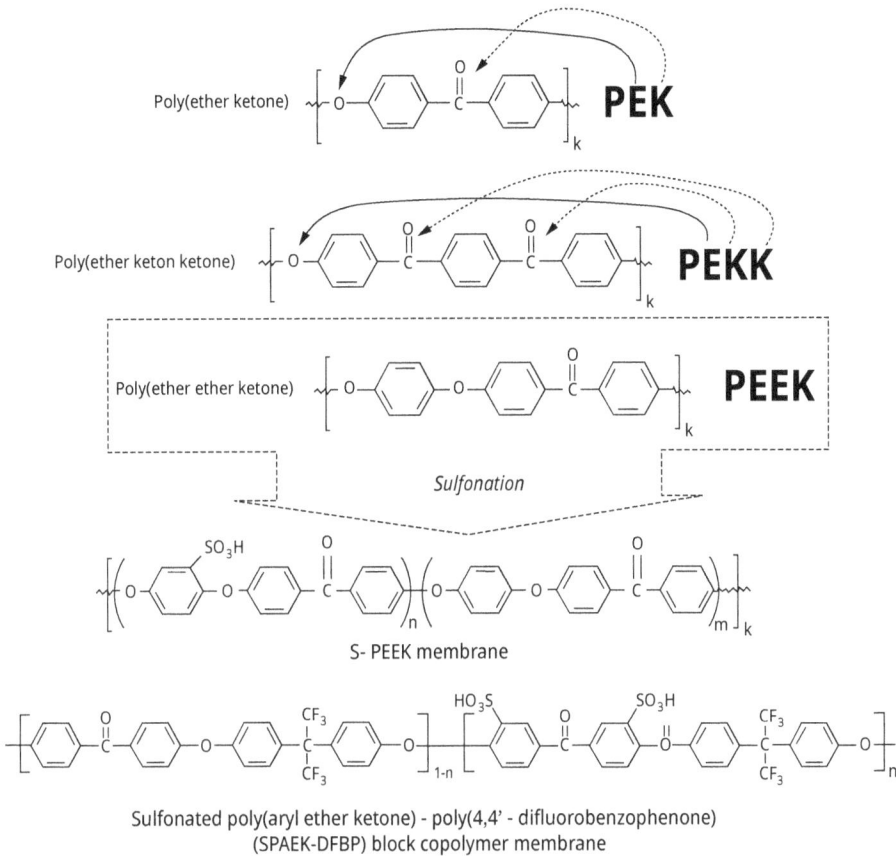

Figure 6: Evolution of PEKK, PEEK, SPEEK, and SPAEK from the base PEK [12].

rioration of FC components is dominated by the reduced durability of the membrane [86]. Furthermore, membranes degrade at high temperatures due to the formation of radicals [87]. The repercussions of membrane degradation aggravate in large-scale operations in which FC stacks are employed. The emergence of hotspots leads to mechanical failures, which creates cell obstructions that ultimately affect the entire cell stack. The cells are exposed to nonhomogeneous reaction conditions because of nonuniform flow of reactants leading to uneven reaction kinetics, ultimately, causing thermal issues [86]. Accurate flow field design, on the other hand, is a rather cumbersome task because of the complexity involved in assuring uniform flow rate and pressure drop across the stacks [88]. Therefore, research on developing membranes that are resistant to high-temperature operation problems and are fortified from potential errors in flow field designs is the current pressing priority. Modified Nafion membranes and non-Nafion composite membranes offer potential solutions to the high-temperature operation problems and have been elaborated upon in the preceding sections. The SPEEK membranes should be promoted as Nafion alternatives due to their low initial cost, facile acceptance of chemical modifications, and expansively available research. Although SPEEK membranes display better mechanical stabilities, upon chemical modification, compared to other polymeric membranes, future development should focus on strengthening their backbone structure since the main chain is sometimes susceptible to scission reducing its durability [89]. Besides this, utilization of PBI-based PEM for high-temperature operations should also be advanced, given their commercial availability, comprehensive and wide-ranging research [90, 91], reasonable thermal and mechanical stabilities, and low cost [17]. Incorporation of ILs into the PBI matrix has been extensively researched. High proton conductivities at elevated temperatures (150–200 °C) were achieved for PBI/IL composite membranes [92–94]. However, the voids between the polymeric chains of PBI were expanded due to the presence of ILs, which weakened the intermolecular forces and mechanical strength [17]. Hence, it is essential to optimize the amount of ILs incorporated into the PBI to ensure escalated proton conductivity at high temperatures with uncompromised mechanical strength. Figure 7(a)–(d) shows the fresh PBI membrane, SEM image of the fresh PBI membrane, PBI-[dema][TfO] composite membrane, and SEM image of the PBI/[Dema][TfO] composite membrane, respectively.

Figure 7: (a) Fresh PBI membrane, (b) SEM image of the fresh PBI membrane, (c) PBI-[dema][TfO] composite membrane, and (d) SEM image of the PBI/[Dema][TfO] composite membrane (adapted from [95]).

Besides this, monitoring of the FC stack parameters by development of dynamic PEMFC models can provide an added layer of protection for the membrane [96]. Appropriate mitigation measures can be implemented upon fluctuations in performance indicator parameters to safeguard the membrane from excessive damage. Additionally, the development of performance degradation models for HT-PEMFCs that account for the operation deterioration causes will help ascertain the feasibility of FC operations in various applications [97]. Developed composite membranes can be tested using the model to study the performance of the FC and membrane degradation patterns in varying conditions.

On the other hand, natural polymers are not only cheap and cost-effective but also promote the cause of sustainability. As stated earlier, one of the goals of Nafion-free membranes is to counter the challenge of increased capital cost of Nafion membranes; therefore, natural polymers offer an attractive alternative to serve as Nafion-free or Nafion-modified membranes. Cellulose is a cheap, biodegradable, and abundantly available natural polymer found in plants [98, 99]. They demonstrate high mechanical strength due to the strong inter- and intramolecular hydrogen bonding network and are potentially serviceable as proton conductors after appropriate modifications [100]. Besides this, chitosan, a derivative of the natural polymer chitin, is another widely explored polymer for utilization in FC applications because of its high glass transition temperature, ability to retain water, and behavior as exceptional fuel barriers [101]. The same merits apply for cellulose as well. Lignin, which is the second most abundant biopolymer on the Earth, has also been utilized as a filler and that is mainly because of its biodegradability, abundance, high thermal stability, abundant hydroxyl groups within its structure, and low cost [22, 102].

Glass transition temperatures indicates the strength of a material and signifies the temperature at which a polymer transitions from soft state to a glassy hard state or vice versa. Suitable modifications by introducing sulfonic acid group in chitosan [103] or in bacterial cellulose [104] or by adding inorganic fillers to the natural polymers or blending them with synthetic polymers [101] need to be done to enhance their proton conductivity. Besides this, nanocellulose materials have received great attention recently, given their nanoscale morphology, dimensional stability, facile chemical modification capability, sustainability, and wide-ranging dimensional scope [105, 106]. Nanocellulose have been employed in PEMs because they serve as effective gas barriers, possess carboxyl functional groups, and have low cost with the cellulose nanocrystals, a category of nanocellulose, boasting a sale price of as low as $4/lb [107, 108]. Carboxyl acid groups work as suitable proton donors and acceptors and are therefore favorable functional groups to have in applications concerning proton transport [109]. The oxidized form of nanocellulose, such as the Tempo-oxidized cellulose nanofiber, is known to demonstrate a higher proton conductivity compared to the nonoxidized cellulose [110]. Besides this, nanocellulose themselves have been successfully doped with imidazole that enhanced the proton conductivity by four times compared to the normal cellulose while conserving the thermal stability from 110 to 150 °C [111]. Although nanocellulose works as superior gas barriers, the enhancement in proton conductivities, by doping, has resulted in increased water uptake causing a reduction in dimensional stability [112]. The reduction in stabilities can be enhanced by utilizing sustainable methods by crosslinking the molecules with environmentally friendly materials such as citric acid [107]. To conclude, natural polymers are a promising alternative to serve as effective membrane materials in PEMFCs; however, the effect of high temperatures on proton conductivity and enhancement in proton conductivity itself while maintaining stability is a prevailing priority in research studies.

5 Conclusions

Membranes are vital components of PEMFC that functions as proton conductors. Deterioration in membrane performance drastically degrades the entire FC performance. A good PEM must exhibit high proton conductivities, superior water retention abilities, resistance to fuel crossover, good mechanical stabilities, and low cost among a host of other desirable characteristics. Conventional perfluorinated sulfonated membranes, for example, Nafion, have high proton conductivities and good electrochemical stabilities. Therefore, Nafion membranes are widely employed membranes in FCs for low-temperature operations (80 °C). However, their performance diminishes at high temperatures (>100 °C) due to their membrane degradation and water evaporation that, eventually, causes membrane dehydration. Researchers attempted to unravel this issue by modifying the Nafion membrane by introducing inorganic materials as fillers, such as Al_2O_3, zirconia, MOFs, and silica, or doping them with ILs to produce composite modified Nafion membranes with enhanced proton conductivities at higher temperatures (above 110 °C), and upgraded membrane properties such as mechanical strength, water retention, resistance to fuel crossover, and durability. However, Nafion-modified membranes are expensive. Hence, Nafion-free membranes have also been researched with cost-effective and easily modifiable polymeric substrates, such as PBI, PEEK, and SPEEK. These polymers can be doped with IL, inorganic materials, and nanomaterials to produce composites capable of producing enhanced conductivity at higher temperatures with low cost. Current research is focused on developing natural polymer-based composite membranes that will serve as low-cost and eco-friendly alternatives to synthetic ones. However, the impact of elevated temperatures on proton conductivities and upgrading the proton conductivity itself are crucial areas of research today.

References

[1] San Martín, I., Ursúa, A. and Sanchis, P. Modelling of PEM fuel cell performance: Steady-state and dynamic experimental validation. Energies, Feb 2014, **7**(2): 670–700. doi: 10.3390/en7020670.

[2] Martis, R., Al-Othman, A., Alkasrawi, M. and Tawalbeh, M. Fuel cells for carbon capture and power generation: Simulation studies. Int. J. Hydrogen Energy, Jan 2021, **46**(8): 6139–6149. doi 10.1016/j.ijhydene.2020.10.208.

[3] Youssef, M. E. S., Amin, R. S. and El-Khatib, K. M. Development and performance analysis of PEMFC stack based on bipolar plates fabricated employing different designs. Arab. J. Chem., Jul 2018, **11**(5): 609–614. doi: 10.1016/j.arabjc.2015.07.005.

[4] Osman, A. I., et al. Hydrogen production, storage, utilisation and environmental impacts: A review. Environ. Chem. Lett., Feb 2022, **20**(1): 153–188. doi: 10.1007/s10311-021-01322-8.

[5] Tawalbeh, M., Alarab, S., Al-Othman, A. and Javed, R. M. N. The operating parameters, structural composition, and fuel sustainability aspects of pem fuel cells: A mini review. Fuels, Aug 2022, **3**(3): 449–474. doi 10.3390/fuels3030028.

[6] Mohammed, H., Al-Othman, A., Nancarrow, P., Tawalbeh, M. and El Haj Assad, M. Direct hydrocarbon fuel cells: A promising technology for improving energy efficiency. Energy, Apr 2019, **172**: 207–219. doi 10.1016/j.energy.2019.01.105.

[7] Anson, C. W. and Stahl, S. S. Mediated fuel cells: Soluble Redox Mediators and their applications to electrochemical reduction of O_2 and Oxidation of H_2, alcohols, biomass, and complex fuels. Chem. Rev., Apr 2020, **120**(8): 3749–3786. doi: 10.1021/acs.chemrev.9b00717.

[8] MA, S., CHEN, Q., Jogensen, F., Stein, P. and Skou, E. 19F NMR studies of Nafion^{TM} ionomer adsorption on PEMFC catalysts and supporting carbons. Solid State Ion., Dec 2007, **178**(29–30): 1568–1575. 10.1016/j.ssi.2007.10.007.

[9] Chopade, S. A., So, S., Hillmyer, M. A. and Lodge, T. P. Anhydrous proton conducting polymer electrolyte membranes via polymerization-induced microphase separation. ACS Appl. Mater. Interfaces, Mar 2016, **8**(9): 6200–6210. doi: 10.1021/acsami.5b12366.

[10] Lim, D.-W., Sadakiyo, M. and Kitagawa, H. Proton transfer in hydrogen-bonded degenerate systems of water and ammonia in metal–organic frameworks. Chem. Sci., 2019, **10**(1): 16–33. doi: 10.1039/C8SC04475A.

[11] Zhu, L. Y., Li, Y. C., Liu, J., He, J., Wang, L. Y. and Du Lei, J. Recent developments in high-performance Nafion membranes for hydrogen fuel cells applications. Petroleum Science, Jun 2022, **19**(3): 1371–1381. China University of Petroleum Beijing doi: 10.1016/j.petsci.2021.11.004.

[12] Ahmad, S., Nawaz, T., Ali, A., Orhan, M. F., Samreen, A. and Kannan, A. M. An overview of proton exchange membranes for fuel cells: Materials and manufacturing. Int. J. Hydrog. Energy, Elsevier Ltd 2022, **47**(44): 19086–19131. May doi: 10.1016/j.ijhydene.2022.04.099.

[13] Wang, Y., Chen, K. S., Mishler, J., Cho, S. C. and Adroher, X. C. A review of polymer electrolyte membrane fuel cells: Technology, applications, and needs on fundamental research. Appl. Energy, 2011, **88**(4): 981–1007. doi: 10.1016/j.apenergy.2010.09.030.

[14] Zhang, H. and Shen, P. K. Recent development of polymer electrolyte membranes for fuel cells. Chem. Rev., May 2012, **112**(5): 2780–2832. doi: 10.1021/cr200035s.

[15] Liu, Y.-R., Chen, -Y.-Y., Zhuang, Q. and Li, G. Recent advances in MOFs-based proton exchange membranes. Coord. Chem. Rev., Nov 2022, **471**: 214740. doi: 10.1016/j.ccr.2022.214740.

[16] Ogungbemi, E., et al. Fuel cell membranes – Pros and cons. Energy, Apr 2019, **172**: 155–172. doi: 10.1016/j.energy.2019.01.034.

[17] Alashkar, A., Al-Othman, A., Tawalbeh, M. and Qasim, M. A critical review on the use of ionic liquids in proton exchange membrane fuel cells. Membranes (Basel)., Feb 2022, **12**(2): 178. doi: 10.3390/membranes12020178.

[18] Fernandez Bordín, S. P., Andrada, H. E., Carreras, A. C., Castellano, G. E., Oliveira, R. G. and Galván Josa, V. M. Nafion membrane channel structure studied by small-angle X-ray scattering and Monte Carlo simulations. Polymer (Guildf), Oct 2018, **155**: 58–63. doi: 10.1016/j.polymer.2018.09.014.

[19] Frühwirt, P., Kregar, A., Törring, J. T., Katrašnik, T. and Gescheidt, G. Holistic approach to chemical degradation of Nafion membranes in fuel cells: Modelling and predictions. Phys. Chem. Chem. Phys., 2020, **22**(10): 5647–5666. doi: 10.1039/C9CP04986J.

[20] Liu, L., Chen, W. and Li, Y. An overview of the proton conductivity of Nafion membranes through a statistical analysis. J. Membr. Sci., Apr 2016, **504**: 1–9. doi: 10.1016/J.MEMSCI.2015.12.065.

[21] Maiyalagan, T. and Pasupathi, S. Components for PEM fuel cells: An overview. Mater. Sci. Forum, Jul 2010, **657**: 143–189. doi: 10.4028/www.scientific.net/MSF.657.143.

[22] Tawalbeh, M., Al-Othman, A., Ka'ki, A., Farooq, A. and Alkasrawi, M. Lignin/zirconium phosphate/ ionic liquids-based proton conducting membranes for high-temperature PEM fuel cells applications. Energy, Dec 2022, **260**: 125237. doi: 10.1016/j.energy.2022.125237.

[23] Banerjee, R., et al. Identifying in operando changes in membrane hydration in polymer electrolyte membrane fuel cells using synchrotron X-ray radiography. Int. J. Hydrog. Energy, 2018, **43**(20): 9757–9769. May doi: 10.1016/j.ijhydene.2018.03.224.

[24] Sudaroli, B. M. and Kolar, A. K. An experimental study on the effect of membrane thickness and PTFE (polytetrafluoroethylene) loading on methanol crossover in direct methanol fuel cell. Energy, Mar 2016, **98**: 204–214. doi: 10.1016/j.energy.2015.12.101.

[25] Paidar, M., Mališ, J., Bouzek, K. and Žitka, J. Behavior of Nafion membrane at elevated temperature and pressure. Desalination Water Treat, 2010, **14**(1–3): 106–111. doi: 10.5004/dwt.2010.1015.

[26] Sun, X., Simonsen, S., Norby, T. and Chatzitakis, A. Composite membranes for high temperature PEM fuel cells and electrolysers: A critical review. Membranes (Basel), Jul 2019, **9**(7): 83. doi: 10.3390/membranes9070083.

[27] Prykhodko, Y., Fatyeyeva, K., Hespel, L. and Marais, S. Progress in hybrid composite Nafion®-based membranes for proton exchange fuel cell application. Chem. Eng. J., 2021, **409**: 127329. doi https://doi.org/10.1016/j.cej.2020.127329.

[28] Okonkwo, P. C., Ben Belgacem, I., Emori, W. and Uzoma, P. C. Nafion degradation mechanisms in proton exchange membrane fuel cell (PEMFC) system: A review. Int. J. Hydrog. Energy, Aug 2021, **46**(55): 27956–27973. doi: 10.1016/j.ijhydene.2021.06.032.

[29] Meng, K., Chen, B., Zhou, H., Shen, J., Shen, Z. and Tu, Z. Investigation on degradation mechanism of hydrogen–oxygen proton exchange membrane fuel cell under current cyclic loading. Energy, Mar 2022, **242**: 123045. doi: 10.1016/j.energy.2021.123045.

[30] Shi, G., et al. Unparalleled mitigation of membrane degradation in fuel cells: Via a counter-intuitive approach: Suppression of H2O2 production at the hydrogen anode using a Ptskin-PtCo catalyst. J. Mater. Chem. A, 2020, **8**(3): 1091–1094. doi: 10.1039/c9ta12023h.

[31] Aoki, M., Uchida, H. and Watanabe, M. Decomposition mechanism of perfluorosulfonic acid electrolyte in polymer electrolyte fuel cells. ElectroChem. Commun., Sep 2006, **8**(9): 1509–1513. doi: 10.1016/j.elecom.2006.07.017.

[32] Li, X., et al. Highly conductive and mechanically stable imidazole-rich cross-linked networks for high-temperature proton exchange membrane fuel cells. Chem. Mater., Feb 2020, **32**(3): 1182–1191. doi: 10.1021/acs.chemmater.9b04321.

[33] Al-Othman, A., Zhu, Y., Tawalbeh, M., Tremblay, A. Y. and Ternan, M. Proton conductivity and morphology of new composite membranes based on zirconium phosphates, phosphotungstic acid, and silicic acid for direct hydrocarbon fuel cells applications. J. Porous. Mater., 2017, **24**(3): 721–729. doi: 10.1007/s10934-016-0309-6.

[34] Liu, L., Pu, Y., Lu, Y., Li, N., Hu, Z. and Chen, S. Superacid sulfated SnO2 doped with CeO2: A novel inorganic filler to simultaneously enhance conductivity and stabilities of proton exchange membrane. J. Membr. Sci., Mar 2021, **621**: 118972. doi: 10.1016/j.memsci.2020.118972.

[35] Rosli, R. E., *et al.* A review of high-temperature proton exchange membrane fuel cell (HT-PEMFC) system. Int. J. Hydrogen Energy, 2017, **42**(14): 9293–9314. doi: 10.1016/j.ijhydene.2016.06.211.

[36] Ying, Y. P., Kamarudin, S. K. and Masdar, M. S. Silica-related membranes in fuel cell applications: An overview. Int. J. Hydrog. Energy, Aug 2018, **43**(33): 16068–16084. doi: 10.1016/j.ijhydene.2018.06.171.

[37] Su, D. Y., Ma, J. and Pu, H. K. The research of Nafion/PTFE/inorganic composite membrane used in direct methanol fuel cell. Adv. Mater. Res., Jan 2014, **881–883**: 927–930. doi: 10.4028/www.scientific.net/AMR.881-883.927.

[38] Sigwadi, R., Mokrani, T., Dhlamini, S. and Msomi, P. F. Nafion® reinforced with polyacrylonitrile/ZrO$_2$ nanofibers for direct methanol fuel cell application. J. Appl. Polym. Sci., Mar 2021, **138**(10): 49978. doi: 10.1002/app.49978.

[39] Alvarez, A., *et al.* Influence of silica morphology in composite Nafion membranes properties. Int. J. Hydrogen Energy, Nov 2011, **36**(22): 14725–14733. doi: 10.1016/j.ijhydene.2011.07.031.

[40] Velayutham, P., Sahu, A. and Parthasarathy, S. A Nafion-Ceria composite membrane electrolyte for reduced methanol crossover in direct methanol fuel cells. Energies, Feb 2017, **10**(2): 259. doi: 10.3390/en10020259.

[41] Sood, R., Cavaliere, S., Jones, D. J. and Rozière, J. Electrospun nanofibre composite polymer electrolyte fuel cell and electrolysis membranes. Nano Energy, Aug 2016, **26**: 729–745. doi: 10.1016/j.nanoen.2016.06.027.

[42] Jin, Y., *et al.* Novel Nafion composite membranes with mesoporous silica nanospheres as inorganic fillers. J. Power Sources, Dec 2008, **185**(2): 664–669. doi: 10.1016/j.jpowsour.2008.08.094.

[43] Al-Othman, A., Tawalbeh, M., Temsah, O. and Al-Murisi, M. Industrial Challenges of MOFs in Energy Applications. In: Abdul-Ghani, O. (Ed.) Encyclopedia of Smart Materials, Elsevier, 2022, pp. 535–543. doi: 10.1016/B978-0-12-815732-9.00030-9.

[44] Wang, H., Zhu, Q.-L., Zou, R. and Xu, Q. Metal-organic frameworks for energy applications. Chem., Jan 2017, **2**(1): 52–80. doi: 10.1016/j.chempr.2016.12.002.

[45] Liu, Q., *et al.* Metal Organic Frameworks Modified Proton Exchange Membranes for Fuel Cells. Front. Chem., Frontiers Media S.A. 2020, **8**. Aug doi: 10.3389/fchem.2020.00694.

[46] Matoga, D., Oszajca, M. and Molenda, M. Ground to conduct: Mechanochemical synthesis of a metal–organic framework with high proton conductivity. Chem. Commun., 2015, **51**(36): 7637–7640. doi: 10.1039/C5CC01789K.

[47] Li, Z., *et al.* Enhanced proton conductivity of Nafion Hybrid membrane under different humidities by incorporating metal–organic frameworks with high phytic acid loading. ACS Appl. Mater. Interfaces, Jun 2014, **6**(12): 9799–9807. doi: 10.1021/am502236v.

[48] Li, Z., *et al.* Constructing efficient ion nanochannels in alkaline anion exchange membranes by the in situ assembly of a poly(ionic liquid) in metal–organic frameworks. J. Mater. Chem. A, 2016, **4**(6): 2340–2348. doi: 10.1039/C5TA10452A.

[49] Dong, X.-Y., Li, -J.-J., Han, Z., Duan, P.-G., Li, L.-K. and Zang, S.-Q. Tuning the functional substituent group and guest of metal–organic frameworks in hybrid membranes for improved interface compatibility and proton conduction. J. Mater. Chem. A, 2017, **5**(7): 3464–3474. doi: 10.1039/C6TA07761G.

[50] Hernández-Fernández, F. J., De Los Ríos, A. P., Mateo-Ramírez, F., Juarez, M. D., Lozano-Blanco, L. J. and Godínez, C. New application of polymer inclusion membrane based on ionic liquids as proton exchange membrane in microbial fuel cell. Sep. Purif. Technol., Feb 2016, **160**: 51–58. doi: 10.1016/j.seppur.2015.12.047.

[51] Rynkowska, E., Fatyeyeva, K. and Kujawski, W. Application of polymer-based membranes containing ionic liquids in membrane separation processes: A critical review. Rev. Chem. Eng., Apr 2018, **34**(3): 341–363. doi: 10.1515/revce-2016-0054.

[52] Sun, P. and Armstrong, D. W. Ionic liquids in analytical chemistry. Analytica Chimica Acta., Feb 2010, **661**(1): 1–16. doi: 10.1016/j.aca.2009.12.007.

[53] Asgari, M. S., Nikazar, M., Molla-abbasi, P. and Hasani-Sadrabadi, M. M. Nafion®/histidine functionalized carbon nanotube: High-performance fuel cell membranes. Int. J. Hydrogen Energy, May 2013, **38**(14): 5894–5902. doi: 10.1016/j.ijhydene.2013.03.010.

[54] Hudson, R., et al. Evaluation of carboxylic, phosphonic, and sulfonic acid protogenic moieties on tunable poly(meta -phenylene oxide) ionomer scaffolds. J. Polym. Sci. Part A: Polym. Chem., Nov 2019, **57**(22): 2209–2213. doi: 10.1002/pola.29503.

[55] Ng, W. W., Thiam, H. S., Pang, Y. L., Chong, K. C. and Lai, S. O. A state-of-art on the development of Nafion-based membrane for performance improvement in direct methanol fuel cells. Membranes, MDPI 2022, **12**(5): May doi: 10.3390/membranes12050506.

[56] Son, B., Oh, K., Park, S., Lee, T., Lee, D. H. and Kwon, O. Study of morphological characteristics on hydrophilicity-enhanced SiO_2/Nafion composite membranes by using multimode atomic force microscopy. Int. J. Energy Res., Jul 2019, **43**(9): 4157–4169. doi: 10.1002/er.4528.

[57] Vinothkannan, M., Ramakrishnan, S., Kim, A. R., Lee, H.-K. and Yoo, D. J. Ceria stabilized by titanium carbide as a sustainable filler in the Nafion Matrix improves the mechanical integrity, electrochemical durability, and hydrogen impermeability of proton-exchange membrane fuel cells: Effects of the filler content. ACS Appl. Mater. Interfaces, Feb 2020, **12**(5): 5704–5716. doi: 10.1021/acsami.9b18059.

[58] Yang, L., Tang, B. and Wu, P. Metal–organic framework–graphene oxide composites: A facile method to highly improve the proton conductivity of PEMs operated under low humidity. J. Mater. Chem. A, 2015, **3**(31): 15838–15842. doi: 10.1039/C5TA03507D.

[59] Tsai, C.-H., Wang, -C.-C., Chang, C.-Y., Lin, C.-H. and Chen-Yang, Y. W. Enhancing performance of Nafion®-based PEMFC by 1-D channel metal-organic frameworks as PEM filler. Int. J. Hydrog. Energy, Sep 2014, **39**(28): 15696–15705. doi: 10.1016/j.ijhydene.2014.07.134.

[60] Rao, Z., Feng, K., Tang, B. and Wu, P. Construction of well interconnected metal-organic framework structure for effectively promoting proton conductivity of proton exchange membrane. J. Membr. Sci., Jul 2017, **533**: 160–170. doi: 10.1016/j.memsci.2017.03.031.

[61] Yang, J., Che, Q., Zhou, L., He, R. and Savinell, R. F. Studies of a high temperature proton exchange membrane based on incorporating an ionic liquid cation 1-butyl-3-methylimidazolium into a Nafion matrix. Electrochim. Acta., Jul 2011, **56**(17): 5940–5946. doi: 10.1016/j.electacta.2011.04.112.

[62] Schmidt, C., Glück, T. and Schmidt-Naake, G. Modification of Nafion membranes by impregnation with ionic liquids. Chem. Eng. Technol., Jan 2008, **31**(1): 13–22. doi: 10.1002/ceat.200700054.

[63] Zhao, Q., Carro, N., Ryu, H. Y. and Benziger, J. Sorption and transport of methanol and ethanol in H^+-Nafion. Polymer (Guildf), Mar 2012, **53**(6): 1267–1276. doi: 10.1016/j.polymer.2012.01.050.

[64] Eisa, A., Al-Othman, A., Al-Sayah, M. and Tawalbeh, M. Novel composite membranes based on polyaniline/ionic liquids for PEM fuel cells applications. Key Eng. Mater., 2020, **865**: 55–60. KEM doi: 10.4028/www.scientific.net/KEM.865.55.

[65] Ka'ki, A., Alraeesi, A., Al-Othman, A. and Tawalbeh, M. Proton conduction of novel calcium phosphate nanocomposite membranes for high temperature PEM fuel cells applications. Int. J. Hydrog. Energy, 2021, **46**(59): 30641–30657. doi: 10.1016/j.ijhydene.2021.01.013.

[66] Nauman Javed, R. M., Al-Othman, A., Nancarrow, P. and Tawalbeh, M. Zirconium silicate-ionic liquid membranes for high-temperature hydrogen PEM fuel cells. Int. J. Hydrog. Energy, May 2022, doi: 10.1016/j.ijhydene.2022.05.009.

[67] Mohammed, H., Al-Othman, A., Nancarrow, P., Elsayed, Y. and Tawalbeh, M. Enhanced proton conduction in zirconium phosphate/ionic liquids materials for high-temperature fuel cells. Int. J. Hydrog. Energy, Jan 2021, **46**(6): 4857–4869. doi: 10.1016/j.ijhydene.2019.09.118.

[68] Al-Othman, A., et al. Novel composite membrane based on zirconium phosphate-ionic liquids for high temperature PEM fuel cells. Int. J. Hydrogen Energy, Jan 2021, **46**(8): 6100–6109. doi: 10.1016/j.ijhydene.2020.02.112.

[69] Al-Othman, A., Tremblay, A. Y., Pell, W., Liu, Y., Peppley, B. A. and Ternan, M. The effect of glycerol on the conductivity of Nafion-free ZrP/PTFE composite membrane electrolytes for direct hydrocarbon fuel cells. J. Power Sources, 2012, **199**: 14–21. doi: 10.1016/j.jpowsour.2011.09.104.

[70] Verdier, N., Foran, G., Lepage, D., Prébé, A., Aymé-Perrot, D. and Dollé, M. Challenges in solvent-free methods for manufacturing electrodes and electrolytes for lithium-based batteries. Polymers, Jan 2021, Vol. **13**(3): 1–26. MDPI AG doi: 10.3390/polym13030323.

[71] Wang, S., Sun, P., Li, Z., Liu, G. and Yin, X. Comprehensive performance enhancement of polybenzimidazole based high temperature proton exchange membranes by doping with a novel intercalated proton conductor. Int. J. Hydrog. Energy, 2018, **43**(21): 9994–10003. doi: 10.1016/j.ijhydene.2018.04.089.

[72] Kozawa, Y., Suzuki, S., Miyayama, M., Okumiya, T. and Traversa, E. Proton conducting membranes composed of sulfonated poly(etheretherketone) and zirconium phosphate nanosheets for fuel cell applications. Solid State Ion., Mar 2010, **181**(5–7): 348–353. doi: 10.1016/j.ssi.2009.12.017.

[73] Zhu, H. SPEEK scaling UP. Joule, Apr 2022, **6**(4): 718–720. doi: 10.1016/j.joule.2022.03.006.

[74] Harun, N. A. M., Shaari, N. and Nik Zaiman, N. F. H. A review of alternative polymer electrolyte membrane for fuel cell application based on sulfonated poly(ether ether ketone). Int. J. Energy Res., Nov 2021, **45**(14): 19671–19708. doi: 10.1002/er.7048.

[75] Yee, R., Zhang, K. and Ladewig, B. The effects of sulfonated poly(ether ether ketone) ion exchange preparation conditions on membrane properties. Membranes (Basel)., Aug 2013, **3**(3): 182–195. doi: 10.3390/membranes3030182.

[76] Di Vona, M. L., et al. Composite polymer electrolytes of sulfonated poly-ether-ether-ketone (SPEEK) with organically functionalized TiO2. J. Membr. Sci., Mar 2011, **369**(1–2): 536–544. doi: 10.1016/j.memsci.2010.12.044.

[77] Doğan, H., Inan, T. Y., Unveren, E. and Kaya, M. Effect of cesium salt of tungstophosphoric acid (Cs-TPA) on the properties of sulfonated polyether ether ketone (SPEEK) composite membranes for fuel cell applications. Int. J. Hydrog. Energy, Aug 2010, **35**(15): 7784–7795. doi: 10.1016/j.ijhydene.2010.05.045.

[78] Deivanayagam, P., Ramanujam Ramamoorthy, A. and Jaisankar, S. N. Synthesis and characterization of sulfonated poly (arylene ether sulfone)/silicotungstic acid composite membranes for fuel cells. Polym. J., Feb 2013, **45**(2): 166–172. doi: 10.1038/pj.2012.102.

[79] Sonpingkam, S. and Pattavarakorn, D. Mechanical properties of sulfonated poly (ether ether ketone) nanocomposite membranes. Int. J. Chem. Eng. Appl., Apr 2014, **5**(2): 181–185. doi: 10.7763/ijcea.2014.v5.374.

[80] Liu, Y.-L. Effective approaches for the preparation of organo-modified multi-walled carbon nanotubes and the corresponding MWCNT/polymer nanocomposites. Polym. J., Apr 2016, **48**(4): 351–358. doi: 10.1038/pj.2015.132.

[81] Sivasubramanian, G., Hariharasubramanian, K., Deivanayagam, P. and Ramaswamy, J. High-performance SPEEK/SWCNT/fly ash polymer electrolyte nanocomposite membranes for fuel cell applications. Polym. J., Oct 2017, **49**(10): 703–709. doi: 10.1038/pj.2017.38.

[82] Zhang, Z., et al. Construction of new alternative transmission sites by incorporating structure-defect metal-organic framework into sulfonated poly(arylene ether ketone sulfone)s. Int. J. Hydrog. Energy, Aug 2021, **46**(53): 27193–27206. doi: 10.1016/j.ijhydene.2021.05.167.

[83] Xu, B., Li, D., Ma, Z., Zheng, M. and Li, Y. Thermodynamic optimization of a high temperature proton exchange membrane fuel cell for fuel cell vehicle applications. Mathematics, Aug 2021, **9**(15): doi: 10.3390/math9151792.

[84] Haider, R., et al. High temperature proton exchange membrane fuel cells: Progress in advanced materials and key technologies. Chem. Soc. Rev., 2021, **50**(2): 1138–1187. doi: 10.1039/D0CS00296H.

[85] Salam, M. A., *et al.* Effect of temperature on the performance factors and durability of proton exchange membrane of hydrogen fuel cell: A narrative review. Mater. Sci. Res. India, Sep 2020, **17**(2): 179–191. doi: 10.13005/msri/170210.

[86] Ali, A., Al-Othman, A. and Tawalbeh, M. Grand challenges in fuel cell technology towards resource recovery. J. Resour. Recovery, 2023, **1**: 1004. doi: 10.52547/jrr.2211.1004.

[87] Nauman Javed, R. M., Al-Othman, A., Tawalbeh, M. and Olabi, A. G. Recent developments in graphene and graphene oxide materials for polymer electrolyte membrane fuel cells applications. Renew. Sustain. Energy Rev., Oct 2022, **168**: 112836. doi: 10.1016/j.rser.2022.112836.

[88] Wang, Y., Wang, S., Wang, G. and Yue, L. Numerical study of a new cathode flow-field design with a sub-channel for a parallel flow-field polymer electrolyte membrane fuel cell. Int. J. Hydrog. Energy, Jan 2018, 2359–2368. doi: 10.1016/j.ijhydene.2017.11.172.

[89] Jawad, N. H., *et al.* Fuel cell types, properties of membrane, and operating conditions: A review. Sustainability, Nov 2022, **14**(21): 14653. doi: 10.3390/su142114653.

[90] Escorihuela, J., Olvera-Mancilla, L., Alexandrova, L., Del Castillo, F. and Compañ, V. Recent progress in the development of composite membranes based on polybenzimidazole for high temperature proton exchange membrane (PEM) fuel cell applications. Polymers (Basel), Aug 2020, **12**(9): 1861. doi: 10.3390/polym12091861.

[91] Nimir, W., *et al.* Approaches towards the development of heteropolyacid-based high temperature membranes for PEM fuel cells. Int. J. Hydrog. Energy, Feb 2023, **48**(17): 6638–6656. doi: 10.1016/j.ijhydene.2021.11.174.

[92] Eguizábal, A., Lemus, J. and Pina, M. P. On the incorporation of protic ionic liquids imbibed in large pore zeolites to polybenzimidazole membranes for high temperature proton exchange membrane fuel cells. J. Power Sources, Jan 2013, **222**: 483–492. doi: 10.1016/j.jpowsour.2012.07.094.

[93] van de Ven, E., Chairuna, A., Merle, G., Benito, S. P., Borneman, Z. and Nijmeijer, K. Ionic liquid doped polybenzimidazole membranes for high temperature proton exchange membrane fuel cell applications. J. Power Sources, Jan 2013, **222**: 202–209. doi: 10.1016/j.jpowsour.2012.07.112.

[94] Eguizábal, A., Lemus, J., Roda, V., Urbiztondo, M., Barreras, F. and Pina, M. P. Nanostructured electrolyte membranes based on zeotypes, protic ionic liquids and porous PBI membranes: Preparation, characterization and MEA testing. Int. J. Hydrog. Energy, Apr 2012, **37**(8): 7221–7234. doi: 10.1016/j.ijhydene.2011.11.074.

[95] Niu, B., *et al.* Polybenzimidazole and ionic liquid composite membranes for high temperature polymer electrolyte fuel cells. Solid State Ion., 2021, **361**: 115569. doi: https://doi.org/10.1016/j.ssi.2021.115569.

[96] Taieb, A., Mukhopadhyay, S. and Al-Othman, A. Adaptive estimation of PEMFC stack model parameters – An experimental verification. Int. J. Hydrog. Energy, Jul 2022, doi: 10.1016/J.IJHYDENE.2022.05.215.

[97] Zhou, M., *et al.* Modeling the performance degradation of a high-temperature PEM fuel cell. Energies, Aug 2022, **15**(15). doi: 10.3390/en15155651.

[98] Wang, S., Lu, A. and Zhang, L. Recent advances in regenerated cellulose materials. Prog. Polym. Sci., Feb 2016, **53**: 169–206. doi: 10.1016/j.progpolymsci.2015.07.003.

[99] Sharma, P. R., *et al.* High aspect ratio carboxycellulose nanofibers prepared by nitro-oxidation method and their nanopaper properties. ACS Appl. Nano Mater., Aug 2018, **1**(8): 3969–3980. doi: 10.1021/acsanm.8b00744.

[100] Lee, C. M., Kubicki, J. D., Fan, B., Zhong, L., Jarvis, M. C. and Kim, S. H. Hydrogen-bonding network and oh stretch vibration of cellulose: Comparison of computational modeling with polarized IR and SFG Spectra. J. Phys. Chem. B, Dec 2015, **119**(49): 15138–15149. doi: 10.1021/acs.jpcb.5b08015.

[101] Muhmed, S. A., *et al.* Emerging chitosan and cellulose green materials for ion exchange membrane fuel cell: A review. Energy, Ecol. Environ., 2020, **5**(2): 85–107. doi: 10.1007/s40974-019-00127-4.

[102] Tawalbeh, M., Al-Othman, A., Ka'ki, A., Farooq, A. and Alkasrawi, M., "Highly proton conductive membranes based on lignin/ZrP/PTFE composite for high temperature PEM fuel cells." in *2022 Advances in Science and Engineering Technology International Conferences, ASET 2022*, Feb. 2022, 1–5. doi: 10.1109/ASET53988.2022.9734834.

[103] Shirdast, A., Sharif, A. and Abdollahi, M. Effect of the incorporation of sulfonated chitosan/sulfonated graphene oxide on the proton conductivity of chitosan membranes. J. Power Sources, Feb 2016, **306**: 541–551. doi: 10.1016/j.jpowsour.2015.12.076.

[104] Yue, L., *et al.* Sulfonated bacterial cellulose/polyaniline composite membrane for use as gel polymer electrolyte. Compos. Sci. Technol., Jun 2017, **145**: 122–131. doi: 10.1016/j.compscitech.2017.04.002.

[105] Thomas, B., *et al.* Nanocellulose, a versatile green platform: From biosources to materials and their applications. Chem. Rev., 2018, **118**(24): 11575–11625. Dec doi: 10.1021/acs.chemrev.7b00627.

[106] Sharma, P. R. and Varma, A. J. Functionalized celluloses and their nanoparticles: Morphology, thermal properties, and solubility studies. Carbohydr. Polym., Apr 2014, **104**: 135–142. doi: 10.1016/j.carbpol.2014.01.015.

[107] Li, S., *et al.* Sustainable plant-based biopolymer membranes for PEM fuel cells. Int. J. Mol. Sci., 2022, **23**(23): 15245. Dec doi: 10.3390/ijms232315245.

[108] Nelson, K., *et al.* American process: Production of low cost nanocellulose for renewable, advanced materials applications. Springer Ser. Mater. Sci., 2016, **224**: 267–302 Springer Verlag doi: 10.1007/978-3-319-23419-9_9.

[109] Bayer, T., *et al.* High temperature proton conduction in nanocellulose membranes: Paper fuel cells. Chem. Mater., Jul 2016, **28**(13): 4805–4814. doi: 10.1021/acs.chemmater.6b01990.

[110] Jankowska, I., Pankiewicz, R., Pogorzelec-Glaser, K., Ławniczak, P., Łapiński, A. and Tritt-Goc, J. Comparison of structural, thermal and proton conductivity properties of micro- and nanocelluloses. Carbohydr. Polym., Nov 2018, **200**: 536–542. doi: 10.1016/j.carbpol.2018.08.033.

[111] Smolarkiewicz, I., *et al.* Proton-conducting microcrystalline cellulose doped with imidazole. thermal and electrical properties. Electrochim. Acta, Feb 2015, **155**: 38–44. doi: 10.1016/j.electacta.2014.11.205.

[112] Lu, Y., Armentrout, A. A., Li, J., Tekinalp, H. L., Nanda, J. and Ozcan, S. A cellulose nanocrystal-based composite electrolyte with superior dimensional stability for alkaline fuel cell membranes. J. Mater. Chem. A, 2015, **3**(25): 13350–13356. doi: 10.1039/C5TA02304A.

Wafa Suwaileh*, Soheil Zarghami, Saima Farooq, Nipa Roy,
Mohammad Boshir Ahmed

Chapter 5
Polymer membranes for catalysis

Abstract: Polymeric membranes are a promising class of materials for catalysis applications. They offer a number of advantages over traditional inorganic membranes, including lower cost, easier fabrication, and better compatibility with biocatalysts. However, polymeric membranes are also more susceptible to chemical and thermal degradation, and they typically have lower permselectivity than inorganic membranes. In this chapter, we discuss the different types of polymeric membranes that are used for catalysis applications, and we review the recent advances in the development of new polymeric membranes with improved catalytic performance. We also discuss the challenges and opportunities for the use of polymeric membranes in catalysis applications.

Keywords: polymeric membranes, catalysis applications, degradation susceptibility, permselectivity, recent advances

1 Introduction

In the past two decades, with the growing demand for higher conversions, increased yields, and enhanced selectivity in producing appetite products while meeting energy-saving requirements, polymeric membranes (PMs) have led to the milestone of innovative configurations and the desired design of catalytic membrane materials [1]. Toward this end, which enables simultaneous reaction and separation, they have received great attention in both industrial and academic research. Catalytic membranes are a particular kind of versatile materials, and these are currently being used in many chemical reactions worldwide. The most popular method of employing catalytic membranes is the removal of a certain product to alter the existing thermodynamic constraints on the high yield and selectivity of the desired product and hence accelerate the equilibrium-limited reactions toward completion. The potential benefits of the catalytic membrane include: (i) enabling an extended range of allowable temperature

---

*Corresponding author: Wafa Suwaileh**, Chemical Engineering Program, Texas A&M University at Qatar, Education City, Doha, Qatar, e-mail: wafa.suwaileh@qatar.tamu.edu
Soheil Zarghami, Department of Chemical Engineering, Isfahan University of Technology, Iran
Saima Farooq, Department of Biological Sciences and Chemistry, College of Arts and Science, University of Nizwa, Oman
Nipa Roy, Department of Physics, Yeungnam University, Gyeongsan, 38541, Republic of Korea
Mohammad Boshir Ahmed, School of Engineering, Edith Cowan University, Joondalup, Australia

https://doi.org/10.1515/9783110796032-005

and pressure for a chemical reaction [1], (ii) reduced occurrence of undesired side reactions, (iii) increased maximum residence time for a given reactor volume [2], and (iv) enhanced purification efficiency. The selection of membrane material for catalytic membranes is primarily determined by the reaction system, particularly when it involves hydrogen (e.g., hydrogenation/dehydrogenation reactions) and/or oxygen (e.g., synthesis of oxy-organic compounds).

A membrane is known as a "semipermeable barrier", which functions as a boundary or dividing layer or partition layer, selectively allowing the passage of certain things (e.g., ions, molecules, and small particles) while impeding the permeation flux of others [3]. The permeability and selectivity of catalytic membranes are critical parameters that are primarily governed by the membrane's layer configurations and porous structure, including factors such as porosity, tortuosity, and pore size. Other variables, including molecular weight, temperature, pressure, and transport mechanisms, also significantly impact the membrane's performance in terms of these properties. Membrane materials are a versatile technology with potential applications across various fields, including biotechnology, pharmaceutical, water and air purification, petrochemical sectors, energy, and environmental applications. In contrast, catalytic membranes are a combination of membranes with chemical reactions which have received great attention to offer advantages in several different instances [4], and separation of the undesired materials may also be carried out by their dematerialization via a chemical reaction [5, 6]. The vast array of catalytic membranes offers numerous advantages with respect to the polymeric materials, which can be used for membrane design, and many of them are less expensive. Hence, it is necessary for the proper selection of the material for a desired application [6]. Furthermore, the catalytic membrane can mitigate thermodynamic equilibrium limitations, achieve higher conversion rates, and obviate the need for additional costs associated with separation or recycling. They have also some limitations like requiring higher capital costs to install in some cases; hence, developing proper catalytic materials is still under development [1]. However, polymeric materials have potential applications in different industrial fields such as the pharmaceutical sector (e.g., for steam reforming of methane), biotechnology, chemical plants, energy (e.g., water splitting), and environmental fields (e.g., nitrate removal), and more sophisticated catalytic amperometric material development.

The utilization of membranes has been widespread across various fields [6], including the use of ceramic catalytic membrane reactors as a green and sustainable technology for heterogeneous catalytic processes, such as methane conversion, where the catalytic membrane functions as both the catalyst and separation material [7, 8]. In addition, membranes can also be used for catalytic degradation of different contaminants as well as pollutants. Therefore, it is necessary to choose a proper material combination to fabricate catalytic materials by considering several parameters like productivity, separation efficacy, selective separation, improved catalytic performance and separation performance, the lifetime of the catalytic membrane, reducing the operating costs, as well as mechanical and chemical perfections.

Considering the aforementioned factors, the development of a catalytic membrane catalyst and their composite remains challenge, despite the well-established use of membranes for several decades [2]. Catalytic membranes can be prepared through different processes so that the main strategy to prepare different kinds of catalytic membrane materials is to enhance the performance of the membrane as well as to degrade different kinds of pollutants simultaneously. For example, in membrane-embedded catalysts, the membrane is prepared in a way that contains the catalyst in its backbone. Another type is membrane-assisted catalysts where the catalyst is applied on the surface of the membrane using different techniques such as precipitation, grafting, and electrodeposition [5]. Based on these two approaches, many catalytic membranes can be fabricated. For example, metal nanoparticles (NPs) are incorporated into the membrane via an intermatrix synthesis process (i.e., a technique based on the immobilization of metal NPs) followed by chemical or electrochemical reduction methods [9]. Similarly, Domènech et al. [6] fabricated a porous membrane via an inner matrix synthesis process, where a palladium (Pd) NP-based catalytic membrane was developed for *p*-nitrophenol to *p*-aminophenol degradation.

The objective of this chapter is to provide a detailed discussion on various membrane materials for catalytic applications, their potential applications, and the possible strategies to enhance membrane catalytic performance.

2 Polymeric membranes for catalysis

A membrane typically acts as a selective barrier, allowing some substances or particles to flow through while blocking others. This behavior is linked to the inherent characteristics of membranes, including their pore shape and size, surface charge, porosity and hydrophobicity, as well as the membranes' geometry and dimensions [10]. The use of membranes in catalytic applications receives a dual blend of catalysis and selective separation simultaneously [11].

PMs are an emerging class of functionalized materials with a wide range of applications in diverse fields of environmental and chemical engineering, and biotechnology [12]. Recently, PMs have been increasingly popular in catalysis as they offer numerous benefits compared to their traditional catalytic counterpart systems, that is, enhanced surface/volume ratio, customizable surface properties, and adjustable pore sizes. PMs in catalysis potentially act as a catalytic membrane reactor (CMR) due to their capability to provide a suitable reaction environment by providing catalyst support and effective separation of reactants and/or products by diffusion-sorption mechanism [12, 13].

PMs are versatile due to their peculiar characteristics such as elastic nature, good sorption and diffusion coefficients, incorporation of a variety of catalysts such as nanosized metallic clusters and complexes, activated carbon, and zeolites [14]. They are privileged over inorganic membranes due to their ability to be used in chemical synthesis involving low reaction temperatures (room temperature to ≤170 °C) for both liquid- or gas-phase reaction mixtures [15].

They can also be produced in a variety of shapes, for example, spiral, hollow, or sheets. Their pertinent ability to create a well-defined porous matrix that may act as a catalyst support is helping to construct single-site catalysts with active sites that are closely similar to one another [13, 16].

The physical nature of the catalyst divides the catalysis into homogeneous and heterogeneous [12]. Homogeneous catalysis, a similar physical state of reaction mixture and catalyst, imparts incredible benefits such as effective contact of reactants with catalyst, high catalytic selectivity, and remarkable catalytic efficiency by working at the molecular level. But at the same time, catalyst separation becomes a soaring issue in reaction engineering that needs serious attention. On the other hand, heterogeneous catalysis having different phases of the catalysts and reactants offers benefits of easy catalyst separation, profound catalyst regeneration and recyclability, and good stability but suffers from reduced catalytic efficiency as well as catalyst selectivity. Despite these pertinent disadvantages of heterogeneous catalysis, it is by far the most widely practiced catalytic system in reaction engineering and industry [17].

Generally, the PMs in catalysis are utilized as catalytic PM reactors (CPMRs), where it acts as a single unit incorporating both separation and reaction simultaneously. This integration of both processes also names the CPMR strategy as *reactive separation* [15, 17]. It usually works in two broad categories: (i) catalytic reactions taking place on the membrane surface and (ii) catalytic reactions within the membrane matrix. By having catalytic species affixed to the membrane surface in the surface-catalyzed reactions, PM can increase the surface area for catalytic reactions and selectively adsorb reactants and/or desorb the products. Alternatively, PMs serve as a supporting platform to immobilize and stabilize the catalyst for reactions to take place within their matrix which boosts both selectivity and activity of reactions. This approach also effectively contributes to customizing both the recycling and usage of catalysts, rendering them an incredible choice to the traditional catalytic reaction systems [18].

PM as a catalytic support offers numerous benefits in terms of its affinity for reagents, which is the primary characteristic of polymers [19]. A hydrophobic environment is created for (substrate) binding by the hydrophobic membrane's barrier function, which isolates and regulates both the substrate and oxidant's access to the active site. This prevents the accumulation of extra polar substances nearby. As a result, metalloporphyrins immobilized on PMs can function as shape-selective catalysts by isolating the catalytic complex and preventing its inactivation by accumulation or self-oxidation [17, 19].

The catalyst immobilization on PMs represents a sophisticated hybrid approach to garnering the positive aspects of both homogeneous and heterogeneous catalyses [20]. The outstanding features of CPMRs operating in a continuous-flow reaction mode have proven them to be the best alternative to the traditional catalytic system. In continuous-flow reaction mode, as the name implies, the input reactants continuously get converted into products by passing through the membrane having an immobilized catalyst. Here the catalyst, despite permanently residing in the reaction medium, does not get consumed during the reaction; hence, a minimal catalyst quantity is used. Therefore, the catalyst efficiency drastically improves and also the waste production potentially reduces [19].

2.1 Catalytic polymeric membranes in organic synthesis

2.1.1 One-step production of phenol by direct hydroxylation of benzene

One-step phenol production by direct hydroxylation of benzene has been considered one of the most difficult oxidation reactions and challenging methods of chemical transformation [21].

Because phenol is more highly reactive to oxidation than benzene, this results in the formation of a wide range of oxidative by-products leading to reduced selectivity and potentially poor reaction yields. Thanks to CPMR methodology that has generously facilitated such difficult chemical transformations by providing controlled long-term contact time of substrate with the catalyst resulting in dramatically high selectivity and reaction yields. The contact time can efficiently be controlled by passing the reactant mixture (substrate and oxidant) through the membrane filled with the catalyst at variable permeate flow rates [19]. In the presence of hydrophobic membrane support, the reaction occurs at the membrane's aqueous side interface, whereas by employing a hydrophilic catalyst, oxidation occurs at the membrane's organic side interface. The crunch of this process is the CMR, having a hybrid role of residing in the catalyst and facilitating/monitoring the reaction in the presence of the catalyst [22].

Recently, an ultrafiltration (UF) polyvinylidene fluoride (PVDF) immobilized with copper oxide (CuO) nanocatalyst membrane reactor was employed by Molinari et al. [14] as a continuous-flow mode for liquid-phase direct hydroxylation of benzene to produce phenol in the presence of hydrogen peroxide (H_2O_2) as oxidant and dimethylacetamide as solvent. High phenol concentration (2.4 g/L) and efficient benzene conversion to phenol (2.3%) were observed at a contact time of 19.4 s at 35 °C using CuO-filled PVDF membrane as compared to CuO catalyst or PVDF membrane only. Traces of reaction by-products such as biphenyl and benzoquinone were detected by only implying the potential of catalyst-filled PMs. Not only the contact time of the catalyst with the reaction media was enhanced by the addition of CuO catalyst but the permeability of the hydrophobic membrane was significantly improved by its hydrophilic character. Catalytic leaching and deactivation of the catalytic membrane were also observed after 1 h post reaction. This was overcome in their next study [22] using vanadium-filled PVDF membrane, where vanadyl acetylacetonate was used as a source of vanadium(IV) whose hydrophobic organic ligand had a higher affinity with the PVDF. This stable catalytic membrane made by 3% w/w catalyst had asymmetric structure with an average narrow pore diameter (26–32 nm). The results showed efficient phenol selectivity (100%) and 3.6% yield within the contact time of 5.3 s at transmembrane pressure of 0.01–0.03 MPa in the presence of H_2O_2 oxidant.

Several factors such as membrane type, types of oxidants (nitrous oxide, oxygen, and H_2O_2), and reducing agents such as ascorbic acid, acetic acid, and trifluoroacetic acid are also considered as dominating factors to obtain the best performance. Another study [23] reported 98% phenol selectivity, 1.2% benzene conversion, and 38.2 mmol/g_{cat}/h phenol productivity by employing iron(II) sulfate catalyst immobilized with polypropylene (PP) membrane within 4 h contact time in the presence of hydrogen peroxide oxidant (18 mmol). These studies emphasized the potential of CPMRs as an advancement to the phenol production reaction schemes, wherein further studies are in progress to enhance the reaction yield and selectivity by overcoming the reaction drawbacks.

2.1.2 Hydrogenation reactions

Hydrogenation reactions refer to the saturation of the olefins under catalytic condition [10]. Catalytic Pd NPs embedded in poly(acrylic acid)/polyethyleneimine (PAA/PEI) membranes were fabricated using layer-by-layer (LBL) deposition technique by Kidambi and Bruening [24] for hydrogenation of short-chain, unsaturated alcohols. The high turnover frequency (TOF) for hydrogenation reaction using reduced PAA/PEI/Pd(0)/$_n$PAA membrane demonstrated that selective diffusion of the reactant to the active site is the controlling factor for the catalytic selectivity.

Singh et al. [25] successfully performed hydrogenation of vegetable (soybean) oil with minimal transfatty acids (TFAs) over asymmetric polyetherimide (PEI) embedded with Pt metal catalyst. The PM reactor catalyzed oil with only 4 wt.% of total TFAs at 70 °C and 65 psi which is incredibly promising as compared to traditional Pt/C slurry reactor systems producing more than 10% TFA at harsh operating conditions (110–190 °C at 30–70 psi). The results proved the authenticity of utilizing a polymer/catalyst membrane reactor for catalyzed hydrogenation application at a wider scale.

The hydrogenation of methylene cyclohexane was investigated by Bottino et al. [26] over Pd NPs deposited on PVDF and ceramic alumina oxide (γ-Al_2O_3) membranes. Both membranes Pd-PVDF/PVP (polyvinylpyrrolidone) and Pd-Al_2O_3 showed better conversion selectivity but the ceramic membrane did not show catalytic activity due to a higher amount of catalyst loading to its thickness.

2.1.3 C–C Suzuki–Miyaura cross-coupling reactions

CPMRs have widely facilitated difficult organic transformations, one of which is the carbon–carbon (C–C) cross-coupling reaction known as the Suzuki–Miyaura reaction [27]. The utilization of CPMRs in a continuous-flow reaction mode provides enormous

benefits; for instance, effective mixing and temperature control, scalability, waste reduction, and energy efficiency.

Gu et al. [28] fabricated microfiltration (MF) polyethylsulfonate (PES) immobilized
with Pd NPs using imidazolium-based ionic liquid acting as a stabilizer and immobi-
lizer for the catalyst. Complete C–C Suzuki–Miyaura cross-coupling reaction was
achieved over catalyzed PES membrane reactor in a flow-through reaction mode
within 10 s contact time without catalyst leaching and reaction by-product formation.
The comparison of reaction mode revealed that the flow-through mode of reaction
resulted in three times higher apparent reaction rate compared to the batch reaction
mode. The results affirmed the possibility of a one-step C–C cross-coupling reaction
without any need for a catalyst separation step from the reaction mixture.

Faria et al. [29] successfully carried out Suzuki–Miyaura cross-coupling reactions
of phenylboronic acid coupling with haloarenes with excellent yield over Pd NP-
supported cellulose acetate polymeric membrane (CA/Pd) under green reaction conditions, that is, aqueous reaction media and phosphine-free. CA/Pd(0) membranes resulted in 96% reaction yield at 100 °C using potassium carbonate and water as a
solvent; however, the TOF progressively reduced to 17% conversion after the sixth reaction cycle due to thermal deformation of the membrane after prolonged reaction time.

The catalyst leaching problem was found to be overcome by the modification of
polymeric support. Fayyazi et al. [30] modified polysulfone (PS) membrane support by
chloromethylation and amination followed by incorporation of Pd NPs, which were
anchored by the amine groups to ensure their stability and immobilization. The membrane was tested for the coupling of iodobenzene with phenylboronic acid reaction
and was found to produce 63% coupling yield at 85 °C at 0.1 mL/cm^2/s flux. The membrane resulted in a negligible reduction of yield (<5%) after 10 cycles, implying its potential for Suzuki coupling reactions.

2.2 Catalytic polymeric membranes in environmental applications

2.2.1 Photocatalytic reactions

Recently, advanced oxidation processes, also known as photocatalytic degradation reactions, have received tremendous attention for the decontamination of environmental pollutants. Photocatalytic PMs have been playing a vital role in this realm both in
aqueous and gaseous media [5, 31]. Table 1 demonstrates a comparison of catalytic
polymeric membrane reactors for photocatalytic applications in organic synthesis.

Li et al. [32] fabricated a nanofiber photocatalytic polymeric membrane in a sandwich manner consisting of polyvinyl alcohol (PVA) layer as a middle layer covered by
polymethylmethacrylate (PMMA) and $H_4SiW_{12}O_{40}$ as top and bottom layers, respec-

Table 1: Comparison of catalytic polymer membrane reactors for organic synthesis applications.

Catalytic polymeric membrane reactor type	Application	Reference
PVDF immobilized with CuO nanocatalyst	Hydroxylation of benzene	[14]
PVDF immobilized with CuO nanocatalyst	Hydroxylation of benzene	[22]
PAA/PEI/Pd(0)/$_n$PAA Pd NPs embedded in PAA/PEI membranes	Hydrogenation of allyl alcohols	[24]
Asymmetric PEI embedded with Pt metal catalyst	Hydrogenation of vegetable (soybean) oil with minimal transfatty acids (TFAs)	[25]
PES/Pd NP microfiltration PES immobilized with Pd NPs	C–C cross-coupling Suzuki–Miyaura reaction	[28]
CA/Pd(0) Pd NP-supported cellulose acetate polymeric membrane	Suzuki–Miyaura cross-coupling reactions of haloarenes coupled with phenylboronic acid	[29]
Pd/PS membranes Modified PS membrane coupled with Pd NPs	Suzuki–Miyaura cross-coupling reactions of haloarenes coupled with phenylboronic acid	[30]

PVDF, polyvinylidene fluoride; CuO, copper oxide; PAA/PEI, poly(acrylic acid)/polyethyleneimine; PEI, polyetherimide; PES, polyethylsulfonate; Pd NPs, palladium nanoparticles; PS, polysulfone.

tively. The as-fabricated membrane provided 94% photocatalytic degradation of methyl orange (MO) dye with 30 min of UV irradiation, outperforming $H_4SiW_{12}O_{40}$/PMMA membrane (13.8%) or $H_4SiW_{12}O_{40}$ fiber (4.6%) owing to considerable water tolerance of the sandwiched membrane.

Ma et al. [33] prepared a novel visible-light active photocatalytic membrane of PP grafted with poly-ionic liquid (PIL) followed by complexation with polyoxometalate (POM). The nonwoven fabric (NWF) membrane of PP was initially modified with poly (4-vinylbenzene)-3-methylimidazolium chloride (PVBMC) and embedded with POM. The as-synthesized PP-g-PVBMC-POM membrane demonstrated efficient photodegradation (95%) of 20 mg/L acid red (AO7) within 120 min visible-light irradiation time with good recyclability and stability up to five cycles.

Dekkouche et al. [34] fabricated photocatalytic PMs by embedding anatase-TiO_2 photocatalyst on three different kinds of PMs such as polytetrafluoroethylene (PTFE), PS, and PVDF and tested them for the removal and photocatalytic degradation of 17α-ethinylestradiol (EE2) and diclofenac (DAF). All three membranes (T-PS, T-PVDF, and T-PTFE) resulted in more than 90% adsorptive removal and UV-LED (light-emitting diode) irradiated photodegradation efficiencies for both pollutants. The hydrophilicity of the membranes was found to enhance by the incorporation of titania which led to higher activities. The membrane recyclability and stability were investigated for three consecutive cycles, and T-PVDF was found to have maximum stability.

Cruz et al. [35] grafted PVDF membrane with polyacrylic acid followed by coating with TiO_2 NP as PVDF/PAA/TiO_2 and tested for photocatalytic degradation of reactive black 5 (RB5) dye, and self-cleaning and antifouling properties. The as-synthesized membrane exhibited 98.7% photocatalytic RB5 degradation within 5 h of UV irradiation at pH 1.5. PVDF/PAA/TiO_2/pH 1.5 showed 72% higher permeate flux as compared to neat titania. Fischer et al. [36] reported a simple method for impregnation of TiO_2 onto multi-layered catalytic PMs (a combination of two hydrophilic membranes, such as PVDF and PES, and a hydrophobic PVDF membrane) to avoid NP leaching and agglomeration by tetraisoperoxide hydrolysis. The as-synthesized membrane showed good photodegradation efficiency for methylene blue (MB), ibuprofen, and DAF (25 mg/L each) within 2.5 h of UV irradiation span implying the potential of this simple method of membrane fabrication for a wide range of applications.

2.2.2 Gas cleaning

Polymeric catalytic membranes have been widely utilized in various environmental applications, as elaborated in Table 2. A polymeric catalytic filter was prepared by Zheng et al. [37] by decorating polyphenylene sulfide (PPS) with a nanocoating of manganese oxide (MnO_2)/polypyrrole (PPy), where PPy acted as catalyst dispersant and binder. MnO_2/PPy/PPS membranes were used for selective catalyzed reduction of nitric oxide (NO) (500 ppm; 5% oxygen in NO) with ammonia (500 ppm), and 79.8–90% reduction efficiency was achieved at 140–180 °C.

Another study by Chen et al. [38] successfully deposited MnO_2-Fe_2O_3 cocatalysts over PPS membranes and tested them for NO conversion in the presence of ammonia.

Table 2: Comparison of catalytic polymer membrane reactors for environmental applications.

Catalytic polymeric membrane reactor type	Application	Process details	Reference
Sandwich photocatalytic polymeric nanofibrous membrane	Photocatalytic degradation of methyl orange dye	Complete degradation of 10 mg/L dye solution	[32]
PVA layer as a middle layer covered by PMMA and $H_4SiW_{12}O_{40}$ as the top and bottom layers			

Table 2 (continued)

Catalytic polymeric membrane reactor type	Application	Process details	Reference
PP-g-PVBMC-POM membrane Visible light active photocatalytic membrane of PP grafted with PIL followed by complexation with POM	Acid red (AO7) photocatalytic degradation	95% degradation of 20 mg/L acid red (AO7) within 120 min visible light irradiation time with good recyclability and stability up to five cycles	[33]
PVDF/PAA/TiO$_2$ PVDF membrane grafted with PAA followed by coating with TiO$_2$ nanoparticles	Photocatalytic degradation of RB5 dye, and self-cleaning and antifouling properties	The as-synthesized membrane exhibited 98.7% photocatalytic RB5 degradation within 5 h of UV irradiation at pH 1.5. PVDF/PAA/TiO$_2$/pH 1.5 showed 72% higher permeate flux as compared to neat titania	[35]
PVDF/PES/PVDF/TiO$_2$ Impregnation of TiO$_2$ onto multilayered catalytic polymeric membranes (combination of two hydrophilic membranes such as PVDF and PES, and a hydrophobic PVDF membrane)	Photocatalytic degradation of methylene blue dye, ibuprofen, and diclofenac drugs	50% degradation of (25 mg/L each) within 2.5 h of UV irradiation span	[36]
(T-PS, T-PVDF, T-PTFE) Anatase TiO$_2$ embedded onto PTFE, PS, and PVDF	Adsorptive removal and photocatalytic degradation of EE2 and DAF	90% adsorptive removal and UV LED-irradiated photodegradation efficiencies for both pollutants	[34]
MnO$_2$/PPy/PPS membranes PPS decoration with a nanocoating of MnO$_2$/PPy, where PPy acted as catalyst dispersant and binder	Selective catalyzed reduction of NO	79.8–90% reduction efficiency was achieved at 140–180 °C 500 ppm NO; 5% oxygen in NO with ammonia (500 ppm)	[37]
MnO$_2$–Fe$_2$O$_3$/PPS MnO$_2$–Fe$_2$O$_3$ cocatalysts over PPS membranes	Selective catalyzed reduction of NO in the presence of ammonia	99% reduction efficiency was achieved at 140–180 °C 500 ppm NO; 5% oxygen in NO with ammonia (500 ppm)	[38]

PVA, polyvinyl alcohol; PMMA, polymethylmethacrylate; PP, polypropylene; PIL, poly-ionic liquid; POM, polyoxometalate; PVDF, polyvinylidene fluoride; PAA, polyacrylic acid; RB5, reactive black 5; PES, polyethersulfone; PTFE, polytetrafluoroethylene; PS, polysulfones; EE2, 17α-ethinylestradiol; DAF, diclofenac; LED, light-emitting diode; PPS, p; MnO$_2$, manganese oxide; PPy, polypyrrole; NO, nitric oxide.

Efficient conversion (99%) of 500 ppm of NO was achieved at 140–180 °C over the membrane filter demonstrating its capability for air cleaning.

2.3 Nanoparticle (NP) decorated catalytic polymeric membrane reactors

NPs or their oxides (Au, Ag, Fe, Al, Mg, Ti, Zr, and Si) have been extensively incorporated to modify the surface or matrix of PMs, which impact their permeability, thermal and mechanical stability, selectivity, conductivity, and antimicrobial activity making them efficient catalytic reactors. These membrane reactors have widely been employed in organic synthesis, wastewater decontamination, fuel cells, and so on [39, 40], and this information is further elaborated in Table 3.

Zhou et al. [41] prepared Au-NP-immobilized nanoporous DMAEM-b-PS membrane, a block copolymer membrane made up of poly(2-dimethylaminoethylmethacrylate) (PDMAEM) and polystyrene (PS). The selective swelling and subsequent evaporation of block copolymer membrane in hot ethanol resulted in a nanoporous structure with adjustable pore size and well-defined geometry. The Au-NP-immobilized block copolymer membrane reactor (PDMAEM-b-PS/Au-NPs) was tested for degradation of toxic MO and rhodamine B (RB) dyes and hydrogenation of 4-nitrophenol (4-NP) in a continuous flow-through reaction mode. The catalytic degradation efficiency of MO and RB was achieved up to 88% and 91% after 6 h, whereas 4-NP was 100% converted to 4-aminophenol at a flow rate of 0.5 mL/min with remarkable recyclability of up to eight cycles. Silver NP-immobilized microporous PP membranes (MPPM/Ag-NPs) were fabricated in the presence of polydopamine (PDA) as a reducing agent by Hu et al. [42] and tested for the degradation of MB dye solution in a continuous-flow reactor mode. MPPM/Ag-NPs exhibited excellent catalytic activity by completely decolorizing MB solution within 2 s at a flow rate of 210 L/m^2/h with good stability and recyclability up to five cycles.

Fang et al. [43] employed an in situ reduction and blending method to fabricate Ag-NP-immobilized PES UF catalytic membranes (UCMs) and evaluated them for catalytic reduction of 4-NP and separation of macropollutants, that is, humic acid (HA) and bovine serum albumin (BSA). Polyphenol tannic acid (TA) acted as both a blending reducing agent and stabilizing agent for immobilized Ag-NPs on the membrane surface. UF PES-TA/Ag-NP membranes having a porosity in the range of 93.2–94.7 and pore size of 14.2–15.8 nm exhibited pure water permeability (239.8 L/m^2.h), excellent rejection for HA and BSA as 87.3% and 96.1%, respectively, and efficient catalytic degradation ability (98%) of 4-NP in a dynamic catalytic reaction mode. UF PES-TA/Ag-NP membranes proved to have outstanding separation and catalytic abilities with enhanced stability and recyclability up to seven consecutive cycles as compared to neat PES or PES/TA membranes.

Ozone-resistant PVDF hollow-fiber MF catalytic membrane and UCMs immobilized with iron oxide NPs supported over powdered activated carbon were fabricated by Li and Yeung [44]. The CMR is packed up in a small volume with a large (7,000 m^{-1}) surface/volume ratio consisting of three different compartments as membrane distributor, membrane contractor, and membrane separator. This compact catalytic reactor exhibited outstanding *N,N*-diethyl-*meta*-toluamide (DEET) conversion (60%) within 4 min contact time and 30% total organic content (TOC) reduction compared to a semi-batch ozone reactor with 20% DEET conversion and 5% TOC reduction.

Ouyang et al. [45] decorated MF membranes of PS and PES LBL coated with poly (sodium 4-styrene sulfonate) (PSS) and poly(allylamine hydrochloride) (PAH) with Au-NPs and investigated their performance for the catalytic reduction of 4-NP with sodium borohydride. The catalytic activity and stability of synthesized MF hollow-fiber PS and PES membranes modified with PSS/PAH/Au-NPs were evaluated. It was found that PS-modified PSS/PAH/Au-NP membrane showed a better catalytic reduction of 4-NP (99%) and good stability (95–98%) for up to 4 h post reaction with negligible Au leaching as compared to PES-modified PSS/PAH/Au-NP membrane due to its high permeability. Hence, LBL deposition of Au-NPs is a useful technique in membrane technology for catalytic hollow-fiber preparation.

Table 3: Comparison of NP-decorated catalytic polymer membrane reactors for different applications.

Catalytic polymeric reactor type	Application	Process details	Reference
PDMAEM-b-PS/Au-NPs PDMAEM and PS	Catalytic degradation of toxic MO and RB dyes and hydrogenation of 4-NP	Continuous flow-through reaction mode Flow rate: 0.5 mL/min MO and RB were degraded up to 88% and 91%, whereas 4-NP was 100% converted to 4-AP Recyclability: up to 8 cycles	[41]
MPPM/Ag-NPs Microporous polypropylene membrane-immobilized silver NPs	Catalytic degradation of MB dye solution	Complete discoloration of MB solution within 2 s at a flow rate of 210 L/m^2/h Good stability and recyclability up to 5 cycles	[42]
Ultrafiltration PES/Ag-NP membranes	Catalytic reduction of 4-nitrophenol and separation of macropollutants, i.e., HA and BSA	Pure water permeability (239.8 L/m^2.h), excellent rejection for HA and BSA at 87.3% and 96.1%, respectively, and efficient catalytic degradation ability (98%) of 4-NP in a dynamic catalytic reaction mode	[43]
Ozone-resistant PVDF microfiltration and ultrafiltration catalytic membranes immobilized with iron oxide NPs supported over powdered activated carbon	Ozone treatment of DEET in water	Compact catalytic reactor DEET conversion (60%) 4 min contact time and 30% TOC reduction	[44]

Table 3 (continued)

Catalytic polymeric reactor type	Application	Process details	Reference
PS/PAH/Au-NPs PS and PES layer-by-layer coated with PSS and PAH with Au-NPs	Catalytic hydrogenation of 4-NP	99% conversion of 4-NP Catalytic activity (95–98%) and negligible Au leaching <5 ppb 4 h post-reaction Permeate flux is 1.4 ± 0.2 mL/(cm^2 min) under a pressure of <0.2 bar	[45]

PDMAEM, poly(2-dimethylaminoethyl methacrylate); PS, polystyrene; MO, methyl orange; RB, rhodamine B; 4-NP, 4-nitrophenol; 4-AP, 4-aminophenol; MB, methylene blue; MPPM, microporous PP membranes; PES, polyethylsulfonate; HA, humic acid; BSA, bovine serum albumin; PVDF, polyvinylidene fluoride; DEET, *N,N*-diethyl-*meta*-toluamide; TOC, total organic content; PS, polysulfone; PES, polyethersulfone; PSS, poly(sodium 4-styrene sulfonate); PAH, poly(allylamine hydrochloride).

3 Membrane materials for catalytic applications

The development of the chemical industries with respect to reducing costs in manufacturing and obtaining a sustainable high-quality product with fewer environmental impacts became a merging of economic interests worldwide. Catalytic membrane materials are emerging as alternative solutions to catalyst challenges in different industrial fields, where the reaction and separation can be combined into a single unit. In addition, catalytic membranes follow the process strengthening strategy, which is a state-of-the-art alternative for sustainable growth depending on the design strategy to obtain profits in manufacturing and processing. The benefits of this strategy are based on several factors, including the reduction of industrial equipment size, energy consumption, environmental impact, decrease in capital costs, environmental impact, increase in the efficacy of the industrial plant, safety controls, and industrial automatization improvement [46]. The selection of the catalytic membrane is influenced by several parameters, such as selectivity, permeability, productivity, membrane lifespan, mechanical and chemical optimization under operating conditions, and cost [47].

As specified, the catalytic membranes can be classified into two categories: (i) organic (polymeric) and (ii) inorganic (ceramic and metal) based on the chemical process requirements. Generally, PMs function at temperatures below 300 °C, whereas inorganic membranes can operate at high temperatures. They indicate a broad range of pH tolerance and high resistance to chemical deterioration. Inorganic membranes have primarily been used in reactions containing small molecules or for high-temperature decomposition reactions [48]. Furthermore, catalytic membranes can be categorized based on their roles as extractor, distributor, and contactor. Extractor catalytic membranes are designed to collect and/or remove one of the products from the reaction medium, thereby increasing the reaction yield. Distributor catalytic membranes' role is to enter one of the reactants, especially gases, in dosing which may react in successive reactions. Also, the contactor catalytic membrane's role is to intimate excellent contact between reactants and catalysts, where reactants are fed separately on either side of the membrane or by forcing them to pass together through catalytic pores of membranes [28, 48]. Despite the excellent qualities that distinguish catalytic membranes, there are some drawbacks such as high costs associated with the fabrication of catalytic membranes. The advantages and disadvantages of catalytic membranes are illustrated in Table 4.

Table 4: Some advantages and disadvantages of catalytic membranes [5].

Catalytic membrane type	Advantages	Disadvantages
Organic catalytic membranes (polymeric catalytic membranes)	– Compared to inorganic and metallic membranes, the technology to fabricate catalytic polymeric membranes is much more advanced – The cost of polymeric membranes is lower than that of metallic or ceramic membranes – Polymeric membranes mainly operate at low temperatures – Less restrictive requirements for the materials used in the construction of the modules	– A membrane with a limited life span – The cost of the membrane preparation and modules – Modeling and prediction may be more challenging depending on the complexity of the process and its chemical reactions – Polymeric membranes have less resistance to high temperatures and aggressive chemicals than inorganic and metallic membranes
Inorganic catalytic membranes (ceramic and metallic)	– High thermal stability – Resistance to high pressure – Resistance to aggressive environments	– The cost of capital is very high – Brittleness – At high temperatures, difficult to membrane-to-module sealing – At medium temperatures, the low permeability of the highly selective (dense) membranes

PMs can be made from rubbery or glassy polymers. The polymer passes from the glassy state to the rubbery state when an amorphous polymer is heated to a temperature above the rubbery transition temperature, T_g. Glassy polymers have high selectivity and lower permeability, whereas rubbery polymers typically have high permeabilities for gases, but relatively low selectivity, which allows them to obtain high purity [49]. A list of commercial polymers that are used as membrane materials with potential applications is given in Table 5.

Polymeric catalytic membranes are mainly of two types: (i) membrane-embedded catalysts and (ii) membrane-assisted catalysts. In the first type, the catalysts are directly embedded in the membrane's backbone, whereas in the second type, catalysts are applied to the membrane's surface through grafting or precipitation (deposition) processes, and the membrane serves as a catalyst carrier. In the case of membrane-embedded, the catalyzation of the PM can be achieved via the blending process. Using various blending techniques, the blending process can be carried out between polymers with a liquid catalyst; for example, sulfuric acid, or solid catalysts such as nanomaterials, like TiO_2, Al_2O_3, and different kinds of resins. Blending membranes with nanomaterials can provide several benefits, including [50, 51]: (a) controlled nanosize of membrane pores due to different interactions between nanomaterials and polymers; (b) lower agglomeration of the polymeric/nanomaterial blend membrane; (c) modification of the electronic and optical responses of the membrane due to the polymer forming a dielectric environment around the nanomaterials; and (d) increased transport of molecules through the membrane and improved mechanical stability due to the embedding of nanomaterials into the membrane matrix. For example, by blending polymeric solution with Fe_2O_3 and multiwalled carbon nanotubes (NTs), a porous PVDF catalytic membrane was developed, which efficiently removed HA through the catalytic oxidation of organic pollutants [52].

Table 5: Different polymeric membranes and their potential applications [5, 15].

Materials	Membrane formation	Potential application
CA and its derivatives	Casting/melt extraction	RO, UF, MF, GS, FO
CN	Casting	UF
CTA	Casting	RO
PP	Melt extraction	PV
PTFE	Melt extraction	UF, MF, ED, MD
PAN	Casting	MF, MD
PE	Melt extraction	UF
PVA	Casting	D, PV
PVDF	Casting	UF, MF, ED
PVC	Casting	MF
PMMA	Casting	D
PP	Melt extraction	MF, MD
PET	Melt extraction	MF
PMP	Melt extraction	GS
PC	Casting	D, MF
PPO	–	UF, GS
PA	Melt extraction	UF
PDMS	Casting	GS, PV
Polysulfone	Casting	UF
PEI	Casting	GS

CA, cellulose acetate; CN, cellulose nitrate; CTA, cellulose triacetate; PP, polypropylene; PTFE, polytetrafluoroethylene; PAN, polyacrylonitrile; PE, polyethylene; PVA, poly(vinyl alcohol); PVDF, poly(vinylidene fluoride); PVC, poly(vinyl chloride); PMMA, polymethyl methacrylate; PP, polypropylene; PET, poly(ethylene terephthalate); PMP, poly(methyl pentenal); PC, polycarbonate; PPO, poly(phenylene oxide); PA, polyamide; pDMS, polydimethylsiloxane; PEI, poly(ether imide); RO, reverse osmosis; UF, ultrafiltration; MF, microfiltration; GS, gas separation; FO, forward osmosis; PV, pervaporation; ED, electrodialysis; MD, membrane distillation; D, desalination.

In the case of membrane-assisted catalysts, the catalyzation of the PM can be achieved by two processes: the grafting process or the precipitation (deposition) process [49]. Grafting is a process that modifies the surface of the polymer by covalently bonding monomers onto the polymer chain. Features that can influence grafting include the nature of the backbone, solvent, monomer additives, initiator, and temperature. The nature of the polymer backbone is a significant factor in determining the success of grafting, as it can either facilitate or hinder the grafting reaction. Grafting techniques can be divided into two basic types: (a) grafting with a single monomer, where the grafting occurred in a single step, and (b) grafting with a mixture of two (or more) monomers, in this case, the two various monomers are grafted side by side to get the required specification, and this type of grafting can produce the bipolar membranes [49]. The precipitation process is based on the polymerization of an oligomer mixture to form a coating that adheres to the substrate by physical forces. For example, wang et al. [53] examined a novel method to deposit the NPs on the PM surface using the roll-to-roll process [53]. Polymer membranes have low fabricating costs due to the use of the roll-to-roll fabrication process [54, 55]. Also, this process can be applied for NP alignment on membrane support and pattern printing on membrane films [56]. Mashentseva et al. [57] studied the effect of the deposition of Au- and Ag-based NTs on the polyethylene terephthalate track-etched membranes [57]. Recently, embedding noble gold and silver NTs in porous membranes has become a significant trend in functional composites to their unique properties and potential application in sensors [58] or electrochemical electrodes [59], in heterogeneous catalysts [60], and in flow-through reactors, also in medicine [57, 61].

4 Fabrication strategies

4.1 Catalysts incorporated into the membrane matrix

Incorporation of NPs into different polymeric matrix is one of the most popular method for enhancing membrane properties such as surface roughness, surface wettability, and fouling resistance. Also embedding NPs inside the polyamide (PA) active layer for fabrication of nanomaterial-modified PA thin-film composite hybrid membranes is a promising approach for designing high-performance membranes [62, 63].

The dispersion of catalysts in the membrane polymer matrix was found to be effective in degrading effluents and in the esterification of acetic acid by ethanol. Ndlovu et al. [64] immobilized Pd-NPs in beta-cyclodextrin-modified PVDF membranes and completely degraded MO azo dye in sodium borohydride.

Mixed matrix membrane reactors for esterification reaction with ethanol have been made from entrapment of metal-organic framework MIL-101(Cr) and 4A zeolite particles within polyimide Matrimid® and sodium alginate matrix membrane, respectively [65, 66]. The fabrication of mixed matrix membranes for esterification reactions assisted by pervaporation has been a topic of interest, and readers are referred to a published review paper [67].

Besides the use of NP fillers, progress has also been made in using polymers as a hydrophilicity-enhancer filler for esterification of ethanol and acetic acid in a membrane reactor. Recently, the blending of sulfonated cross-linked chitosan solution with the sulfonated PES (SPES) solution was used to prepare high hydrophilic catalytic pervaporation membrane reactor through nonsolvent-induced phase inversion method for esterification of acetic acid by ethanol [68].

Wang et al. [69] fabricated a self-cleaning catalytic membrane for water treatment. First, they applied a combination of Cu(0)-mediated reversible deactivation radical polymerization and dicyclohexylcarbodiimide coupling reaction to synthesize ferrocene-containing polymers. They prepared a casting solution of PS and added PS-b-PHEA and PSFb-PFc as hydroxyl-containing block polymer and ferrocene-containing copolymer, respectively. The solution was cast on a glass plate and then immersed into a tank filled with deionized (DI) water. The addition of PSF-b-PHEA into PS casting solution increased the hydrophilicity of the membranes. The PSFb-PFc was also used as the Fenton reagent, and the obtained catalytic membranes showed high activity in catalyzing Fenton's reactions.

4.2 Catalyst coating on the membrane surface

Didaskalou et al. [70] used this method to fabricate polybenzimidazole (PBI)-anchored catalytic membrane prepared via an azide–alkyne cycloaddition-integrated synthesis –separation. In this regard, a PBI nanofiltration (NF) membrane was prepared by phase inversion. Then the presynthesized organocatalyst (azido-derivatized cinchona-squaramide bifunctional) was covalently grafted onto the NF membrane surface.

Researchers also utilized the surface grafting method for preparation of functionalized catalytic membranes for esterification by pervaporation. Zhang et al. [71] developed ionic liquid-functionalized PVA catalytic composite membranes. They used free radical polymerization method for grafting between PVA and acidic ionic liquids.

Coating of catalytic NPs onto the membrane surface was found to be promising for developing catalytic membranes. Membrane surface coating with hydrophilic manganese dioxide (MnO_2) NPs through in situ cocasting resulted in dual-layer MnO_2-PVDF composite UF membrane with antifouling and catalytic properties (Figure 1) [72].

Novel catalytic PMs with antimicrobial capability were developed by high-voltage electrospraying of a polymeric solution (N-methyl-2-pyrrolidinone (NMP), PVP–PSF containing Ag NPs (60–150 nm) on a commercial PES MF membrane [73].

It is worth noting that fabrication of mussel-inspired PDA intermediate layer was found to be effective in introducing NPs onto membrane surfaces [74, 75]. So a combina-

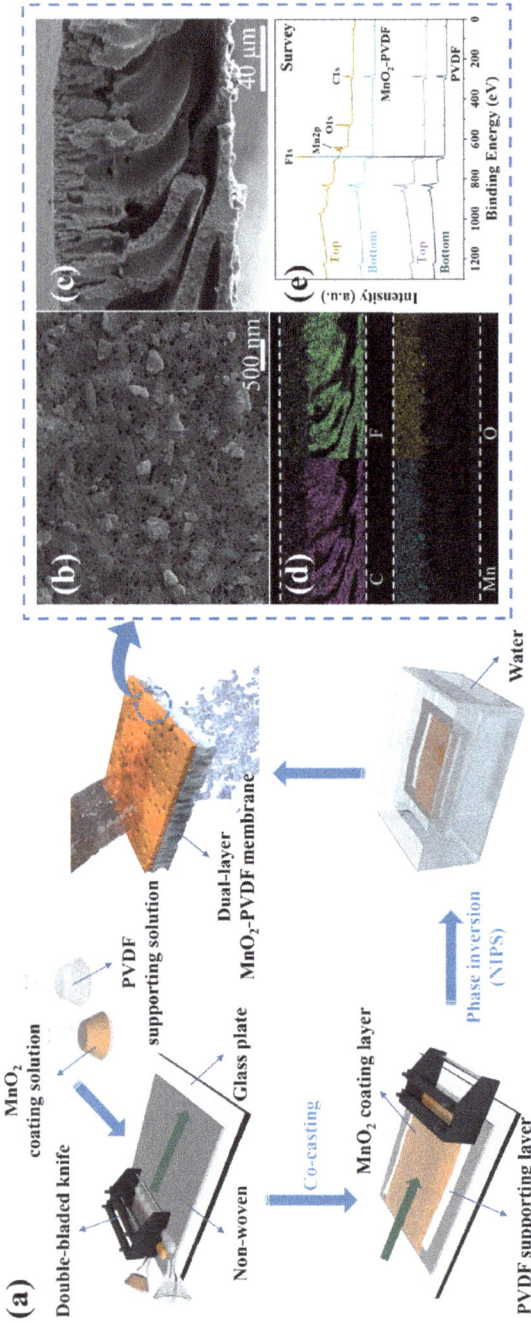

Figure 1: Synthesis and characterization of dual-layer MnO_2-PVDF composite membrane. (a) Preparation route of MnO_2-PVDF membrane. (b) SEM top surface image of MnO_2-PVDF membrane. (c) SEM cross-sectional image of MnO_2-PVDF membrane. (d) Elemental mapping of carbon, fluorine, manganese, and oxygen within the cross section of MnO_2-PVDF membrane. (e) XPS full survey spectra of PVDF and MnO_2-PVDF membranes [72].

tion of high-voltage electrospinning technique and dopamine chemistry was also used to develop electrospun nanofibrous membrane for catalytic removal of contaminants [76].

Wang et al. [76] fabricated conventional polyacrylonitrile (PAN) nanofibrous membranes. Then they constructed PDA microspheres on the PAN nanofibrous membrane. Finally, they immobilized silver NPs on PDA-coated PAN nanofibrous membrane surface via immersion in $AgNO_3$ aqueous solution with continuous shaking.

Following the progress in PDA coating application for development of the catalytic PMs, Sun et al. [77] coated the surface of commercial PVDF membranes (pore size 0.22 μm) with dopamine by immersion in dopamine/tris-HCl buffer solution. Subsequently, vacuum filtration of $FeOCl/MoS_2$ solution followed by aqueous solution of polyethylene glycol/glutaraldehyde was applied.

Recently, Yang et al. [78] fabricated PDA/PEI intermediate layer on the pristine hydrophilic NWF MF membrane surface. Then, the copper sulfate ($CuSO_4$) solution was filtered through the PDA/PEI-NWF membrane. After filtration, the obtained membrane was immersed in $CuSO_4$ solution and ammonia was added dropwise. Finally, the obtained collection was put in oven (at 80 °C for 3 h) for covering copper oxide (CuO) NPs onto the membrane surface and the membrane pores' walls through hydrothermal reaction.

A two-step fabrication approach was utilized for the synthesis of UF Pd-immobilized catalytic membrane. Zhang et al. fabricated PDA coating on commercial PES membranes (effective pore size 0.03 μm). Then, the PDA-modified PES membranes were immersed in $PdCl_2$ solution (1 g/L) for 30 min, followed by soaking in 2 mL $NaBH_4$ solution (5.33 g/L) to fabricate Pd-immobilized membranes [79].

Bioinspired coating with other materials containing catechol groups such as TA, levodopa (L-dopa), and proanthocyanidins has been used for preparing membranes with special wettability and antifouling properties [74, 75]. These materials can be used for development of catalytic PMs.

5 Strategies to enhance membrane performance

Development of postfunctionalized membranes is also found to be attractive for various applications. However, examples of modified catalytic PMs are relatively rare.

Wang et al. [80] developed catalytic UF and NF PMs for naproxen removal. They used the Non-solvent induced phase separation (NIPS) technique to prepare PES@CoFe$_2$O$_4$ catalytic UF membranes. They cast PES polymeric casting solution (CoFe$_2$O$_4$-based catalytic NPs, SPES, glycerol, and NMP) on a clean glass plate using a steel casting knife with 150 μm clearance gap. The obtained film was immersed in DI water coagulation bath at 20 °C. After that, for preparation of a stable selective layer on top of UF membranes, layer-by-layer polyelectrolyte coating was done, which includes: (1) immersion of the UF membranes in a poly(diallyldimethylammonium chloride) solution (0.1 g/L) in a 50 mM/L

NaCl concentration, (2) rinsing in 50 mM/L NaCl solution (three times), (3) immersion of the membranes in a PSS solution (0.1 g/L) in 50 mM/L NaCl concentration, and (4) repeating the rinsing process. The catalytic NF membranes showed higher naproxen rejection in comparison to the catalytic UF membranes.

In another study, Zhai et al. [81] developed double-layer PVDF composite membranes with a top layer of semi-interpenetrating polymer network and a bottom layer of PMAA-Pd/PVDF (catalytic layer) for one-step process including separation and catalytic degradation (Figure 2). So the undesired pollutants (such as Congo red, direct black 38, direct red 23 dyes, and BSA) were separated by the top coating layer. On the other hand, p-nitrophenol as an organic pollutant controllably penetrated through the top layer and then entered into the catalytic layer. Finally, the product (p-aminophenol) was in situ separated and collected. The in situ separation of the product improved the reaction efficiency.

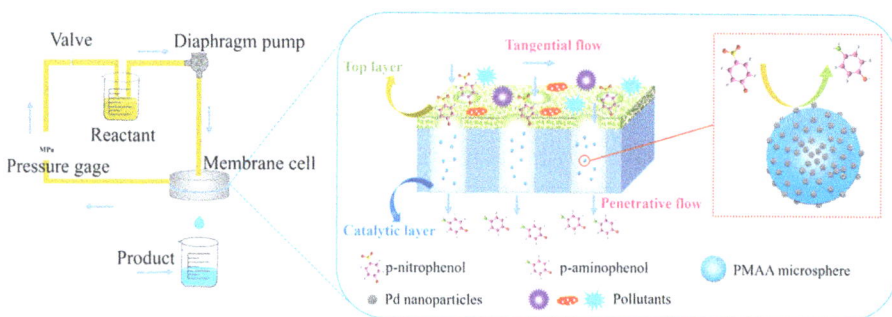

Figure 2: A double-layer composite membrane for continuous cross-flow catalytic degradation and in situ separation. Adapted with permission from Zhai et al. [81].

6 Future perspectives

PMs in catalysis primarily serve as reactors, owing to the dual blend of the separation process and catalysis simultaneously. It is a promising area of research that offers considerable advantages over traditional catalytic reaction systems such as improved selectivity and sensitivity, high yields, low energy consumption, and waste minimization.

In light of the reviewed literature, it is believed that the incorporation of a catalyst onto the polymer membrane shares the qualities of both homogeneous and heterogeneous catalyses. Nonetheless, the catalytic polymer membrane reactor system should exhibit all the expected advantages such as efficient selectivity and activity, high reaction yield with the least reaction by-products, considerably high TOF, and easy separation from the reaction mixture. Lastly, the catalyst recyclability and stability must be ensured as well.

7 Conclusion

Over the past decade, the study and development of polymeric catalytic membranes have gained significant attention as a promising research area. This chapter presents a review of the applications of polymeric catalytic membranes, as well as various membrane materials and fabrication methods used for catalytic purposes. By incorporating specific materials into catalytic membranes, it becomes possible to degrade various types of effluents, thus promoting the advancement of novel catalytic membranes. Recent research has shown notable advancements in the development of catalytic PMs achieved through the incorporation of catalysts into the membrane matrix or via catalyst coating on the membrane surface. Nevertheless, the exploration of novel strategies aimed at enhancing membrane separation performance remains a crucial avenue for future research in this field.

The exploration of green chemistry approaches for membrane surface coating that utilizes safer chemicals and solvents represents an active and promising research area in the field.

References

[1]   Stankiewicz, A. I., and Moulijn, J. A. Process intensification: Transforming chemical engineering. Chem. Eng. Prog., 2000, **96**: 22–34.

[2]   Nestoridis, V., Andreou, I., and Vayenas, C. G. Optimal residence time policy for product yield maximization in chemical reactors. J. Optim. Theory Appl., 1986, **49**: 271–287. https://doi.org/10. 1007/BF00940760.

[3]   Yang, D., Tang, Y., Zhu, B., Pang, H., Rong, X., Gao, Y., Du, F., Cheng, C., Qiu, L., and Ma, L. Engineering cell membrane-cloaked catalysts as multifaceted artificial peroxisomes for biomedical applications. Adv. Sci., 2023, **10**: 2206181.

[4]   Liu, Q. L., and Chen, H. F. Modeling of esterification of acetic acid with n-butanol in the presence of Zr(SO4)2·4H2O coupled pervaporation. J. Memb. Sci., 2002, **196**: 171–178. https://doi.org/https://doi. org/10.1016/S0376-7388(01)00543-9.

[5]   Abdallah, H. A review on catalytic membranes production and applications. Bull. Chem. React. Eng. Catal., 2017, **12**: 136–156. https://doi.org/10.9767/bcrec.12.2.462.136-156.

[6]   Domènech, B., Muñoz, M., Muraviev, D. N., and Macanás, J. Catalytic membranes with palladium nanoparticles: From tailored polymer to catalytic applications. Catal. Today, 2012, **193**: 158–164. https://doi.org/https://doi.org/10.1016/j.cattod.2012.02.049.

[7]   Julbe, A., Farrusseng, D., and Guizard, C. Porous ceramic membranes for catalytic reactors – Overview and new ideas. J. Memb. Sci., 2001, **181**: 3–20. https://doi.org/https://doi.org/10.1016/ S0376-7388(00)00375-6.

[8]   Zhang, G., Jin, W., and Xu, N. Design and fabrication of ceramic catalytic membrane reactors for green chemical engineering applications, engineering. 2018, **4**: 848–860. https://doi.org/https://doi. org/10.1016/j.eng.2017.05.001.

[9]   Ruiz, P., Muñoz, M., Macanás, J., and Muraviev, D. N. Intermatrix synthesis of polymer-stabilized PGM@Cu core–shell nanoparticles with enhanced electrocatalytic properties. React. Funct. Polym., 2011, **71**: 916–924. https://doi.org/https://doi.org/10.1016/j.reactfunctpolym.2011.05.009.

[10] Algieri, C., Coppola, G., Mukherjee, D., Shammas, M. I., Calabro, V., Curcio, S., and Chakraborty, S. Catalytic membrane reactors: The industrial applications perspective. Catalysts, 2021, **11**. https://doi.org/10.3390/catal11060691.

[11] Buonomenna, M. G., Choi, S. H., and Drioli, E. Catalysis in polymeric membrane reactors: The membrane role. Asia-Pac. J. Chem. Eng., 2010, **5**: 26–34. https://doi.org/https://doi.org/10.1002/apj.379.

[12] Qing, W., Li, X., Shao, S., Shi, X., Wang, J., Feng, Y., Zhang, W., and Zhang, W. Polymeric catalytically active membranes for reaction-separation coupling: A review. J. Memb. Sci., 2019, **583**: 118–138. https://doi.org/https://doi.org/10.1016/j.memsci.2019.04.053.

[13] Bet-moushoul, E., Mansourpanah, Y., Farhadi, K., and Tabatabaei, M. TiO2 nanocomposite based polymeric membranes: A review on performance improvement for various applications in chemical engineering processes. Chem. Eng. J., 2016, **283**: 29–46. https://doi.org/https://doi.org/10.1016/j.cej.2015.06.124.

[14] Molinari, R., Poerio, T., and Argurio, P. Liquid-phase oxidation of benzene to phenol using CuO catalytic polymeric membranes. Desalination, 2009, **241**: 22–28. https://doi.org/https://doi.org/10.1016/j.desal.2007.11.081.

[15] Ozdemir, S. S., Buonomenna, M. G., and Drioli, E. Catalytic polymeric membranes: Preparation and application. Appl. Catal. A Gen., 2006, **307**: 167–183. https://doi.org/https://doi.org/10.1016/j.apcata.2006.03.058.

[16] Dolatshah, M., Zinatizadeh, A. A., Zinadini, S., and Zangeneh, H. Preparation, characterization and performance assessment of antifouling L-lysine (C, N codoped)-TiO2/WO3-PES photocatalytic membranes: A comparative study on the effect of blended and UV-grafted nanophotocatalyst. J. Environ. Chem. Eng., 2022, **10**: 108658. https://doi.org/https://doi.org/10.1016/j.jece.2022.108658.

[17] Gotardo, M. C. A. F., Guedes, A. A., Schiavon, M. A., José, N. M., Yoshida, I. V. P., and Assis, M. D. Polymeric membranes: The role this support plays in the reactivity of the different generations of metalloporphyrins. J. Mol. Catal. A Chem., 2005, **229**: 137–143. https://doi.org/https://doi.org/10.1016/j.molcata.2004.11.014.

[18] Qing, W., Liu, F., Yao, H., Sun, S., Chen, C., and Zhang, W. Functional catalytic membrane development: A review of catalyst coating techniques. Adv. Colloid Interface Sci., 2020, **282**: 102207. https://doi.org/https://doi.org/10.1016/j.cis.2020.102207.

[19] Biniaz, P., Makarem, M. A., and Rahimpour, M. R. Membrane reactors. Catal. Immobil., 2020, 307–324. https://doi.org/https://doi.org/10.1002/9783527817290.ch9.

[20] Tsotsis, T. T., Minet, R. G., Champagnie, A. M., and Liu, P. K. T. Catalytic Membrane Reactors. In: Comput. Des. Catal, CRC Press, 2020, pp. 471–552.

[21] Al-Megrenb, H. A., Poerio, T., Brunettia, A., Barbieria, G., Driolia, E., AL-Hedaibb, B. S. A., Al-Hamdanb, A. S. N., and Al-Kinanyb, M. C. A continuous membrane reactor for benzene hydroxylation to phenol. Chem. Eng. Trans., 2013, **32**: 679–684.

[22] Molinari, R., Argurio, P., and Poerio, T. Vanadyl acetylacetonate filled PVDF membranes as the core of a liquid phase continuous process for pure phenol production from benzene. J. Memb. Sci., 2015, **476**: 490–499. https://doi.org/https://doi.org/10.1016/j.memsci.2014.12.006.

[23] Molinari, R., and Poerio, T. Remarks on studies for direct production of phenol in conventional and membrane reactors. Asia-Pacific J. Chem. Eng., 2010, **5**: 191–206. https://doi.org/https://doi.org/10.1002/apj.369.

[24] Kidambi, S., and Bruening, M. L. Multilayered polyelectrolyte films containing palladium nanoparticles: Synthesis, characterization, and application in selective hydrogenation. Chem. Mater., 2005, **17**: 301–307. https://doi.org/10.1021/cm048421t.

[25] Singh, D., Rezac, M. E., and Pfromm, P. H. Partial hydrogenation of soybean oil with minimal trans fat production using a Pt-decorated polymeric membrane reactor. J. Am. Oil Chem. Soc., 2009, **86**: 93–101. https://doi.org/10.1007/s11746-008-1321-z.

[26] Bottino, A., Capannelli, G., Comite, A., and Di Felice, R. Polymeric and ceramic membranes in three-phase catalytic membrane reactors for the hydrogenation of methylenecyclohexane. Desalination, 2002, **144**: 411–416. https://doi.org/https://doi.org/10.1016/S0011-9164(02)00352-1.

[27] Gu, Y., Bacchin, P., Lahitte, J.-F., Remigy, J.-C., Favier, I., Gómez, M., Gin, D. L., and Noble, R. D. Catalytic membrane reactor for Suzuki-Miyaura C−C cross-coupling: Explanation for its high efficiency via modeling. AIChE J., 2017, **63**: 698–704. https://doi.org/https://doi.org/10.1002/aic.15379.

[28] Gu, Y., Favier, I., Pradel, C., Gin, D. L., Lahitte, J.-F., Noble, R. D., Gómez, M., and Remigy, J.-C. High catalytic efficiency of palladium nanoparticles immobilized in a polymer membrane containing poly(ionic liquid) in Suzuki–Miyaura cross-coupling reaction. J. Memb. Sci., 2015, **492**: 331–339. https://doi.org/https://doi.org/10.1016/j.memsci.2015.05.051.

[29] Faria, V. W., Oliveira, D. G. M., Kurz, M. H. S., Gonçalves, F. F., Scheeren, C. W., and Rosa, G. R. Palladium nanoparticles supported in a polymeric membrane: An efficient phosphine-free "green" catalyst for Suzuki–Miyaura reactions in water. RSC Adv., 2014, **4**: 13446–13452. https://doi.org/10.1039/C4RA01104J.

[30] Fayyazi, F., Ahmadi Feijani, E., and Mahdavi, H. Chemically modified polysulfone membrane containing palladium nanoparticles: Preparation, characterization and application as an efficient catalytic membrane for Suzuki reaction. Chem. Eng. Sci., 2015, **134**: 549–554. https://doi.org/https://doi.org/10.1016/j.ces.2015.05.008.

[31] Kumari, P., Bahadur, N., and Dumée, L. F. Photo-catalytic membrane reactors for the remediation of persistent organic pollutants – A review. Sep. Purif. Technol., 2020, **230**: 115878. https://doi.org/https://doi.org/10.1016/j.seppur.2019.115878.

[32] Li, T., Zhang, Z., Li, W., Liu, C., Wang, J., and An, L. H4SiW12O40/polymethylmethacrylate/polyvinyl alcohol sandwich nanofibrous membrane with enhanced photocatalytic activity. Colloids Surfaces A Physicochem. Eng. Asp., 2016, **489**: 289–296. https://doi.org/https://doi.org/10.1016/j.colsurfa.2015.10.061.

[33] Ma, S., Meng, J., Li, J., Zhang, Y., and Ni, L. Synthesis of catalytic polypropylene membranes enabling visible-light-driven photocatalytic degradation of dyes in water. J. Memb. Sci., 2014, **453**: 221–229. https://doi.org/https://doi.org/10.1016/j.memsci.2013.11.021.

[34] Dekkouche, S., Morales-Torres, S., Ribeiro, A. R., Faria, J. L., Fontàs, C., Kebiche-Senhadji, O., and Silva, A. M. T. In situ growth and crystallization of TiO2 on polymeric membranes for the photocatalytic degradation of diclofenac and 17α-ethinylestradiol. Chem. Eng. J., 2022, **427**: 131476. https://doi.org/https://doi.org/10.1016/j.cej.2021.131476.

[35] Cruz, N. K. O., Semblante, G. U., Senoro, D. B., You, S.-J., and Lu, S.-C. Dye degradation and antifouling properties of polyvinylidene fluoride/titanium oxide membrane prepared by sol–gel method. J. Taiwan Inst. Chem. Eng., 2014, **45**: 192–201. https://doi.org/https://doi.org/10.1016/j.jtice.2013.04.011.

[36] Fischer, K., Grimm, M., Meyers, J., Dietrich, C., Gläser, R., and Schulze, A. Photoactive microfiltration membranes via directed synthesis of TiO2 nanoparticles on the polymer surface for removal of drugs from water. J. Memb. Sci., 2015, **478**: 49–57. https://doi.org/https://doi.org/10.1016/j.memsci.2015.01.009.

[37] Zheng, Y., Zhang, Y., Wang, X., Xu, Z., Liu, X., Lu, X., and Fan, Z. MnO2 catalysts uniformly decorated on polyphenylene sulfide filter felt by a polypyrrole-assisted method for use in the selective catalytic reduction of NO with NH3. RSC Adv., 2014, **4**: 59242–59247. https://doi.org/10.1039/C4RA07168A.

[38] Chen, X., Zheng, Y., and Zhang, Y. MnO2-Fe2O3 catalysts supported on polyphenylene sulfide filter felt by a redox method for the low-temperature NO reduction with NH3. Catal. Commun., 2018, **105**: 16–19. https://doi.org/https://doi.org/10.1016/j.catcom.2017.09.006.

[39] Kudaibergenov, S. E., and Dzhardimalieva, G. I. Flow-through catalytic reactors based on metal nanoparticles immobilized within porous polymeric gels and surfaces/hollows of polymeric membranes. Polymers (Basel), 2020, **12**. https://doi.org/10.3390/polym12030572.

[40] Ng, L. Y., Mohammad, A. W., Leo, C. P., and Hilal, N. Polymeric membranes incorporated with metal/ metal oxide nanoparticles: A comprehensive review. Desalination, 2013, **308**: 15–33. https://doi.org/ https://doi.org/10.1016/j.desal.2010.11.033.

[41] Zhou, J., Zhang, C., and Wang, Y. Nanoporous block copolymer membranes immobilized with gold nanoparticles for continuous flow catalysis. Polym. Chem., 2019, **10**: 1642–1649. https://doi.org/10. 1039/C8PY01789A.

[42] Hu, M. X., Guo, Q., Li, J. N., Huang, C. M., and Ren, G. R. Reduction of methylene blue with Ag nanoparticle-modified microporous polypropylene membranes in a flow-through reactor. New J. Chem., 2017, **41**: 6076–6082. https://doi.org/10.1039/C7NJ01068K.

[43] Fang, X., Li, J., Ren, B., Huang, Y., Wang, D., Liao, Z., Li, Q., Wang, L., and Dionysiou, D. D. Polymeric ultrafiltration membrane with in situ formed nano-silver within the inner pores for simultaneous separation and catalysis. J. Memb. Sci., 2019, **579**: 190–198. https://doi.org/https://doi.org/10.1016/j. memsci.2019.02.073.

[44] Li, Y., and Yeung, K. L. Polymeric catalytic membrane for ozone treatment of DEET in water. Catal. Today, 2019, **331**: 53–59. https://doi.org/https://doi.org/10.1016/j.cattod.2018.06.005.

[45] Ouyang, L., Dotzauer, D. M., Hogg, S. R., Macanás, J., Lahitte, J.-F., and Bruening, M. L. Catalytic hollow fiber membranes prepared using layer-by-layer adsorption of polyelectrolytes and metal nanoparticles. Catal. Today, 2010, **156**: 100–106. https://doi.org/https://doi.org/10.1016/j.cattod. 2010.02.040.

[46] Lalia, B. S., Kochkodan, V., Hashaikeh, R., and Hilal, N. A review on membrane fabrication: Structure, properties and performance relationship. Desalination, 2013, **326**: 77–95. https://doi.org/https://doi. org/10.1016/j.desal.2013.06.016.

[47] Westermann, T., and Melin, T. Flow-through catalytic membrane reactors – Principles and Applications. Chem. Eng. Process. Process Intensif, 2009, **48**: 17–28. https://doi.org/https://doi.org/ 10.1016/j.cep.2008.07.001.

[48] Westermann, T., Kopriwa, N., Schröder, A., and Melin, T. Effective dispersion model for flow-through catalytic membrane reactors combining axial dispersion and pore size distribution. Chem. Eng. Sci., 2010, **65**: 1609–1615. https://doi.org/https://doi.org/10.1016/j.ces.2009.10.023.

[49] Bhattacharya, A., and Misra, B. N. Grafting: A versatile means to modify polymers: Techniques, factors and applications. Prog. Polym. Sci., 2004, **29**: 767–814. https://doi.org/https://doi.org/10. 1016/j.progpolymsci.2004.05.002.

[50] Vanherck, K., Verbiest, T., and Vankelecom, I. Comparison of two synthesis routes to obtain gold nanoparticles in polyimide. J. Phys. Chem. C., 2012, **116**: 115–125. https://doi.org/10.1021/jp207244y.

[51] Smuleac, V., Varma, R., Sikdar, S., and Bhattacharyya, D. Green synthesis of Fe and Fe/Pd bimetallic nanoparticles in membranes for reductive degradation of chlorinated organics. J. Memb. Sci., 2011, **379**: 131–137. https://doi.org/https://doi.org/10.1016/j.memsci.2011.05.054.

[52] Alpatova, A., Meshref, M., McPhedran, K. N., and Gamal El-Din, M. Composite polyvinylidene fluoride (PVDF) membrane impregnated with Fe2O3 nanoparticles and multiwalled carbon nanotubes for catalytic degradation of organic contaminants. J. Memb. Sci., 2015, **490**: 227–235. https://doi.org/ https://doi.org/10.1016/j.memsci.2015.05.001.

[53] Wang, B., Jackson, E. A., Hoff, J. W., and Dutta, P. K. Fabrication of zeolite/polymer composite membranes in a roller assembly. Microporous Mesoporous Mater., 2016, **223**: 247–253. https://doi. org/https://doi.org/10.1016/j.micromeso.2015.11.003.

[54] Merkel, T. C., Lin, H., Wei, X., and Baker, R. Power plant post-combustion carbon dioxide capture: An opportunity for membranes. J. Memb. Sci., 2010, **359**: 126–139. https://doi.org/https://doi.org/10. 1016/j.memsci.2009.10.041.

[55] Sawamura, K., Furuhata, T., Sekine, Y., Kikuchi, E., Subramanian, B., and Matsukata, M. Zeolite membrane for dehydration of isopropylalcohol–water mixture by vapor permeation. ACS Appl. Mater. Interfaces, 2015, **7**: 13728–13730. https://doi.org/10.1021/acsami.5b04085.

[56] Cakmak, M., Batra, S., and Yalcin, B. Field assisted self-assembly for preferential through thickness ("z-direction") alignment of particles and phases by electric, magnetic, and thermal fields using a novel roll-to-roll processing line. Polym. Eng. Sci., 2015, **55**: 34–46. https://doi.org/https://doi.org/10.1002/pen.23861.

[57] Mashentseva, A., Borgekov, D., Kislitsin, S., Zdorovets, M., and Migunova, A. Comparative catalytic activity of PET track-etched membranes with embedded silver and gold nanotubes. Nucl. Instruments Methods Phys. Res. Sect. B Beam Interact. With Mater. Atoms, 2015, **365**: 70–74. https://doi.org/https://doi.org/10.1016/j.nimb.2015.07.063.

[58] Yang, M., Lee, K. G., Kim, J. W., Lee, S. J., Huh, Y. S., and Choi, B. G. Highly ordered gold-nanotube films for flow-injection amperometric glucose biosensors. RSC Adv., 2014, **4**: 40286–40291. https://doi.org/10.1039/C4RA05273K.

[59] Bahari Mollamahalle, Y., Ghorbani, M., and Dolati, A. Electrodeposition of long gold nanotubes in polycarbonate templates as highly sensitive 3D nanoelectrode ensembles. Electrochim. Acta, 2012, **75**: 157–163. https://doi.org/https://doi.org/10.1016/j.electacta.2012.04.119.

[60] Mashentseva, A., Borgekov, D., Zdorovets, M., and Russakova, A. Synthesis, structure, and catalytic activity of Au/poly(ethylene terephthalate) composites. Acta Phys Pol A, 2014, **125**: 1263–1266.

[61] Yu, Y., Kant, K., Shapter, J. G., Addai-Mensah, J., and Losic, D. Gold nanotube membranes have catalytic properties. Microporous Mesoporous Mater., 2012, **153**: 131–136. https://doi.org/https://doi.org/10.1016/j.micromeso.2011.12.011.

[62] Akther, N., Phuntsho, S., Chen, Y., Ghaffour, N., and Shon, H. K. Recent advances in nanomaterial-modified polyamide thin-film composite membranes for forward osmosis processes. J. Memb. Sci., 2019, **584**: 20–45. https://doi.org/https://doi.org/10.1016/j.memsci.2019.04.064.

[63] Shabani, Z., Zarghami, S., and Mohammadi, T. Nanomaterials for Fouling-resistant RO Membranes. In: Nanotechnol. Beverage Ind, Elsevier, 2020, pp. 151–184. https://doi.org/10.1016/B978-0-12-819941-1.00006-7.

[64] Ndlovu, L. N., Malatjie, K. I., Donga, C., Mishra, A. K., Nxumalo, E. N., and Mishra, S. B. Catalytic degradation of methyl orange using beta cyclodextrin modified polyvinylidene fluoride mixed matrix membranes imbedded with in-situ generated palladium nanoparticles. J. Appl. Polym. Sci., 2023, **140**: e53270. https://doi.org/10.1002/app.53270.

[65] Bhat, S. D., and Aminabhavi, T. M. Pervaporation-aided dehydration and esterification of acetic acid with ethanol using 4A zeolite-filled cross-linked sodium alginate-mixed matrix membranes. J. Appl. Polym. Sci, 2009, **113**: 157–168. https://doi.org/10.1002/app.29545.

[66] de la Iglesia, Ó., Sorribas, S., Almendro, E., Zornoza, B., Téllez, C., and Coronas, J. Metal-organic framework MIL-101(Cr) based mixed matrix membranes for esterification of ethanol and acetic acid in a membrane reactor. Renew. Energy, 2016, **88**: 12–19. https://doi.org/10.1016/j.renene.2015.11.025.

[67] Castro-Muñoz, R., De la Iglesia, Ó., Fíla, V., Téllez, C., and Coronas, J. Pervaporation-assisted esterification reactions by means of mixed matrix membranes. Ind. Eng Chem Res., 2018, **57**: 15998–16011. https://doi.org/10.1021/acs.iecr.8b01564.

[68] Yahya, R., and Elshaarawy, R. F. M. Highly sulfonated chitosan-polyethersulfone mixed matrix membrane as an effective catalytic reactor for esterification of acetic acid. Catal. Commun., 2023, **173**: 106557. https://doi.org/10.1016/j.catcom.2022.106557.

[69] Wang, Y., Zhang, J., Bao, C., Xu, X., Li, D., Chen, J., Hong, M., Peng, B., and Zhang, Q. Self-cleaning catalytic membrane for water treatment via an integration of Heterogeneous Fenton and membrane process. J. Memb. Sci., 2021, **624**: 119121. https://doi.org/10.1016/j.memsci.2021.119121.

[70] Didaskalou, C., Kupai, J., Cseri, L., Barabas, J., Vass, E., Holtzl, T., and Szekely, G. Membrane-grafted asymmetric organocatalyst for an integrated synthesis–separation platform. ACS Catal., 2018, **8**: 7430–7438. https://doi.org/10.1021/acscatal.8b01706.

[71] Zhang, L., Li, Y., Liu, Q., Li, W., and Xing, W. Fabrication of ionic liquids-functionalized PVA catalytic composite membranes to enhance esterification by pervaporation. J. Memb. Sci., 2019, **584**: 268–281. https://doi.org/10.1016/j.memsci.2019.05.006.

[72] Li, P., Liang, H., Luo, X., Cheng, X., Ding, J., Wu, D., Liu, L., Gao, X., and Li, G. Organic-inorganic composite ultrafiltration membrane with anti-fouling and catalytic properties by in-situ co-casting for water treatment. J. Memb. Sci., 2022, **662**: 120984. https://doi.org/10.1016/j.memsci.2022.120984.

[73] Wan, Z., Gan, L., Wang, W.-N., and Jiang, Y. Rapid membrane surface functionalization with Ag nanoparticles via coupling electrospray and polymeric solvent bonding for enhanced antifouling and catalytic performance: Deposition and interfacial reaction mechanisms. J Colloid Interface Sci., 2023, **639**: 203–213. https://doi.org/10.1016/j.jcis.2023.02.047.

[74] Zarghami, S., Mohammadi, T., Sadrzadeh, M., and Van der Bruggen, B. Superhydrophilic and underwater superoleophobic membranes – A review of synthesis methods. Prog. Polym. Sci., 2019, **98**: 101166. https://doi.org/10.1016/j.progpolymsci.2019.101166.

[75] Zarghami, S., Mohammadi, T., and Sadrzadeh, M. Superhydrophobic/Superhydrophilic Polymeric Membranes for Oil/Water Separation. In: Oil−Water Mix. Emuls. Vol. 1 Membr. Mater. Sep. Treat., American Chemical Society, 2022, pp. 119–184. SE–4 https://doi.org/10.1021/bk-2022-1407.ch004.

[76] Wang, J., Pei, X., Liu, G., Bai, J., Ding, Y., Wang, J., and Liu, F. Gravity-driven catalytic nanofibrous membrane with microsphere and nanofiber coordinated structure for ultrafast continuous reduction of 4-nitrophenol. J. Colloid Interface Sci., 2019, **538**: 108–115. https://doi.org/10.1016/j.jcis.2018.11.086.

[77] Sun, X., Zheng, H., Jiang, S., Zhu, M., Zhou, Y., Wang, D., Fan, Y., Zhang, D., and Zhang, L. Fabrication of FeOCl/MoS2 catalytic membranes for pollutant degradation and alleviating membrane fouling with peroxymonosulfate activation. J. Environ. Chem. Eng., 2022, **10**: 107717. https://doi.org/10.1016/j.jece.2022.107717.

[78] Yang, J., Huang, Y., Cheng, Y., Wu, X., Lu, J., Wan, Q., Feng, J., Zeng, Q., Zhao, S., Yu, L., and Xiong, Z. Long-acting removal of high-toxic p-nitrophenol in wastewater via peroxymonosulfate activation by cyclic membrane catalysis. J Clean Prod., 2023, **401**: 136739. https://doi.org/10.1016/j.jclepro.2023.136739.

[79] Zhang, N., Wu, Y., Yuen, G., and De Lannoy, C.-F. Ultrafiltration Pd-immobilized catalytic membrane microreactors continuously reduce nitrophenol: A study of catalytic activity and simultaneous separation. Sep. Purif. Technol., 2023, **312**: 123318. https://doi.org/10.1016/j.seppur.2023.123318.

[80] Wang, T., De Vos, W. M., and De Grooth, J. CoFe2O4-peroxymonosulfate based catalytic UF and NF polymeric membranes for naproxen removal: The role of residence time. J. Memb. Sci., 2022, **646**: 120209. https://doi.org/10.1016/j.memsci.2021.120209.

[81] Zhai, X., Chen, X., Shi, X., Wang, S., Wang, S., Wu, Q., Ma, Y., Wang, J., Wan, D., and Pan, J. Simultaneously enhancing purification, catalysis and in situ separation in a continuous cross-flow catalytic degradation process of multi-component organic pollutants by a double-layer PVDF composite membrane. J. Environ. Chem. Eng., 2022, **10**: 107160. https://doi.org/10.1016/j.jece.2022.107160.

Neelesh Ashok*, Sruthi M. S., Taniya Rose Abraham, Sabu Thomas

Chapter 6
Polymer membranes for pervaporation

Abstract: Pervaporation is a membrane separation technique that has gained significant attention due to its potential in various industrial applications. This abstract provides a concise overview of the book chapter on pervaporation, focusing on the key aspects. It begins by introducing the principles and mechanisms of pervaporation and subsequently highlights the importance of polymer membranes in achieving efficient separation. The chapter explores the classification of polymeric membranes and discusses their morphology, structure, and the factors influencing polymer selection. The synthesis and modification of polymer membranes are addressed, including various fabrication techniques and methods for enhancing membrane performance. Crosslinking and surface functionalization techniques are explored, emphasizing their role in improving the stability, selectivity, and permeability of membranes. The chapter also highlights characterization techniques for evaluating membrane properties, such as morphological analysis, thermal and mechanical properties, and spectroscopic techniques.

Transport mechanisms in polymer membranes, including diffusion and sorption, are discussed, with a focus on the solution–diffusion model and its relevance to pervaporation. The concepts of selectivity and permeability are explained, along with the factors influencing membrane performance. Performance evaluation and optimization are explored, providing insights into pervaporation performance metrics, factors affecting membrane performance, and strategies for performance enhancement. Furthermore, the article explores recent advances and emerging trends in the field, including advanced polymer membrane materials, hybrid and composite membranes, modification techniques for performance enhancement, and novel applications. These advancements hold promise for addressing current challenges and expanding the scope of pervaporation in diverse industries. This article provides a comprehensive overview of pervaporation, encompassing principles, polymer materials, membrane synthesis and modification, characterization techniques, transport mechanisms, performance evaluation, and recent advances. It serves as a valuable resource for researchers and professionals

*Corresponding author: **Neelesh Ashok**, Amrita School of Artificial Intelligence, Amrita Vishwa Vidyapeetham, Coimbatore, Tamil Nadu, India, e-mail: neeleshashok@gmail.com
Sruthi M. S., Amrita School of Artificial Intelligence, Amrita Vishwa Vidyapeetham, Coimbatore, Tamil Nadu, India
Taniya Rose Abraham, Sabu Thomas, School of Energy Materials, Mahatma Gandhi University, Kottayam, Kerala, India

https://doi.org/10.1515/9783110796032-006

working in membrane separation and pervaporation, facilitating a deeper understanding of the topic and inspiring further advancements in this field.

Keywords: polymer membranes, pervaporation, selectivity, permeability, separation

1 Overview of pervaporation

Pervaporation is a membrane separation process that has gained significant attention in recent years for its ability to efficiently separate liquid mixtures based on their vapor pressure differences. It is a versatile and energy-efficient technique that offers advantages over conventional separation methods such as distillation or absorption. Pervaporation has found applications in various industries, including petrochemical, pharmaceutical, food and beverage, and environmental sectors. The concept of pervaporation involves the use of a selective membrane that allows certain components of a liquid mixture to selectively permeate through the membrane, while blocking others [1]. This separation is driven by the vapor pressure difference between the feed side and the permeate side of the membrane. The membrane acts as a barrier, selectively sorbing and diffusing one or more components of the liquid mixture, thereby separating them from the feed solution. The selectivity and permeability of the membrane play a crucial role in determining the separation efficiency of pervaporation [2]. Different membrane materials and designs can be tailored to exhibit specific selectivity toward target components, allowing for the separation of complex mixtures with high precision. Polymer membranes, in particular, have been extensively studied and utilized in pervaporation due to their tunable properties, ease of fabrication, and cost-effectiveness. One of the key advantages of pervaporation is its ability to operate under mild conditions of temperature and pressure [3]. Unlike distillation, which requires high energy input, pervaporation can achieve separation at lower temperatures, reducing energy consumption. This makes it a more environmentally friendly and economically viable option for separation processes. Pervaporation has a wide range of applications. It is commonly used for the dehydration of organic solvents or alcohols, such as the removal of water from ethanol in biofuel production. It is also employed in the purification of industrial wastewater, recovery of valuable chemicals or solvents, and concentration of liquid mixtures [4]. Pervaporation has shown promise in the separation of azeotropic or close-boiling mixtures, which are challenging to separate using traditional methods. Advancements in membrane materials, fabrication techniques, and modification methods have further improved the performance and efficiency of pervaporation. Researchers are actively exploring novel materials, such as hybrid or composite membranes, and developing strategies to enhance selectivity, permeability, and stability. These advancements are paving the way for the implementation of pervaporation in new areas and opening doors to innovative separation solutions [5–7].

1.1 Importance of polymer membranes

Polymer membranes play a crucial role in various scientific and industrial applications. Their importance stems from their unique properties, versatility, and tunability [8, 9]. Membranes can be classified as microfiltration, ultrafiltration, and nanofiltration membranes on the basis of pore sizes (Figure 1).

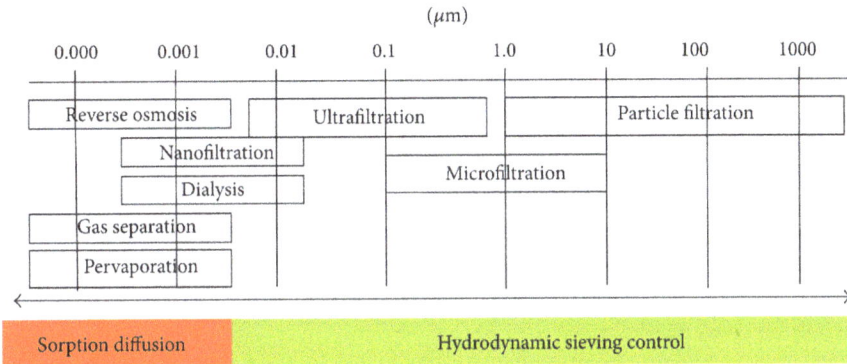

Figure 1: The choice of membrane with respect to the size of particles encountered. Copyright © 2015 Ghoshna Jyoti et al.

The following are some key points highlighting the importance of polymer membranes:
–   **Separation and purification:** Polymer membranes are extensively used in separation and purification processes. They enable the selective passage of certain molecules or ions, while retaining others, based on factors such as size, charge, and solubility. This makes them valuable in applications such as water treatment, gas separation, desalination, and purification of chemicals and pharmaceuticals.
–   **Membrane-based technologies:** Polymer membranes serve as the foundation for various membrane-based technologies, such as reverse osmosis, ultrafiltration, nanofiltration, and pervaporation. These technologies have wide-ranging applications in industries like food and beverage, pharmaceuticals, biotechnology, and energy production. Polymer membranes enable efficient and cost-effective separation, concentration, and purification of desired components.
–   **Environmental sustainability:** Polymer membranes contribute to environmental sustainability by facilitating resource conservation and waste reduction. For example, in water treatment, membranes can remove pollutants, contaminants, and microorganisms, providing safe drinking water and reducing the strain on freshwater resources. In industrial processes, membrane separation can minimize the generation of harmful byproducts, making the processes more environmental friendly.

– **Energy efficiency:** Polymer membranes offer energy-efficient separation solutions compared to traditional techniques. They operate at lower pressures and temperatures, reducing energy consumption. Membrane-based technologies like pervaporation and membrane distillation provide alternatives to energy-intensive processes such as distillation, offering significant energy savings and promoting sustainability.
– **Versatility and customization:** Polymer membranes exhibit diverse properties that can be tailored to specific applications. They can be designed with different pore sizes, surface chemistries, and thicknesses to achieve the desired selectivity, permeability, and mechanical stability. This versatility allows for customization according to the requirements of different separation processes.
– **Scalability and cost-effectiveness:** Polymer membranes are highly scalable and cost-effective compared to other separation techniques. They can be mass-produced using various fabrication methods, including phase inversion, electrospinning, and interfacial polymerization. The scalability and cost-effectiveness of polymer membranes make them practical for large-scale industrial applications.

Pervaporation stands out as a unique category among membranes due to its distinct separation criteria. Pervaporation membranes, which are nonporous, have found extensive use in the separation of liquid mixtures. The separation mechanism relies on the affinity between the molecules and the membrane materials. As a result, molecules with higher affinity are adsorbed and diffuse through the membrane, while molecules with lower affinity are retained by the membrane.

1.2 Principles and mechanisms

The principles and mechanisms underlying membrane-based separations, including those involving polymer membranes, are key to understanding their performance and applications (Table 1).

The pervaporation process involves the transport of components across the membrane, which can be described by a solution desorption model. This model takes into account a series of processes: (i) diffusion of the component through the liquid boundary layer to the membrane surface, (ii) sorption/diffusion into the membrane, (iii) transport through the membrane, and (iv) diffusion through the vapor phase boundary layer into the bulk of the permeance (as shown in Figure 2).

By manipulating the membrane material – its structure, and its surface properties – researchers can enhance selectivity, permeability, and separation efficiency in various membrane-based applications.

Table 1: Fundamental principles and mechanisms.

Principle/ mechanism	Description
Selective permeation	The membrane selectively allows certain components to permeate while rejecting others, based on physicochemical properties and interactions with the membrane material. Polymer membranes can be engineered for specific selectivity in separation processes.
Sorption	It is the process of solutes or molecules being adsorbed onto the membrane surface or penetrating into the membrane structure. Polymer membranes interact differently with solutes, based on factors such as polarity, size, charge, and affinity, influencing transport behavior.
Diffusion	It is the movement of solutes or molecules from areas of high to low concentration. Solutes diffuse through the membrane matrix or along the membrane surface. Factors like solute size, solute–membrane interactions, and concentration gradients impact the rate of diffusion.
Solution–diffusion model	The solute dissolves into the membrane and then diffuses through it. The solute's ability to dissolve and diffuse depends on factors such as size, solute–membrane interactions, and polymer chain mobility. Solution–diffusion model explains solute transport through polymer membranes.
Size exclusion	Polymer membranes act as molecular sieves, permitting smaller molecules or ions to pass through, while blocking larger ones. Size-exclusion mechanism relies on pore size and structure of the membrane, enabling size-based separation.
Donnan exclusion	This occurs in charged membranes or in the presence of charged solutes, preferentially transporting ions or charged species, based on electrical forces and ion–ion interactions. Donnan exclusion plays a role in separating charged solutes or rejecting ions in specific applications.

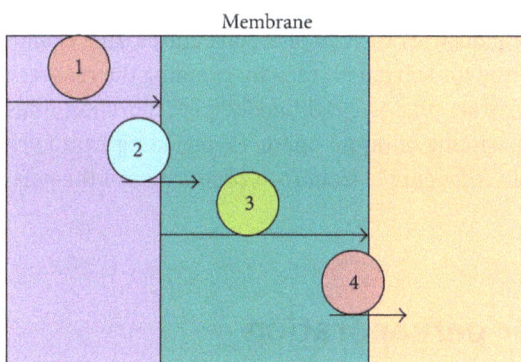

Figure 2: Major steps involved in the transport of a component through a pervaporation membrane. Copyright © 2015 Ghoshna Jyoti et al.

1.3 Advantages and applications

Pervaporation offers several advantages and finds diverse applications in various industries. One of the key advantages is its energy efficiency, as it operates at lower temperatures and pressures compared to conventional separation techniques like distillation [10]. This results in significant energy savings, making pervaporation a sustainable option for separation processes. Additionally, pervaporation provides high selectivity, enabling the separation of components with similar boiling points or azeotropic compositions, which are challenging to separate using traditional methods [11]. Its versatility is evident in its application to different liquid mixtures, including water–organic solvent systems, organic–organic mixtures, and liquid–gas separations, making it applicable in industries such as petrochemicals, pharmaceuticals, food and beverage, and environmental sectors [12, 13]. Pervaporation systems are relatively simple in design and operation, requiring minimal equipment and space. They can be easily integrated into existing processes, making them cost-effective and practical for industrial applications. Pervaporation also plays a crucial role in solvent recovery, allowing the separation and concentration of solvents from liquid mixtures. This is particularly valuable in industries where solvent recycling and reuse are essential for cost reduction and environmental sustainability [14, 15]. The advantages of pervaporation, including energy efficiency, selective separation, versatility, simplicity, and solvent recovery, make it a valuable separation technique with diverse applications across various industries. Its continued development and implementation contribute to efficient and sustainable separation processes.

In terms of applications, pervaporation is commonly employed in the dehydration of organic solvents, such as removing water from ethanol in biofuel production or drying solvents in chemical processes. It is also utilized in wastewater treatment, facilitating the purification and removal of volatile organic compounds (VOCs) from industrial wastewater. Moreover, pervaporation is employed in the separation and purification of specialty chemicals in industries such as pharmaceuticals, fine chemicals, and fragrance production. It is used for solvent extraction, enabling the recovery and concentration of solvents from extract phases. Additionally, pervaporation finds application in the concentration and refining of liquid mixtures, including fruit juice concentration, removal of undesirable components from food products, and the purification of essential oils [16].

2 Polymer materials for pervaporation

Polymer materials play a vital role in pervaporation due to their tunable properties, ease of fabrication, and cost-effectiveness. These materials can be tailored to exhibit specific characteristics, such as selectivity, permeability, stability, and compatibility

with different separation environments [17]. Here are some key aspects regarding
polymer materials for pervaporation:

- **Classification of polymeric membranes:** Polymer membranes used in pervaporation can be broadly classified into two categories: glassy polymers and rubbery
polymers. Glassy polymers, such as polysulfone and polyimide, exhibit a rigid
and amorphous structure, providing high selectivity but lower permeability. Rubbery polymers, such as polydimethylsiloxane (PDMS) and polyisoprene, possess a
flexible and rubber-like structure, offering high permeability but relatively lower
selectivity [18].
- **Membrane morphology and structure:** The morphology and structure of polymer
membranes significantly influence their separation performance. Common membrane structures include dense films, composite membranes, and asymmetric membranes. Dense films have a homogeneous structure and are typically used for gas
separation. Composite membranes consist of a thin selective layer, supported by a
porous substrate, providing a balance between selectivity and permeability. Asymmetric membranes have a graded structure with a dense top layer and a porous
support layer, combining the advantages of both dense films and porous materials.
- **Polymer selection:** The selection of the polymer material depends on the specific
separation requirements. Factors to consider include solute size, solute–membrane
interactions, solvent compatibility, thermal stability, and chemical resistance. Various
polymers have been investigated for pervaporation, including polymeric blends, copolymers, and functionalized polymers. Examples of commonly used polymer materials include poly(vinyl alcohol) (PVA), polyetherimide (PEI), polyethylene glycol
(PEG), and poly(acrylonitrile) (PAN).
- **Membrane fabrication techniques:** Polymer membranes can be fabricated
using various techniques, including solution casting, phase inversion, electrospinning, and layer-by-layer assembly. Solution casting involves dissolving the polymer in a solvent and casting the solution onto a substrate, followed by solvent
evaporation to form a solid membrane. Phase inversion involves precipitating
the polymer from a casting solution by changing the solvent composition or temperature. Electrospinning produces nanofiber membranes by electrostatically
drawing polymer solutions or melts into fine fibers. Layer-by-layer assembly involves depositing alternating layers of oppositely charged polymers to create a
multilayer membrane.
- **Modification techniques:** Polymer membranes can be modified to enhance their
separation performance. Surface modification techniques, such as plasma treatment, chemical functionalization, and coating with selective materials, can improve selectivity, hydrophilicity, and fouling resistance. Crosslinking techniques,
including chemical crosslinking and physical crosslinking, can enhance the stability and mechanical properties of polymer membranes.

Polymer materials provide a versatile platform for designing and fabricating membranes for pervaporation. The choice of polymer, membrane structure, and fabrication techniques can be tailored to meet specific separation requirements, leading to the development of efficient and selective membranes for various applications. Ongoing research aims to explore novel polymer materials, advanced fabrication methods, and surface modification techniques to further enhance the performance of polymer membranes in pervaporation [19].

2.1 Classification of polymeric membranes

Classification of polymeric membranes is a systematic categorization of these membranes, based on their properties, structure, and performance characteristics. It helps in understanding the different types of membranes and their suitability for specific separation processes (Table 2).

Table 2: Classification of polymeric membranes.

Classification	Description
Glassy polymers vs. rubbery polymers	Glassy polymers, such as polysulfone (PSF) and polyimide (PI), have a rigid and amorphous structure, offering high selectivity but lower permeability. Rubbery polymers, such as polydimethylsiloxane (PDMS) and polyisoprene (PI), have a flexible and rubber-like structure, providing high permeability but relatively lower selectivity.
Dense films	They are homogeneous, nonporous membrane layers, made from glassy polymers. They are suitable for gas separation applications where small gas molecules permeate, while larger molecules are blocked.
Composite membranes	They are composed of a thin selective layer, supported by a porous substrate. They provide a balance between selectivity and permeability, and are widely used in applications such as reverse osmosis and nanofiltration.
Asymmetric membranes	They are graded structures, with a dense top layer and a porous support layer. The dense layer provides selective separation, while the porous support enhances mechanical strength and permeability. They are fabricated using phase-inversion techniques.
Nanocomposite membranes	They are membranes incorporating nanoparticles or nanofillers into the polymer matrix to enhance performance. Nanoparticles can alter selectivity, permeability, and stability, offering improved separation efficiency.
Ion-exchange membranes	They contain fixed-charge groups, allowing selective transport of ions, based on charge and size. They are widely used in applications involving ion transport and separation, such as electrodialysis and fuel cells.

2.2 Membrane morphology and structure

Membrane morphology and structure play a crucial role in determining the performance and properties of polymeric membranes. Table 3 outlines different membrane morphologies and their structures.

Table 3: Types of membrane morphologies and their structures.

Morphology/structure	Description
Dense films	They are uniform, nonporous membranes, with a dense structure throughout. These films have a homogeneous cross section without any voids or pores. Dense films provide high selectivity but may have lower permeability compared to other membrane structures.
Composite membranes	They consist of a thin selective layer, supported by a porous substrate. The selective layer provides separation functionality, while the porous support layer enhances the mechanical strength and facilitates mass transfer. Composite membranes offer a balance between selectivity and permeability.
Asymmetric membranes	They have a graded structure, with a dense top layer and a porous support layer. The dense top layer serves as the active separation layer, while the porous support layer provides mechanical stability. Asymmetric membranes combine the advantages of dense films and porous materials.
Porous membranes	They possess a network of interconnected pores throughout the membrane. The pore size, distribution, and morphology can vary, ranging from microporous to macroporous structures. Porous membranes offer high permeability but may have lower selectivity, compared to dense films.
Composite/porous hybrid membranes	They combine the characteristics of composite membranes and porous membranes. These membranes consist of a selective layer with a mixed morphology, such as a dense selective layer on a porous support or a composite layer with embedded porous particles. Hybrid membranes offer a combination of selectivity, permeability, and mechanical stability.

These different membrane morphologies and structures allow for a range of separation capabilities, permeability rates, and mechanical properties. The choice of membrane morphology and structure depends on the specific application requirements and the desired balance between selectivity and permeability [20–22].

2.3 Polymer selection

Polymer selection is a critical aspect in designing polymeric membranes for various separation processes. The choice of polymer material influences the membrane's selectivity, permeability, stability, and compatibility with the separation environment [23, 24]. Here are some key considerations for polymer selection in membrane applications:

- **Solubility and swelling:** The polymer should be soluble in a suitable solvent to form a homogeneous membrane casting solution. Additionally, the polymer should exhibit minimal swelling in the presence of solvents or solutes to maintain the membrane's integrity and separation performance.
- **Chemical compatibility:** The polymer should be chemically compatible with the solvents, solutes, and operating conditions of the separation process. It should resist chemical degradation, fouling, and swelling when exposed to the separation environment.
- **Selectivity:** The polymer should possess the necessary selectivity to separate the desired components effectively. Different polymers exhibit varying affinities and interactions with specific solutes, allowing for tailored selectivity in separation processes.
- **Permeability:** The polymer should have sufficient permeability to allow for the desired flux or throughput of the separated components. The polymer's molecular structure, chain mobility, and free volume play a role in determining its permeability characteristics.
- **Mechanical strength and stability:** The polymer should possess adequate mechanical strength to withstand handling, operating pressures, and temperature variations. It should maintain its structural integrity and separation performance over the desired lifespan of the membrane.
- **Thermal stability:** The polymer should exhibit thermal stability, remaining functional, and maintaining its properties under the intended operating temperature range. Thermal stability is particularly crucial for applications involving high temperatures or thermal cycling.
- **Cost and availability:** Considerations of cost and availability are important in practical applications. The polymer should be cost-effective and readily available in the desired form and quantity for membrane fabrication.

Commonly used polymers in membrane applications include polymeric blends, copolymers, and functionalized polymers. Examples of popular polymer materials for pervaporation and other separation processes include poly(vinyl alcohol) (PVA), polysulfone (PSF), polyimide (PI), polyethylene glycol (PEG), and poly(acrylonitrile) (PAN), among others. The specific polymer selection depends on the target separation, compatibility with the separation environment, and the desired membrane properties [25, 26]. Polymer selection is a crucial step in membrane design, as it directly impacts the membrane's selectivity, permeability, stability, and compatibility. Considerations of solubility, chemical compatibility, selectivity, permeability, mechanical strength, thermal stability, and cost are vital for successful polymer selection in membrane applications.

3 Synthesis and modification of polymer membranes

The synthesis and modification of polymer membranes play a crucial role in tailoring their properties and optimizing their performance for specific separation processes. Polymer membranes can be synthesized using various fabrication techniques, and their properties can be further enhanced through modification methods. This section provides an overview of the synthesis and modification approaches employed in the development of polymer membranes.

3.1 Membrane fabrication techniques

Membrane fabrication techniques are essential for producing polymeric membranes with the desired structures and properties. These techniques determine the membrane's morphology, thickness, pore size, and overall performance. Here are some commonly used membrane fabrication techniques:

- **Solution casting:** Solution casting involves dissolving the polymer in a suitable solvent to form a casting solution. The solution is then cast onto a solid substrate or a rotating drum. As the solvent evaporates, a thin polymeric film or membrane is formed. Solution casting allows for precise control of the membrane thickness and is suitable for producing dense films or composite membranes.
- **Phase inversion:** Phase inversion is a widely used technique to fabricate membranes with different structures, including dense films, composite membranes, and asymmetric membranes. It involves the conversion of a homogeneous polymer–solvent casting solution into a solid membrane by inducing a phase separation. This phase separation can be achieved by changing the solvent composition, temperature, or by employing nonsolvents. Phase inversion allows for the formation of membranes with controlled porosity and tailored structures.
- **Electrospinning:** Electrospinning is a method used to produce polymer fibers, with diameters ranging from nanometers to micrometers. It involves applying a high electric field to a polymer solution or melt, causing the solution to form a charged jet. The jet is then stretched and collected on a grounded or rotating collector, resulting in a fibrous membrane. Electrospinning enables the fabrication of nanofiber membranes with high surface-area-to-volume ratio and use in potential applications such as filtration, tissue engineering, and energy storage.
- **Interfacial polymerization:** Interfacial polymerization is employed to fabricate thin-film composite membranes. It involves the reaction between monomers at the interface of two immiscible phases, typically a liquid and a gas or a liquid and a solid. The monomers diffuse into the opposite phase and react to form a thin

selective layer on a support material. Interfacial polymerization allows for the production of membranes with a highly selective layer for specific separations.
– **Breath figures:** Breath figures are formed by the condensation of water droplets on a polymer film surface. This technique involves casting a polymer solution onto a substrate and exposing it to a humid environment. Water droplets condense on the surface, creating a regular pattern of spherical pores. By controlling the environmental conditions, such as temperature and humidity, membranes with well-defined pore structures can be fabricated. Breath figures offer a simple and cost-effective approach to produce porous membranes.
– **Template methods:** Template methods involve the use of preformed templates or sacrificial materials to create porous structures in membranes. For example, sacrificial templating uses a sacrificial material that is incorporated into the polymer matrix and later removed, leaving behind a porous structure. Template methods provide control over pore size, distribution, and shape, enabling the fabrication of membranes with specific pore architectures.

These membrane fabrication techniques offer versatility in producing membranes with various structures and properties. The choice of technique depends on the desired membrane morphology, porosity, thickness, and the targeted application requirements. Researchers continue to explore new fabrication techniques and combinations to develop innovative membranes for improved separation performance.

3.2 Polymer modification methods

Polymer modification methods are employed to enhance the properties and performance of polymeric membranes for specific separation applications. These methods involve modifying the structure, surface, or properties of the polymer membranes through various techniques. Here are some commonly used polymer modification methods:
– **Surface functionalization:** Surface functionalization involves introducing specific functional groups or molecules onto the surface of the polymer membrane. This can be achieved through surface grafting, plasma treatment, chemical reactions, or adsorption of functional molecules. Surface functionalization can enhance membrane properties such as selectivity, hydrophilicity, fouling resistance, and interaction with target solutes.
– **Crosslinking:** Crosslinking refers to the formation of chemical bonds between polymer chains in the membrane matrix. It improves the membrane's mechanical strength, stability, chemical resistance, and thermal stability. Crosslinking can be achieved through chemical crosslinking agents, UV irradiation, heat, or other crosslinking methods. Crosslinked membranes are less prone to swelling and deformation, leading to improved separation performance and long-term stability.

- **Blending and copolymerization:** Blending involves the incorporation of additional polymers or additives into the membrane matrix. This can be done by physically blending polymer solutions or by melt blending polymers. Blending allows for the modification of membrane properties such as selectivity, permeability, mechanical strength, and thermal stability. Copolymerization involves the polymerization of multiple monomers to form a copolymer membrane with tailored properties.
- **Nanocomposite formation:** Nanocomposite membranes involve the incorporation of nanoparticles or nanofillers into the polymer matrix. The nanoparticles can be organic or inorganic and provide unique properties such as enhanced selectivity, mechanical strength, thermal stability, and fouling resistance. Nanocomposite membranes are prepared through techniques like in situ polymerization, solution blending, or impregnation of nanoparticles into the polymer matrix.
- **Surface coating:** Surface coating involves applying a thin layer of a specific material onto the membrane surface. This can be achieved through techniques such as dip coating, spin coating, or chemical vapor deposition. Surface coatings can improve membrane properties such as selectivity, fouling resistance, hydrophilicity, or antimicrobial properties. Examples include coating with selective polymers, metals, metal oxides, or graphene-based materials.
- **Molecular imprinting:** Molecular imprinting involves creating specific recognition sites or cavities within the polymer matrix that can selectively bind to target molecules. This technique is based on the formation of a template-monomer complex, followed by polymerization and removal of the template molecule. Molecularly imprinted membranes offer selective separation capabilities and are used in applications such as molecular recognition, sensors, and drug delivery.

These polymer modification methods provide means to tailor the properties and performance of polymeric membranes to meet specific separation requirements. By applying appropriate modification techniques, researchers can enhance the selectivity, permeability, stability, and compatibility for improved membrane performance in various separation processes.

3.3 Crosslinking and surface functionalization

Crosslinking and surface functionalization are two important methods used for modifying polymeric membranes to enhance their properties and performance. Crosslinking involves the formation of covalent bonds between polymer chains, creating a three-dimensional network within the membrane matrix [27]. This method improves the mechanical strength, stability, chemical resistance, and thermal stability of the membrane. Crosslinked membranes are less prone to swelling and deformation, making them suitable for demanding separation environments [28]. Crosslinking can be achieved through various techniques:

- **Chemical crosslinking:** It involves the use of crosslinking agents, such as multifunctional monomers or crosslinking agents, which react with the polymer chains. Common crosslinking agents include glutaraldehyde, divinylbenzene, or polyfunctional isocyanates. Chemical crosslinking can be performed through solution immersion, in situ polymerization, or surface grafting methods.
- **Physical crosslinking:** Physical crosslinking relies on non-covalent interactions to form crosslinks between polymer chains. This can be achieved through methods such as thermal annealing, UV irradiation, or ionizing radiation. Physical crosslinking offers the advantage of reversible crosslinks, allowing for the regeneration or reshaping of the membrane.

Crosslinking improves the stability and durability of polymeric membranes, making them resistant to chemical attack, fouling, and physical stresses. It also helps maintain the membrane's structural integrity, enhancing its separation performance and prolonging its lifespan.

Surface functionalization involves modifying the surface of the polymeric membrane by introducing specific functional groups or molecules. This method aims to improve membrane properties such as selectivity, hydrophilicity, fouling resistance, and interaction with target solutes. Surface functionalization techniques include:

- **Grafting:** Grafting involves attaching functional molecules or polymer chains onto the membrane surface through covalent bonding. It can be achieved through various methods, including grafting-from, grafting-to, or grafting-through techniques. Grafted molecules can enhance the membrane's surface properties, such as hydrophilicity, ion selectivity, or antifouling characteristics.
- **Plasma treatment:** Plasma treatment involves subjecting the membrane surface to a low-temperature plasma discharge. This process alters the surface chemistry and morphology, improving hydrophilicity, wettability, and surface energy. Plasma treatment can also introduce reactive functional groups onto the surface, enabling subsequent chemical reactions or grafting.
- **Chemical functionalization:** Chemical functionalization involves modifying the membrane surface through chemical reactions. This can include techniques such as diazotization, acylation, or silanization. Chemical functionalization allows for the introduction of specific functional groups onto the membrane surface, altering its chemical and physical properties.

Surface functionalization can enhance the membrane's selectivity toward specific solutes, reduce fouling tendencies, improve antifouling properties, and enable interactions with target molecules. It provides control over surface characteristics and allows customization of membranes for specific separation applications [29]. Both crosslinking and surface functionalization methods offer valuable means to modify polymeric membranes and optimize their properties for improved separation perfor-mance. These techniques enable researchers to develop membranes with enhanced stability, selectivity, permeability, and compatibility, catering to the requirements of various separation processes [30].

4 Characterization techniques

Characterization techniques play a vital role in understanding and evaluating the properties and performance of polymeric membranes. These techniques provide valu-able insights into membrane morphology, structure, chemical composition, mechani-cal properties, transport properties, and surface characteristics. Here is an overview of commonly used characterization techniques for polymeric membranes [31, 32]:

Morphological analysis:
- Scanning electron microscopy (SEM): SEM provides high-resolution imaging of the membrane surface and cross-section, allowing for the examination of mem-brane morphology, pore structure, and surface roughness.
- Transmission electron microscopy (TEM): TEM offers detailed imaging at the nanoscale, providing information about the membrane's ultrastructure, including nanopores, particle distribution, and interfacial characteristics.
- Atomic force microscopy (AFM): AFM enables the measurement of surface topog-raphy and the characterization of membrane roughness at the nanoscale.

Thermal and mechanical properties:
- Differential scanning calorimetry (DSC): DSC measures the heat flow associated with thermal transitions in the membrane, providing information about glass transition temperature, melting point, crystallinity, and thermal stability.
- Thermogravimetric analysis (TGA): TGA measures the weight changes of the membrane as a function of temperature, allowing for the evaluation of thermal stability, degradation temperatures, and thermal decomposition behavior.
- Mechanical testing: Mechanical properties such as tensile strength, elasticity, and flexibility can be determined through techniques like tensile testing, flexural test-ing, or indentation testing.

Transport properties and permeability measurement:
- Permeability testing: Permeability measurements determine the membrane's ability to allow the passage of specific substances. Techniques such as gas perme-ation tests, liquid permeation tests, and pervaporation tests are used to evaluate the permeability and selectivity of membranes.

– Bubble point/pore size distribution: Bubble point testing measures the minimum pressure required to force liquid through the membrane pores, providing information about the membrane's pore size distribution and pore structure.

Spectroscopic techniques for membrane analysis:
– Fourier transform infrared spectroscopy (FTIR): FTIR analyzes the vibrational modes of chemical bonds, allowing for the identification of functional groups, polymer structure, and chemical composition.
– X-ray photoelectron Spectroscopy (XPS): XPS provides information about the elemental composition, surface chemistry, and chemical bonding states of the membrane surface.
– Nuclear magnetic resonance (NMR): NMR spectroscopy provides insights into the molecular structure, polymer chain dynamics, and diffusion behavior within the membrane.

The selection of appropriate techniques depends on the specific properties and parameters of interest. By employing a combination of characterization techniques, researchers can gain a comprehensive understanding of the membrane structure, and its properties and performance, facilitating the development and optimization of membranes for various separation applications.

5 Transport mechanisms in polymer membranes

Transport mechanisms in polymer membranes play a fundamental role in various separation processes. These mechanisms govern the movement of molecules or ions through the membrane, determining the selectivity and permeability of the separation system. Understanding the transport mechanisms is essential for optimizing membrane performance and designing membranes tailored for specific applications [33, 34]. Polymer membranes exhibit different transport mechanisms, including diffusion and sorption, which are governed by factors such as membrane morphology, polymer properties, solute characteristics, and operating conditions. By unraveling the intricacies of these transport mechanisms, researchers can enhance membrane efficiency, selectivity, and durability, paving the way for advancements in fields such as water purification, gas separation, and membrane-based processes for chemical and pharmaceutical industries.

5.1 Diffusion and sorptionin membranes

Diffusion and sorption are two key transport mechanisms that occur in polymeric membranes. These mechanisms play a crucial role in the separation and transport of molecules across the membrane. Diffusion refers to the movement of molecules from an area of high concentration to an area of low concentration. In polymeric membranes, diffusion occurs as solute molecules or gas molecules dissolve into the membrane and move through the polymer matrix. The rate of diffusion is influenced by factors such as the solute concentration gradient, membrane thickness, porosity, and the diffusion coefficient of the solute in the membrane material [35]. The diffusion process in membranes can be described by Fick's laws, which relate the diffusion flux to the concentration gradient [36]. Diffusion is generally driven by the concentration difference across the membrane, and occurs in both dense films and porous membranes. In dense films, diffusion typically dominates the transport mechanism, while in porous membranes, both diffusion and convection contribute to mass transport.

Sorption is the process by which solute molecules are adsorbed or absorbed onto the surface or within the membrane material. Sorption can occur through physical interactions, such as van der Waals forces or hydrogen bonding, or through chemical interactions such as specific molecular recognition or ion exchange. Sorption plays a significant role in the separation process, particularly in membranes with selective layers. The sorption capacity and selectivity of the membrane depend on factors such as the chemical nature of the solute and membrane, surface properties, and the presence of functional groups or pores. Sorption can be reversible, where solutes are weakly bound and easily desorb, or irreversible, where solutes form strong bonds with the membrane material [37].

The combination of diffusion and sorption processes determines the transport behavior of solutes across polymeric membranes. Understanding and controlling these mechanisms is essential for optimizing membrane performance, achieving high selectivity, and improving separation efficiency. Various factors, such as membrane structure, polymer properties, solute characteristics, and operating conditions, influence the relative contributions of diffusion and sorption in membrane transport. By manipulating these factors, researchers can design membranes with enhanced selectivity, permeability, and stability, catering to the specific requirements of diverse separation applications.

5.2 Solution–diffusion model

The solution–diffusion model is a widely used theoretical framework for understanding and describing the transport behavior of solutes through polymeric membranes. It is based on the concept that the transport process occurs through two sequential steps: solution and diffusion [38]. In the solution step, solutes dissolve into the membrane matrix, typically through sorption or permeation into the polymer. The solutes first interact with the membrane surface and then penetrate the polymer structure. The solubility and sorption characteristics of the solutes in the membrane material play a crucial role in this step. Diffusion: Following the solution step, the solutes diffuse through the polymer matrix, driven by a concentration gradient. The diffusion process involves the random motion of solutes within the polymer, where they move from regions of high concentration to regions of low concentration. Diffusion is influenced by factors such as the diffusion coefficient of the solute, membrane thickness, porosity, and the nature of the polymer matrix.

The solution–diffusion model assumes that the solute transport occurs primarily through these two steps. It provides a framework for predicting the permeability and selectivity of polymeric membranes by considering the solubility and diffusivity of solutes in the membrane material. The model assumes that the solute molecules are independent and do not interact with each other during transport [39]. The solution–diffusion model is often used in the design and optimization of membranes for various separation processes. By understanding the solubility and diffusivity properties of different solutes and membranes, researchers can predict the membrane's performance in terms of permeability, selectivity, and separation efficiency. However, it is important to note that the model is a simplified representation of the complex transport behavior in membranes, and deviations from the model may occur in real-world scenarios due to phenomena such as concentration polarization, sorption kinetics, and interactions between solute molecules and the polymer matrix [40].

5.3 Selectivity and permeability

Selectivity and permeability are two essential properties of polymeric membranes that determine their effectiveness in separation processes. Selectivity refers to the membrane's ability to preferentially allow the passage of certain solutes or molecules, while restricting or rejecting others. It is a measure of the membrane's ability to separate components based on their size, shape, charge, or affinity. Selectivity is typically quantified by the separation factor, which is the ratio of the permeabilities of two different solutes through the membrane. The selectivity of a membrane depends on various factors, including:

- Molecular size and shape: Membranes with smaller pores or restricted pathways can selectively allow smaller molecules to pass while blocking larger ones.
- Charge and polarity: Membranes with charged functional groups or ion-exchange properties can exhibit selectivity toward specific ions or polar molecules.
- Solubility and sorption: Membranes with specific interactions or sorption sites can preferentially adsorb or dissolve certain solutes, leading to selectivity.

Permeability: Permeability refers to the rate at which solutes can pass through the membrane. It is a measure of the membrane's ability to allow the transport of solutes, and is often quantified by the permeability coefficient, which represents the rate of solute diffusion through the membrane per unit area and per unit concentration gradient. The permeability of a membrane is influenced by several factors, including:

- Membrane thickness: Thinner membranes generally exhibit higher permeability due to shorter diffusion paths.
- Porosity: Membranes with higher porosity or larger pore sizes can offer higher permeability but may sacrifice selectivity.
- Polymer properties: The inherent diffusivity and solubility characteristics of the polymer matrix impact the membrane's permeability.
- Transport mechanism: Different transport mechanisms, such as diffusion or convection, contribute to the overall permeability.

Balancing selectivity and permeability is crucial in membrane design. While high selectivity is desirable for specific separations, it often comes at the expense of reduced permeability. Finding the optimal trade-off between these properties depends on the target separation application and the desired performance requirements.

Polymeric membranes can be tailored to achieve the desired selectivity and permeability by carefully selecting the membrane material, modifying the membrane structure, introducing functional groups, or incorporating nanoparticles. The selection and optimization of membrane properties play a key role in meeting the specific needs of separation processes in various industries, including water treatment, gas separation, pharmaceutical purification, and chemical processing.

6 Performance evaluation and optimization

Performance evaluation and optimization are crucial aspects in the development and utilization of polymeric membranes for separation processes. Performance evaluation involves assessing the membrane's efficiency, selectivity, permeability, stability, and other relevant parameters. Various performance metrics are used, including flux, separation factor, rejection rate, and fouling resistance. Evaluation techniques such as permeability testing, solute transport measurements, and fouling studies provide valuable insights into membrane performance. Optimization strategies aim to enhance membrane performance by adjusting key parameters such as membrane composition, structure, surface modification, and operating conditions. These strategies can involve material selection, membrane fabrication optimization, modification techniques, and process optimization. By evaluating and optimizing membrane performance, researchers and engineers can improve separation efficiency, reduce energy consumption, increase membrane lifespan, and address challenges such as fouling and selectivity limitations. This continuous improvement process leads to the development of more efficient and sustainable membrane-based separation technologies [41, 42].

Table 4: Various performance metrics.

Performance metric	Description
Separation factor	Quantifies the membrane's ability to separate components in the liquid mixture
Flux	Measures the rate of permeate passage through the membrane per unit area
Permeability	Represents the membrane's ability to allow the transport of specific components
Pervaporation separation index (PSI)	Comprehensive metric, combining separation factor and flux, to assess overall pervaporation performance
Stability	Evaluates the membrane's ability to maintain separation performance, over time, without significant degradation
Energy efficiency	Measures the energy consumption required to achieve the desired separation

These performance metrics provide a comprehensive assessment of the effectiveness and efficiency of polymeric membranes in pervaporation processes. They aid in the comparison, optimization, and development of membranes for various separation applications as shown in Table 4 above.

6.1 Factors affecting membrane performance

These factors play a crucial role in determining the efficiency, selectivity, and stability of the membrane. Table 5 defines the important factors that influence membrane performance:

Considering and optimizing these factors is essential for achieving optimal membrane performance in terms of selectivity, permeability, stability, and longevity. By carefully selecting the membrane material, adjusting the membrane structure, optimizing operating conditions, addressing fouling concerns, implementing appropriate pretreatment and conditioning techniques, and ensuring membrane durability, researchers and engineers can design membranes that meet the specific requirements of separation processes.

Table 5: Factors affecting membrane performance.

Factor	Description
Membrane material	The choice of polymer material, considering properties such as solubility, diffusivity, selectivity, thermal stability, and mechanical strength
Membrane structure	The membrane's morphology, pore size, and structure, which impact selectivity and permeability
Membrane thickness	The thickness of the membrane, influencing mass transfer rates, diffusion, selectivity, and flux
Operating conditions	Parameters such as temperature, pressure, and feed composition, which affect solute diffusion, selectivity, and flux
Fouling and membrane cleaning	The accumulation of unwanted substances on the membrane surface, impacting flux, selectivity, and overall performance
Membrane pretreatment and conditioning	Techniques such as rinsing, soaking, and chemical treatments to modify surface properties and enhance membrane stability
Membrane aging and durability	Changes in membrane properties over time, influenced by chemical degradation, mechanical stresses, and exposure to harsh conditions

6.2 Strategies for performance optimization

To optimize the performance of polymeric membranes in separation processes, various strategies can be employed. The following strategies play a major role to enhance selectivity, permeability, stability, and overall separation efficiency.

- Membrane material selection: Careful selection of the membrane material is crucial for achieving the desired separation performance. Different polymers possess unique properties that can be tailored to specific separation needs. Considering factors such as solubility, diffusivity, selectivity, thermal stability, and mechanical strength helps in identifying the most suitable membrane material for the target application.
- Membrane structure and morphology: Modifying the membrane structure and morphology can significantly impact performance. Techniques such as controlling pore size, adjusting membrane thickness, and optimizing the membrane architecture (e.g., dense films or porous membranes) allow for fine-tuning of selectivity and permeability. Membrane fabrication methods, including phase inversion, electrospinning, and self-assembly, can be utilized to achieve the desired membrane structure.
- Surface modification: Surface modification techniques alter the surface properties of the membrane to improve selectivity, fouling resistance, and interaction with the target solutes. Surface functionalization, coating, or grafting of specific

molecules or functional groups onto the membrane surface can enhance selectivity, hydrophilicity, or antifouling properties. Surface roughness and charge can also be tailored to optimize performance.

– Polymer blending and Composite membranes: Blending two or more polymers or incorporating additives into the membrane matrix can improve performance. Blending polymers with complementary properties can enhance selectivity and permeability. Adding nanoparticles, such as zeolites, carbon nanotubes, or graphene oxide, can improve selectivity, mechanical strength, and stability through enhanced dispersion and interaction with the polymer matrix.

– Process optimization: Optimizing operating conditions such as temperature, pressure, and feed composition can significantly impact membrane performance. Adjusting these parameters to match the separation requirements and solute characteristics can improve selectivity, permeability, and energy efficiency. Optimization may involve finding the optimal temperature, pressure, or concentration gradient to maximize separation performance while minimizing energy consumption.

– Fouling mitigation and cleaning: Fouling can adversely affect membrane performance. Implementing strategies to mitigate fouling, such as pretreatment, backwashing, or the use of antifouling coatings, can improve membrane longevity and maintain performance. Regular cleaning protocols, such as chemical cleaning or membrane rejuvenation techniques, help restore membrane performance and extend its operational lifespan.

– Advanced characterization and modeling: Utilizing advanced characterization techniques, such as molecular simulations or advanced imaging methods, can provide a deeper understanding of membrane structure, transport mechanisms, and performance. Incorporating modeling and simulation tools can aid in predicting and optimizing membrane performance, guiding membrane design, and facilitating the selection of operating conditions.

These strategies for performance optimization empower researchers and engineers to enhance the efficiency, selectivity, stability, and longevity of polymeric membranes in the separation processes. By employing a combination of these strategies, tailored membranes can be developed to meet the specific requirements of various industries, including water treatment, gas separation, pharmaceuticals, and chemical processing.

7 Advanced polymer membrane materials

Advanced polymer membrane materials refer to the development and utilization of innovative polymer-based membranes with enhanced properties and performance characteristics. These materials are designed to overcome the limitations of traditional membranes and meet the increasing demands of various separation processes [43, 44].

Table 6: Details about advanced polymer membrane materials.

Advanced polymer membrane material	Description	Examples
High-performance polymers	Polymers with exceptional properties such as high thermal stability, chemical resistance, and selectivity	Polyimides, polybenzimidazoles, polyetherimides
Functionalized polymers	Polymers with specific functional groups, incorporated to enhance selectivity and sorption properties	Ion-exchange polymers, affinity-based polymers
Polymer nanocomposites	Hybrid materials, combining polymers with nanoscale fillers to enhance mechanical strength and selectivity	Polymer-graphene oxide composites, polymer-nanoparticle hybrids
Responsive and Smart polymers	Polymers that exhibit changes in properties in response to external stimuli, allowing for switchable behavior	Temperature-responsive polymers, pH-responsive polymers
Membranes with tunable pore sizes	Membranes engineered to have adjustable pore sizes for precise molecular sieving and size-based separations	Microporous membranes, nanoporous membranes with controlled pore size distributions

These advanced polymer membrane materials demonstrate significant advancements in membrane technology, offering improved selectivity, permeability, stability, and responsiveness. They contribute to the development of more efficient and versatile membrane systems for various separation processes as depicted in Table 6 above.

7.1 Hybrid and composite membranes

Hybrid and composite membranes have gained significant attention in membrane technology due to their unique properties and enhanced separation performance. These membranes are created by combining different materials, such as polymers, inorganic materials, or nanoparticles, to take advantage of their synergistic effects [45].

These hybrid and composite membranes demonstrate the potential to overcome limitations of traditional polymeric membranes by synergistically combining the advantages of different materials. They offer improved selectivity, permeability, stability, and mechanical strength, opening up new possibilities for separation processes in diverse industries such as water treatment, gas separation, and chemical processing. Ongoing research continues to explore novel hybrid and composite membrane designs to further enhance their performance and expand their applications [46] as summarized in Table 7.

Table 7: Hybrid and Composite membranes.

Membrane	Description
Polymer–inorganic hybrids	Membranes that combine polymers with inorganic materials, such as zeolites, metal-organic frameworks (MOFs), or silica, to enhance performance; these hybrids offer improved selectivity and stability.
Mixed matrix membranes (MMMs)	Membranes consisting of a polymeric matrix embedded with inorganic fillers, such as zeolites, MOFs, or carbon nanotubes; these fillers enhance selectivity, permeability, and mechanical strength.
Nanocomposite membranes	Membranes that incorporate nanoparticles, such as graphene oxide, metal oxides, or carbon nanotubes, into a polymer matrix; these nanoparticles improve selectivity, mechanical strength, and transport properties.
Polymeric–inorganic thin-film composite (TFC) membranes	Membranes composed of a thin polymeric selective layer deposited onto an inorganic support or porous substrate; TFC membranes offer high selectivity, low resistance, and improved separation efficiency.
Interfacial polymerization composite membranes (IPC)	Membranes formed by the in-situ polymerization of monomers at the interface of a support material; IPC membranes exhibit excellent selectivity, stability, and performance in various separation processes.
Hybrid ceramic membranes	Membranes that combine ceramic materials with polymer components to create hybrid structures; these membranes offer high temperature resistance, chemical stability, and enhanced separation performance.

7.2 Modification techniques for performance enhancement

Modification techniques play a crucial role in enhancing the performance of polymeric membranes. These techniques involve altering the membrane's surface properties or structure to improve selectivity, permeability, fouling resistance, and stability as tabulated in Table 8 below.

These modification techniques allow researchers to tailor membrane properties to specific separation needs and optimize performance. By functionalizing the membrane surface, blending different polymers, crosslinking, applying thin film coatings, or modifying surface roughness or charge, membranes can exhibit enhanced selectivity, permeability, fouling resistance, and stability. These modifications enable the development of tailored membranes for various applications, including water treatment, gas separation, pharmaceutical purification, and chemical processing [47].

Table 8: Common modification techniques for performance enhancement in polymeric membranes.

Modification technique	Description
Surface functionalization	Chemical modification of the membrane surface to introduce functional groups or coatings with specific properties
Polymer blending	Incorporation of different polymers to create blends with improved properties, such as enhanced selectivity or stability
Nanocomposite membranes	Membranes that incorporate nanoparticles, such as graphene oxide, metal oxides, or carbon nanotubes, into a polymer matrix; these nanoparticles improve selectivity, mechanical strength, and transport properties
Crosslinking	Formation of covalent bonds within the polymer matrix to increase membrane stability, mechanical strength, and fouling resistance
Surface roughening	Creating a rough surface texture to enhance hydrodynamic flow, reduce fouling, or increase active surface area for enhanced separation
Molecular sieving	Introduction of selective barriers, such as pore-blocking agents or size-exclusion materials, to achieve precise molecular sieving properties
Surface roughening	Creating a rough surface texture to enhance hydrodynamic flow, reduce fouling, or increase active surface area for enhanced separation
Surface charge modification	Alteration of the membrane surface charge to influence solute–surface interactions, selectivity, and fouling resistance

7.3 Novel applications and future directions

Novel applications and future directions in the field of polymeric membranes explore new areas where these membranes can be utilized, and provide insights into potential advancements. Table 9 provides an overview of novel applications and future directions in the field of polymeric membranes [48, 49].

These novel applications and future directions highlight the expanding use of polymeric membranes across various industries and research fields. They reflect the growing demand for sustainable and efficient separation technologies and the need to address emerging challenges. Continued advancements in membrane materials, design, and process optimization, along with the exploration of new applications, will pave the way for improved membrane performance, reduced energy consumption, and increased sustainability in the future.

Table 9: Overview of novel applications and future directions.

Novel applications and future directions	Description
Water desalination	Utilizing polymeric membranes for desalination processes, including reverse osmosis and membrane distillation
Wastewater treatment	Application of polymeric membranes for efficient removal of contaminants and pollutants from wastewater
Gas separation	Employing polymeric membranes for the separation and purification of gases, such as CO_2 capture and natural gas processing
Renewable energy	Integration of polymeric membranes in energy-related processes, including fuel cells, batteries, and hydrogen production
Biomedical applications	Utilizing membranes for biomedical applications, such as drug delivery, tissue engineering, and blood filtration
Emerging contaminant removal	Addressing the removal of emerging contaminants, such as pharmaceuticals and microplastics, from water sources
Membrane reactors	Combining membrane separation and reaction processes in membrane reactors for simultaneous separation and conversion
Sustainability and Environmental impact reduction	Developing eco-friendly membranes, optimizing energy efficiency, and reducing environmental impact in membrane processes
Advanced modeling and Simulation techniques	Employing advanced modeling and simulation tools to enhance understanding, design, and optimization of membrane systems
Multifunctional membranes	Developing membranes with multifunctional capabilities, such as simultaneous separation and catalysis

8 Conclusion

In conclusion, pervaporation is a promising membrane separation technique that offers numerous advantages and applications in various industries. This book chapter provided an in-depth overview of pervaporation, including its principles, mechanisms, and the importance of polymer membranes. The classification of polymeric membranes, their morphology, and structure were discussed, highlighting the role of polymer selection in achieving desired separation performance. The article also explored the synthesis and modification of polymer membranes, including fabrication techniques, polymer modification methods, and crosslinking and surface functionalization approaches. Characterization techniques for evaluating membrane properties were presented, encompassing morphological analysis, thermal and mechanical properties, transport properties, and spectroscopic techniques.

Transport mechanisms in polymer membranes, such as diffusion and sorption, as well as the solution–diffusion model, were discussed in detail. The concepts of selectivity and permeability were explained, shedding light on the factors influencing membrane performance. The chapter also delved into performance evaluation and optimization, presenting pervaporation performance metrics, factors affecting membrane performance, and strategies for performance enhancement. Moreover, recent advances and emerging trends in the field were outlined, including advanced polymer membrane materials, hybrid and composite membranes, modification techniques for performance enhancement, and novel applications and future directions. These advancements hold great potential for addressing current challenges and expanding the scope of pervaporation in various fields.

References

[1]   Ma, X., Ji, Y., Wang, T., Wang, J., and Qiu, J. Pervaporation membranes for liquid mixture separation: Principles, materials, and applications. Chem. Eng. J., 2020, **391**: 123499. doi: 10.1016/j. cej.2020.123499.

[2]   Goh, P. S., Ismail, A. F., and Ng, B. C. Pervaporation membranes for liquid mixture separation: Selectivity and permeability considerations. J. Membr. Sci., 2019, **573**: 211–230. doi: 10.1016/j. memsci.2018.12.071.

[3]   Wang, X., Gao, C., and Li, G. Polymer membranes for pervaporation: Materials, structures, and applications. Chem. Eng. J., 2020, **400**: 125947. doi: 10.1016/j.cej.2020.125947.

[4]   Wang, J., Song, H., Wang, T., Liu, W., and Xu, N. Pervaporation membranes for energy-efficient separation: Applications and future prospects. Sep. Purif. Technol., 2021, **273**: 118975. doi: 10.1016/j. seppur.2021.118975.

[5]   Li, X., Wang, Z., Sun, Z., Li, X., and Yu, S. Recent advances in pervaporation separation of azeotropic and close-boiling mixtures: Materials, techniques, and applications. Sep. Purif. Technol., 2021, **276**: 119118. doi: 10.1016/j.seppur.2021.119118.

[6]   Yang, C., Chen, J., Xu, N., and Wang, Y. Pervaporation separation of azeotropic mixtures: Challenges and opportunities. Ind. Eng. Chem. Res., 2021, **60**(20): 7115–7134. doi: 10.1021/acs.iecr.1c00804.

[7]   Ghosh, A. K., Mandal, S., Das, S., Sujith, R., and Banerjee, S. Recent advances in pervaporation separation of azeotropic mixtures using polymeric membranes: A comprehensive review. Sep. Purif. Rev., 2021, **50**(4): 381–419. doi: 10.1080/15422119.2020.1763033.

[8]   Ghasemzadeh, K., Ismail, A. F., Matsuura, T., Rana, D., and Jaafar, J. Recent advances in polymer–based membranes for pervaporation process. Sep. Purif. Rev., 2019, **48**(1): 1–24. doi: 10.1080/15422119.2018.1442739.

[9]   Nunes, S. P., Peinemann, K. V., and Geise, G. M. Polymer membranes for membrane distillation: Recent advances and future prospects. J. Membr. Sci., 2019, **572**: 72–94. doi: 10.1016/j. memsci.2018.10.017.

[10] Pabby, A. K., Rizvi, S. S. H., and Sastre A. M. Handbook of Nembrane Separations Chemical, Pharmaceutical, Food, and Biotechnological Applications, Elsevier, 2000.

[11] Goh, P. S., Ismail, A. F., Ng, B. C., and Hilal, N. Pervaporation separation of organic–organic mixtures: Recent advances and challenges. Sep. Purif. Technol., 2018, **199**: 459–482. doi: 10.1016/j. seppur.2018.02.056.

[12]   Li, J., Li, S., and Liao, C. Recent advances in pervaporation separation membranes for bioalcohol production. Renewable Energy, 2019, **139**: 1203–1214. doi: 10.1016/j.renene.2019.03.073.

[13]   Kim, S. J., and Hoek, E. M. V. Pervaporation membranes for biofuel production: Membrane development, modification, and module design. J. Membr. Sci., 2016, **499**: 480–502. doi: 10.1016/j.memsci.2015.10.055.

[14]   Dong, L., Li, Y., Huang, H., Wang, Z., and Li, X. Pervaporation: A membrane–based technology for solvent dehydration and recovery. Chem. Eng. J., 2019, **356**: 870–885. doi: 10.1016/j.cej.2018.09.213.

[15]   Surya Murali, R., and Bhaumik, D. Pervaporation: A membrane–based separation technology for solvent recovery and wastewater treatment. Environ. Sci. Pollut. Res., 2020, **27**(9): 8759–8777. doi: 10.1007/s11356-020-07807-0.

[16]   Ye, X., and Zhang, R. Pervaporation dehydration of organic solvents for sustainable processes. Ind. Eng. Chem. Res., 2018, **57**(14): 4779–4792. doi: 10.1021/acs.iecr.7b05197.

[17]   Goh, P. S., Ismail, A. F., and Ng, B. C. Polymeric membranes for pervaporation: Preparation, characterization, and applications. Polymers, 2018, **10**(1): 30. doi: 10.3390/polym10010030.

[18]   Yin, J., and Xu, T. Polymer membranes for pervaporation and vapor permeation processes: Recent advances and perspectives. J. Membr. Sci., 2017, **524**: 446–469. doi: 10.1016/j.memsci.2016.11.065.

[19]   Li, Y., Li, J., and Liao, C. Advanced fabrication techniques for polymeric membranes in pervaporation. J. Membr. Sci., 2020, **610**: 118243. doi: 10.1016/j.memsci.2020.118243.

[20]   Hilal, N., Ismail, A. F., Matsuura, T and Oatley-Radcliffe, D. (Eds.). Membrane Characterization, Elsevier, 2017, pp. 1–41.

[21]   Nunes, S. P., and Peinemann, K. V. Membrane Technology in the Chemical Industry: Principles and Applications, Wiley–VCH, 2010, **4**.

[22]   Wang, Z., Xu, T., Li, J. Polymeric membranes with engineered morphology for pervaporation: Recent advances and future directions. J. Membr. Sci., 2020, **603**: 118002. doi: 10.1016/j.memsci.2020.118002.

[23]   Rautenbach, R., Friedrich, H. B., and Schlosser, E. Handbook of Membrane Separations: Chemical, Pharmaceutical, Food, and Biotechnological Applications, CRC Press, 2016.

[24]   Lueptow, R. M., and Feng, X. Polymer membrane materials for the pervaporation separation of organic liquid mixtures: A review. J. Membr. Sci., 2017, **523**: 596–614. doi: 10.1016/j.memsci.2016.10.025.

[25]   Ribeiro, M. C., Pereira, C. M., and Pereira, M. F. R. Poly(vinyl alcohol)-based membranes for pervaporation: A review. J. Membr. Sci., 2018, **549**: 142–160. doi: 10.1016/j.memsci.2017.11.064.

[26]   Kujawa, J., Oliveira, M. S., Gorgojo, P., and Nunes, S. P. Polysulfone membranes for pervaporation: A review. J. Membr. Sci., 2020, **597**: 117678. doi: 10.1016/j.memsci.2019.117678.

[27]   Qiu, W., and Li, H. Crosslinked polymeric membranes for pervaporation: A review. J. Membr. Sci., 2017, **527**: 2–27. doi: 10.1016/j.memsci.2016.11.050.

[28]   Ji, Y., Li, Z., Zhang, B., and Chung, T. S. Crosslinked polymeric membranes for pervaporation and vapor permeation separation: Recent advances and future prospects. Chem. Eng. Sci., 2019, **194**: 274–293. doi: 10.1016/j.ces.2018.08.026.

[29]   Fang, Y., Liu, Y., Ma, Y., and Xu, T. Surface modification of polymeric membranes for enhanced pervaporation performance: A review. J. Membr. Sci., 2019, **584**: 245–264. doi: 10.1016/j.memsci.2019.03.001.

[30]   Zhao, Y., and Yan, Y. Surface modification of polymeric membranes for antifouling performance in membrane distillation: A review. J. Membr. Sci., 2019, **582**: 297–318. doi: 10.1016/j.memsci.2019.03.063.

[31]   Basu, S., Maity, S., and Bhattacharya, P. K. Techniques for characterization of polymeric membranes for gas separation. Sep. Purif. Rev., 2017, **46**(1): 1–46. doi: 10.1080/15422119.2016.1216902.

[32]   Hwang, K. J., Kim, S. H., Park, H. B., and Lee, Y. M. Recent advances in characterization techniques for polymeric membranes. Chem. Eng. Sci., 2017, **166**: 135–153. doi: 10.1016/j.ces.2017.03.055.

[33] Paul, D. R., and Yampolskii, Y. (Eds.). Sodium chloride sorption in sulfonated polymers for membrane applications. J. Membr. Sci., 2012, **423**: 195–208.

[34] Mulder, M. Basic Principles of Membrane Transport, Springer Science & Business Media, 2012.

[35] Robeson, L. M. Correlation of separation factor versus permeability for polymeric membranes. J. Membr. Sci., 1991, **62**(2): 165–185. doi: 10.1016/0376-7388(91)80060-L.

[36] Mulder, M., and Smolders, C. A. Introduction to Membranes: Basic Principles of Membrane Technology (2nd ed.), Kluwer Academic Publishers, 2001.

[37] Van den Berg, M., Wessling, M., and Lammertink, R. G. H. Membrane Technology: Volume 4: Membranes for Water Treatment, Wiley-VCH, 2011.

[38] Baker, R. W. Membrane Technology and Applications (3rd ed.), Wiley, 2004.

[39] Chung, T. S., and Jiang, L. Y. Interfacially polymerized membranes for reverse osmosis/ nanofiltration: A review. Polymer, 2007, **48**(26): 6917–6929. doi: 10.1016/j.polymer.2007.09.032.

[40] Paul, D. R., and Yampolskii, Y. Transport in Polymer Membranes, Royal Society of Chemistry, 2012.

[41] Hwang, K. J., Jeon, S., Kim, J., Kim, Y. J., Kang, Y., and Lee, Y. M. Performance evaluation and optimization of polymeric membranes for pervaporation separation of water/ethanol mixtures. Ind. Eng. Chem. Res., 2016, **55**(47): 12031–12040. doi: 10.1021/acs.iecr.6b02309.

[42] Barczak, M., and Krzeminski, A. Performance evaluation and optimization of membrane distillation processes: A comprehensive review. J. Membr. Sci., 2019, **588**: 117208. doi: 10.1016/j. memsci.2019.117208.

[43] Duong, H. D., and Nunes, S. P. Advanced polymeric materials for membranes in liquid and gas separations: Recent progress and future prospects. Prog. Polym. Sci., 2020, **100**: 101186. doi: 10.1016/j.progpolymsci.2019.101186.

[44] Marconnet, C., Chaouki, J., and Roudier, S. Advanced polymeric membranes for solvent recovery: Materials, properties, and separation performance. J. Membr. Sci., 2019, **573**: 476–498. doi: 10.1016/ j.memsci.2018.11.051.

[45] Khayet, M., and Matsuura, T. Preparation and characterization of hybrid organic-inorganic membranes for nanofiltration and forward osmosis processes. J. Membr. Sci., 2011, **369**(1–2): 187–198. doi: 10.1016/j.memsci.2010.11.023.

[46] Huang, L., Zhu, L., Li, C., Cui, Z., Yang, H., and Li, J. Recent advances in composite membranes for pervaporation separation. J. Membr. Sci., 2019, **580**: 178–200. doi: 10.1016/j.memsci.2019.03.031.

[47] Liao, Y., Xu, Z. L., and Wang, K. Recent advances in surface modification of polymeric membranes for water treatment. J. Membr. Sci., 2021, **627**: 119247. doi: 10.1016/j.memsci.2021.119247.

[48] Figoli, A., Marino, T., and Simone, S. Recent advances and future perspectives of polymer membranes for food and beverage industry. Food Res. Int., 2021, **145**: 110425. doi: 10.1016/j. foodres.2021.110425.

[49] Carta, M., Malpass-Evans, R., Croad, M., and Rogan, Y. Permeation and selectivity in solvent resistant nanofiltration: A review. J. Mater. Chem. A, 2013, **1**(33): 9002–9021. doi: 10.1039/c3ta11949b.

Asmaa Selim

Chapter 7
Advances in polymer membranes
for pervaporation

Abstract: Recent years have seen a lot of interest in membrane separation technology from the areas of chemical engineering, food science, analytical science, and environmental science. Electrodialysis, ultrafiltration, and reverse osmosis are some of the most common ways that membranes are used to separate substances. Pervaporation (PV)-based membrane separation offers distinct advantages such as low cost, high energy efficiency, and separation efficiency, in addition to shared benefits such as simplicity, flexibility, and a small floor area. Additional benefits encompass reciprocal advantages such as limited spatial requirements, uncomplicated design, and adaptability. PV has been acknowledged as a feasible technique for liquid separation not only for azeotropic mixtures but also for solvents with comparable boiling temperatures and more particularly organic–organic combinations and thermally sensitive compounds. Additionally, it is effective in eliminating dilute organics from aqueous solutions. Furthermore, PV has been employed for the purpose of eliminating low-concentration organic compounds. The theoretical framework and practical implementations of PV have undergone noteworthy progress and stimulating innovations in recent decades. The efficacy of the separation process is significantly influenced by various parameters, among which the PV membrane holds paramount importance. The objective of this chapter is to present a comprehensive summary of the potential of polymeric PV membranes in terms of their application for PV, fabrication techniques, and materials. This article provides a comprehensive explanation of PV including the underlying principles and separation mechanisms of polymer membranes as well as the various factors that impact their performance. The aim of this chapter is to facilitate reader comprehension.

Keywords: polymer membranes, pervaporation, sustainability, energy efficiency

1 Introduction and historical background

In recent years all throughout the world, people are interested in developing new separation technologies that might reduce waste and energy consumption. This is all in the

Asmaa Selim, Renewable Energy Research Group, Institute of Materials and Environmental Chemistry, Research Centre for Natural Science, 1117, Budapest, Hungary; Chemical Engineering and Pilot Plat Department, Engineering and Renewable Energy Research Institute National Research Centre, 12622 Giza, Egypt, e-mail: asmaa.selim@ttk.hu

https://doi.org/10.1515/9783110796032-007

name of a greener, more sustainable future. It is essential to integrate energy/process integration, recycling, and the recovery of valuable chemicals from waste streams in order to achieve this goal. Because of their potential advantages over more traditional separation methods, membrane technologies are increasingly being considered in a variety of commercial settings. Emerging and cutting-edge membrane-based separation methods may find widespread use in various industries such as water treatment, energy production, food processing, chemical manufacturing, and healthcare [1–3]. The membrane can be considered as permselective barrier because it allows just some species through while acting as a barrier to others. The rapidly evolving membrane technologies are state-of-the-art separation methods with potential applications in the environmental remediation, green energy, culinary, chemical, and pharmaceutical industries. Microfiltration, ultrafiltration, reverse osmosis (RO), electrodialysis, gas separation (GS), and pervaporation (PV) are the six most common membrane technologies used in these contexts [4–6].

With its low energy requirements, low cost, minimal impact on the environment, and high separation efficiency, PV stands out as a crucial approach among those available. RO and membrane GS are two other processes that share similarities with PV. Similarities between PV and steam permeation exist in that they both utilize gaseous constituents on the feed side of a membrane. The flow of PV is independent of the inlet pressure, in contrast to steam permeation, where the productivity of steam is heavily reliant on the pressure of the inlet gas [7]. Therefore, PV is a membrane separation procedure useful in a variety of contexts, including solvent dehydration and organic mixture separation. In azeotropic systems, where pure solvents can only be recovered with the use of entertainers, which in turn must be removed via another separation step, this method offers substantial advantages [8, 9]. Distillation, adsorption, and extraction are all tried and true methods of system separation, but PV is being tested on those systems that are particularly challenging to purify. PV technology has demonstrated promise as a feasible approach for eliminating low-concentration volatile organic compounds (VOCs) from wastewater, retrieving volatile fragrance compounds from fruit juices, and segregating azeotropic and closely boiling points solvents as well as organic mixtures, while preserving heat-sensitive materials. Unlike other methods, PV doesn't require the use of a third component that could lead to unintended consequences like hydrolysis when separating azeotropic mixtures [10]. In PV, the membrane exhibits selective permeability, effectively segregating the liquid feed and vapor permeation phases and facilitating the transportation of only the desired components of the feed liquid through evaporation. Hence, the permeation resistance is contingent upon the equilibrium of adsorption and the motion of the penetration components through the membrane, the performance of PV is virtually unaffected by the vapor-liquid equilibrium.

While PV membranes and processes have received much attention in recent years, their origins may be traced back to 1906, when Kahlenberg investigated the PV of an alcohol mixture via a rubber membrane [10, 11]. Kober coined the word "pervaporation" in 1907 while investigating the permeability of film consist of cellulose nitrate toward water vapor from albumin and toluene aqueous solutions. In 1956, Heisler initially reported using a cellulose membrane to separate water and ethanol. In 1965, an American oil corporation collaborated with Néel and his French colleagues to investigate the potential of the PV process [12, 13]. Sulzer Chemtech has been engaged in the development of thin-film composite membranes since 1997, which utilize the high-permeable Leob and Sourirajan membrane with its asymmetric structure as a substrate for thin-selective polymer layer. GFT Gesellschaft fur Trenntechnik GmBH first developed industrial-scale PV membranes in 1980. For commercial applications, the first composite PV membrane was PVA/polyacrylonitrile (PAN). The first commercial applications of inorganic membranes appeared in the year 2000. NaA zeolite membranes were used for ethanol dehydration in the commercial field, making this novel technology accessible to the PV process [14, 15]. PV plants, on the other hand, used a hydrophilic membrane to dewater organic solvents. Additionally, membranes with preferential to specific organic solvent (organoselective) were pioneered in the United States for the purpose of treating groundwater by eliminating trichloroethylene. Membrane Technology and Research developed a spiral-wound polydimethylsiloxane (PDMS) membrane in 1993. Industrially manufactured PDMS membranes provide a method for distilling most organic solvents out of water [16–18]. A third category of PV membrane has emerged by the end of the twentieth century. The new membrane, termed as "oragnoselective membrane," is highly selective toward specific solvent at the expense of others. In Europe, the value of the final membrane has been proven, as it is responsible for the abundance of ethyl *tert*-butyl ether, which is preferred to lead and its derivatives when advertising high octane fuel. Sulzer Chemtech (PERVAP® 2256) developed the first commercial organoselective membrane for removing methanol or ethanol from organic solutions [19–21].

Currently, PV membranes utilized in industry are composed of diverse materials, including two main classes of materials. Polymeric membranes that are composed of organic polymer chains are interconnected through cross-linking and create minuscule pores, enabling the diffusion of molecules. Inorganic membranes which are composed of ceramics or zeolites despite the challenges associated with large-scale industrial production and high costs, inorganic membranes may provide several benefits over polymeric membranes including enhanced solvent resistance and the capacity to function at elevated temperatures. Subsequently, an additional category of membranes has been identified, denoted as mixed matrix or hybrid membranes, which are composed of a blend of organic polymeric membrane and inorganic particles distributed throughout the polymeric matrix. Polymeric membranes have been widely adopted by the membrane separation industry due to their exceptional cost-effectiveness and versatility in various applications. Polymeric membranes are required to exhibit several crucial properties, with the foremost being an inclination toward a particular constituent. The regulation of pore size in polymeric membranes can be readily achieved during the production process. This confers a noteworthy benefit. The process of installation necessitates a significant level of adaptability while also requiring minimal spatial requirements. Therefore, polymer-based membranes are often regarded as the most flexible

and economically viable alternative to inorganic membranes because of their low production costs and simplicity of manufacture [22].

Therefore, in this chapter, a through view on PV principles and performance, the
different polymeric membranes used PV for various applications, and different fabrication techniques perspectives will be attempted.

2 Definition, fundamentals, and separation performance of PV process

2.1 Definition and theory

PV is a method of separating liquid mixtures through a nonporous membrane that is
powered by the partial vapor pressure gradient. The PV technique is the ideal choice
for separating combinations with close boiling points or that are azeotropic because
of its high selectivity and energy efficiency. During the PV process, the driving force is
the difference in chemical potential for the selective permeation of a particular component of a liquid mixture through a dense membrane. This occurs at the direct contact between the surface of the membrane and feed mixture. Application of sweep gas
or the use of a vacuum pump creates the pushing force across the membrane's sides.
The permeate, the desirable liquid component, changes phase from liquid to gas from
the feed side to the permeate side of the membrane, respectively, making it different
from conventional membrane separation processes. Condensation using dry ice or liquid nitrogen is commonly used to collect the permeate solution [23, 24].

The well-known mechanism for the dense called solution–diffusion has been applied to the process of PV. Diffusion is the mechanism that carries the transport over
the membrane in the following four steps: The process of membrane filtration comprises four distinct steps, namely: (1) the migration of the feed toward the membrane
surface, (2) the adsorption of the desired and preferential compound, (3) the transport
of the preferential compound via the membrane by diffusion, and (4) the detachment
of the preferential compound from the membrane and subsequent converted to gas
at the permeate side [25, 26]. According to Böddeker's findings, the membranes exhibit
a preference for molecules that are more condensable or have specific interactions
with the membrane materials, resulting in a sorption selectivity [27]. The Hildebrand
and Hanson solubility parameters are frequently utilized to determine the interaction
strength or solute molecule affinity toward membrane materials. This is done to identify the aforementioned factors. In general, the degree of preferential sorption of a
solute into a membrane is positively correlated with the proximity of the solubility
parameters of the membrane material and the solute. In order to maintain selectivity,
it is necessary for elastomeric membranes to have a greater thickness compared to
low-swelling glassy membranes, as they are more susceptible to swelling by the feed

component mixture. According to literature, the utilization of membranes with reduced thickness may lead to an increase in flux. However, this may also result in challenges related to the regulation of concentration polarization and maintenance of low permeate pressure within the system [19, 28].

2.2 Separation performance

Flux and permeance as well as selectivity and separation factors are the conventional performance metrics which commonly employed for the evaluation of PV's efficacy. The PV flux is defined as the quantity of substance that passes via a given area of membrane within a specific period of time. The measurement of permeate flux is usually expressed in units of weight; however, it can also be in mole as well as volume units. The relationship between both partial flux and difference of partial vapor pressure of a component across the membrane, commonly referred to as the driving force, has been established. To assess the effectiveness of PV, it is necessary to employ standard metrics commonly utilized in the membrane sector as aforementioned. The assessment of PV efficacy was conducted through the implementation of total flux (J) and separation factor (β). The flux, represented by the symbol J (g/m^2 h), was determined to be dependent on the permeate weight, W (g), the effective area of the membrane, A (m^2), and the duration of the experiment, t (h). The aforementioned correlation is established via utilization of the subsequent mathematical expression:

$$J = \frac{W}{A \times t} \tag{1}$$

The calculation of the separation factor (β) involves the utilization of the subsequent equation:

$$\beta_{i,j} = \frac{Y_i/Y_j}{X_i/_j} \tag{2}$$

where Y_i and X_i denote the mass fractions of components i and j in the permeate and feed, respectively. The efficiency of membranes during the PV process can be quantified by both the pervaporation separation index (PSI) and the enrichment factor:

$$\text{PSI} = J(\beta - 1) \tag{3}$$

$$\text{Enrichment factor} = Y_w/X_w \tag{4}$$

Baker et al. [29] suggest that permeance and selectivity are more appropriate measures for comprehending the intrinsic properties of a membrane, as opposed to the latter two, which are heavily influenced by operational factors such as temperature. To assess the intrinsic characteristics of the membrane, the calculation of two component permeability was conducted utilizing the subsequent equation:

$$P_i = \frac{J_i \delta}{\left(x_i \gamma_i P_i^{\text{sat}} - y_i P^p\right)} \tag{5}$$

where P_i (g/m h kPa) is utilized to denote the permeability of component i, while δ (m) represents the thickness of the membrane, J_i (g/m^2 h) is the individual flux, and γ_i denotes the activity coefficient. Additionally, P_i^{sat} (kPa) is the saturated vapor pressure, and x_i and y_i are the mole fraction in the feed and permeate side, respectively. The variable P^p (kPa) represents the pressure downstream. The Wilson equation is utilized to determine the activity coefficients, whereas the Antoine equation is employed to calculate the P_i^{sat}. The calculation of selectivity (α_{ij}) involved determines the ratio between the permeability of i and j:

$$\alpha_{ij} = \frac{P_i}{P_j} \tag{6}$$

2.3 Factors affect pervaporation performance

The PV's overall performance is subject to the influence of multiple factors including the unique attributes of the feed constituents, the membrane, and the operational parameters of the process. The efficacy of separation in PV is reliant not only on the operational parameters such as feed concentration and flow rate and operating temperature but also on the characteristics of the material of the membrane, which encompass, but are not restricted to, crystallinity, degree of cross-linking, and hydrophobicity.

2.4 Feed concentration

The concentration of the feed is a significant factor that exerts a considerable impact on the efficacy of the PV technique. Theoretically, photovoltaic technology has the potential to effectively fractionate any liquid mixture across a broad spectrum of concentration levels. In general, the presence of a high concentration of selective constituents on the membrane induces an increase in volume due to swelling. As a consequence, there is an augmentation of the void volume within the polymer network. Consequently, it enables the penetration of both low and high affinity components through the membrane, resulting in an increased total permeate flux [30, 31]. Furthermore, for binary mixture, the permeation flux of through the membrane is ascribed to the plasticization phenomenon induced by the two-feed component [32].

2.5 Feed temperature

The prevailing characteristic in the PV process may be regarded as the operational temperature [33]. As a result, variations in temperature can exert a direct influence on the driving force, feed component permeability, and their ability to diffuse through the membrane and consequently on permeation flux. Moreover, alterations in temperature exert a significant influence on feed component vapor pressures as well as on the thermodynamic characteristics of the feed [34]. The PV performance is commonly characterized by Arrhenius equations to demonstrate the impact of temperature on the process [35]:

$$J = J_o \exp\left(-\frac{E_J}{RT}\right) \tag{7}$$

The aforementioned equation pertains to the flux of membranes, where J represents the variable of interest and J_o denotes the pre-exponential factor, while E_J signifies the apparent activation energy, measured in o kJ/mol. R represents the gas constant, measured in units of kJ/mol K, and T represents the temperature in K.

The aforementioned equation can also be utilized to determine the ability of the feed component to diffuse across the membrane by substituting the permeation fluxes with the diffusion coefficients. The diffusion coefficient can be estimated by applying Fick's law to the individual flux of component i through the membrane:

$$J_i = -D_i\left(\frac{dC_i}{dx}\right) \tag{8}$$

The permeation flux (kg/m^2 s), diffusion coefficient (m^2/s), and concentration (kg/m^3) of component i in the membranes are represented by J_i, D_i, and C_i, respectively. For analysis, it is postulated that a linear model can describe the concentration profile through the membrane. The determination of D_i based on average concentration can be achieved through the utilization of the modified Fick equation, where δ represents the thickness of the membrane. This assertion has been documented in the literature [36, 37]:

$$D_i = \frac{J_i * \delta}{C_i} \tag{9}$$

2.6 The molecular size of the permeate component

The initial stages of the permeation mechanism entail the solvation of molecules within the polymer membrane, followed by their subsequent transport through the membrane via diffusion. Preferential permeation occurs due to variations in either solubility or diffusivity. The solubility of a substance is primarily influenced by variations in the chemical properties of the species that are permeating, while the diffusivity is predominantly determined by the size and shape of these molecules as well as the extent to which the diffusing species aggregate within the polymer. To summarize, three overarching trends were identified: In the context of binary permeation within a homologous series, it has been observed that the species with a lower molecular weight exhibits a preference for permeation. Furthermore, it has been found that molecules possessing a smaller diameter tend to permeate at a faster rate than their bulkier counterparts. Additionally, it has been noted that shape and size effects play a dominant role in the permeation behavior of chemically similar molecules. Molecules exhibiting significant dissimilarities in chemical properties are more susceptible to the influence of factors such as solubility, as opposed to their shape and dimensions [30].

2.7 Membrane swelling

The degree of membrane swelling is a crucial parameter in the PV mechanism. The preferred solvent initiates a swelling process in the membranes, resulting in enhanced flexibility of the polymer chain and, consequently, an escalated rate of permeation through the membrane. Hence, the extent of swelling is a crucial factor that necessitates consideration in the PV procedure. The swelling of a polymer is primarily influenced by three key factors, namely polymer crystallinity degree, ability of solvent to diffuse through membrane, and its rate. This assertion is supported by extant research [38]. The phenomenon of membrane swelling may arise as a consequence of the strong attraction between a permeating species and the selective polymeric membrane during PV. The phenomenon of swelling has been observed to exert an influence on both flux and selectivity, as reported in literature. It is imperative to exercise restraint or regulation over the extent of membrane swelling, as it has been found to have a detrimental effect on membrane selectivity and integrity [11, 39]. A balance between the processes of sorption and swelling must be achieved. For preferential permeation to occur, it is essential that a considerable degree of chemical affinity is present between a specific constituent and the membrane. In cases where the affinity is excessively high, the membrane may undergo swelling and consequent loss of integrity [40].

2.8 Membrane-free volume

The free volume of a polymer refers to the cumulative interstitial voids present between the polymer chains in amorphous, noncrystalline substances. The presence of side groups in polymer chains results in imperfect packing, leading to the existence of empty space or free volume. The determination of a polymer's free volume can be achieved through the measurement of its specific volume, followed by the subsequent calculation of the volume occupied by the constituent groups that comprise the polymer. The minuscule interstices between polymer chains that arise due to the thermal agitation of polymer molecules are referred to as free volumes in amorphous polymeric membranes. The temporal occurrence of these volume elements is commensurate with the movements of the permeants across the membrane. The probability of pore-flow characteristics in the membrane increases with the presence of individual free volume elements (pores) of larger dimensions [2].

2.9 Membrane material

The selection of membrane material in PV is primarily dependent on the specific application, which will be discussed in the following section. This has led to the development of numerous membrane materials that are capable of performing the most fundamental applications of liquid separations, such as the separation of organic chemicals from water and the separation of organic substances from one another. Organic compounds can be separated from water in two different ways: either by using hydrophilic polymer for removing water from the mixture such as esterification reactions or by hydrophobic materials to eliminate the organic residues. When categorizing polymeric substances based on their physical properties, they are typically divided into the three groupings: polymers with glassy behavior, rubbery or elastomeric behaving polymers, and polymers with ionic characterization. While first and the last groups of polymers are often preferred for the creation of hydrophilic membranes for the purpose of water removal, elastomeric polymers are commonly used for hydrophobic membranes in order to separate organic compounds from water.

3 Polymeric membranes in pervaporation application

Generally speaking, to date, the viable application of PV process can be classified into three main processes: (1) Hydrophilic PV, mainly dehydration of aqueous solution, which is considered the most studied application PV, (2) hydrophobic PV, and (3) organophilic PV or organic–organic mixture separation using organoselective membrane [18, 41].

As outlined in Section 1, this chapter is dedicated to the examination of the latest advancements in polymer-based membranes for various PV applications. Therefore, this section is divided into three categories: (i) hydrophilic PV, (ii) hydrophobic PV, and (iii) organophilic PV.

3.1 Hydrophilic pervaporation

Among the three applications, hydrophilic PV particularly dehydration of organics is one of the early and best developed membranes for PV process. Generally, hydrophilic polymers are employed for the membrane preparation, which could encourage the water solubility and transport through the membranes. Additionally, based on the synergic effect, as water has smaller molecule size compared to the organic solvents this will enhance its dissolution and diffusion in the hydrophilic membranes through hydrogen bond [7, 42, 43]. There are several prevalent polymers utilized for hydrophilic membranes which include poly(vinyl alcohol) (PVA), chitosan (CS), alginate (SA), polysulfone (PS), polyimide (PI), polyamide (PA), polyaniline, perfluoropolymers, and polybenzimidazole (PBI) [19, 44–46]. PV dehydration is mainly reported for azeotropic aqueous solution such as ethanol, isopropanol, n-propanol, 2-butanol, n-butanol, and tetrahydrofuran, where the water azeotrope is at 4, 12.6, 28.3, 26.8, 42.5, and 5.3 wt.%, respectively. Research has been conducted on the dehydration of significant organic solvents, including acetic acid and acetone aqueous solution, due to their crucial role in the chemical industry. Moreover, investigations have been carried out on the azeotrope mixture [8]. PVA has gained significant popularity as a PV membrane for dehydration due to its remarkable hydrophilicity, ability to form films, and thermal stability. The hydrophilicity of polyvinyl alcohol (PVA) is ascribed to the hydroxyl groups present in its pendant structure, whereas its semicrystalline nature is a result of the presence of two carbon atoms in its backbone. Despite PVA membrane has excellent perm-selectivity toward water, its permeation flux is relatively low. Conversely, PVA exhibits water solubility, thereby inducing instability within the aqueous medium, thereby impacting membrane swelling and, consequently, the efficacy of separation [7, 47, 48]. Several strategies were suggested to prevent the swelling of PVA membranes during ethanol dehydration process, including blending, cross-linking, and grafting. Initial research primarily centered on cross-linking, which appeared to be the most effective through chemical or physical cross-linking methods [20, 49]. Chemical cross-linking can be achieved through the use of different cross-linking agents, including but not limited to glutaraldehyde, maleic acid, and citric acid. The process of cross-linking entails the consumption of certain hydroxyl groups present in the PVA chain, thereby causing a reduction in the hydrophilicity of the membrane. This reduction in hydrophilicity leads to an increase in selectivity toward water molecules and a subsequent decrease in the permeation flux [50–52]. In contrast, PV application has received relatively less attention with regards to cross-linking by the physical methods such as freezing, heat treatment, and irradiation. According to various reports, it has been observed that PVA can undergo effective cross-linking with 30 min heating above its glass transition temperature. Moreover, the application of heat treatment resulted in a noteworthy enhancement in the degree of crystallinity of the polymer chain, leading to an increase in density. The elimination of water from the polymer chain causes a reduction in the affinity of PVA toward water, ultimately resulting in a decrease in the water content of the swollen material. Consequently, the extent of swelling exhibited by PVA membranes and their corresponding separation efficiency can be regulated through thermal processing [38, 53–57]. Furthermore, the incorporation of various hydrophilic polymers with less compact structures, such as CS and SA, has been widely reported as a successful method to modify PVA molecular structure [58–62]. This approach effectively reduces its crystallinity, ultimately enhancing the membrane permeation flux. Polymeric membranes, including the aforementioned modified membranes, typically experienced a compromise between permeability (or flux) and selectivity (or separation factor). Additionally, for PV dehydration, in addition to PVA, sodium alginate (SA) is also widely studied. SA was reported as standing polymeric membrane for ethanol/ water solution [63]. Kalyani et al. [64] reported the use of SA blended with PVP for the same solution. The blend membranes show good selectivity of 364 and extensive flux of 0.5 kg/(m^2 h) for separating azeotropic ethanol–water solution [64]. Additionally, PVA-blended SA membrane was reported for dehydration of isopropanol/water [60]. The pure SA membrane exhibits the highest separation selectivity of 365. On the contrary of the separation selectivity, the permeation flux of the blended membranes shows increase when increasing the PVA percentage. Blend membrane with 40% PVA show the highest permeation flux of 260 g/(m^2 h). Previously, Yeom and Lee [65] reported the use of cross-linked SA with GA for dehydration of IPA/water. The study investigates that increasing the cross-linker content results in reducing both separation factor and permeation flux. For the dehydration of ethanol solution with 70–90%, cross-linked membrane yield fluxes in the range of 60–1,000 g/(m^2 h), while the separation factor was in the range of 450–2,600. Subsequently, SA was used for the fabrication of composite membrane comprising alginate layer supported by a CS layer atop poly(vinylidine fluoride). They conducted tests for three distinct membranes by altering selective layers using cobalt alginate and alginic acid as cross-linkers. Despite exhibiting a lower separation factor compared to the SA membrane, the alginic acid membrane demonstrated superior mechanical strength, long-term stability, and a higher overall flux. The composite membranes were applied for the dehydration of ethanol-water and isopropanol water solutions. The alginic acid cross-linked membrane showed higher separation factor when dehydrating ethanol than for isopropanol comparing to the SA membrane. While the cobalt cross-linked membranes exhibit lower fluxes and higher separation for both solutions compared to pure SA membrane [66]. Following the same concept, blended SA with polyaniline (PAni) casted on two different porous support (PAN and PES) was reported for acetic acid dehydration by Moulik et al. [67]. The PAN supported membrane exhibits higher flux of 70 g/(m^2 h) and 441 separation factor compared to 40 g/(m^2 h) and ~360 for the PES supported membrane when using 2 wt.% water in the feed. SA-based membrane has been reported to show an outstanding performance through review on SA as PV membrane for the dehydration application and its modification has been published by Aminabhavi et al. [68]. The utilization of CS, as a biopolymer, has been extensively researched and suggested as a viable means to attain the replacement of synthetic polymers in membrane preparation for various PV applications, particularly in distinct azeotropic water-organic systems. Various CS-based membrane concepts have been devised, such as material blending, composite, and mixed matrix membranes, and have undergone testing for diverse separation applications. The application of CS in PV application has been extensively reviewed in the literature [69–72]. Besides the hydrophilic polymers, PS is a hydrophobic polymer with high mechanical and chemical stability. The use of PS in PV dehydration is usually accompanied by surface modification such as sulfonating or with the formation of thin film composite membranes where PS is used as support layer. Sodium-sulfonated PS showed improvement in both the flux and separation factor when tested for water/ethanol mixture separation, and the sulfonation process enhances the membrane ability to swell even at elevated temperature. The separation factor shows an extreme increase from ~500 to ~2,000 when increasing the ethanol concentration in the feed from 10 to 90 wt.% at 25 °C. Conversely, the permeation flux wasn't significantly affected by increasing the ethanol feed concentration. However, the separation factor observed a drastically decrease when increasing the temperature, concluding that the sulfonated PS membrane was optimum at low temperature [73]. Moreover, sulfonated PS that is used for the preparation of asymmetric membranes by using wet phase inversion method was reported for ethanol/water dehydration by using PV. The degree of sulfonation was found to be one of the main factors affecting the membrane hydrophilicity and therefore the separation performance. The highest PSI was achieved when highest sulfonation degree was applied if 0.92. The increase in the ethanol concentration in the feed results in slight decrease in the flux, while the separation factor is noticeably increased [74]. Additionally, the use of PS as support layer for hollow-fiber composite PVA-SA blend membrane was reviewed by Lipnizki et al. [75] for the dehydration of several binary solutions such as ethanol, isopropanol, *n*-butanol, and tertiary butanol using PV process. Moreover, in addition to the above-mentioned polymers and to overcome their limitation for the dehydration of corrosive liquids by PV at elevated operating temperatures new polymers were explored and developed. A new series of promising polymers with stiff and rigid chains such as polycarboxylic acids (e.g., poly(acrylic acid (PAA)) and amorphous perfluoropolymer (e.g., Teflon) have been considered; for example, the use of blended PAA-nylon 6 membranes for the PV dehydration of acetic acid-water. Using 40% PAA in the blend membrane led to separation factor of 82.3 and approximately 96 mol% water in the permeate at 15 °C and ~76 mol% of water in the feed [76]. Consequently, the application of perfluoropolymer-based membranes for the dehydration of wide range of aqueous solution as ethanol, isopropanol, and ethyl acetate [44]. In addition to their application for the dehydration of *N,N*-dimethyl formamide (DMF), *N,N*-dimethylsulfoxide (DMSO), and *N,N*-dimethylacetamide (DMAc) [77], PAs represent a class of polymers with notable heat resistance that exhibit performance of PA-based membranes in PV dehydration. Among these, nylon is the most widely recognized and is available in various forms, depending on the specific monomers employed in the polymerization process. Since its inception, a diverse array of polymeric amides with varying properties has been uncovered. Lee et al. [78] reported the use of two different aromatic PA membranes prepared by direct condensation of F-aramide and H-aramide for PV dehydration of ethanol aqueous solution. The authors concluded that increasing the ethanol concentration results in increasing both the flux and the separation factor of both membranes. However, the F-aramide membranes showed a lower separation factor of water/ethanol compared to the H-aramide membrane. The highest flux of 293 g/(m²h) and separation factor of 60 were obtained using the H-aramide membranes at 25 °C and 10 wt.% water in the feed [78]. In addition to the abovementioned studies, Table 1 summarizes the additional reports for PV performance for different organic solution dehydration using different membranes.

Table 1: Polymeric membranes for hydrophilic pervaporation.

Feed mixture	Mass ratio (wt./wt.%)	Membrane material	T (°C)	Flux (g/m²h)	Separation factor	Reference
MeOH/water	85/15	PAI-PEI	60	1,030	4.71	[79]
MeOH/water	90/10	SA/PVA	60	30	135	[80]
EtOH/water	95/5	PVA-PAA	50	260	50	[81]
EtOH/water	90/10	Modified CS	70	80	52	[82]
IPA/water	85/15	Ultem	60	7	585	[83]
H₂O₂/water	43/57	PDD-TFE (CMS-3)	25	70	12	[84]
DMAc/water	90/10	Perfluoropolymer	60	8	1000	[77]
Acetone/water	85/15	PBI	50	300	490	[85]
EG/water	64/36	PBI/PEI	60	758	592	[86]
Ac.Ac/water	50/50	PBI (2.5 wt.% H₂SO₄)	60	207	5461	[87]

PAI, polyamideimide; PEI, polyetherimide; PDDTFE, perfluorodimethyldioxole-tetrafluoroethylene; DMAc, dimethylacetamide; PDD-TFE, PBI, polybenzimidazole; EG, ethylene glycol; Ac.Ac, acetic acid.

More recently, apart from its primary application in solvent dehydration, hydrophilic PV membrane has the potential to be integrated with reactions that produce water as a byproduct, such as esterification, as depicted in Figure 1 [88]. This integration can lead to an improvement in the conversion rate through the in situ removal of water, as reported in the literature [89, 90].

A recent study reported the fabrication of spongy porous layer of PVA and 4-sulfophthalic acid which act as catalytic porous layer for an active separation layer of composite PVA and SA for the sake of the fabrication of catalytic composite membrane [91]. Figure 2 shows the structure of both the PVA-SA layer and the spongy porous catalytic layer. The membrane was found to impressively enhance the acetic acid conversion even after several cycles of reaction. The acetic acid conversion reaches 95.9% after continuous process for half day at 75 °C and only decreased to 80% after manifold cycles. Simultaneously, the PVA-based membrane facilitates the removal of the byproduct, water, through a PV process. The utilization of catalytic membrane technology has the potential to broaden the range of applications for PVA membranes beyond separation processes and into the realm of process intensification [92, 93].

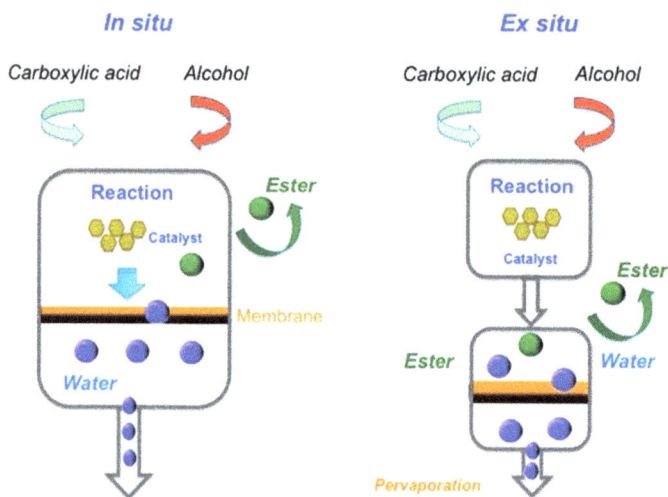

Figure 1: Esterification reaction assisted with pervaporation membranes in both in situ and ex situ arrangement (reprinted with permission from [88]. Copyright © 2023, American Chemical Society).

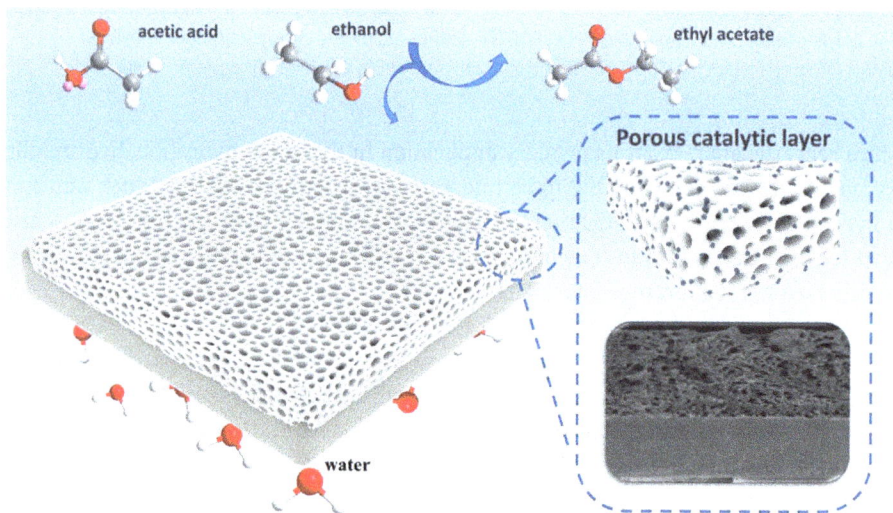

Figure 2: Catalytic porous spongy layer of PVA and 4-sulfophthalic for acetic acid separation (reprinted with permission from [91]. Copyright © 2022, American Chemical Society).

3.2 Hydrophobic pervaporation

As previously noted, PV exhibits a competitive advantage over adsorption technology in situations where there are multiple or relatively low concentrations of organic compounds in water. This is due to the fact that PV is a continuous process and offers greater operational convenience compared to the intermittent process of adsorption. PV has been found to be a viable method for eliminating organic compounds from aqueous solutions, particularly in instances where the concentration of such compounds is relatively low. This technique has been applied in various processes, including the removal of organic solvents from process side streams, the recovery of organics from fermentation broths, the elimination of aroma compounds, the dealcoholization of wine or beer, and the treatment of wastewater [94–96]. In contrast to the phenomenon of dehydration, the PV process involves the selective permeation of organic compounds with larger molecular sizes through a membrane in order to eliminate dilute organic compounds from aqueous solutions. Therefore, it is necessary for the membrane to consist of organophilic or hydrophobic substances to enhance the attraction of organic compounds to it [97–99]. As a result, the hydrophobic or organophilic membrane is dependent solely on its sorption selectivity rather than diffusion selectivity to achieve the selective permeation of organic compounds over water. This is due to the fact that the molecular size of organics results in a lower diffusivity compared to water. The thermodynamic sorption of components within a membrane is a crucial factor in the identification and design of hydrophobic membrane materials. This can be determined by examining the solubility parameters. In general, a higher degree of similarity between the solubility parameter of a given component and that of the membrane material results in a greater level of attraction between the membrane and the component. The Hansen solubility parameter theory can be utilized to determine the solubility parameters of small molecules by taking into account their hydrogen bonding, polar, and dispersive force contributions. Similarly, a technique for determining the solubility parameter of membrane materials was suggested through a collective contribution approach [95, 100, 101]. In addition to poly(dimethylsiloxane) or PDMS and poly(octyl methyl siloxane) or POMS, poly(1-trimethylsilyl-1-propyne) or PTMSP, poly(ether-*block*-amide), nitrile-butadiene rubber, and styrene-butadiene rubber are among the commonly employed membranes for the purpose of separating organic solvents [102–110]. Among the mentioned polymers, PDMS-based membranes are extensively utilized due to their cost-effectiveness and favorable processability. The potential of PDMS for the separation of organic compounds such as butanol, ethanol, methanol, and isopropanol has been investigated in various studies. Therefore, PDMS membrane is widely regarded as the standard for membranes with alcohol selectivity [95, 111–114]. This is attributed to the close solubility parameters of ethanol (26.2 MPa$^{0.5}$) and butanol (23.3 MPa$^{0.5}$) to that of PDMS (14.9 MPa$^{0.5}$) opposed to water (47.9 MPa$^{0.5}$). Consequently, they can be dissolved more readily into a PDMS matrix as compared to water. Additionally, the structural framework of PDMS is composed of Si–O units that exhibit a high degree of flexibility.

This characteristic confers significant flexibility to the PDMS chains. This increases the ability of substances to pass through the membrane. Moreover, the organophilic nature of PDMS is attributed to the hydrophobic methyl substituents present on the silicon atom [115].

Commercial pristine PDMS membranes are available such as the commercial Pervap 1060, Pervap 2200, and Pervap 4060 PDMS membranes. In fact, PDMS PV membranes are usually fabricated from two most commonly utilized commercial PDMS precursors distinguished based on their terminal groups. First, through condensation reaction of hydroxyl-terminated PDMS which is frequently cross-linked using tetraethoxysilane resulting in the formation of a three-dimensional network structure. Second, it involves the utilization of vinyl-terminated PDMS which undergoes an additional reaction with a hydrosilyl-containing cross-linker, resulting in the formation of a linear architecture. Therefore, precursor's molecular weight and cross-linker quantities are critical factors in optimizing the transport and mechanical characteristics of PDMS membranes [116, 117].

The potential of hydrophobic PV membranes in a fermentation broth has been found to be significant [118], with applicability not only as an ex situ but also as an in situ process. However, PV separation of bioalcohols presents a significant challenge due to the trade-off effect between the anti-biofouling property and the PV performance of the membrane. Due to their hydrophobic nature, protein-based biofoulants exhibit a preference for adsorption onto the membrane surface. The occurrence of membrane fouling can lead to a decline in separation efficiency, which can be either temporary or permanent in nature, ultimately reducing the operational lifespan of the membrane [119, 120]. In general, reducing the surface hydrophobicity appears to be a viable approach for improving antifouling properties. However, the solution–diffusion model will have a detrimental impact on the alcohol flux. Extensive research has been carried out to enhance the efficacy of PV membranes by altering the membrane fabrication process, owing to the scarcity of hydrophobic materials for recovery applications. In order to improve the anti-biofouling properties of PDMS membranes while maintaining their separation performance, measures must be taken. A novel variety of anti-biofouling PDMS membrane was produced through a straightforward cross-linking process involving fluorosilane and PDMS. The PDMS membrane was endowed with an ultra-low surface energy by introducing fluoroalkyl groups, which maintained its excellent hydrophobicity. The preparation scheme of the PDMS membrane and its fluorination process is described in Figure 3. As a result, microbial adhesion onto the membrane surface was alleviated [121]. The incorporation of a fluoroalkyl moiety resulted in a notable enhancement of hydrophobicity and a decrease in microbial adhesion on the surface of the membrane. The results indicate that the fluorinated PDMS membrane demonstrated superior performance in terms of stabilized flux of 740 g/(m^2 h) and separation factor of 21.8 compared to 360 g/(m^2 h) and 7.1 separation factor for pristine PDMS membrane when utilized in conjunction with ABE fed-batch fermentation. An alternative approach involved the utilization of PDMS as the selective-layer material, which was applied as a coating on different substrates. The overall performance of composite membranes is determined by the viscosity of PDMS solutions and the properties of substrates such as pore size and porosity [122–125]. Hecke et al. [126] utilized a PDMS/PI composite membrane to extract butanol from a solution that underwent ABE fermentation. The permeate was enriched to 195 g/L after the application of in situ butanol recovery membrane. Additionally, a clear decrease in the butanol toxicity in the fermentor was recognized which result in enhancing the glucose concentration to 126 g/L and improve productivity to 0.3 g/(L h). The coupling of the fermentation process with PV results in elongating the process from 475 h to 1,172 h with an average flux of ~0.37 kg/(m^2 h).

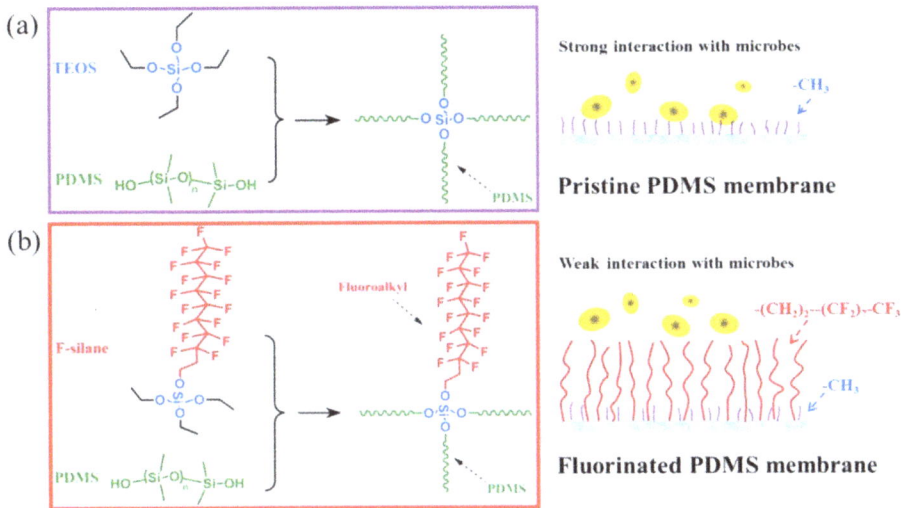

Figure 3: Preparation of PDMS membrane using silane cross-linker (a) and fluorination of PDMS through condensation reaction using fluorosilan (reprinted with permission from [121]. Copyright © 2023, Elsevier 555867147816).

On the other hand, PEBA is a prospective class of polymers that exhibits potential for the development of hydrophobic PV membranes. The potential of PEBA for the removal of diverse organic constituents from aqueous solutions is considerable. The aforementioned statement pertains to the distinct chemical composition of the PEBA polymer [127, 128]. PEBA is a copolymer that is thermoplastic in nature and comprises rigid PA segments and flexible polyether segments. The commercially accessible trade name for the PEBA polymer is Pebax. PEBA copolymers are synthesized using different types of poly(amides) including PA6, PA66, PA12, and poly(ethers) such as ethylene oxide and poly(tetramethylene oxide). According to the literature, the copolymer's mechanical stability is regulated by the crystal structure of its PA segments, whereas the amorphous structure of its polyether segments is responsible for its high permeability to organic substances [127, 129]. Pebax 2533 is a PEBA copolymer that exhibits hydrophobic properties due to its significant proportion of polyether segments (80%). PEBA membranes have gained significant popularity in PV applications due to their favorable characteristics, including high permeability, high selectivity, and long-term stability, as reported in the previous section. Pebax 2533 based membrane is more effective in achieving optimal membrane performance for the recovery of organic compounds due to its maximized organophilicity and was used for several pervaporative removal of organic compounds (e.g., acetone, butanol, and ethanol) from aqueous solutions [129–131]. However, in comparison to the PDMS membrane, the PEBA membrane exhibits a significantly lower water contact angle due to the existence of the hydrophilic segment [132, 133]. Additionally, PEBA has lower separation factor for alcohol/water observed in the PEBA membrane (3–4 for ethanol/water and approximately 20 for butanol/water) in contrast to the higher separation factor (for ethanol/water around 7–8 and for butanol/water approximately 40) [109, 110]. Conversely, the PEBA membrane demonstrated a significantly greater phenol/water separation factor in comparison to the PDMS membrane (30 vs. 6). This indicates a high level of organophilicity in PEBA membranes, which makes them advantageous for selectively permeating larger organic molecules, such as phenol.

Cao et al. [134] reported the PV separation of phenol, *p*-cersol, *p*-chlorophenol, and *p*-nitrophenol aqueous solutions using PEBA (2533) membrane at temperate ranges from 30 to 70 °C and up to 0.6 wt.% feed concentration. The coupling effects among phenolic solutes, summary of the total permeation flux, and description of the process are presented in Figure 4.

Among the four different solutions, phenol solution observes the highest total and phenol fluxes over the full range of the feed concentration. While *p*-cersol achieves the highest permeate concentration and enriching factor of the same range of the feed concentration. For the four solutions, increasing the feed concentration results in increasing the total flux, except *p*-chlorophenol. In addition to the above-reviewed literature, Table 2 presents more results for hydrophobic PV performance for different solutions under different conditions.

3.3 Organophilic pervaporation

Nowadays the most challenging area of research in PV is the organic–organic separation, due to the necessity of membranes possessing satisfactory separation properties and long-term stability during the process. Furthermore, the enlargement of polymers in a solvent consisting solely of organic compounds may diminish the membrane's capacity for size discrimination and stability. Moreover, certain blends containing organic compounds, specifically aliphatic and aromatic compounds, demonstrate comparable physicochemical characteristics, leading to inadequate distinctions in membrane solubility and diffusivity. In general the major application for the organic–organic separa-

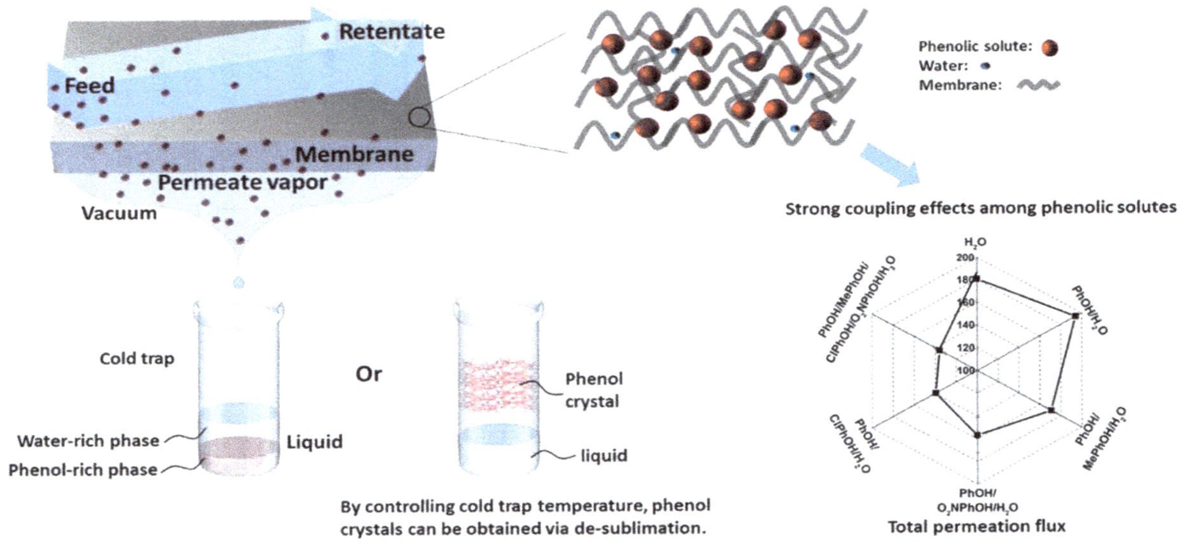

Figure 4: Summary of the separation process, coupling effect among phenolic solutes and PEBA membrane, and resulted total flux (reprinted with permission from [134]. Copyright © 2023 Elsevier 5574210164790).

Table 2: Polymeric membranes for hydrophobic pervaporation.

Feed mixture	Organic wt.%	Membrane material	T (°C)	Flux (g/m^{2}h)	Separation factor	Reference
n-Butanol/water	1	PDMS	60	1,000	41	[135]
EtOH/water	5	POSS/PEBA	65	427	5.7	[131]
Furfural/water	1	PTFPMS	80	1,041	45.6	[136]
Furfural/water	2	Polyurethaneurea.	75	28	638	[137]
THF/water	0.5	HG30	30	26.3	170	[138]
Diaceytl/water	0.02	Peba2533	70	700	40	[139]
S-MTB/water	0.01	Peba2533	70	800	110	

PTFPMS, poly[(3,3,3-trifluoropropyl)methylsiloxane]; THF, tetrahydrofuran; HG30, homo-P4VP-grafted PVC; MTB, methylthiobutanoate.

tion could be divided to four section or types or organics: (i) polar/nonpolar, (ii) aromatic/aliphatic, (iii) aromatic/alicyclic, and (iv) isomers [140, 141].

Contemporary studies have focused on alcohols and tertiary butyl ethers as prevalent substitutes for tetraethyl-lead in order to enhance the octane rating of gasoline. Illustrations of such substitutions comprise ethanol/ethyl-*tert*-butyl ether (ETBE) and methanol/ methyl-*tert*-butyl ether (MTBE) as polar/nonpolar solvents. Hydrophilic polymers are the preferred material for PV membranes due to their ability to selectively allow polar molecules to permeate the membrane, while hindering the passage of nonpolar molecules. Hydrophilic polymers, including PVA, CS, and cellulose acetate (CA), as well as moderately hydrophilic polymers like polyetheretherketone (PEEK), have been utilized for the separation of polar solvents from nonpolar solvents. For the purpose of separating methanol and MTBE, Rhim and Kim [142] utilized PVA/PAA and PVA/sulfosuccinic acid (SSA). The blend membranes show increase and then decrease in the separation factor when increasing the PVA content to reach maximum of 4,000 at 85 wt.% PVA when separating MTBE/MeOH = 80/20 wt.% at 50 °C, while the permeation flux acts vice versa with a minimum value of 10.6 g/(m^2 h) at the same condition. On the other side, the observed results for the cross-linked SSA/PVA membrane show that increasing the SSA % in the PVA membrane results in decrease in the flux and increase in the separation factor up to 5 wt.%. At 50 °C and for MTBE/MeOH = 80/20 solution, the membrane with 5 wt.% shows a separation factor of ~1500 and flux around 26 g/(m^2 h). While, Madandar and Mohammadi [143] reported the use of PVA-based membranes supported on a microporous support of PAN and studied the influence of the permeate vacuum pressure on the PV performance. The work present that decreasing the permeate pressure will lead to increase in the PV performance of the membrane. A PV separation index of 54 kg/m^2 was achieved at 15 mbar when separating MTBE/MeOH = 70/30 wt.% at 35 °C. The enhancement of the separation factor using modified PEEK with cardo (lactone) groups using chloroform as a solvent was also reported by Zereshki et al. [144]. The modification of PEEK with cardo (lactone) group decreases its crystallinity and facilitates its solubility in organic solvents. The modified PEEK membrane faced trade-off behavior between separation factor and permeation flux when increasing the MeOH concentration, with a maximum flux of 0.113 kg/(m^2 h) at 87.2 wt.% MeOH and maximum separation factor of 250 at 1 wt.% MeOH [144].

Nam [145] has reported the use of CS composite membrane modified by H$_2$SO$_4$ and four different surfactants for the PV of 20% methanol/MTBE. The study shows that the use of surfactant improved both the separation factor and the flux at room temperature [145]. The modification of CS with surfactant enhanced the PV performance to achieve 470 g/(m^2 h) permeation flux and MeOH concentration higher than 98 wt.% in the permeate compared to 100 g/(m^2 h) as flux and only 70 wt.% MeOH in the permeate for the membrane modified with sulfuric acid, when separating 20% methanol/MTBE at 25 °C.

Additionally, CA could be considered as the most studied hydrophilic polymer for both ethanol and methanol separation from their tertiary butyl ethers [146]. Several studies presented a set of membranes composed of CA and modified CA, which were utilized for the purpose of separating MeOH/MTBE and EtOH/ETBE mixtures. The studies revealed that the incorporation of CA in conjunction with CHAP has resulted in a notable enhancement in flux, with an approximate 25-fold increase. However, this has been accompanied by a reduction in the separation factor to half its original value during the separation of 30% methanol/MTBE [147, 148]. Additionally, the influence of acetylation degree of MeOH/MTBE separation was studied by Cao et al. [149]. The higher the acetylation degree results in higher MeOH concentration in the permeate and is over a range between 2 and 40 wt.% MeOH in the feed operating at 50 °C. On the other hand, Qian et al. [150] studied the influence of the viscosity of CA and its blends in dilute solutions on the PV performance for separating MeOH/MTBE. The work investigates the performance of CA dissolved in tetraheydrofuran, dioxane, acetone, and butanone and the performance of blend CA with poly vinylbutyral in dioxane. It was observed that the permeation flux is inversely proportional with the viscosity of the solution for the separation of 3 wt.% MeOH in the feed.

Blend of CA with poly(vinyl pyrrolidone-*co*-vinyl acetate) was reported for the separation of ethanol/ETBE [151]. In addition, the utilization of a CA membrane featuring grafted poly(methyl diethylene glycol methacrylate) copolymer of varying lengths was implemented in the PV separation process of an ethanol (EtOH) /ETBE mixture. The researchers noted that the PV efficacy is contingent upon both the quantity and composition of the graft copolymer present within the membrane. Furthermore, it was observed that the short-grafted copolymer exhibited twice the selectivity and lower permeability in comparison to the long-grafted copolymer. The results showed enhancement in the separation factor. For separating azeoptropic EtOH/ETBE solution at 50 °C, the lowest selectivity of ethanol weight fraction in the permeate of 0.876 was achieved with 44 wt.% long-graft CA. The aforementioned phenomenon can be ascribed to the inflexible configuration and partitioning of phases within the CA substrate. This renders the transportation of the bulky ETBE molecules across the membrane. However, the diffusion of ethanol is facilitated by these same properties [152]. On the contrary, hydrophobic polymer such as cellulose triacetate membrane show an impressive separation factor of 9480 and moderate flux for the separation of 5% MeOH/MTBE at 30 °C [153].

Conversely, the differentiation between aromatic and aliphatic substances, such as toluene and iso-octane, as well as toluene and *n*-heptane, is being considered. Moreover, the separation of aromatic and alicyclic compounds, such as benzene and cyclohexane, has garnered significant interest for the purpose of separating cyclohexane from mixtures containing benzene and cyclohexane. This process is particularly relevant in the context of production plants for benzene, toluene, and xylene, as it has the potential to improve the overall production capacity of these chemicals [154]. To achieve this segregation, several polymers such as PIs and their copolymers were heavily employed. The synthesis of these polymers was achieved using carboxylic acid anhydrides and primary diamines as the fundamental constituents. The utilization of PIs as membrane materials for organic/organic separations was facilitated by their chemical and thermal stability, which can be attributed to the rigid and stiff nature of their chains. Typically, it is observed that PI exhibits a greater inclination toward aromatic as opposed to aliphatic, owing to the favorable interactions that occur between the polar imide and/or benzene groups and the electrons present in aromatic rings. This phenomenon can be attributed to the advantageous intermolecular interactions involving the polar imide and/or benzene moieties. As a result, PI membranes exhibit the capacity to differentially isolate aromatic substances from their amalgamations with aliphatic compounds [155–160].

Moreover, it has been noted that membranes containing supramolecular entities enhance flow rate or enable permselectivity by generating inclusion complexes or clathrate/cage compounds. The aforementioned compounds are present within cellular membranes. The utilization of supramolecular compounds such as cyclodextrins, calixerene, and calixarenes as transport carriers in PV membranes has been investigated in various studies [161–163]. The molecular recognition capabilities of these classes of supramolecules are widely acknowledged. The presence of calixerene in poly (methyl methacrylate)-*graft*-poly(dimethylsiloxane), PVA, and poly(vinyl acetal) membranes leads to a significant increase in benzene selectivity from its aqueous solution and over cyclohexane, respectively [164–166].

Furthermore, block copolymers containing soft-hard segments are significant membrane materials with the ability to separate organic/organic mixtures. The block copolymers under consideration exhibit the ability to mitigate excessive swelling through the presence of hard segments, while simultaneously displaying an affinity toward organic compounds via the soft segments. Comprehensive separation performance in organic/organic mixtures can be achieved by modifying the size and chemistry of both the soft and hard segments. The evaluation of this particular performance can be quantified through the utilization of metrics such as permeation flux, separation factor, and structural stability. As an illustration, apart from the accessibility of PEBA and its ability to create a high-quality layer, it exhibits a notable inclination toward aromatic compounds owing to the selective interaction between these compounds and the polyether component. The PV process has garnered significant interest for its ability to separate aromatic and aliphatic compounds. In this regard, the thermally cross-linked PEBA composite membrane supported on tubular ceramic substrate has been found to exhibit exceptional and consistent performance when used in conjunction with toluene/n-heptane [167]. The simple preparation technique of the thermal cross-linked composite membrane is illustrated in Figure 5. It was found that increasing the PEBA concentration from 4 to 7 wt.% increases the separation factor from ~ 2 to 4.75 of 50 wt.% toluene/n-heptane at 40 °C while decreasing the flux from 96 to 27 g/(m^2 h).

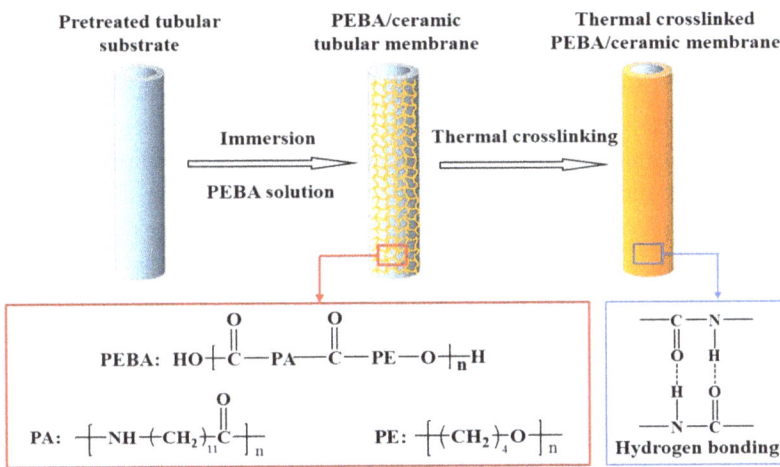

Figure 5: Preparation of thermal cross-linked PEBA membrane supported on tubular ceramic substrate (reprinted with permission from [167]. Copyright 2015 Elsevier 5577001409716).

4 Fabrication techniques

Generally, polymeric membranes for PV are flat sheet membranes and either one dense layer or asymmetric with a thin-selective dense layer. In laboratory settings, dense membranes are commonly utilized to assess the inherent characteristics of materials. However, their implementation in industrial separation procedures is impeded by the comparatively limited transmembrane flow. Asymmetric membranes are characterized by a selective layer that is thin and supported by a substrate that is relatively thick and microporous. The substrate's microporous structure significantly decreases the resistance of the substructure for the transportation of penetrants, thereby augmenting the trans-membrane flux. There exist multiple techniques for the fabrication of the PV membrane. This section provides an overview of several prominent methods for fabricating polymeric membranes, and the most used ones are solution casting, hollow fiber spinning, and solution coating.

4.1 Solution casting

The technique of solution casting is widely utilized for the production of flat-sheet membranes that are intended for diverse applications. Initially, the polymer and any potential additives are dissolved in a solvent to create a homogeneous solution. This solution is subsequently applied to a level surface, after which the solvent is eliminated through either evaporation or phase inversion mechanisms. Multilayered membranes can be fabricated through multisolution castings either with or without the use of porous supports. The production of dense membranes involves a gradual and thorough evaporation process to remove solvents. On the other hand, the creation of asymmetric membranes with interconnected cell structures is achieved through a phase inversion process, which entails immersion in a nonsolvent bath during solvent removal. The incorporation of high-volatility solvents into the casting solution, followed by an evaporation process prior to phase inversion, facilitates the development of a dense upper layer.

4.2 Solution coating

Solution coating is yet another method that is routinely utilized in the process of producing thin composite membranes. The synthesis of composite membranes is commonly achieved through the application of a thin-selective layer onto a microporous substrate or support, which can be in the shape of a flat sheet, hollow fiber, or tubular structure. The aforementioned action is undertaken with the aim of attaining the intended amalgamation of composite membranes. In point of fact, the degree of porosity of the substrate needs to be sufficiently high such that the coated selective layer can exert the majority of its influence over the resistance of the membrane. In order to prevent the infiltration of the coating solution, it is necessary for the substrate surface's pore size distribution to be tightly controlled and devoid of significant defects. Prior to the application of the coating, the substrate is prewetted with a solvent that has a low boiling point and is not compatible with the coating solvent. This process serves to reduce the likelihood of intrusion. Consequently, the prewetting solvent is extracted through the process of drying in order to acquire the coated film. Achieving uniform coating on small-diameter hollow fibers presents a significant challenge, as uneven coating can adversely impact the separation process. The utilization of spin coating is a viable method in the production of thin films that exhibit uniformity. The present technique involves the application of a small quantity of polymer dispersion onto the central region of the substrate, which is simultaneously rotating at a predetermined velocity. The utilization of spin coating technology by various researchers has facilitated the production of consistently improved membranes. They enhanced the permselectivity of the PV membrane through the utilization of alternate self-assembly and spin coating techniques in synthesizing the active layer were reported [168–172].

4.3 Hollow fiber spinning

Hollow fiber polymeric membranes, which were initially developed by Mahon in 1966, have emerged as a significant and crucial type of synthetic membranes due to their numerous advantageous characteristics [173]. The elevated demand for hollow fiber membranes, in comparison to alternative membranes, can be attributed to their heightened packing density and greater effective membrane area per unit volume of the separation device, or membrane module, and a self-containing vacuum channel when utilizing the shell feed mode. In addition, the use of hollow fiber spinning membrane offers substantial self-mechanical reinforcement and facilitates the manipulation of modules during both fabrication and operational processes. Most of the advancements and alterations in the synthesis of hollow fiber spinning membrane have been accomplished through heuristic approaches, supported by prior knowledge, experiential evidence, and qualitative scientific comprehension. The development of membrane technology has arisen as a means to offer sustainable solutions to meet the worldwide need for clean energy, water, and health management. To produce new hollow fiber membrane with superior separation capabilities, researchers must examine the inherent physicochemical characteristics of the membranes, manipulate phase inversion procedures, and regulate dope rheological responses during the membrane creation process. The synthesis of high-performance hollow fiber membrane that possesses both desirable morphology and separation performance is a formidable task. The presence of macrovoids and irregular shapes in hollow fibers can lead to a decline in the mechanical properties of hollow fiber membranes when subjected to extended periods of high pressure or vibration. The enhancement of separation performance through alterations in polymer molecular structure is constrained by the established compromise between permeability and selectivity. Specifically, an augmentation in permeability is typically concomitant with a reduction in selectivity and vice versa. The development of an effective formulation, understanding of coagulation chemistry, appropriate spinneret design, and optimization of spinning conditions including air gap, temperature, and take-up speed are crucial factors in the production of high-quality fibers. During the process of spinning, the membrane is generated via phase inversion. Inversion occurs upon contact between the nascent fiber and the coagulant. Upon extrusion of the polymer dope and bore fluid in the lumen side of the nascent fiber, immediate coagulation takes place at its internal surface subsequent to its emergence from the spinneret. Partial coagulation commences on the external surface of the nascent fiber as it traverses the air gap region due to the existence of moist air. The phase inversion process culminates upon the complete precipitation of the fiber in the external coagulation bath. The selective layer's thickness and morphology can be altered through the manipulation of spinning dope, bore fluid, external coagulant, and take-up speed. The intricacy of hollow fiber spinning escalates with the progression of spinning techniques from single-layer to dual-layer coextrusion. The utilization of dual-layer hollow fibers presents several benefits including reduced costs and increased flexibility in material selection and morphology for both the selective and supporting layers [174–177].

5 Conclusion

In the realm of segregating azeotropic mixtures, PV has been identified as a more energy-efficient alternative to conventional separation techniques such as distillation and absorption processes. The present chapter provides an overview of the historical evolution of materials employed in the fabrication of polymeric PV membranes. The pervaporation technique has been employed for the separation of organic–organic mixtures using organophilic PV, dehydration of organic–water mixtures using hydrophilic PV, and recovery of organics from aqueous solutions using hydrophobic PV. To date, these three categories of PV separation systems have been employed. This article comprehensively addresses the historical context of PV, the separation performance, the fundamental principles, the process performance influencing factors, and the transport mechanism of PV membrane materials. The subsequent account provides a concise overview of the primary achievements and challenges pertaining to each category of material.

PV has been a significant contributor to the process of solvent dehydration and separation. Its potential can be further enhanced by integrating it with other feasible liquid separation technologies and identifying suitable materials for certain corrosive solvents. PV has been instrumental in facilitating the hydrophilic PV process, which involves the removal of water from organic–water mixtures. PV utilizing hydrophobic membranes presents a highly auspicious avenue for application. Nevertheless, further investigation is required to enhance the hydrodynamics of the PV system, thereby augmenting the selectivity and stability of the membranes. These advancements hold significant potential for implementation in industries such as chemical, petrochemical, food, and beverage. The cost-effectiveness, safety, and eco-friendliness of various membranes currently available make them a viable option for organophilic PV, potentially competing favorably with established separation technologies due to their unique benefits. However, in the past few decades, there has been relatively minimal progress in the separation of organic compounds through PV as compared to solvent dehydration and the elimination of organics from water. The increasing prevalence of patents and scientific publications pertaining to PV systems utilized in organic separations serves as evidence of the widespread adoption of this system within the organic separations domain. Several studies have indicated that PV methodologies are technically feasible for the separation of various mixtures, including but not limited to MeOH/MTBE, EtOH/ETBE, benzene/cyclohexane, and toluene/n-heptane.

Nevertheless, the implementation of this method of separation in industrial contexts is currently at an early stage of development. The plausibility of this assertion could be attributed to the absence of membrane modules in the current market that exhibit both durability and affordability suitable for deployment in industrial contexts. The operational conditions necessitate persistent contact of the membranes with solvent-containing feed solutions, while operating in an environment characterized by a relatively elevated temperature. Furthermore, an examination and discussion have been conducted on the fabrication procedures employed in the manufacture of PV membranes, including solution casting and hollow fiber spinning solution coating.

References

[1] Abels, C., Carstensen, F., and Wessling, M. Membrane processes in biorefinery applications. J. Membr. Sci., 2013, **444**: 285–317.

[2] Baker, R. W. Membrane Technology and Applications (2nd ed.), John Wiley & Sons, Ltd, Chichester, 2004. https://doi.org/https://doi.org/10.1002/9781118359686.

[3] Murder, M. Basic Principle of Membrane Technology, Kluwer Academic publishers, 1996.

[4] Ren, J., and Wang, R. Preparation of polymeric membranes. In: Wang, L. K., Chen, J. P., Hung, Y.-T., Shammas, N. K. (Eds.) Membrane and Desalination Technologies, Humana Press, Springers Science, New York, 2011, pp. 47–100.

[5] Sirkar, K. K. Membrane Handbook, Springer Science+Business Media, New York, 1992.

[6] (Tony) Fane, A. G., Wang, R., and Jia, Y. Membrane technology: Past, present and future. In: Wang, L. K., Chen, J. P., Hung, Y.-T., Shammas, N. K. (Eds.) Membrane and Desalination Technologies, Humana Press, Springers Science, New York, 2011, pp. 1–45.

[7] Chapman, P. D., Oliveira, T., Livingston, A. G., and Li, K. Membranes for the dehydration of solvents by pervaporation. J. Membr. Sci., 2008, **318**: 5–37. https://doi.org/10.1016/j.memsci.2008.02.061.

[8] Castro-Muñoz, R., Galiano, F., Fíla, V., Drioli, E., and Figoli, A. Mixed matrix membranes (MMMs) for ethanol purification through pervaporation: Current state of the art. Rev. Chem. Eng., 2018, **35**: 565–590.

[9] Kober, P. A. Pervaporation, perstillation and percrystallization. 1. J. Am. Chem. Soc., 1917, **39**: 944–948. https://doi.org/10.1021/ja02250a011.

[10] Feng, X., and Huang, R. Y. M. Liquid separation by membrane pervaporation: A review. Ind. Eng. Chem. Res., 1997, **36**: 1048–1066. https://doi.org/10.1021/ie960189g.

[11] Feng, X., and Huang, R. Y. M. Liquid separation by membrane pervaporation. Ind. Eng. Chem. Res., 1997, **36**: 1048–1066.

[12] Kujawski, W. Application of pervaporation and vapour permeation in environmental protection. J. Environ. Stud., 2000, **9**: 13–26.

[13] Jonquières, A., Clément, R., Lochon, P., Néel, J., Dresch, M., and Chrétien, B. Industrial state-of-the-art of pervaporation and vapour permeation in the western countries. J. Membr. Sci., 2002, **206**: 87–117.

[14] Binning, R. C., Lee, R., Jennings, J., and Martin, E. C. Separation of liquid mixtures by permeation. Ind. Eng. Chem., 1961, **53**: 45–50.

[15] Morigami, Y., Kondo, M., Abe, J., Kita, H., and Okamoto, K. The first large-scale pervaporation plant using tubular-type module with zeolite NaA membrane. Sep. Purif. Technol., 2001, **25**: 251–260.

[16] József, A. T., Toth, A. J., and József, A. T. Liquid Waste Treatment with Physicochemical Tools for Environmental Protection, Budapest University of Technology and Economics, 2015.

[17] Kujawa, J., Cerneaux, S., and Kujawski, W. Removal of hazardous volatile organic compounds from water by vacuum pervaporation with hydrophobic ceramic membranes. J. Membr. Sci., 2015, **474**: 11–19. https://doi.org/https://doi.org/10.1016/j.memsci.2014.08.054.

[18] Smitha, B., Suhanya, D., Sridhar, S., and Ramakrishna, M. Separation of organic–organic mixtures by pervaporation-a review. J. Membr. Sci., 2004, **241**: 1–21.

[19] Shao, P., and Huang, R. Y. M. Polymeric membrane pervaporation. J. Membr. Sci., 2007, **287**: 162–179.

[20] Peng, M., Vane, L. M., and Liu, S. X. Recent advances in VOCs removal from water by pervaporation. J. Hazard Mater., 2003, **98**: 69–90.

[21] Lipnizki, F., Hausmanns, S., Ten, P.-K., Field, R. W., and Laufenberg, G. Organophilic pervaporation: Prospects and performance. Chem. Eng. J., 1997, **3**: 113–129.

[22] Jiang, L. Y., Chung, T. S., and Rajagopalan, R. Dehydration of alcohols by pervaporation through polyimide Matrimid® asymmetric hollow fibers with various modifications. Chem. Eng. Sci., 2008, **63**: 204–216.

[23] Crespo, J. G., and Böddeker, K. W. Energy requirements. In: Membrane Processes in Separation and Purificatio, Springer Science & Business Media, Dordrecht, 1994, pp. 186–192.

[24] Huang, R. Y. M., and Yeom, C. K. Pervaporation separation of aqueous mixtures using crosslinked poly(vinyl alcohol) (PVA). II. Permeation of ethanol-water mixtures. J. Membr. Sci., 1990, **51**: 213–292.

[25] Rautenbach, R., Herion, C., and Meyer-Blumentoth, U. Pervaporation membrane separation processes. In: Huang, R. Y. M. (Ed.) Membrane Science and Technology Series, Elsevier, Amsterdam, 1990, pp. 181–191.

[26] Wijmans, J. G., and Baker, R. W. The solution-diffusion model: A review. J. Membr. Sci., 1995, **107**. https://doi.org/10.1016/0376-7388(95)00102-I.

[27] Böddeker, K. W., Bengtson, G., and Bode, E. Pervaporation of low volatility aromatics from water. J. Membr. Sci., 1990, **53**: 143–158. https://doi.org/10.1016/0376-7388(90)80010-J.

[28] Böddeker, K. W. Terminology in pervaporation. J. Membr. Sci., 1990, **51**: 259–272. https://doi.org/10.1016/S0376-7388(00)80350-6.

[29] Baker, R. W., Wijmans, J. G., and Huang, Y. Permeability, permeance and selectivity: A preferred way of reporting pervaporation performance data. J. Membr. Sci., 2010, **348**: 346–352. https://doi.org/10.1016/j.memsci.2009.11.022.

[30] Hyder, M. N. Preparation, Characterization and Performance of Poly(vinyl alcohol) based Membranes for Pervaporation Dehydration of Alcohols, University of Waterloo, 2008.

[31] Shirazi, Y. Effects of CNTs content on physicochemical and pervaporation separation properties of PVA membranes. Sep. Sci. Technol., 2013, **48**: 716–727.

[32] Yeom, C. K., and Huang, R. Y. M. Modelling of the pervaporation separation of ethanol-water mixtures through crosslinked poly(vinyl alcohol) membrane. J. Membr. Sci., 1992, **67**: 39–55. https://doi.org/10.1016/0376-7388(92)87038-Y.

[33] Schaetzel, P., Vauclair, C., Nguyen, Q. T., and Bouzerar, R. A simplified solution – Diffusion theory in pervaporation: The total solvent volume fraction model. J. Membr. Sci., 2004, **244**: 117–127.

[34] Liang, B., Zhan, W., Qi, G., Lin, S., Nan, Q., Liu, Y., Cao, B., and Pan, K. High performance graphene oxide/polyacrylonitrile composite pervaporation membranes for desalination applications. J. Mater. Chem. A, 2015, **3**: 5140–5147.

[35]   Liu, G. G., Shen, J., Liu, Q., Liu, G. G., Xiong, J., Yang, J., and Jin, W. Ultrathin two-dimensional MXene membrane for pervaporation desalination. J. Membr. Sci., 2018, **548**: 548–558.

[36]   Qureshi, N., and Gj, M. Bioconversion of renewable resources into ethanol: An economic evaluation of selected hydrolysis, fermentation, and membrane technologies. Energy Sources, 1995, **17**: 241–265.

[37]   Kim, S., and Dale, B. E. Global potential bioethanol production from wasted crops and crop residues. Biomass Bioenergy, 2004, **26**: 361–375.

[38]   Katz, M. G., and Wydeven, T. Selective permeability of PVA membranes. II. Heat-treated membranes. J. Appl. Polym. Sci., 1982, **27**: 79–87.

[39]   Sharma, A., Thampi, S. P., Suggala, S. V., and Bhattacharya, P. K. Pervaporation from a dense membrane: Roles of permeant-membrane interactions, Kelvin effect, and membrane swelling. Langmuir, 2004, **20**: 4708–4714.

[40]   Garcia Villaluenga, J. P., and Tabe-Mohammadi, A. A review on the separation of benzene/cyclohexane mixtures by pervaporation processes. J. Membr. Sci., 2000, **169**: 159–174. https://doi.org/10.1016/S0376-7388(99)00337-3.

[41]   Wang, L., Wang, Y., Wu, L., and Wei, G. Fabrication, properties, performances, and separation application of polymeric pervaporation membranes: A review. Polymers (Basel), 2020, **12**: 1466. https://doi.org/10.3390/polym12071466.

[42]   Tusel, G. F., and Brüschke, H. E. A. Use of pervaporation systems in the chemical industry. Desalination, 1985, **53**: 327–338. https://doi.org/10.1016/0011-9164(85)85070-0.

[43]   Semenova, S. I., Ohya, H., and Soontarapa, K. Hydrophilic membranes for pervaporation: An analytical review. Desalination, 1997, **110**: 251–286. https://doi.org/10.1016/S0011-9164(97)00103-3.

[44]   Smuleac, V., Wu, J., Nemser, S., Majumdar, S., and Bhattacharyya, D. Novel perfluorinated polymer-based pervaporation membranes for the separation of solvent/water mixtures. J. Membr. Sci., 2010, **352**: 41–49. https://doi.org/10.1016/j.memsci.2010.01.058.

[45]   Abdul Wahab, M. S., Rahman, S. A., and Samah, R. A. Hydrophilic enhancement of polysulfone membrane via graphene oxide embedded thin film nanocomposite for isopropanol dehydration. Vacuum, 2020, **180**: 109569. https://doi.org/10.1016/j.vacuum.2020.109569.

[46]   Wang, K. Y., Weber, M., and Chung, T.-S. Polybenzimidazoles (PBIs) and state-of-the-art PBI hollow fiber membranes for water, organic solvent and gas separations: A review. J. Mater. Chem. A, 2022, **10**: 8687–8718. https://doi.org/10.1039/D2TA00422D.

[47]   Chaudhari, S., Kwon, Y., Moon, M., Shon, M., Nam, S., and Park, Y. Poly(vinyl alcohol) and poly(vinylamine) blend membranes for isopropanol dehydration. J. Appl. Polym. Sci., 2017, **134**. https://doi.org/https://doi.org/10.1002/app.45572.

[48]   Chaudhri, S. G., Rajai, B. H., and Singh, P. S. Preparation of ultra-thin poly(vinyl alcohol) membranes supported on polysulfone hollow fiber and their application for production of pure water from seawater. Desalination, 2015, **367**: 272–284.

[49]   Vane, L. M. Separation technologies for recovery and dehydration of alcohols from fermentation broths. Biofuels Bioprod. Biorefining, 2008, **2**: 553–588.

[50]   Birck, C., Degoutin, S., Tabary, N., Miri, V., and Bacquet, M. New crosslinked cast films based on poly (vinyl alcohol): Preparation and physico-chemical properties. Express Polym. Lett., 2014, **8**: 941–952. https://doi.org/10.3144/expresspolymlett.2014.95.

[51]   Meng, P., Chen, C., Yu, L., Li, J., and Jiang, W. Crosslinking of PVA pervaporation membrane by maleic acid. Tsinghua Sci. Technol., 2000, **5**: 172–175.

[52]   Park, J.-S., Park, J.-W., and Ruckenstein, E. On the viscoelastic properties of poly(vinyl alcohol) and chemically crosslinked poly(vinyl alcohol). J. Appl. Polym. Sci., 2001, **82**: 1816–1823. https://doi.org/https://doi.org/10.1002/app.2023.

[53]   Bolto, B., Tran, T., Hoang, M., and Xie, Z. Crosslinked poly(vinyl alcohol) membranes. Prog. Polym. Sci., 2009, **24**: 969–981.

[54] Bolto, B., Hoang, M., and Xie, Z. A review of membrane selection for the dehydration of aqueous ethanol by pervaporation. Chem. Eng. Process, 2011, **50**: 227–235.

[55] Mg, K., and Wydeven, T. Selective permeability of PVA membranes, I. Radiation-crosslinked membranes. J. Appl. Polym. Sci., 1981, **26**: 2935–2946.

[56] Gohil, J. M., Bhattacharya, A., and Ray, P. Studies on the cross-linking of poly(vinyl alcohol). J. Polym. Res., 2006, **13**: 161–169. https://doi.org/https://doi.org/10.1007/s10965-005-9023-9.

[57] Mallapragada, S. K., and Peppas, N. A. Dissolution mechanism of semicrystalline poly(vinyl alcohol) in water. J. Polym. Sci., Part B: Polym. Phys., 1996, **34**: 1339–1346.

[58] Li, B.-B., Xu, Z.-L., Qusay, F. A., and Li, R. Chitosan-poly (vinyl alcohol)/poly (acrylonitrile) (CS–PVA/PAN) composite pervaporation membranes for the separation of ethanol–water solutions. Desalination, 2006, **193**: 171–181.

[59] Lee, Y. M. Modified chitosan membranes for pervaporation. Desalination, 1993, **90**: 277–290.

[60] Kurkuri, M. D., Toti, U. S., and Aminabhavi, T. M. Syntheses and characterization of blend membranes of sodium alginate and poly(vinyl alcohol) for the pervaporation separation of water + isopropanol mixtures. J. Appl. Polym. Sci., 2002, **86**: 3642–3651. https://doi.org/10.1002/app.11312.

[61] Toti, U. S., and Aminabhavi, T. M. Different viscosity grade sodium alginate and modified sodium alginate membranes in pervaporation separation of water + acetic acid and water + isopropanol mixtures. J. Membr. Sci., 2004, **228**: 199–208. https://doi.org/10.1016/j.memsci.2003.10.008.

[62] Kuila, S. B., and Ray, S. K. Dehydration of dioxane by pervaporation using filled blend membranes of polyvinyl alcohol and sodium alginate. Carbohydr. Polym., 2014, **101**: 1154–1165. https://doi.org/10.1016/j.carbpol.2013.09.086.

[63] Yeom, C. K., Jegal, J. G., and Lee, K. H. Characterization of relaxation phenomena and permeation behaviors in sodium alginate membrane during pervaporation separation of ethanol-water mixture. J. Appl. Polym. Sci., 1996, **62**: 1561–1576. https://doi.org/10.1002/(SICI)1097-4628(19961205) 62:10<1561::AID-APP8>3.0.CO;2-M.

[64] Kalyani, S., Smitha, B., Sridhar, S., and Krishnaiah, A. Separation of ethanol–water mixtures by pervaporation using sodium alginate/poly(vinyl pyrrolidone) blend membrane crosslinked with phosphoric acid. Ind. Eng. Chem. Res., 2006, **45**: 9088–9095. https://doi.org/10.1021/ie060085y.

[65] Yeom, C. K., and Lee, K.-H. Characterization of sodium alginate membrane crosslinked with glutaraldehyde in pervaporation separation. J. Appl. Polym. Sci., 1998, **67**: 209–219. https://doi.org/10.1002/(SICI)1097-4628(19980110)67:2<209::AID-APP3>3.0.CO;2-Y.

[66] Huang, R. Y., Pal, R., and Moon, G. Pervaporation dehydration of aqueous ethanol and isopropanol mixtures through alginate/chitosan two ply composite membranes supported by poly(vinylidene fluoride) porous membrane. J. Membr. Sci., 2000, **167**: 275–289. https://doi.org/10.1016/S0376-7388(99)00293-8.

[67] Moulik, S., Nazia, S., Vani, B., and Sridhar, S. Pervaporation separation of acetic acid/water mixtures through sodium alginate/polyaniline polyion complex membrane. Sep. Purif. Technol., 2016, **170**: 30–39. https://doi.org/10.1016/j.seppur.2016.06.027.

[68] Bhat, S. D., and Aminabhavi, T. M. Pervaporation separation using sodium alginate and its modified membranes: A review. Sep. Purif. Rev., 2007, **36**: 203–229. https://doi.org/10.1080/15422110701539061.

[69] Castro-Muñoz, R., González-Valdez, J., and Ahmad, M. Z. High-performance pervaporation chitosan-based membranes: New insights and perspectives. Rev. Chem. Eng., 2021, **37**: 959–974. https://doi.org/10.1515/revce-2019-0051.

[70] Jyothi, M. S., Reddy, K. R., Soontarapa, K., Naveen, S., Raghu, A. V., Kulkarni, R. V., Suhas, D. P., Shetti, N. P., Nadagouda, M. N., and Aminabhavi, T. M. Membranes for dehydration of alcohols via pervaporation. J. Environ. Manage., 2019, **242**: 415–429. https://doi.org/10.1016/j.jenvman.2019.04.043.

[71] Bolto, B., Hoang, M., and Xie, Z. A review of membrane selection for the dehydration of aqueous ethanol by pervaporation. Chem. Eng. Process. Process Intensif., 2011, **50**: 227–235. https://doi.org/10.1016/j.cep.2011.01.003.

[72] Chakrabarty, T., Kumar, M., and Shahi, V. K. Chitosan based membranes for separation, pervaporation and fuel cell applications: Recent developments. In: Elnashar, M. (Ed.) Biopolymers, InTechOpen, London, UK, 2010, pp. 201–226.

[73] Hung, M.-Y., Chen, S.-H., Liou, R.-M., Hsu, C.-S., and Lai, J.-Y. Pervaporation separation of a water/ethanol mixture by a sodium sulfonate polysulfone membrane. J. Appl. Polym. Sci., 2003, **90**: 3374–3383. https://doi.org/10.1002/app.12947.

[74] Chen, S.-H., Liou, R.-M., Lin, Y.-Y., Lai, C.-L., and Lai, J.-Y. Preparation and characterizations of asymmetric sulfonated polysulfone membranes by wet phase inversion method. Eur. Polym. J., 2009, **45**: 1293–1301. https://doi.org/10.1016/j.eurpolymj.2008.11.030.

[75] Lipnizki, F., Field, R. W., and Ten, P.-K. Pervaporation-based hybrid process: A review of process design, applications and economics. J. Membr. Sci., 1999, **153**: 183–210. https://doi.org/10.1016/S0376-7388(98)00253-1.

[76] Huang, R. Y. M., Moreira, A., Notarfonzo, R., and Xu, Y. F. Pervaporation separation of acetic acid-water mixtures using modified membranes. I. Blended polyacrylic acid (PAA)-nylon 6 membranes. J. Appl. Polym. Sci., 1988, **35**: 1191–1200. https://doi.org/10.1002/app.1988.070350506.

[77] Tang, J., and Sirkar, K. K. Perfluoropolymer membrane behaves like a zeolite membrane in dehydration of aprotic solvents. J. Membr. Sci., 2012, **421–422**: 211–216. https://doi.org/10.1016/j.memsci.2012.07.015.

[78] Lee, K.-R., Wang, Y.-H., Teng, M.-Y., Liaw, D.-J., and Lai, J.-Y. Preparation of aromatic polyamide membrane for alcohol dehydration by pervaporation. Eur. Polym. J., 1999, **35**: 861–866. https://doi.org/10.1016/S0014-3057(98)00056-1.

[79] Wang, Y., HongGoh, S., Chung, T. S., and Na, P. Polyamide-imide/polyetherimide dual-layer hollow fiber membranes for pervaporation dehydration of C1–C4 alcohols. J. Membr. Sci., 2009, **326**: 222–233.

[80] Bano, S., Mahmood, A., and Lee, K.-H. Vapor permeation separation of methanol-water mixtures: Effect of experimental conditions. Ind. Eng. Chem. Res., 2013, **52**: 10450–10459.

[81] Ruckensien, E., and Liang, L. Pervaporation of ethanol-water through polyvinylalcohol–polyacrylamide interpenetrating polymer network membranes unsupported and supported on polyethersulfone ultrafiltration membranes: A comparison. J. Membr. Sci., 1996, **110**: 99–107. https://doi.org/https://doi.org/10.1016/0376-7388(95)00240-5.

[82] Izquierdo-Gil, M. A., and Jonsson, G. Factors affecting flux and ethanol separation performance in vacuum membrane distillation (VMD). J. Membr. Sci., 2003, **214**: 113–130.

[83] Wang, Y., Jiang, L., Matsuura, T., Chung, T.-S., and Goh, S. H. Investigation of the fundamental differences between polyamide-imide (PAI) and polyetherimide (PEI) membranes for isopropanol dehydration via pervaporation. J. Membr. Sci., 2008, **318**: 217–226.

[84] Roy, S., Thongsukmak, A., Tang, J., and Sirkar, K. K. Concentration of aqueous hydrogen peroxide solution by pervaporation. J. Membr. Sci., 2012, **389**: 17–24. https://doi.org/10.1016/j.memsci.2011.09.046.

[85] Shi, G. M., Wang, Y., and Chung, T.-S. Dual-layer PBI/P84 hollow fibers for pervaporation dehydration of acetone. AIChE J., 2012, **58**: 1133–1145. https://doi.org/10.1002/aic.12625.

[86] Wang, Y., Gruender, M., and Chung, T. S. Pervaporation dehydration of ethylene glycol through polybenzimidazole (PBI)-based membranes. 1. Membrane fabrication. J. Membr. Sci., 2010, **363**: 149–159. https://doi.org/10.1016/j.memsci.2010.07.024.

[87] Wang, Y., Shung Chung, T., and Gruender, M. Sulfonated polybenzimidazole membranes for pervaporation dehydration of acetic acid. J. Membr. Sci., 2012, **415–416**: 486–495. https://doi.org/10.1016/j.memsci.2012.05.035.

[88] Castro-Muñoz, R., De la Iglesia, Ó., Fíla, V., Téllez, C., and Coronas, J. Pervaporation-assisted esterification reactions by means of mixed matrix membranes. Ind. Eng. Chem. Res., 2018, **57**: 15998–16011. https://doi.org/10.1021/acs.iecr.8b01564.

[89] Lv, B., Liu, G., Dong, X., Wei, W., and Jin, W. Novel reactive distillation–pervaporation coupled process for ethyl acetate production with water removal from reboiler and acetic acid recycle. Ind. Eng. Chem. Res., 2012, **51**: 8079–8086. https://doi.org/10.1021/ie3004072.

[90] Zhang, L., Li, Y., Liu, Q., Li, W., and Xing, W. Fabrication of ionic liquids-functionalized PVA catalytic composite membranes to enhance esterification by pervaporation. J. Membr. Sci., 2019, **584**: 268–281. https://doi.org/10.1016/j.memsci.2019.05.006.

[91] Han, J., Liang, Y., He, C., Tong, Y., and Li, W. Porous PVA-g-SPA/PVA-SA catalytic composite membrane via lyophilization for esterification enhancement. Langmuir, 2022, **38**: 2660–2667. https://doi.org/10.1021/acs.langmuir.1c03381.

[92] Sun, H., Qu, Z., Yu, J., Ma, H., Li, B., Sun, D., and Ge, Y. Asymmetric 5-sulfosalicylic acid-PVA catalytic pervaporation membranes for the process intensification in the synthesis of ethyl acetate. Sep. Purif. Technol., 2022, **282**: 120113. https://doi.org/10.1016/j.seppur.2021.120113.

[93] Díaz, E. G., Álvarez-García, S., Luque, S., and Álvarez, J. R. Low-temperature hydrophilic pervaporation of lactic acid esterification reaction media. Membranes (Basel), 2022, **12**: 96. https://doi.org/10.3390/membranes12010096.

[94] Jiang, H., Shi, W., Liu, Q., Wang, H., Li, J., Wu, C., Li, Y., and Wei, Z. Intensification of water/ethanol separation by PVA hybrid membrane with different functional ligand UiO-66-X nanochannels in pervaporation process. Sep. Purif. Technol., 2021, **256**: 117802. https://doi.org/https://doi.org/10.1016/j.seppur.2020.117802.

[95] Chen, J., Li, J., Qi, R., Ye, H., and Chen, C. Pervaporation performance of crosslinked polydimethylsiloxane membranes for deep desulfurization of FCC gasoline. J. Membr. Sci., 2008, **322**: 113–121. https://doi.org/10.1016/j.memsci.2008.05.032.

[96] Zhang, H., and Wang, Y. Poly(vinyl alcohol)/ZIF-8-NH 2 mixed matrix membranes for ethanol dehydration via pervaporation. AIChE J., 2016, **62**: 1728–1739. https://doi.org/10.1002/aic.15140.

[97] Zhang, Q. G., Liu, Q. L., Jiang, Z. Y., and Chen, Y. Anti-trade-off in dehydration of ethanol by novel PVA/APTEOS hybrid membranes. J. Membr. Sci., 2007, **287**: 237–245. https://doi.org/10.1016/j.memsci.2006.10.041.

[98] Pereira, C. C., Ribeiro, C. P., Nobrega, R., and Borges, C. P. Pervaporative recovery of volatile aroma compounds from fruit juices. J. Membr. Sci., 2006, **274**: 1–23. https://doi.org/10.1016/j.memsci.2005.10.016.

[99] Castro-Muñoz, R. Pervaporation-based membrane processes for the production of non-alcoholic beverages. J. Food Sci. Technol., 2019, **56**: 2333–2344. https://doi.org/10.1007/s13197-019-03751-4.

[100] Catarino, M., and Mendes, A. Dealcoholizing wine by membrane separation processes, Innov. Food Sci. Emerg. Technol., 2011, **12**: 330–337. https://doi.org/10.1016/j.ifset.2011.03.006.

[101] Yi, S., and Wan, Y. Volatile organic compounds (VOCs) recovery from aqueous solutions via pervaporation with vinyltriethoxysilane-grafted-silicalite-1/polydimethylsiloxane mixed matrix membrane. Chem. Eng. J., 2017, **313**: 1639–1646. https://doi.org/https://doi.org/10.1016/j.cej.2016.11.061.

[102] Chapman, P. D. D., Oliveira, T., Livingston, A. G., and Li, K. Membranes for the dehydration of solvents by pervaporation. J. Membr. Sci., 2008, **318**: 5–37.

[103] Castro-Muñoz, R. Pervaporation: The emerging technique for extracting aroma compounds from food systems. J. Food Eng., 2019, **253**: 27–39. https://doi.org/10.1016/j.jfoodeng.2019.02.013.

[104] Peng, P., Shi, B., and Lan, Y. A review of membrane materials for ethanol recovery by pervaporation. Sep. Sci. Technol., 2010, **46**: 234–246. https://doi.org/10.1080/01496395.2010.504681.

[105] Borisov, I. L., Malakhov, A. O., Khotimsky, V. S., Litvinova, E. G., Finkelshtein, E. S., Ushakov, N. V., and Volkov, V. V. Novel PTMSP-based membranes containing elastomeric fillers: Enhanced 1-

butanol/water pervaporation selectivity and permeability. J. Membr. Sci., 2014, **466**: 322–330. https://doi.org/https://doi.org/10.1016/j.memsci.2014.04.037.

[106] Claes, S., Vandezande, P., Mullens, S., De Sitter, K., Peeters, R., and Van Bael, M. K. Preparation and benchmarking of thin film supported PTMSP-silica pervaporation membranes. J. Membr. Sci., 2012, **389**: 265–271. https://doi.org/http://dx.doi.org/10.1016/j.memsci.2011.10.035.

[107] Van Hecke, W., and De Wever, H. High-flux POMS organophilic pervaporation for ABE recovery applied in fed-batch and continuous set-ups. J. Membr. Sci., 2017, **540**: 321–332. https://doi.org/ https://doi.org/10.1016/j.memsci.2017.06.058.

[108] Rom, A., and Friedl, A. Investigation of pervaporation performance of POMS membrane during separation of butanol from water and the effect of added acetone and ethanol. Sep. Purif. Technol., 2016, **170**: 40–48. https://doi.org/http://dx.doi.org/10.1016/j.seppur.2016.06.030.

[109] Liu, Q., Li, Y., Li, Q., Liu, G. G., Liu, G. G., and Jin, W. Mixed-matrix hollow fiber composite membranes comprising of PEBA and MOF for pervaporation separation of ethanol/water mixtures. Sep. Purif. Technol., 2019, **214**: 2–10. https://doi.org/https://doi.org/10.1016/j.seppur.2018.01.050.

[110] Li, Y., Shen, J., Guan, K., Liu, G., Zhou, H., and Jin, W. PEBA/ceramic hollow fiber composite membrane for high-efficiency recovery of bio-butanol via pervaporation. J. Membr. Sci., 2016, **510**: 338–347. https://doi.org/http://dx.doi.org/10.1016/j.memsci.2016.03.013.

[111] Mao, H., Li, S.-H., Zhang, A.-S., Xu, L.-H., Lu, -J.-J., and Zhao, Z.-P. Novel MOF-capped halloysite nanotubes/PDMS mixed matrix membranes for enhanced n-butanol permselective pervaporation. J. Memb. Sci., 2020, **595**: 117543. https://doi.org/https://doi.org/10.1016/j.memsci.2019.117543.

[112] León, J. A., and Fontalvo, J. PDMS modified membranes by 1-dodecanol and its effect on ethanol removal by pervaporation. Sep. Purif. Technol., 2019, **210**: 364–370.

[113] Kujawska, A., Knozowska, K., Kujawa, J., Li, G., and Kujawski, W. Fabrication of PDMS based membranes with improved separation efficiency in hydrophobic pervaporation. Sep. Purif. Technol., 2020, **234**: 116092. https://doi.org/https://doi.org/10.1016/j.seppur.2019.116092.

[114] Liu, G., Wei, W., Wu, H., Dong, X., Jiang, M., and Jin, W. Pervaporation performance of PDMS/ceramic composite membrane in acetone butanol ethanol (ABE) fermentation–PV coupled process. J. Membr. Sci., 2011, **373**: 121–129.

[115] Mao, H., Zhen, H.-G., Ahmad, A., Zhang, A.-S., and Zhao, Z.-P. In situ fabrication of MOF nanoparticles in PDMS membrane via interfacial synthesis for enhanced ethanol permselective pervaporation. J. Membr. Sci., 2019, **573**: 344–358. https://doi.org/10.1016/j.memsci.2018.12.017.

[116] Jin, M.-Y., Liao, Y., Tan, C.-H., and Wang, R. Development of high performance nanofibrous composite membranes by optimizing polydimethylsiloxane architectures for phenol transport. J. Membr. Sci., 2018, **549**: 638–648. https://doi.org/10.1016/j.memsci.2017.10.051.

[117] Wei, W., Xia, S., Liu, G., Dong, X., Jin, W., and Xu, N. Effects of polydimethylsiloxane (PDMS) molecular weight on performance of PDMS/ceramic composite membranes. J. Membr. Sci., 2011, **375**: 334–344. https://doi.org/10.1016/j.memsci.2011.03.059.

[118] Cheng, X., Pan, F., Wang, M., Li, W., Song, Y., Liu, G., Yang, H., Gao, B., Wu, H., and Jiang, Z. Hybrid membranes for pervaporation separations. J. Membr. Sci., 2017, **541**: 329–346. https://doi.org/10. 1016/j.memsci.2017.07.009.

[119] Rana, D., and Matsuura, T. Surface modifications for antifouling membranes. Chem. Rev., 2010, **110**: 2448–2471. https://doi.org/10.1021/cr800208y.

[120] Zhu, H., Liu, G., and Jin, W. Recent progress in separation membranes and their fermentation coupled processes for biobutanol recovery. Energy Fuels, 2020, **34**: 11962–11975. https://doi.org/10. 1021/acs.energyfuels.0c02680.

[121] Zhu, H., Li, X., Pan, Y., Liu, G., Wu, H., Jiang, M., and Jin, W. Fluorinated PDMS membrane with anti-biofouling property for in-situ biobutanol recovery from fermentation-pervaporation coupled process. J. Membr. Sci., 2020, **609**: 118225. https://doi.org/10.1016/j.memsci.2020.118225.

[122] Niemistö, J., Kujawski, W., and Keiski, R. L. Pervaporation performance of composite poly(dimethyl siloxane) membrane for butanol recovery from model solutions. J. Membr. Sci., 2013, **434**: 55–64. https://doi.org/10.1016/j.memsci.2013.01.047.

[123] Rozicka, A., Niemistö, J., Keiski, R. L., and Kujawski, W. Apparent and intrinsic properties of commercial PDMS based membranes in pervaporative removal of acetone, butanol and ethanol from binary aqueous mixtures. J. Membr. Sci., 2014, **453**: 108–118. https://doi.org/10.1016/j.memsci.2013.10.065.

[124] Zhan, X., Li, J., Huang, J., and Chen, C. Enhanced pervaporation performance of multi-layer PDMS/PVDF composite membrane for ethanol recovery from aqueous solution. Appl. Biochem. Biotechnol., 2010, **160**: 632–642. https://doi.org/10.1007/s12010-008-8510-y.

[125] Inthavee, W., Kanchanatawee, S., and Boontawan, A. Development of a composite tubular membrane for separation of acetone-butanol-ethanol (ABE) from fermentation broth by using pervaporation technique. Thai J. Agric. Sci., 2011, **44**: 400–407.

[126] Hecke, W. V., Vandezande, P., Claes, S., Vangeel, S., Beckers, H., Diels, L., and Wever, H. D. Integrated bioprocess for long-term continuous cultivation of *Clostridium acetobutylicum* coupled to pervaporation with PDMS composite membranes. Bioresour. Technol., 2012, **111**: 368–377. https://doi.org/10.1016/j.biortech.2012.02.043.

[127] Thanakkasaranee, S., Kim, D., and Seo, J. Preparation and characterization of poly(ether-*block*-amide)/polyethylene glycol composite films with temperature-dependent permeation. Polymers (Basel), 2018, **10**: 225. https://doi.org/https://doi.org/10.3390/polym10020225.

[128] Soloukipour, S., Mousavi, S. M., Saljoughi, E., and Pourafshari Chenar, M. PEBA/PS blend pervaporation membranes: Preparation, characterization and performance investigation. Desalin. Water Treat., 2019, **153**: 24–35. https://doi.org/10.5004/dwt.2019.23951.

[129] Aburabie, J., Peinemann, K.-V., and Peinemann, K.-V. Crosslinked poly(ether block amide) composite membranes for organic solvent nanofiltration applications. J. Membr. Sci., 2017, **523**: 264–272. https://doi.org/https://doi.org/10.1016/j.memsci.2016.09.027.

[130] Wu, G., Li, Y., Geng, Y., and Jia, Z. In situ preparation of COF-LZU1 in poly(ether-*block*-amide) membranes for efficient pervaporation of *n*-butanol/water mixture. J. Membr. Sci., 2019, **581**: 1–8. https://doi.org/10.1016/j.memsci.2019.03.044.

[131] Le, N. L., Wang, Y., and Chung, T.-S. Pebax/POSS mixed matrix membranes for ethanol recovery from aqueous solutions via pervaporation. J. Membr. Sci., 2011, **379**: 174–183. https://doi.org/10.1016/j.memsci.2011.05.060.

[132] Liu, F., Liu, L., Feng, X., Liu, L., Feng, X., Liu, F., Liu, L., and Feng, X. Separation of acetone–butanol–ethanol (ABE) from dilute aqueous solutions by pervaporation. Sep. Purif. Technol., 2005, **42**: 273–282. https://doi.org/http://dx.doi.org/10.1016/j.seppur.2004.08.005.

[133] Liu, S., Liu, G., Zhao, X., and Jin, W. Hydrophobic-ZIF-71 filled PEBA mixed matrix membranes for recovery of biobutanol via pervaporation. J. Membr. Sci., 2013, **446**: 181–188. https://doi.org/10.1016/j.memsci.2013.06.025.

[134] Cao, X., Wang, K., and Feng, X. Removal of phenolic contaminants from water by pervaporation. J. Membr. Sci., 2021, **623**: 119043. https://doi.org/10.1016/j.memsci.2020.119043.

[135] Bai, Y., Dong, L., Lin, J., Zhu, Y., Zhang, C., Gu, J., Sun, Y., and Xu, Y. High performance polydimethylsiloxane pervaporative membranes with hyperbranched polysiloxane as a crosslinker for separation of n-butanol from water. RSC Adv., 2015, **5**: 52759–52768. https://doi.org/10.1039/C5RA06886J.

[136] Yang, Y., Si, Z., Cai, D., Teng, X., Li, G., Wang, Z., Li, S., and Qin, P. High-hydrophobic CF3 groups within PTFPMS membrane for enhancing the furfural pervaporation performance. Sep. Purif. Technol., 2020, **235**: 116144. https://doi.org/10.1016/j.seppur.2019.116144.

[137] Ghosh, U. K., Pradhan, N. C., and Adhikari, B. Separation of furfural from aqueous solution by pervaporation using HTPB-based hydrophobic polyurethaneurea membranes. Desalination, 2007, **208**: 146–158. https://doi.org/10.1016/j.desal.2006.04.078.

[138] Choudhury, S., and Ray, S. K. Poly(4-vinylpyridine) and poly(vinyl acetate-co-4-vinylpyridine) grafted polyvinyl chloride membranes for removal of tetrahydrofuran from water by pervaporation. Sep. Purif. Technol., 2021, **254**: 117618. https://doi.org/10.1016/j.seppur.2020.117618.

[139] Zou, Y., Liu, Y., Muhammad, Y., Tong, Z., and Feng, X. Experimental and modelling studies of pervaporative removal of odorous diacetyl and S-methylthiobutanoate from aqueous solutions using PEBA membrane. Sep. Purif. Technol., 2018, **200**: 1–10. https://doi.org/10.1016/j.seppur.2018.01.069.

[140] Smitha, B. Separation of organic? Organic mixtures by pervaporation? A review*1. J. Membr. Sci., 2004, **241**: 1–21. https://doi.org/10.1016/j.memsci.2004.03.042.

[141] Ong, Y. K., Shi, G. M., Le, N. L., Tang, Y. P., Zuo, J., Nunes, S. P., and Chung, T.-S. Recent membrane development for pervaporation processes, Prog. Polym. Sci., 2016, **57**: 1–31. https://doi.org/https://doi.org/10.1016/j.progpolymsci.2016.02.003.

[142] Rhim, J.-W., and Kim, Y.-K. Pervaporation separation of MTBE-methanol mixtures using cross-linked PVA membranes. J. Appl. Polym. Sci., 2000, **75**: 1699–1707. https://doi.org/10.1002/(SICI)1097-4628(20000401)75:14<1699::AID-APP3>3.0.CO;2-O.

[143] Madandar, A., and Mohammadi, T. Effect of permeate pressure on pervaporation of methyl tert-butyl ether/methanol mixtures. Desalination, 2006, **200**: 390–392. https://doi.org/10.1016/j.desal.2006.03.387.

[144] Zereshki, S., Figoli, A., Madaeni, S. S., Simone, S., Esmailinezhad, M., and Drioli, E. Pervaporation separation of MeOH/MTBE mixtures with modified PEEK membrane: Effect of operating conditions. J. Membr. Sci., 2011, **371**: 1–9. https://doi.org/10.1016/j.memsci.2010.11.068.

[145] Yong Nam, S. Pervaporation separation of methanol/methyl t-butyl ether through chitosan composite membrane modified with surfactants. J. Membr. Sci., 1999, **157**: 63–71. https://doi.org/10.1016/S0376-7388(98)00368-8.

[146] Ma, X., Hu, C., Guo, R., Fang, X., Wu, H., and Jiang, Z. HZSM5-filled cellulose acetate membranes for pervaporation separation of methanol/MTBE mixtures. Sep. Purif. Technol., 2008, **59**: 34–42. https://doi.org/10.1016/j.seppur.2007.05.023.

[147] Niang, M., Luo, G., and Schaetzel, P. Pervaporation separation of methyl tert-butyl ether/methanol mixtures using a high-performance blended membrane. J. Appl. Polym. Sci., 1997, **64**: 875–882. https://doi.org/10.1002/(SICI)1097-4628(19970502)64:5<875::AID-APP7>3.0.CO;2-M.

[148] Yang, J. S., Kim, H. J., Jo, W. H., and Kang, Y. S. Analysis of pervaporation of methanol-MTBE mixtures through cellulose acetate and cellulose triacetate membranes. Polymer (Guildf)., 1998, **39**: 1381–1385. https://doi.org/10.1016/S0032-3861(97)00416-3.

[149] Cao, S., Shi, Y., and Chen, G. Influence of acetylation degree of cellulose acetate on pervaporation properties for MeOH/MTBE mixture. J. Membr. Sci., 2000, **165**: 89–97. https://doi.org/10.1016/S0376-7388(99)00222-7.

[150] Qian, H., Miao, J., Zhang, Y., and Chen, L. Influence of viscosity slope coefficient of CA and its blends in dilute solutions on permeation flux of their films for MeOH/MTBE mixture. J. Membr. Sci., 2002, **203**: 167–173. https://doi.org/10.1016/S0376-7388(02)00004-2.

[151] Nguyen, Q.-T., Clément, R., Noezar, I., and Lochon, P. Performances of poly(vinylpyrrolidone-co-vinyl acetate)-cellulose acetate blend membranes in the pervaporation of ethanol–ethyl tert-butyl ether mixtures. Sep. Purif. Technol., 1998, **13**: 237–245. https://doi.org/10.1016/S1383-5866(98)00046-X.

[152] Billy, M., Da Costa, A. R., Lochon, P., Clément, R., Dresch, M., and Jonquières, A. Cellulose acetate graft copolymers with nano-structured architectures: Application to the purification of bio-fuels by pervaporation. J. Membr. Sci., 2010, **348**: 389–396. https://doi.org/10.1016/j.memsci.2009.11.027.

[153] Niang, M. A triacetate cellulose membrane for the separation of methyl *tert*-butyl ether/methanol mixtures by pervaporation. Sep. Purif. Technol., 2001, **24**: 427–435. https://doi.org/10.1016/S1383-5866(01)00143-5.

[154] Garcia Villaluenga, J. P., and Tabe-Mohammadi, A. A review on the separation of benzene/cyclohexane mixtures by pervaporation processes. J. Membr. Sci., 2000, **169**: 159–174. https://doi.org/10.1016/S0376-7388(99)00337-3.

[155] Liu, H.-X., Wang, N., Zhao, C., Ji, S., and Li, J.-R. Membrane materials in the pervaporation separation of aromatic/aliphatic hydrocarbon mixtures – A review. Chin. J. Chem. Eng., 2018, **26**: 1–16. https://doi.org/10.1016/j.cjche.2017.03.006.

[156] Jiang, L. Y., Wang, Y., Chung, T.-S., Qiao, X. Y., and Lai, J.-Y. Polyimides membranes for pervaporation and biofuels separation, Prog. Polym. Sci., 2009, **34**: 1135–1160. https://doi.org/10.1016/j.progpolymsci.2009.06.001.

[157] Kung, G., Jiang, L. Y., Wang, Y., and Chung, T.-S. Asymmetric hollow fibers by polyimide and polybenzimidazole blends for toluene/iso-octane separation. J. Membr. Sci., 2010, **360**: 303–314. https://doi.org/10.1016/j.memsci.2010.05.030.

[158] Katarzynski, D., and Staudt, C. Temperature-dependent separation of naphthalene/*n*-decane mixtures using 6FDA–DABA-copolyimide membranes. J. Membr. Sci., 2010, **348**: 84–90. https://doi.org/10.1016/j.memsci.2009.10.043.

[159] Ribeiro, C. P., Freeman, B. D., Kalika, D. S., and Kalakkunnath, S. Aromatic polyimide and polybenzoxazole membranes for the fractionation of aromatic/aliphatic hydrocarbons by pervaporation. J. Membr. Sci., 2012, **390–391**: 182–193. https://doi.org/10.1016/j.memsci.2011.11.042.

[160] Ribeiro, C. P., Freeman, B. D., Kalika, D. S., and Kalakkunnath, S. Pervaporative separation of aromatic/aliphatic mixtures with poly(siloxane-*co*-imide) and poly(ether-*co*-imide) membranes. Ind. Eng. Chem. Res., 2013, **52**: 8906–8916. https://doi.org/10.1021/ie302344z.

[161] Böhmer, V. Calixarenes, macrocycles with(almost) unlimited possibilities. Angew. Chem. Int. Ed. English, 1995, **34**: 713–745. https://doi.org/10.1002/anie.199507131.

[162] Yamasaki, A., Iwatsubo, T., Masuoka, T., and Mizoguchi, K. Pervaporation of ethanol/water through a poly(vinyl alcohol)/cyclodextrin (PVA/CD) membrane. J. Membr. Sci., 1994, **89**: 111–117. https://doi.org/10.1016/0376-7388(93)E0217-8.

[163] Kusumocahyo, S. P., Kanamori, T., Sumaru, K., Iwatsubo, T., and Shinbo, T. Pervaporation of xylene isomer mixture through cyclodextrins containing polyacrylic acid membranes. J. Membr. Sci., 2004, **231**: 127–132. https://doi.org/10.1016/j.memsci.2003.11.011.

[164] Dubey, V., Pandey, L. K., and Saxena, C. Pervaporation of benzene/cyclohexane mixtures through supramolecule containing poly(vinyl acetal) membranes. Sep. Purif. Technol., 2006, **50**: 45–50. https://doi.org/10.1016/j.seppur.2005.11.005.

[165] Pandey, L. K., Saxena, C., and Dubey, V. Modification of poly(vinyl alcohol) membranes for pervaporative separation of benzene/cyclohexane mixtures. J. Membr. Sci., 2003, **227**: 173–182. https://doi.org/10.1016/j.memsci.2003.08.024.

[166] Uragami, T., Meotoiwa, T., and Miyata, T. Effects of the addition of calixarene to microphase-separated membranes for the removal of volatile organic compounds from dilute aqueous solutions. Macromolecules, 2001, **34**: 6806–6811. https://doi.org/10.1021/ma010693j.

[167] Wu, T., Wang, N., Li, J., Wang, L., Zhang, W., Zhang, G., and Ji, S. Tubular thermal crosslinked-PEBA/ceramic membrane for aromatic/aliphatic pervaporation. J. Membr. Sci., 2015, **486**: 1–9. https://doi.org/10.1016/j.memsci.2015.03.037.

[168] Shieh, J.-J., Chung, T.-S., and Paul, D. R. Study on multi-layer composite hollow fiber membranes for gas separation. Chem. Eng. Sci., 1999, **54**: 675–684. https://doi.org/10.1016/S0009-2509(98)00256-5.

[169] Yave, W., Car, A., Funari, S. S., Nunes, S. P., and Peinemann, K.-V. CO_2-philic polymer membrane with extremely high separation performance. Macromolecules, 2010, **43**: 326–333. https://doi.org/10.1021/ma901950u.

[170] Chung, T.-S., Kafchinski, E. R., Kohn, R. S., Foley, P., and Straff, R. S. Fabrication of composite hollow fibers for air separation. J. Appl. Polym. Sci., 1994, **53**: 701–708. https://doi.org/10.1002/app.1994.070530520.

[171] Li, P., Chen, H. Z., and Chung, T.-S. The effects of substrate characteristics and pre-wetting agents on PAN–PDMS composite hollow fiber membranes for CO_2/N_2 and O_2/N_2 separation. J. Membr. Sci., 2013, **434**: 18–25. https://doi.org/10.1016/j.memsci.2013.01.042.

[172] Ding, J., Zhang, M., Jiang, Z., Li, Y., Ma, J., and Zhao, J. Enhancing the permselectivity of pervaporation membrane by constructing the active layer through alternative self-assembly and spin-coating. J. Membr. Sci., 2012, **390–391**: 218–225. https://doi.org/10.1016/j.memsci.2011.11.043.

[173] Roy, S., and Singha, N. R. Polymeric nanocomposite membranes for next generation pervaporation process: Strategies, challenges and future prospects. Membranes (Basel), 2017, **7**. https://doi.org/10.3390/membranes7030053.

[174] Ong, Y. K., and Chung, T.-S. Pushing the limits of high performance dual-layer hollow fiber fabricated via I 2 PS process in dehydration of ethanol. AIChE J., 2013, **59**: 3006–3018. https://doi.org/10.1002/aic.14149.

[175] Peng, N., Widjojo, N., Sukitpaneenit, P., Teoh, M. M., Lipscomb, G. G., Chung, T.-S., and Lai, J.-Y. Evolution of polymeric hollow fibers as sustainable technologies: Past, present, and future. Prog. Polym. Sci., 2012, **37**: 1401–1424. https://doi.org/10.1016/j.progpolymsci.2012.01.001.

[176] Liu, Y.-L., Yu, C.-H., Ma, L.-C., Lin, G.-C., Tsai, H.-A., and Lai, J.-Y. The effects of surface modifications on preparation and pervaporation dehydration performance of chitosan/polysulfone composite hollow-fiber membranes. J. Membr. Sci., 2008, **311**: 243–250. https://doi.org/10.1016/j.memsci.2007.12.040.

[177] Hua, D., Kang Ong, Y., Wang, P., and Chung, T.-S. Thin-film composite tri-bore hollow fiber (TFC TbHF) membranes for isopropanol dehydration by pervaporation. J. Membr. Sci., 2014, **471**: 155–167. https://doi.org/10.1016/j.memsci.2014.07.059.

Sadia Bano, Muhammad Altaf, Sajda Bano

Chapter 8
Electrospun nanofibrous membranes (ENMs) for environmental applications

Abstract: For the next 50 years, environmental pollution will top the list of the most important global issues confronting humanity. Nanotechnology is addressing these difficulties by designing and manufacturing functional nanofibers with a focus on energy and environmental applications. Electrospinning is an emerging field that has been regarded as a crucial scientific and economic endeavor in nanotechnology. For practically all polymer solutions, melts, emulsions, and suspensions with suitable viscosity, the electrospinning technique can be utilized to manufacture polymer nanofibers since it is relatively easy to use and convenient. These nanofiber membranes can be used for various environmental applications depending on their properties.

Keywords: environment remediation, electrospinning, polymers, membranes, nanofibers

1 Introduction

Due to the rapid expansion of industrialization, urbanization, and contemporary agricultural development, environmental pollution has grown as a global concern [1, 2]. Technological improvements in the products and procedures have enabled industry to generate new products and new pollutants in abundance that are above the capability of the environment to clean itself [3]. Water and air are two important components of the environment, which are being adversely affected by pollution [1]. There is no doubt that the increasing air and water pollution poses a threat to the ecosystem. Therefore, it is necessary to immediately shift focus on dealing with them. The severity of the situation can be assessed as more than 80% of the world's population is still exposed to air quality levels that exceed World Health Organization (WHO) regulations, according to a WHO report from 2018 [4]. And according to some estimates, industrial water use would rise by 50% annually by 2025 compared to the level in 1995, and between 2 and 7 billion people will experience water scarcity by the third

Sadia Bano, Department of Chemistry, Lahore College for Women University,
e-mail: leochem10@gmail.com
Muhammad Altaf, Department of Chemistry, Chiang Mai University, Thailand,
e-mail: muhammadaltaf11966@gmail.com
Sajda Bano, Department of Chemistry, University of Punjab, Pakistan,
e-mail: sajdabano.com@gmail.com

https://doi.org/10.1515/9783110796032-008

decade of the twenty-first century [5]. So, there is a dire need to work on environmental pollution mitigation and remediation using advanced technologies.

One of the most primitive approaches for creating a better atmosphere and solving pollution is filtration technology. Air filtration equipment must be installed in every industrial unit with possible emission hazard. And to cope with the issue of water scarcity, use of alternate sources of water, such as seawater and rainwater, and filter out any toxins that are present. Membrane separation process has a special place in environmental remediation processes mainly due to its smaller carbon footprint and being efficient, selective and reliable [6]. Membranes have been produced by a number of technologies including sintering, phase inversion, stretching, and track-etching [7, 8]. Recently, electrospinning has been used to make composite polymeric nanofiber membranes quickly and efficiently to be exploited in a number of environmental applications related to nanotechnology [9]. Nanomaterials and nanostructures have a significant impact on several fields, such as the environment, energy, power, national security, military projects, and even health care, due to rapid development in new technologies. Nanomaterials are different from macro-sized materials not just in the size of their characteristic dimensions but also in their ability to offer vastly increased performance and new opportunities for a variety of practical applications. The National Science Foundation defines a nanofiber as having a diameter of 100 nm or less, while generally, the term "nanofiber" refers to a strand with a diameter smaller than one micron [10].

The need to use cutting-edge filtration technology in an energy and economical approach has increased interest in nanostructured membranes, particularly those manufactured using "electrospinning," which produce nanofibrous membranes [11]. The electrospinning technique enables the mass manufacture of thin, nanoscale, extremely practical mesh-like devices. Like electrospinning, electrospraying involves electrostatic acceleration of solution drops toward a target to produce consistently sized particles or thin-film coatings [12]. When a high voltage is supplied to the polymer solution placed in the syringe, a nonwoven nanofibrous membrane is created and the process is termed electrospinning. It is a straightforward but effective method for producing nanofiber membranes. Comparing ENMs to traditional polymeric or ceramic membranes reveals that the former has higher porosity (usually about 80%), a lighter base weight, a greater active surface area (up to 40 m^2/g based on the fiber diameter and porosity) as well as pores that are interconnected [13]. The electrospinning process is very flexible because different modifications can be made to the fiber to enable it to perform a particular function. For instance, nanofibers can be functionalized by binding active species like nanoparticles, different fiber diameters can be obtained by adjusting processing parameters, and they can also be assembled on a variety of matrices by using different alignment, stacking, or folding. By doing this, composite membranes and multilevel structured membranes with good thermal stability and antibacterial properties can be created [14–16].

Its global attention may also be due to simpler process management and lower production costs. ENMs have been acknowledged as competitive contenders for a variety of applications, including energy storage, healthcare, biotechnology, as well as environmental applications [17–19]. The benefits of electrospun filters have drawn industrial firms for more research and production. Table 1 lists some major companies around the world which produce ENMs. The purpose of this chapter is to investigate recent developments in electrospun nanofiber membranes and their environmental applications.

Table 1: Some major companies working on ENMs around the world.

Company	Origin	Website
Finetex	Korea	www.ftene.com
Donaldson	USA	www.donaldson.com
Electrospunra	Singapore	www.electrospunra.com
Elmarco	Czech	www.elmarco.com
Bioinicia	Spain	www.bioinicia.com
Mecc-nanofiber	Japan	https://www.mecc-nano.com/
Nano fiberlabs	China	https://www.nanofiberlabs.com

2 Fabrication of nanofibers by electrospinning

Electrospinning, also known as electrostatic spinning, is not a very new yet is a very powerful nanofiber fabrication technique. It is versatile enough to produce fibers on micro- and nanoscales and is believed to be a variant of the electrospray process. Fiber sizes for electrospun nanofibrous membranes normally vary from 100 to 500 nm, but when the fiber diameter drops below 50 nm, characteristics like surface area and porosity become more important.

2.1 Materials used for ENMs production

Solution, melt, or sol-gel suspension of various polymers have been extensively used to make nanofibers over the years by electrospinning [20]. Polyacrylonitrile (PAN) [21], polyurethane [22], polystyrene (PS) [9, 23], and polyvinylidene fluoride (PVDF) have been widely used as polymers for electrospinning membranes [24].

It is challenging to meet the demand for ENMs since different ENMs with varying properties are needed for various applications. We can resolve the issue by using the technique of coblending polymers, a method that combines polymers with various properties into a single reservoir to enhance the performance of the membranes [25]. The composite formulations can produce electrospun nanofibers with increased specific surface areas, increased porosity, customizable chemical compositions, and variable fiber assembly topologies. Compared to their unaltered counterparts, the modified electrospun nanofibers had better stability and superior catalytic capabilities [26]. Polymer blends are also used to make ENMs such as zein polymer/nylon-6 blend and prepare membranes for chromium removal [27].

To obtain sustainable and effective transformation in ENMs, researchers frequently dope carbon-based nanomaterials, metal nanomaterials, and inorganic nanoparticles into polymer solutions and fix them in polymer nanofibers [28]. This trend has produced many useful examples such as polylactic acid/TiO_2 [29], poly(4-chloro-3-methylphenyl methacrylate) with nano silver [30], PAN-$CuCl_2$ [31], PAN/TiO_2 [32], and polycaprolactone matrix with iron-intercalated montmorillonite filler [33]. Organic solvents are mostly used to make polymer or blend solutions to be introduced in syringes.

Table 2 lists some of the important materials used for ENMs production for environmental applications in recent years.

Table 2: A list of important polymers and solvents used in literature for production of ENMs by electrospinning.

Material	Solvent	Application	References
Polyacrylonitrile	DMF	Anti-*Escherichia coli* membrane for water treatment.	[21]
Polyurethane	50/50 v/v DMF/THF	Nanofiltration	[22]
Polystyrene (PS)	Cyclohexane	CO_2 adsorption and antibacterial	[23]
Polylactic acid/TiO_2	Dichloromethane (DCM)	Removal of Methylene Blue	[29]
PS	Dimethyl formamide (DMF) and chloroform		[9]
Poly(4-chloro-3-methylphenyl methacrylate) and nanosilver	DMF	Water sanitization	[30]
PAN-$CuCl_2$	DMF	High-performance air filters	[31]
PAN/TiO_2	DMF	Particulate matter pollutants filtration	[32]
Polyvinylidene fluoride (PVDF)	DMAc and acetone	Removal of heavy metals from sludge.	[24]
Zein polymer/nylon-6	DMF and formic acid	Hexavalent chromium removal	[27]
Polycaprolactone matrix with iron-intercalated montmorillonite filler	Dichloromethane (DCM)	Removal of arsenic in wastewater	[33]
Porous polyimide (6FDA-TrMPD)	DMF	Oil spill removal	[34]
Fluorinated polyimides	DMF	Oil removal	[35]

Table 2 (continued)

Material	Solvent	Application	References
Ag/sulfonated polyethersulfone/ polyethersulfone	DCM	Removal of heavy metals	[36]
Microporous polyimide + MOF	DMF	Air filtration	[37]
Functionalized polyimides	DMF	Removal of textile dyes	[38]

2.2 Basic principle

A high-voltage power source, a pump for controlling the injection rate of the polymer solution, a syringe connected to an injection needle through which the polymer travels, and a collector (drum) are all components of the electrospinning process. The collector is connected to one of the power supply's electrodes, and the needle tip is connected to the other.

To make the dope solution, a polymer needs to be dissolved in the proper solvent. The polymer is prepared for electrospinning once it has completely dissolved and been degassed. To improve the dope solution's electrospinnability, the electrical conductivity of the dope solution should be high enough. Some polymers, however, may exude unwanted conductivity; therefore, the correct additives (usually salts) should be combined with the polymer solution to increase the conductivity [39].

The dope solution can then be added to the needle to create nanofibers. The procedure of providing high voltages for the polymeric droplets generated at the needle tip is the essential feature of the electrospinning technique. Electrical charge exists among the polymeric molecules in the droplet created by the same voltage. Electrical potential must exceed the surface tension of polymer dope for an electrospinning process to be successful. As a result, the forced electric field must reach a particular value or the minimum necessary high-voltage power. The Taylor cone at the tip of the spinneret then releases a charged polymer jet [40]. The Taylor cone forms in the polymeric droplet because of the electrospinning repulsion. The polymer will eventually travel from the needle tip toward the loaded collector on the other side; after the solvent evaporation, very thin polymeric fibers have been deposited on the collector surface (Figure 1) [41]. Therefore, the electrospinning method has been studied as a quick and flexible substitute for the creation of nanofibers.

If a droplet is placed within a capacitor, the development of a fluid droplet at the syringe's tip constitutes the initial step in fiber production. In comparison to the field-free situation, the charged fluid experiences shape deformations when it interacts with the external electric field [42].

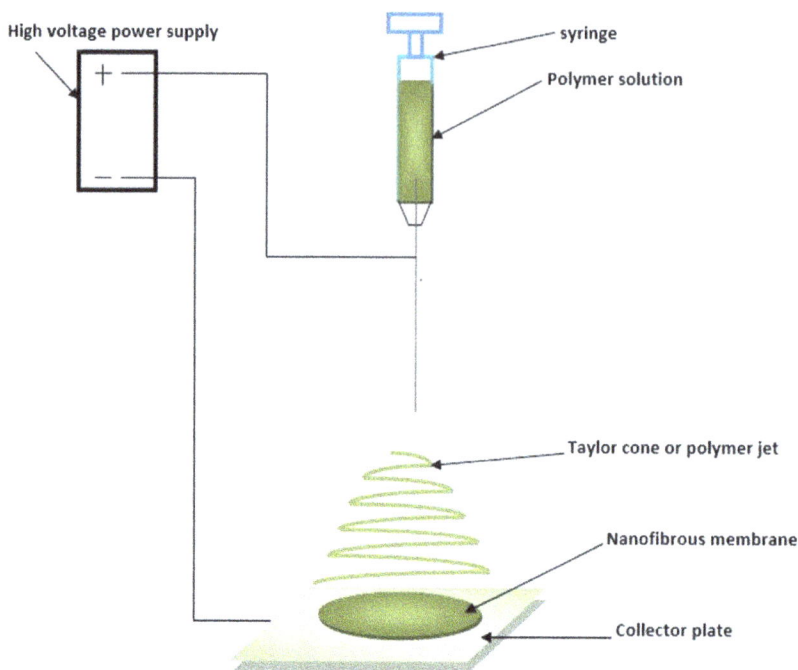

Figure 1: Diagrammatic illustration of an electrospinning setup.

The single-jet electrospinning process has a relatively poor production rate. For instance, it can take many hours to construct an electrospun nanofibrous mat weighing 1 g using a single spinneret. A workable solution to this issue is the multiple-jet electrospinning procedure using numerous spinnerets [43]. The fundamental electrospinning mechanism has developed over time into increasingly intricate, polyvalent devices that can handle a variety of materials in molten or solution forms. The introduction of advancements in the needle design or ejector configuration has made this possible. One of these technologies is coaxial electrospinning, which enables the electrospinning of fibers with a core and shell made of various materials. Using this approach, it is also possible to electrospun hollow fibers [44]. Similar to this, multiaxial electrospinning enables the processing of multiple materials, with the fiber consisting of several laminae or sheaths [45]. The use of needleless electrospinning can serve as an illustration of a unique ejector arrangement. It enables direct electrospinning from an exposed liquid surface. The high production rate that makes this method acceptable for commercial applications is a major benefit [46].

2.3 Postfabrication treatment

The selection of solution ingredients, processing settings, ambient factors, and collector design all affects the mechanical properties of electrospun fibers. The fibrous web needs postfabrication procedures to improve its mechanical properties because once electrospun, it has minimal mechanical integrity. Both chemical and physical methods are used in the treatment strategies. Testing on extracted single fiber specimens or large-scale testing on membrane specimens can be used to determine how various postfabrication processes affect the properties of electrospun membranes. Due to its ease of use and inexpensive cost, the latter scenario appears more frequently in literature [45]. Most important postfabrication techniques are crosslinking, stretching, solvent welding, heat treatment/annealing, and hot stretching.

Due to its special characteristics of a good surface-to-volume ratio and incredible porosity, the nanofiber layer can also be used as the supporting substrate of ENMs [47]. Secondary electrospinning, inorganic deposition, polymer coating, and interfacial polymerization were typically used to create a selective layer on top of the nanofiber layer, which is important for controlling the direction of fluid flow and the effectiveness of separation [48]. Additionally, structural matching should be properly customized by independent adjustment to attain the greatest performance for the desired application. Examples of this would be the thickness and porosities of the supporting layer and the selective layer.

2.4 Effect of different operating conditions on ENMs

The solvent and polymer employed to prepare the polymeric solution affect the wettability/hydrophilicity of membranes, so both must be carefully chosen [49]. When using polymeric solutions, it is advised to use a volatile and low viscosity solvent to avoid the plugging in the syringe [50]. Apart from nature and composition of precursors used, concentration of the polymer solution, the applied voltage, the spacing between the injector and the collector, and the injection rate must all be controlled when manufacturing nanofiber membranes using electrospinning. All these factors affect the properties of final product by electrospinning [29]. There is a lot of work available on operating conditions for electrospinning. Lu et al. [51] specifically gives details on the key operating conditions for producing electrospun fibers from Zeina, cellulose, PV, PAN, and nylon [51].

2.4.1 Polymer concentration

The polymer concentration is crucial to controlling the viscosity during electrospinning, and the fiber creation of the electrospinning process requires a minimal solution concentration, even though other factors including molecular weight, solvent property, and temperature could also affect the viscosity. In a study conducted on solution concentration and its effect tensile strength confirmed that the mechanical characteristics of the nonaligned nanofiber mats were shown to be impacted by solution concentration. The tensile strengths of the CA nanofiber mats increased with increasing solution concentration [52].

2.4.2 Applied voltage

When the electrostatic force in the solution surpasses the surface tension of the solution, the high voltage will induce the appropriate charges on the solution, commencing the electrospinning process. In general, strong positive and negative voltages of more than 6 kV can cause the solution drop at the tip of the needle convert into the shape of a Taylor Cone [53]. A spike in the voltage being used causes the electrostatic force of the polymer solution to increase, which suggests that the jet is stretching, and as a result, the length of the fibers decreases. It has been discovered that altering the applied voltage will alter the initial drop's form, which will alter the fibers' structure and morphology [54].

2.4.3 Tip-to-collector distance

The electrospinning procedure and the resulting fibers are influenced by both the flight time and the electric field strength. The flight time and the intensity of the electric field will both be directly affected by changing the gap between the tip and the collector. The charged jet must be given ample time for most of the solvents to evaporate during fiber production. The jet will travel a shorter distance before it reaches the collection plate when the tip-to-collector distance is low, and the electric field intensity will also increase at the same time (Figure 2). As a result, there may not be enough time for the solvents to evaporate. This distance, which typically ranges from 10 to 15 cm in an electrospinning setup, normally provides enough flight time for the solvent to evaporate and deposit a dry fiber strand [55].

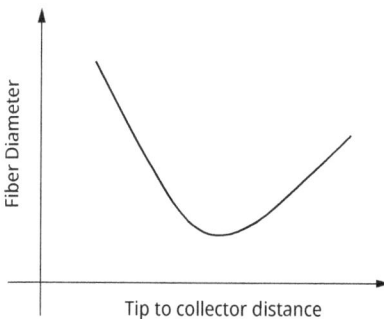

Figure 2: Relationship of fiber diameter and tip-to-collector distance.

2.4.4 Feed rate

The majority of electrospun nanofibers are generally not significantly affected by the solution feed-rate outside of a certain range. The pace at which the solution is dispensed and the rate at which the solution is taken from the nozzle tip during electrospinning, however, must be balanced. The distribution of fiber diameter is at its tightest at this optimal feed rate, and any deviation will lead to a wider range of fiber diameter [25].

2.4.5 Environmental factors

Humidity and temperature are two other factors which affect the features of nanofibers by electrospinning. When electrospinning is done in an atmosphere with an elevated humidity, it's possible that water condenses on the fiber's surface. As a result, this may have an impact on the morphology of the fiber particularly for polymers solutions in volatile solvents. In a study on the effect of humidity and temperature on electrospun nanofiber morphology of CA and PVP conducted by Sander and coworkers concluded that both polymers behave differently. The average fiber diameter of CA nanofibers increases with humidity, but the average diameter of PVP nanofibers falls. Temperature and humidity variations dramatically alter the average diameter of electrospun nanofibers [56].

The temperature of the solution has the dual effects of speeding up its evaporation and lowering the viscosity of the polymer solution, making it possible to electrospin the polymer solution that was easily spinnable at room temperature. According to Demir et al.'s [57] research, higher temperatures may allow electrospun polyurethane fibers to have a more uniform diameter due to the solution's lower viscosity and greater solubility in the solvent, which enable even more stretching of the solution. As a result, the columbic torte can stretch the solution further because the increased temperature has enhanced the mobility of polymer molecules. However, using high temperatures may result in the nanofibers losing their activity when biological components like enzymes and proteins are added to the solution.

Regarding the electrospinning procedure, proper electrospinning settings as well as adjustments are needed for better manufacturing and performance. The focus is on inexpensive, nontoxic, and ecologically beneficial nanomaterials. In addition, ENM preparation methods are typically said to take a long time and have a low yield on a lab scale. As a result, it's essential to create more effective equipment for the broad manufacturing of high-quality ENMs [25].

3 Environmental applications

Due to their effectiveness and energy economy, membrane-based separation methods have assumed a leading position in water purification [58]. Both membrane adsorption and filtering techniques are regarded as efficient ways to remove undesirable species from contaminated water and air or stop dangerous toxins from entering the environment and the human body.

3.1 Water filtration

Currently, freshwater shortage is included among the major crises at the world level. The percentage of freshwater is 0.6 in the whole world's water. While, from the total water of the earth, only 0.2% is potable (safe to drink). Numerous sectors that contaminate natural water sources are industry, household, and agriculture. Unfortunately, the industrial unit is noted as the major contamination source [59]. Although the industrial sector has revolutionized the world, it is badly affecting both the community and the weather owing to the rise in pollution and decline in green urban areas [60]. Various approaches such as chemical (coagulation, oxidation, advanced oxidation processes, and electrochemical oxidation), acoustical (the electric and radiation procedure), and physical (sedimentation, adsorption, and filtration) have been exploited for polluted water purification [61].

The use of electrospun nanofiber membranes (ENMs) and nanocomposites in wastewater treatment is an emerging idea that is progressing more quickly, getting more recognition from different scientists, and raising interest in improving water quality. Water pollutants like germs, dyes, heavy metals, and oil have all been successfully removed from water using electrospun nanofiber membranes [62]. According to Gee et al. [63], the electrospinning approach has been extensively employed as an efficient process to fabricate micro and nanofibers employing numerous materials for the synthesis of water filtration membranes. Not only the careful optimization of traditional needle electrospinning but also the control of limiting factors are the fundamental features that facilitate the preparation of defect-free, efficient, and highly interlinked pore structure nanofiber membranes. A schematic illustration of the filtration membrane and adsorption is shown in Figure 3.

Recently, PAN/TiO$_2$/Ag nanofiber membrane for use in wastewater purification was created successfully. It has high antibacterial action against *Escherichia coli* and *Staphylococcus aureus* and outstanding photocatalytic activity in the degradation of dyes under visible light [34]. A one-step electrospinning method for creating titanium dioxide nanoparticle-modified polyamide 6 (PA6) nanofiber membranes was developed, where the aggregation of these nanoparticles might produce a photocatalytic function. The treatment of contaminated water could be the main potential use for these membranes. Since these membranes' ability to degrade contaminants is dependent on their photocatalytic activity, it is crucial to examine this capability [64].

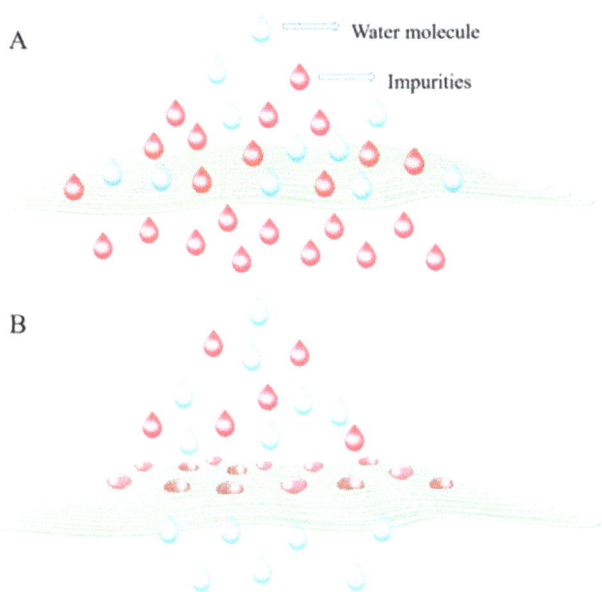

Figure 3: Schematic illustration of (A) adsorption membrane; (B) filtration membrane [62] (copyright Elsevier 2020).

In another study, triple layer membranes were used for water treatment where the middle layer was made of PAN nanofibers by electrospinning. This PAN layer was sandwiched between nylon-6 and polyethylene glycol layer by dry- and wet-phase inversion technique [65]. For use as antibacterial nanofibers, Toniatto et al. [66] developed nanostructured poly(lactic acid) electrospun fibers with high loadings of TiO_2 nanoparticles. In adsorption and filtering processes including MF, UF, nanofiltration (NF), RO, water/oil separation, and others, ENMs have been widely used [7].

3.1.1 Microfiltration

In portable water purification and sewage pretreatment, where the removal of waterborne bacteria, suspended microparticles, algae, and other contaminants is necessary, MF serves as a pressure-driven method of separation that sieves particles ranging from 0.1 to 10 mm in the liquid medium. ENMs have been investigated as novel microfiltration materials for the very efficient and economical removal of contaminants with particle sizes of 0.2 µm (even 0.1 µm) from water. These membranes have a range of fiber diameters as well as varying membrane porosity distributions and thicknesses [67]. Polyethersulfone ENMs, which are helpful as water treatment membranes for water purification, were used as microfiltration medium [68]. The microfiltration activity of electrospun nanofibers membranes was studied by Wang et al. [67] under the consideration of various aspects such as different fiber parameters, membrane thickness, and membrane porosity. They noted that the pore size is significantly impacted by the fiber diameter and the membrane porosity and the hot pressing have a dominant effect to reduce porosity and enhance the rejection performance [67]. In addition, the rejection performance of microfiltration membrane is notably affected by the particle size distribution of suspended particles [69].

For microfiltration with improved water flux and fouling resistance, Li et al. [70] employed electrospinning method for the synthesis of PVA-blended polyvinylidene fluoride membranes. The resulted PVA-blended membranes with characteristic bicontinuous phase and ridge and valley shape revealed the super wettability and the hydrophilic activity exploring the minimal irreversible fouling ratio and ultrahigh flux with a retention rate more than 99% against 3 μm PS microspheres. The novelty of facile preparation method with potential advantages including high water flux under exceptionally low pressure was exhibited. They explored that the relative high flux can be attained without operating chemical cleaning, only by the rinsing with distilled water, which enhances the membranes durability and making as an affordable operating strategy for the filtration [70]. With two to three times higher flux (1.5 L/m^2 h), the PAN/polyester (PET) nanofibrous MF membranes functioned noticeably better than the commercial MF membrane with the same mean pore size (0.22 m). With an LRV of 6, the nanofibrous MF filter was able to maintain an extremely high rejection ratio for bacteria and microparticles. The findings imply that electrospun nanofibrous membranes provide suitable materials for MF applications requiring high flux [71].

3.1.2 Ultrafiltration

UF is a technique that may remove a wide variety of particles from a liquid environment, including viruses, emulsions, proteins, and colloids, with diameters ranging from around 1–100 nm [72]. In the ultrafiltration (UF) process, the membranes typically contain pores with diameters between 0.01 and 0.1 μm. The UF process normally operates between 1 and 10 bars of pressure. Large species, including bacteria, proteins, and particles, are retained using the UF membrane. However, the membrane holes allow for the passage of water, ions, and solutes with low molecular weight [73].

The most used membrane system in the UF process is now thin-film composite (TFC) membranes. The top ultrathin selective layer, the middle porous support layer, and the bottom nonwoven fabric layer make up the three basic layers of such a membrane system. Previously, ENMs were used to replace the traditional middle layer. The TFNC membrane format refers to this entire fibrous membrane structure [10]. It was found that the TFNC membranes' flux rate and rejection rate were significantly greater compared to that of commonly available UF films [74]. Consequently, the use of electrospun fibers as the middle or top layer of TFC membranes has garnered attention on a global scale.

Surface modification like coatings has been made my researchers to construct advanced treated denser layers. For instance, Bahmani et al. [75] synthesized a TFC UF membrane using PAN as coating element over the PAN nanofiber membrane. The TFC membranes are superior in activity of the flux (1.7 times), and refusal of arsenic ions (1.1–1.3 times) in comparison with the UF membranes fabricated via phase inversion method, which attributes to the porous structure of the support layer [75].

It has been noticed that the wettability of UF membrane is improved by the hydrophilic modification, increasing penetrate ability of water and reducing fouling. Vanangamudi et al. [76] employed nylon-6,6/CS as blender on the hydrophilic polyvinylidene fluoride nanofiber support layer for the preparation of a hydrophilic composite ultrafiltration membrane owing to its excellent features of hydrophilicity and mutual solubility [76]. Wang et al. [76] implemented an affordable hot-pressing hypothesis to prepare efficient ultrafiltration membranes employing electrospun nanofibers without the expenditure of chemical methods. In the fabrication process, ENMs containing a skeleton polymer (i.e., PAN) and a thermoplastic polymer (i.e., PVDF) were treated via hotpressing generating the softened/broken of PVDF nanofibers and resultantly led to the fusion of nanofibers. In this way, they obtained membrane with lowest porosities with high values of pure water flux. Further, a complete rejection of PS particles (size is >20 nm) was examined and proved as an innovative approach to develop highly efficient ultrafiltration membranes [77]. A solution for membrane pore wetting and an improve-

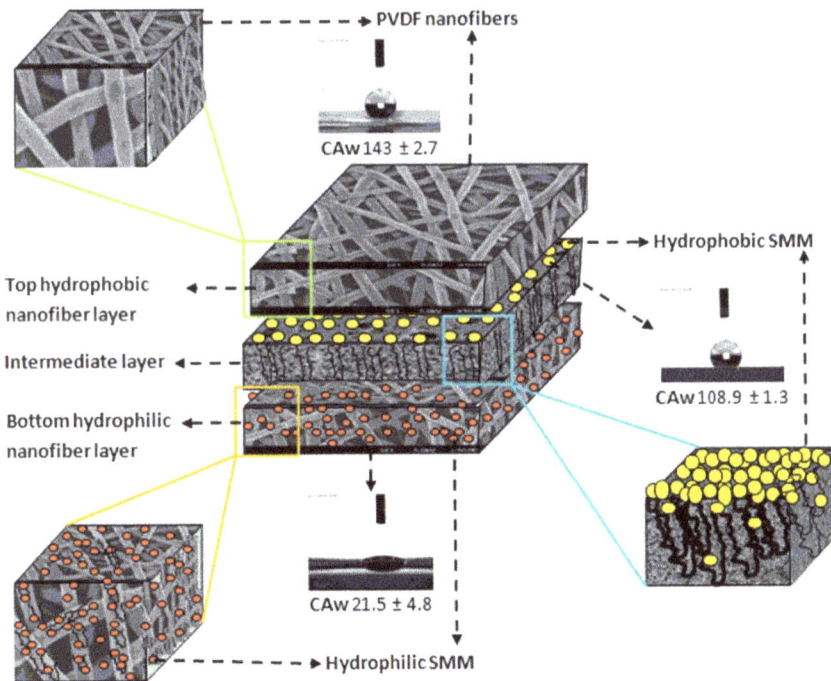

Figure 4: Illustration of a triple layer membrane for ultrafiltration, with electrospun nanofibers as top layer [78] (copyright Springer Nature 2014 CC license).

ment in the pure water flux in membrane distillation are made possible by the creative design and synthesis of nanofiber-based hydro-philic/phobic membranes, which have a thin hydro-phobic nanofiber layer on top and a thin hydrophilic nanofiber layer on the bottom of the conventional casted micro-porous layer (Figure 4).

3.1.3 Nanofiltration

This type of rejection technique is applied to remove particles having 100–1,000 nm in diameter. In water treatment, NF is widely employed for color removal, softening, taste enhancement, odor, and trace organic effluents removal and disinfection [79]. The difference in the activity of NF and reverse osmosis (RO) is the rejection of ions, and RO will be able to reject monovalent ions (i.e., salt of NaCl), while nanofiltration would remove divalent and trivalent species (i.e., $AlCl_3$ and Na_2SO_4) [80].

In present research area, TFC and NF membranes have similarity in their development via IP process using a porous support later. The permeability of NF membranes, like all TFC membranes, is affected by the support layers' thickness and porosity. As the support layer, electrospun nanofibers are being used as an excellent candidate for TFC-NF membrane. In comparison with the conventional UF membranes, TFC-NF membranes are efficient in working due shape of selective layer, higher porosity, interlinked voids, providing the more rejection [81]. For instance, Liu et al. [82] used one-step process to prepare TFC membrane employing PVDF as support layer. In comparison to traditional support membrane, the tree-like prepared ENMs attributed with less tortuosity, possessed high porosity, high-interlinked pore ratio, and less pore size (below 200 nm). They determined the removal efficiency against $MgSO_4$ and NaCl solution and found the rejection rate of 97% and 76%, respectively, exploring as a potential candidate for water purification.

In another study, after electrospinning of PAN precursor solution, heat treatment was used to create free-standing carbon nanofiber (CNF) mats with various fiber size distributions that ranged from 126 to 554 nm in diameter. Tetraethoxyorthosilicate was also used to give the CNF flexibility and boost its specific surface area. The manufactured membranes could successfully filter out NPs of various types (Au, Ag, and TiO_2) and sizes (10–100 nm in diameter) from aqueous solutions. It should be noted that elimination of Ag NP with diameters as tiny as 10 nm was nearly 100% with an incredibly high flux of 47,620 L/m^2 /h/bar [83]. Figure 5 exhibits the rejection features of membrane processes for water purification.

3.1.4 Forward osmosis and reverse osmosis

Forward osmosis (FO) and RO are the techniques employed for the water treatment. Mainly, water treatment includes the desalination of seawater and brackish water, re-

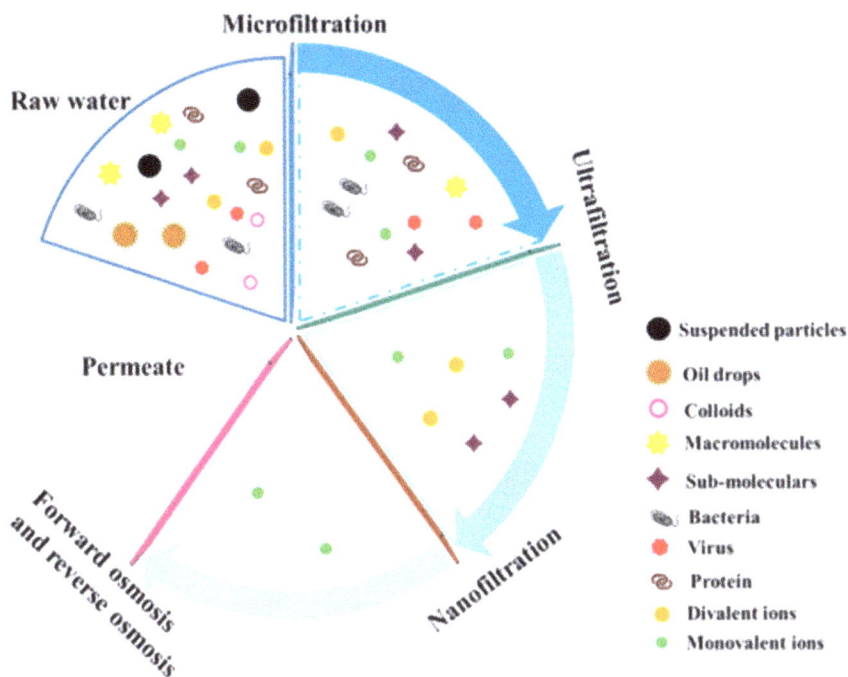

Figure 5: The representation of removal features of membrane processes in water purification [81] copyright Elsevier 2020.

sultantly producing pure water. The estimated amount of salt in seawater and brackish water is around 35,000 ppm and 1,000–5,000 ppm, respectively. These approaches can also be utilized in other sectors such as food industry and metallurgy industry [84]. Figure 6 explores an osmosis process scheme for the desalination process. At present, the removal process with minimal energy usage is FO. Like osmosis process, FO works with the addition of osmotic pressure maintained by the addition of solutes. In FO, an increase in osmotic pressure is attained by the addition of salt concentration to make the solvent flow not the solute to the other phase, simply to say, movement of only water particles [85]. In RO, a high pressure is created to start the solution-diffusion process. Commonly, there are different parameters to categorize membrane such as pore size (0.1–1 nm), driven pressure (15–80 bar), and high rejection ratio (99% for divalent and 98% for monovalent ions). Compared with other membrane processes, RO can remove the tiny pollutants namely Na +, Cl⁻, and monovalent species, making it an excellent and most widely used candidate for desalination and water purification [86].

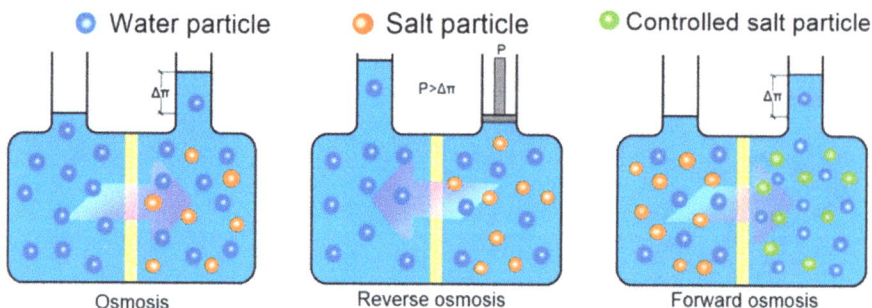

Figure 6: Representation of an osmosis process scheme for the desalination process [84] (Copyright Elsevier 2021 CC license).

3.1.5 Membrane distillation

A great alternative for treating industrial wastewater and desalinating saltwater is membrane distillation (MD) which holds tremendous potential for tackling the freshwater crisis while offering higher than anticipated energy efficiency [87]. The thermal pressure gradient drives the hybrid thermal/membrane process known as MD, in which vapors travels through a hydrophobic membrane from the hot side (320–350 K) and condenses in the cool side [88]. However, low permeate flux and membrane wetting for feeds containing low surface tension chemicals remain MD technology's principal drawbacks.

Over the past 5 years, the performance of ENMs in MD has greatly improved due to new materials and membrane configuration innovations. For instance, it has been demonstrated that polyvinylidene fluoride-*co*-hexafluoropropylene (PH) ENMs are the best and offer significant advantages over others [89]. Additionally, biomimetic ENMs with highly effective super hydrophobicity have been developed [90]. In 2016, Zuo and Chung [91] made a breakthrough by incorporating metal-organic frameworks (MOFs) into membranes for MD. This resulted in new applications for ENMs. As well as, the ENMs may acquire super omniphobic qualities from fluorine-containing compounds that have just come to market, giving them outstanding antiwetting capabilities [92]. It is interesting that, compared to the commercial MD membrane materials, polypropylene and polytetrafluoroethylene, PVDF, and PVDF-HFP are currently the polymers utilized most frequently in ENMs for MD assessment. The electrospinnable polymeric materials with intrinsic hydrophobicity employed for MD, however, are far rarer than hydrophilic filtration membranes. Therefore, the creation of innovative electrospun membrane materials requires considerable attention. Using an eccentric-axial electrospinning process, Li et al. [93] reported a new profiled PAN-PS core-shell nanofiber with an odd C-shaped groove structure. In another study by using heat pressing, the CNT layers were cohered onto the PVDF ENMs. To spray the CNTs without aggregation, ethanol was used to scatter them. In a vacuum MD, the produced superhydrophobic PVDF/CNT ENMs demonstrated the greatest water flux of 28.4 L/m^2/h, with over 26 h of stable performance [94].

3.2 Oil and water separation

Due to their intrinsic negative impacts on the environment, the proper treatment of oily wastewater, which is mostly released by the petrochemical, textile, metallurgical, and food processing sectors, as well as from other oil-contaminated water, is urgently necessary [95]. Skimming, ultrasonic separation, and flotation are examples of conventional procedures, but they are not suitable for oil/water emulsion separation due to their low separation efficiency, high energy costs, and secondary contamination [96]. Due to their significant specific surface area and greater porosity, electrospun nanofibers of wettability are a novel porous material for oil-water separation. Without the use of extra energy, it can naturally separate water and oil [97]. By carefully regulating the diameter of nanofibers with sizes ranging from a few micrometers to several nanometers, the separation effectiveness of oil/water emulsions might be increased [96].

Three different types of ENs membranes with unique wettability have been created so far for the separation of oil from water, including superhydrophobic ENs membranes for removing oil and superhydrophilic ENs membranes for separating water from oil as well as intelligent EN membranes with wettability response to external stimuli that allow for the switchable removal of oil or water [98]. The is a relation of hydrophobicity of material and water contact angle (WCA) [99].

a) The membrane surface is deemed to be hydrophobic if the static WCA is greater than 90°, and it is characterized as superhydrophobic if the WCA is greater than 150° [100].
b) Due to their strong affinity for water molecules and desired anti-oil-fouling efficiency, superhydrophilic and underwater superoleophobic membranes having OCA > 150 and WCA ≤ 5 have attracted significant attention in the oil/water separation recently [101].

Superhydrophilic substances have drawn a lot of interest as potential options for efficient oil/water separation and water treatment [102]. Combination of superhydrophilic and superoleophilic materials for membrane fabrications makes them more efficient for oil and water separation. A variety of superhydrophilic and underwater superoleophobic ENs membranes have been developed for the oil/water separation process in the literature. These membranes were inspired by the distinctive structure as well as surface properties of sand and fish scales [103]. Electrospun nanofibers and an in situ polymerized fluorinated polybenzoxazine functional layer that included NPs (such as silica and alumina NPs) were combined easily to create nanofibrous membranes. The electrospun nanofibers used in this application can be silicon dioxide nanofibers or polymer-based nanofibers, such as cellulose acetate or CA, poly(*m*-phenylene isophthalamide) [95].

The membranes that are created have a sensitive organic PVDF-HFP core wrapped inside an inorganic CuO nanosheet shell. The PVDF-HFP networks give the membranes their super-flexibility, and the CuO nanosheet's incredibly rough surface gives them their superhydrophilic properties. The PVDF-HFP core was created by using electrospinning process [101]. PAN is the most popular hydrophilic polymer currently in use for creating superhydrophilic ENs membranes. Bae et al. [104] produced asymmetric wettability membranes of PAN using electrospinning technique and assessed them for oil/water separation. This asymmetric wettability was produced using different concentration of polymer solution and they showed remarkable separation efficiency [104].

However, the resilience of these membranes in adverse environments, where the coating layer's micro/nanostructure can be harmed, could become a serious problem. Wei et al.'s [105] innovative investigation revealed that an oxidation post-treatment created an anticorrosion surface layer in situ on the hydrophobic poly(arylene sulphide sulfone) ENs membrane's surface. Owing to the different wettability on both sides of the membrane, Janus ENMs (J-ENMs) have gathered a lot of interest as membranes for oil-in-water emulsion separation. These characteristics can provide a capillary effect and enable unidirectional water flow. The capillary impact on J-ENM performance, however, is still unclear. In this study, J-ENMs were created using the electrospinning technique. These J-ENMs feature a dual-layer structure with hydrophilic PAN and hydrophobic PS nanofibrous membranes present at the same time. When compared to single layer membrane, the emulsion flux of J-ENMs was 1.7 times greater, indicating a promising application for the oil/water separation industry [106].

3.3 Air filtration

ENMs stand out from typical fibrous filters in a number of ways including their small diameter, high specific surface area, interlinked porous frameworks, and tunable morphologies [107]. The ENMs have demonstrated promising results in the removal of hazardous gases, volatile organic chemicals, and minute particles from the air. Nanofiber-based membranes are appropriate for PM filtration because thinner fibers result in more surface area per mass [108]. For the electrospun membranes, a wide variety of polymers, such as PET, PU, and PAN, could be chosen.

The theory of air filtration specifies several ways by which particulate matter is filtered. Diffusion, inertial impact, interception, and electrostatic attraction are four of the main mechanisms for air filtration. When a small particle diffuses into the area around the fiber during filtering and does not go forward, diffusion takes place. When a particle strikes fiber directly, it has an inertial impact. When a particle is moving through a flow but makes physical touch with fiber, it slows down and stops; this is known as an "interception mechanism." Charged fibers attract PMs electrostatically, which is different from the mechanical mechanisms [109]. The two most significant particle capture mechanisms of nanofibers may be diffusion and interception. Brownian motion causes small, tiny

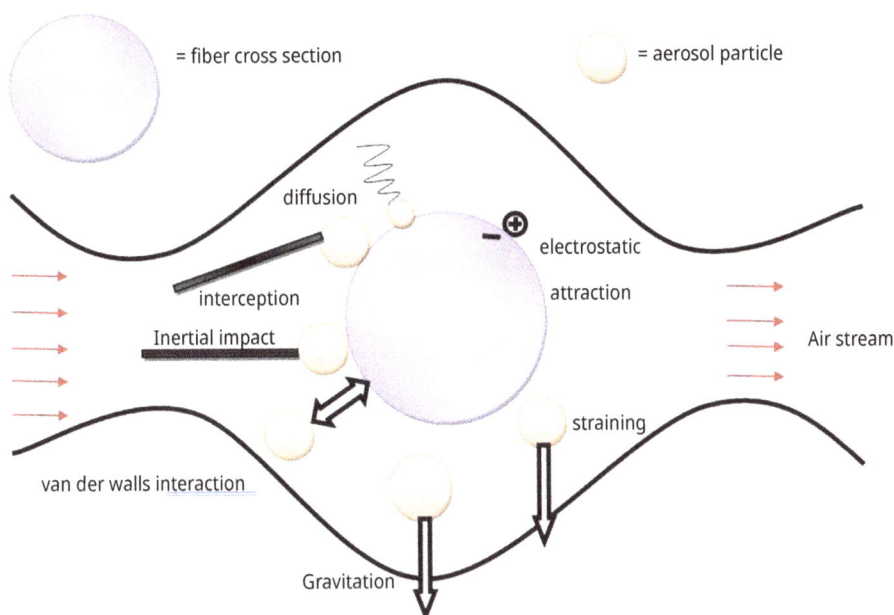

Figure 7: Illustration of different particle capture mechanisms inside a nanofiber.

particles smaller than 0.5 mm to move randomly which can be captured by diffusion. When the distance between the fiber surface and the particle center is equal to or less than the particle radius, particle interception happens. Figure 7 gives an illustration of different particle capture mechanisms.

Different kinds of electrospun nanofiber membranes have been created and developed as the central component of air filters. Here we are categorizing them based on the composition of electrospun nanofibers.

3.3.1 Single component polymer membranes

For PM filtering, several electrospun nanofiber filters made from single-component polymers have been effectively researched and manufactured. Single-component PEBA (poly(ether-*block*-amide)) membranes were created with a diameter of 1.92–3.31 nm. The following parameters must be met for the fabrication of PEBA membranes: melting temperature of 270 °C, 6 cm tip-to-collector distance, 20 kV of applied voltage, and collector speed of 55 rpm. The findings demonstrated that operating factors such voltage, distance, collector speed, and melting temperature had an impact on fiber diameter.

The membranes were meant to be used for hydrophobic packing, separation, evaporation, and osmotic membranes because of their characteristics [110]. In another study, PVDF micro- and nanofibrous membranes were tested for filtration performance in terms of filtration efficiency and air permeability. A positive correlation has been observed between pore size and the fibrous membrane's air permeability and filtering ability. The filtration efficiency testing results were compared to the expected values from the most recent theoretical models based on single fiber filtration efficiency. However, since the fiber diameter was at the nanoscale and the ratio of particle size to fiber diameter was much higher than what the theoretical model calls for, the anticipated values did not correlate well with the experimental observations [111].

In another work on 8% PAN in DMF solution when spun electrically, produced membranes. The membranes were effective at adsorbing formaldehyde and xylene as well as volatile organic compounds (VOCs) and achieved a minimal pressure drop and filtering efficiency of above 95% [112].

The ENMs can also develop antimicrobial abilities. For instance, 12% PA6 in acetic acid/formic acid in a ratio of 2:1 is functionalized with 1% dodecyl trimethyl ammonium bromide (DTAB). Electrospinning was carried out at a temperature of 25 °C with electrical voltages of 60 kV in the needle and 30 kV in the collector, spaced 24 cm apart, and a rate of flow of 30 mm/min. An air blowing test revealed that the DTAB gives the membrane its antibacterial characteristics [113].

3.3.2 Composite membranes

Small diameter and high porosity are two unique characteristics of single-component polymeric nanofiber membranes that make them ideal for air filtration. Most of these membranes still don't have some necessary qualities, though, such as mechanical toughness, chemical resistance, and antifouling capabilities. Composite polymeric nanofiber membranes are created by combining two or more polymers with distinct properties to solve the drawbacks of these materials [50]. To give air filters additional functional qualities, electrospun nanofibers can be modified or post-treated. For instance, electrospinning nanofibers with silver nanoparticles may be used to create air filters with an antimicrobial coating that can help kill viruses and bacteria as well as trap them. In order to add new properties to the nanofiber filters, such as the adsorption of VOCs for the removal of odors, which is the same application as active carbon filters, the electrospinning solution can be changed with additional additives or the process can be progressed [114].

In a recent study, nanofibers of polyvinyl chloride (PVC) and PA6 were used to create the composite membrane for air filtration. These self-powered electrospun composite nanofibers membranes were highly efficient and experienced low pressure drops. Using the triboelectric effect, various electronegative materials were used to prepare nanofibers so that charges could be created on the membrane surface. The influence of the composited nanofiber membrane's capability for charge production on filtration performance enhances the air filtration efficiency of these membranes [115].

Due to their potential uses in gas separation, MOFs have also been incorporated into the polymers for electrospinning. For instance, the zeolitic imidazolate framework-8 serves to change polyacrylic acid [116], polyimide [117], and PAN [118]. It was used in all three instances to filter PM2.5, with respective filtration efficiencies of 99.6, 96.6, and 99.9. An excellent PM removal effectiveness above 99% and the extended lifespan of ENMs made by PAN-CuCl$_2$ composite was observed. At particular CuCl$_2$ concentrations, it is discovered that they seem to have a unique net structure, which significantly boosts PM removal effectiveness compared to pure PAN nanofibrous sheet [31].

It has been demonstrated that the performance of air filtration and the material's photocatalytic properties, which enable us to trap toxic airborne particles while also developing self-cleaning properties when exposed to sunlight, are improved when ammonium tetra thiomolybdate is present in a PAN polymeric matrix when developing membranes. The flow rate was 0.3 mL/h, the collection distance was 18 cm, 13 kV applied voltage, and the operating temperature was 25 °C [119].

4 Conclusion

Electrospinning is a flexible technique for creating nanofibrous porous membranes, using a variety of polymers. The morphology of the produced membranes can be chosen by selecting the polymer, solvent, and electrospinning process parameters. Because of their unique properties, including their high surface-to-mass ratio, excellent pore interconnectivity, and uniform distribution of their dimensions, electrospun nanofiber membranes are used in filtering processes.

ENMs, particularly those reinforced by useful nanoparticles, can effectively replace conventional glass fiber-based filters for air filtration. Nanoparticular additives give the ENM the ability to remove chemical pollutants, giving them an advantage over high-efficiency particulate absorption filters. To create dimensionally stable nanofiber membranes and enhance adhesion between the nanofiber membrane and microfiber medium, more research is necessary. The development of novel sample preparation procedures and sophisticated computational tools will enable more precise and accurate determination of the thickness of nanofiber membrane. The nanocomposite ENMs thus provide advantages including increased filtration effectiveness, increased protection duration, and weight reduction. Technology for treating oily wastewater should undergo a major revolution soon because of the development of ENs membranes with improved oil/water separation capability.

To create dimensionally stable nanofiber membranes and enhance adhesion between the nanofiber membrane and microfiber medium, more research is necessary. The development of novel sample preparation procedures and sophisticated computational tools will enable more precise and accurate determination of the thickness of nanofiber membrane.

References

[1] Ukaogo, P. O., Ewuzie, U., and Onwuka, C. V. Environmental pollution: Causes, effects, and the remedies. In: Microorganisms for Sustainable Environment and Health, Elsevier, 2020, pp. 419–429.

[2] Drioli, E., Macedonio, F., and Tocci, E. Membrane science and membrane engineering for a sustainable industrial development. Sep. Purif. Technol., **275**: 119196, 2021.

[3] Konni, M., et al. Titanium dioxide electrospun nanofibers for dye removal – A review. J. Appl. Nat. Sci., 2022, **14**(2): 450–458.

[4] Osseiran, N., and Chriscaden, K. J. W. G. Switzerland, Air pollution levels rising in many of the world's poorest cities. 2016.

[5] Teow, Y. H., and Mohammad, A. W. J. D. New generation nanomaterials for water desalination: A review. Desalination, 2019, **451**: 2–17.

[6] Abdulhamid, M. A., and Muzamil, K. Recent progress on electrospun nanofibrous polymer membranes for water and air purification: A review. Chemosphere, 2023, **310**: 136886.

[7] Liao, Y., et al. Progress in electrospun polymeric nanofibrous membranes for water treatment: Fabrication, modification and applications. Prog. Polym. Sci., 2018, **77**: 69–94.

[8] Lalia, B. S., et al. A review on membrane fabrication: Structure, properties and performance relationship. Desalination, 2013, **326**: 77–95.

[9] Hussein, A. K., et al. Spun of improvised cis-1, **3**, 4, 6-tetranitrooctahydroimidazo-[4, 5-d]–Imidazole (BCHMX) in polystyrene nanofibrous membrane by electrospinning techniques. BMC Chem., 2022, **16**(1): 1–11.

[10] Wang, X., and Hsiao, B. S. Electrospun nanofiber membranes. Curr. Opin. Chem. Eng., 2016, **12**: 62–81.

[11] Tijing, L. D., et al. 1.16 Electrospinning for membrane fabrication: Strategies and applications. In: Drioli, E., Giorno, L., Fontananova, E. (Eds.) Comprehensive Membrane Science and Engineering (2nd ed.), Elsevier, Oxford, 2017, pp. 418–444.

[12] Manea, L., et al. Electrospun membranes for environmental protection. In: IOP Conference Series: Materials Science and Engineering, IOP Publishing, 2018, **374**: 012081.

[13] Vanangamudi, A., et al., Nanofibers for Membrane Applications, Springer International Publishing, 2018.

[14] Lv, D., et al. Green electrospun nanofibers and their application in air filtration. Macromol. Mater. Eng., 2018, **303**(12): 1800336.

[15] Zhu, M., et al. Electrospun nanofibers membranes for effective air filtration. Macromol. Mater. Eng., 2017, **302**(1): 1600353.

[16] Valencia-Osorio, L. M., and Álvarez-Láinez, M. L. Global view and trends in electrospun nanofiber membranes for particulate matter filtration: A review. Macromol. Mater. Eng., 2021, **306**(10): 2100278.

[17] Qin, X., and Subianto, S. Electrospun nanofibers for filtration applications. In: Electrospun Nanofibers, Elsevier, 2017, pp. 449–466.

[18] Rubio-Valle, J. F., et al. Electrospun nanofibres with antimicrobial activities. In: Antimicrobial Textiles from Natural Resources, Elsevier, 2021, pp. 589–618.

[19] Chang, C., et al. Medical fibers and biotextiles. In: Biomaterial Science, Elsevier, 2020, pp. 575–600.

[20] Barhoum, A., et al. Nanofibers as new-generation materials: From spinning and nano-spinning fabrication techniques to emerging applications. Appl. Mater. Today, 2019, **17**: 1–35.

[21] Ahire, J., et al. Polyacrylonitrile (PAN) nanofibres spun with copper nanoparticles: An anti-*Escherichia coli* membrane for water treatment. Appl. Microbiol. Biotechnol., 2018, **102**: 7171–7181.

[22] Moslehi, M., et al. Preparation and characterization of polyamide thin film composite nanofiltration membrane based on polyurethane nanofibrous support. J. Polym. Environ., 2021, **29**: 2463–2477.

[23] Dolanský, J., Demel, J., and Mosinger, J. Multifunctional polystyrene nanofiber membrane with bounded polyethyleneimine and NO photodonor: Dark-and light-induced antibacterial effect and enhanced CO_2 adsorption. J. Mater. Sci., 2019, **54**: 2740–2753.

[24] Cao, D.-Q., et al. Recovery of extracellular polymeric substances from excess sludge using high-flux electrospun nanofiber membranes. Membranes, 2023, **13**(1): 74.

[25] Xue, J., et al. Electrospinning and electrospun nanofibers: Methods, materials, and applications. Chem. Rev., 2019, **119**(8): 5298–5415.

[26] Xia, C., et al. Recent advances on electrospun nanomaterials for zinc–air batteries. Small Sci., 2021, **1**(9): 2100010.

[27] Ansari, S., et al. Fabrication and characterization of electrospun zein/nylon-6 (ZN6) nanofiber membrane for hexavalent chromium removal. Environ. Sci. Pollut. Res., 2022, **29**: 653–662.

[28] Akther, N., et al. Recent advances in nanomaterial-modified polyamide thin-film composite membranes for forward osmosis processes. J. Membr. Sci., 2019, **584**: 20–45.

[29] Mohammad, N., and Atassi, Y. TiO_2/PLLA electrospun nanofibers membranes for efficient removal of methylene blue using sunlight. J. Polym. Environ., 2021, **29**: 509–519.

[30] Shekh, M. I., et al. Nano silver-embedded electrospun nanofiber of poly (4-chloro-3-methylphenyl methacrylate): Use as water sanitizer. Environ. Sci. Pollut. Res., 2017, **24**: 5701–5716.

[31] Hu, S., et al. Preparation and characterization of electrospun PAN-$CuCl_2$ composite nanofiber membranes with a special net structure for high-performance air filters. Polymers, 2022, **14** (20): 4387.

[32] Ruan, D., et al. Transparent PAN: TiO_2 and PAN-*co*-PMA: TiO_2 nanofiber composite membranes with high efficiency in particulate matter pollutants filtration. Nanoscale Res. Lett., 2020, **15**(1): 1–8.

[33] Dela Peña, E. M. B., et al. The design of a bench-scale adsorbent column based on nanoclay-loaded electrospun fiber membrane for the removal of arsenic in wastewater. Water Environ. J., 2021, **35** (3): 937–942.

[34] Topuz, F., Abdulhamid, M. A., Nunes, S. P., and Szekely, G. Hierarchically porous electrospun nanofibrous mats produced from intrinsically microporous fluorinated polyimide for the removal of oils and non-polar solvents. Environ. Sci. Nano, 2020, **7**(5): 1365–1372.

[35] Topuz, F., Abdulhamid, M. A., and Szekely, G. Superoleophilic oil-adsorbing membranes based on porous and nonporous fluorinated polyimides for the rapid remediation of oil spills. Chem. Eng. J., 2022, **449**, 137821.

[36] Talukder, M. E., et al. Ag nanoparticles immobilized sulfonated polyethersulfone/polyethersulfone electrospun nanofiber membrane for the removal of heavy metals. Sci. Rep., 2022, **12**(1): 5814.

[37] Topuz, F., Abdulhamid, M. A., Hardian, R., Holtzl, T., and Szekely, G. Nanofibrous membranes comprising intrinsically microporous polyimides with embedded metal–organic frameworks for capturing volatile organic compounds. J. Hazard Mater., 2022, **424**, 127347.

[38] Cseri, L., Topuz, F., Abdulhamid, M. A., Alammar, A., Budd, P. M., and Szekely, G. Electrospun adsorptive nanofibrous membranes from ion exchange polymers to snare textile dyes from wastewater. Adv. Mater. Technol., 2021, **6**(10): 2000955.

[39] Topuz, F., Abdulhamid, M. A., Holtzl, T., and Szekely, G. Nanofiber engineering of microporous polyimides through electrospinning: Influence of electrospinning parameters and salt addition. Mater. Des., 2021, **198**, 109280.

[40] Shirazi, M. M. A., and Asghari, M. Electrospun filters for oil–water separation. In: Filtering Media by Electrospinning: Next Generation Membranes for Separation Applications, Springer, 2018, pp. 151–173.

[41] Ramakrishna, S. An Introduction to Electrospinning and Nanofibers, World scientific, 2005.

[42] Wendorff, J. H., Agarwal, S., and Greiner, A. Electrospinning: Materials, Processing, and Applications, John Wiley & Sons, 2012.

[43] Dosunmu, O., et al. Electrospinning of polymer nanofibres from multiple jets on a porous tubular surface. Nanotechnology, 2006, **17**(4): 1123.

[44] Gualandi, C., et al. An innovative co-axial system to electrospin in situ crosslinked gelatin nanofibers. Biomed. Mater., 2016, **11**(2): 025007.

[45] Nauman, S., Lubineau, G., and Alharbi, H. F. Post processing strategies for the enhancement of mechanical properties of ENMS (electrospun nanofibrous membranes): A review. Membranes, 2021, **11**(1): 39.

[46] Yang, W., et al. Optimal spinneret layout in Von Koch curves of fractal theory based needleless electrospinning process. AIP Adv., 2016, **6**(6): 065223.

[47] Su, Q., Zhang, J., and Zhang, L.-Z. Fouling resistance improvement with a new superhydrophobic electrospun PVDF membrane for seawater desalination. Desalination, 2020, **476**: 114246.

[48] Doan, H. N., et al. Recycled PET as a PDMS-Functionalized electrospun fibrous membrane for oil–water separation. J. Environ. Chem. Eng., 2020, **8**(4): 103921.

[49] Kaur, S., et al. The characterization of electrospun nanofibrous liquid filtration membranes. J. Mater. Sci., 2014, **49**: 6143–6159.

[50] Ding, B., Wang, X., and Yu, J. Electrospinning: Nanofabrication and Applications, William Andrew, 2018.

[51] Lu, T., et al. Multistructured electrospun nanofibers for air filtration: A review. ACS Appl. Mater. Interfaces, 2021, **13**(20): 23293–23313.

[52] Tarus, B., et al. Effect of polymer concentration on the morphology and mechanical characteristics of electrospun cellulose acetate and poly (vinyl chloride) nanofiber mats. Alexandria Eng. J., 2016, **55** (3): 2975–2984.

[53] Bagbi, Y., Pandey, A., and Solanki, P. R. Electrospun nanofibrous filtration membranes for heavy metals and dye removal. In: Thomus, S. (Ed.) Nanoscale Materials in Water Purification, Elsevier, 2019, pp. 275–288.

[54] Bakar, S., et al. Effect of voltage and flow rate electrospinning parameters on polyacrylonitrile electrospun fibers. In: IOP Conference Series: Materials Science and Engineering, IOP Publishing, 2018.

[55] Long, Y.-Z., et al. Electrospinning: The setup and procedure. In: Electrospinning: Nanofabrication and Applications, Elsevier, 2019, pp. 21–52.

[56] De Vrieze, S., et al. The effect of temperature and humidity on electrospinning. J. Mater. Sci., 2009, **44**: 1357–1362.

[57] Demir, M., Yilgor, I., Yilgor, E. E. A., and Erman, B. Electrospinning of polyurethane fibers. Polymer, 2002, **43**(11): 3303–3309.

[58] Dickhout, J. M., et al. Produced water treatment by membranes: A review from a colloidal perspective. J. Colloid Interface Sci., 2017, **487**: 523–534.

[59] Baskar, D., et al. Mechanism of nanofibres on removal of water pollutants – A review. NISCAIR-CSIR, 2018, pp. 451–458.

[60] Saeed, M., et al. Synthesis of a visible-light-driven Ag_2O–Co_3O_4 Z-scheme photocatalyst for enhanced photodegradation of a reactive yellow dye. New J. Chem., 2022, **46**(48): 23297–23304.

[61] Alothman, A. A., et al. Facile synthesis and comparative study of the enhanced photocatalytic degradation of two selected dyes by TiO_2-g-C_3N_4 composite. Environ. Sci. Pollut. Res., 2023, **30**(13): 37332–37343.

[62] Cui, J., et al. Electrospun nanofiber membranes for wastewater treatment applications. Sep. Purif. Technol., 2020, **250**: 117116.

[63] Gee, S., Johnson, B., and Smith, A. Optimizing electrospinning parameters for piezoelectric PVDF nanofiber membranes. J. Membr. Sci., 2018, **563**: 804–812.

[64] Blanco, M., et al. TiO_2-doped electrospun nanofibrous membrane for photocatalytic water treatment. Polymers, 2019, **11**(5): 747.

[65] Nallathambi, G., et al. Preparation and characterization of triple layer membrane for water filtration. Environ. Sci. Pollut. Res., 2020, **27**: 29717–29724.

[66] Toniatto, T., et al. Nanostructured poly (lactic acid) electrospun fiber with high loadings of TiO_2 nanoparticles: Insights into bactericidal activity and cell viability. Mater. Sci. Eng. C, 2017, **71**: 381–385.

[67]   Wang, Z., et al. Microfiltration performance of electrospun nanofiber membranes with varied fiber diameters and different membrane porosities and thicknesses. Polymer, 2017, **114**: 64–72.

[68]   Bae, J., Baek, I., and Choi, H. J. Mechanically enhanced PES electrospun nanofiber membranes (ENMs) for microfiltration: The effects of ENM properties on membrane performance. Water Res., 2016, **105**: 406–412.

[69]   Jianxin, L., et al. Numerical investigation of the membrane fouling during microfiltration of semiconductor wastewater. Desalin. Water Treat., 2016, **57**(11): 4756–4768.

[70]   Li, M., et al. Super-hydrophilic electrospun PVDF/PVA-blended nanofiber membrane for microfiltration with ultrahigh water flux. J. Appl. Polym. Sci., 2020, **137**(9): 48416.

[71]   Tang, Y., et al. Electrospun nanofiber-based membranes for water treatment. Polymers, 2022, **14** (10): 2004.

[72]   Yogarathinam, L. T., et al. Electrospun membranes for UF/NF/RO/FO/PRO membranes and processes. In: Electrospun and Nanofibrous Membranes. Elsevier, 2023, pp. 347–369.

[73]   Shirazi, M. M. A., Bazgir, S., and Meshkani, F. Electrospun nanofibrous membranes for water treatment. Adv. Membr. Technol., 2020, **57**(3): 467–504.

[74]   Ma, W., et al. Electrospun fibers for oil–water separation. RCS Adv., 2016, **6**(16): 12868–12884.

[75]   Bahmani, P., et al. High-flux ultrafiltration membrane based on electrospun polyacrylonitrile nanofibrous scaffolds for arsenate removal from aqueous solutions. J. Colloid Interface Sci., 2017, **506**: 564–571.

[76]   Vanangamudi, A., et al. Nanofiber composite membrane with intrinsic Janus surface for reversed-protein-fouling ultrafiltration. ACS Appl. Mater. Interfaces, 2017, **9**(21): 18328–18337.

[77]   Wang, Z., et al. Hot-pressed PAN/PVDF hybrid electrospun nanofiber membranes for ultrafiltration. J. Membr. Sci., 2020, **611**: 118327.

[78]   Prince, J., et al. Nanofiber based triple layer hydro-philic/-phobic membrane-a solution for pore wetting in membrane distillation. Sci. Rep., 2014, **4**(1): 6949.

[79]   Low, Z. X., et al. Fouling resistant 2D boron nitride nanosheet–PES nanofiltration membranes. J. Membr. Sci., 2018, **563**: 949–956.

[80]   Ji, Y., et al. Recent developments in nanofiltration membranes based on nanomaterials. Chin. J. Chem. Eng., 2017, **25**(11): 1639–1652.

[81]   Chen, H., et al. Functionalized electrospun nanofiber membranes for water treatment: A review. Sci. Total Environ., 2020, **739**: 139944.

[82]   Liu, F., et al. Preparation and characterization of novel thin film composite nanofiltration membrane with PVDF tree-like nanofiber membrane as composite scaffold. Mater. Des., 2020, **196**: 109101.

[83]   Faccini, M., et al. Electrospun carbon nanofiber membranes for filtration of nanoparticles from water. J. Nanomater. 2015, **2015**: 2–2.

[84]   Escribá, A., et al. Incorporation of nanomaterials on the electrospun membrane process with potential use in water treatment. Colloids Surf. A: Physicochem. Eng. Asp., 2021, **624**: 126775.

[85]   Yadav, S., et al. Recent developments in forward osmosis membranes using carbon-based nanomaterials. Desalination, 2020, **482**: 114375.

[86]   Asadollahi, M., Bastani, D., and Musavi, S. A. J. D. Enhancement of surface properties and performance of reverse osmosis membranes after surface modification: A review. Desalination, 2017, **420**: 330–383.

[87]   Ray, S. S., et al. A comprehensive review: Electrospinning technique for fabrication and surface modification of membranes for water treatment application. RSC Adv., 2016, **6**(88): 85495–85514.

[88]   Al-Obaidani, S., et al. Potential of membrane distillation in seawater desalination: Thermal efficiency, sensitivity study and cost estimation. J. Membr. Sci., 2008, **323**(1): 85–98.

[89]   An, X., Liu, Z., and Hu, Y. J. D. Amphiphobic surface modification of electrospun nanofibrous membranes for anti-wetting performance in membrane distillation. Desalination, 2018, **432**: 23–31.

[90]   Zhu, Z., et al. Dual-bioinspired design for constructing membranes with superhydrophobicity for direct contact membrane distillation. Environ. Sci. Technol., 2018, **52**(5): 3027–3036.

[91]   Zuo, J., and Chung, T.-S. J. W. Metal–organic framework-functionalized alumina membranes for vacuum membrane distillation. Water, 2016, **8**(12): 586.

[92]   Lee, E.-J., et al. Engineering the re-entrant hierarchy and surface energy of PDMS-PVDF membrane for membrane distillation using a facile and benign microsphere coating. Environ. Sci. Technol., 2017, **51**(17): 10117–10126.

[93]   Li, D., Yan, Y., and Wang, H. Recent advances in polymer and polymer composite membranes for reverse and forward osmosis processes. Prog. Polym. Sci., 2016, **61**: 104–155.

[94]   Yan, K.-K., et al. Superhydrophobic electrospun nanofiber membrane coated by carbon nanotubes network for membrane distillation. Desalination, 2018, **437**: 26–33.

[95]   Wang, X., et al. Electrospun nanofibrous materials: A versatile medium for effective oil/water separation. Mater. Today, 2016, **19**(7): 403–414.

[96]   Chu, Z., Feng, Y., and Seeger, S. Oil/water separation with selective superantiwetting/superwetting surface materials. Angew. Chem. Int. Ed., 2015, **54**(8): 2328–2338.

[97]   Wang, H., et al. Progress in electrospun nanofibrous membranes used for oil-water separation. Mater. Rev., 2017, **31**(10): 144–151.

[98]   Su, R., et al. Recent progress in electrospun nanofibrous membranes for oil/water separation. Sep. Purif. Technol., 2021, **256**: 117790.

[99]   Law, K.-Y. Water–surface interactions and definitions for hydrophilicity, hydrophobicity and superhydrophobicity. Pure Appl. Chem., 2015, **87**(8): 759–765.

[100] Wang, S., et al. Bioinspired surfaces with superwettability: New insight on theory, design, and applications. Chem. Rev., 2015, **115**(16): 8230–8293.

[101] Liu, Z., et al. Efficient oil/water separation membrane derived from super-flexible and superhydrophilic core–shell organic/inorganic nanofibrous architectures. Polymers, 2019, **11**(6): 974.

[102] Zhang, Z.-M., et al. Green and robust superhydrophilic electrospun stereocomplex polylactide membranes: Multifunctional oil/water separation and self-cleaning. J. Membr. Sci., 2020, **593**: 117420.

[103] Zhang, J., et al. Electrospun flexible nanofibrous membranes for oil/water separation. J. Mater. Chem. A, 2019, **7**(35): 20075–20102.

[104] Bae, J., Kim, K. S., and Choi, H. Effect of asymmetric wettability in nanofiber membrane by electrospinning technique on separation of oil/water emulsion. Chemosphere, 2018, **204**: 235–242.

[105] Wei, Z., et al. A novel high-durability oxidized poly (arylene sulfide sulfone) electrospun nanofibrous membrane for direct water-oil separation. Sep. Purif. Technol., 2020, **234**: 116012.

[106] Liang, Y., et al. Capillary effect in Janus electrospun nanofiber membrane for oil/water emulsion separation. Chemosphere, 2019, **221**: 479–485.

[107] Fan, Z.-Y., et al. Folic acid modified electrospun poly (vinyl alcohol)/polyethyleneimine nanofibers for cancer cell capture applications. Chin. J. Polym. Sci., 2016, **34**(6): 755–765.

[108] Sarbatly, R., Sariau, J., and Alam, M. F. I. Advances in nanofiber membrane. Mater. Today Proc., 2021, **46**: 2118–2121.

[109] Kadam, V. V., Wang, L., and Padhye, R. J. Electrospun nanofibre materials to filter air pollutants – A review. J. Ind. Text., 2018, **47**(8): 2253–2280.

[110] Sarwar, Z., et al. Fabrication and characterization of PEBA fibers by melt and solution electrospinning. J. Mater. Res. Technol., 2019, **8**(6): 6074–6085.

[111] Xiao, Y., et al. Performance of electrospun polyvinylidene fluoride nanofibrous membrane in air filtration. Autex Res. J., 2020, **20**(4): 552–559.

[112] Kadam, V., et al. Electrospun polyacrylonitrile/β-cyclodextrin composite membranes for simultaneous air filtration and adsorption of volatile organic compounds. ACS Appl. Nano Mater., 2018, **1**(8): 4268–4277.

[113] Ryšánek, P., et al. Stability of antibacterial modification of nanofibrous PA6/DTAB membrane during air filtration. Mater. Sci. Eng.: C, 2019, **96**: 807–813.

[114] Bortolassi, A. C. C., et al. Efficient nanoparticles removal and bactericidal action of electrospun nanofibers membranes for air filtration. Mater. Sci. Eng.: C, 2019, **102**: 718–729.

[115] Shao, Z., et al. Self-powered electrospun composite nanofiber membrane for highly efficient air filtration. Nanomaterials, 2020, **10**(9): 1706.

[116] Guo, J., et al. PAA@ ZIF-8 incorporated nanofibrous membrane for high-efficiency PM2.5 capture. Chem. Eng. J., 2021, **405**: 126584.

[117] Hao, Z., et al. Electrospun polyimide/metal-organic framework nanofibrous membrane with superior thermal stability for efficient PM2.5 capture. ACS Appl. Mater. Interfaces, 2019, **11**(12): 11904–11909.

[118] Wang, Z., et al. Polymer/MOF-derived multilayer fibrous membranes for moisture-wicking and efficient capturing both fine and ultrafine airborne particles. Sep. Purif. Technol., 2020, **235**: 116183.

[119] Al-Attabi, R., et al. One-pot synthesis of catalytic molybdenum based nanocomposite nano-fiber membranes for aerosol air remediation. Sci. Total Environ., 2019, **647**: 725–733.

Neelesh Ashok*, Sruthi M. S., Taniya Rose Abraham, Sabu Thomas

Chapter 9
Polymer membrane for desalination and distillation

Abstract: Nowadays, there has been growing interest in membrane distillation technology for treating high-salinity water or contaminated wastewater. This hybrid technology combines the advantages of membrane separation and thermal distillation. This process is based on transmembrane partial vapor pressure, which leads to the transport of only water vapor molecules through a hydrophobic microfiltration membrane. This chapter explores the key properties and advantages of polymer membranes in desalination and distillation, including their high selectivity, scalability, and potential for customization. Additionally, it highlights recent advancements in membrane technology, such as the integration of novel materials, surface modifications, and innovative module designs. These developments hold the promise of significantly improving the overall sustainability and cost-effectiveness of desalination and distillation processes, contributing to a more sustainable and water-secure future.

Keywords: water purification, separation technology, nanocomposite membranes, thin-film membranes, reverse osmosis, membrane distillation

1 Introduction

Polymeric membranes have emerged as a promising solution in the fields of desalination and distillation, offering potential advancements in addressing the pressing global water scarcity. These membranes possess unique properties that enable selective separation of salts and impurities from water, making them pivotal in desalination processes. Additionally, their application in membrane distillation has gained attention due to their ability to facilitate vapor transport, while restricting the passage of nonvolatile contaminants. This chapter serves as an introduction to the utilization of polymeric membranes for addressing these pressing water-related issues. It provides an overview of the fundamental principles underlying membrane-based separa-

*Corresponding author: Neelesh Ashok, Amrita School of Artificial Intelligence, Amrita Vishwa Vidyapeetham, Coimbatore, Tamil Nadu, India, e-mail: a_neelesh@cb.amrita.edu
Sruthi M. S., Amrita School of Artificial Intelligence, Amrita Vishwa Vidyapeetham, Coimbatore, Tamil Nadu, India
Taniya Rose Abraham, Sabu Thomas, School of Energy Materials, Mahatma Gandhi University, Kottayam, Kerala, India

https://doi.org/10.1515/9783110796032-009

tion, highlighting the unique properties of polymeric membranes that make them particularly suitable for desalination and distillation applications [1]. Additionally, the chapter outlines the key objectives and scope of this document, aiming to explore the latest advancements, challenges, and future prospects in harnessing polymeric membranes as a sustainable solution for meeting the ever-growing demand for fresh water. By delving into the significance of these membranes in transforming water treatment methodologies, this chapter sets the stage for a comprehensive exploration of their pivotal role in desalination and distillation technologies.

1.1 Overview of the global water crisis and the need for advanced desalination and distillation technologies

The global water crisis is a complex and pressing challenge that stems from the increasing demand for freshwater resources, coupled with the depletion and contamination of existing water sources. With population growth, urbanization, and industrialization, water scarcity has become a serious concern in many regions across the world. According to the United Nations, by 2030, nearly half of the global population could be living in areas with high water stress. One of the key factors contributing to the water crisis is the uneven distribution of freshwater resources. Some regions experience abundant rainfall and have access to ample surface and groundwater, while others face arid conditions and rely heavily on limited water sources. Additionally, overextraction, pollution from industrial and agricultural activities, and climate change further exacerbate the problem, leading to reduced water quality and availability. To address the growing demand for freshwater and mitigate the effects of water scarcity, advanced water treatment technologies have become crucial. Figure 1 depicts the different types of desalination techniques [2]. Desalination and distillation are two such technologies that play a significant role in converting non-potable water into usable and safe drinking water, as well as in meeting the water needs of various industries. Desalination involves the removal of salts and impurities from seawater or brackish water to produce fresh water. This process has gained popularity in regions with abundant seawater resources, but limited freshwater availability. Reverse osmosis, one of the most widely used desalination methods, relies on polymeric membranes to separate salt and other contaminants from water [7]. Distillation, on the other hand, is a separation process that involves heating water to produce steam, which is then condensed back into liquid form, leaving behind impurities. Membrane distillation, a variant of traditional distillation, utilizes polymeric membranes to facilitate vapor transport, while rejecting nonvolatile contaminants, making it an energy-efficient and environmentally friendly option.

As the demand for clean water continues to rise, there is an urgent need for the advancement of desalination and distillation technologies to improve their efficiency, reduce energy consumption, and lower overall costs. Innovative approaches, such as the utilization of polymeric membranes, offer promising solutions to enhance water

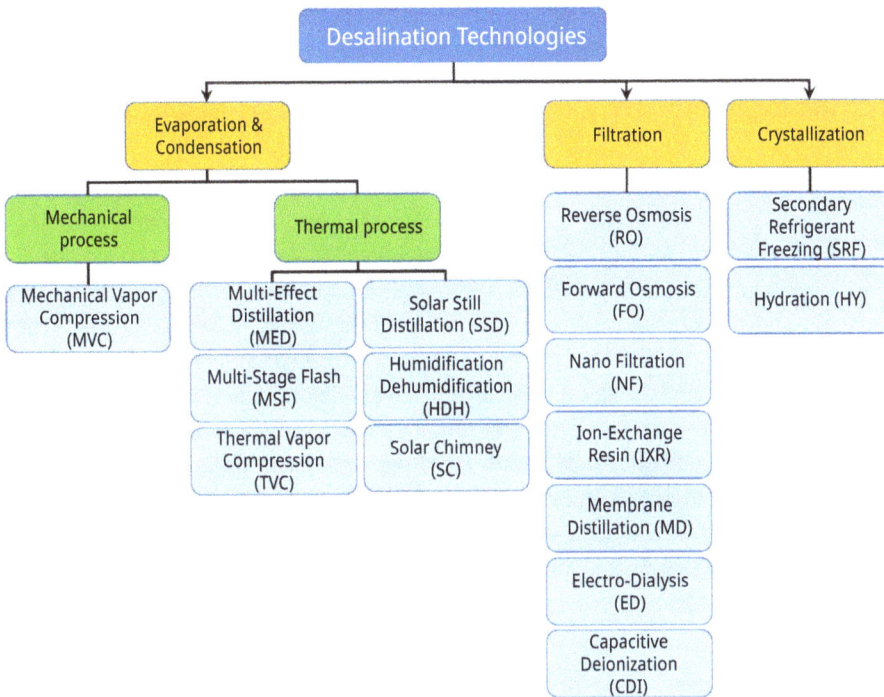

Figure 1: Desalination techniques.
*https://www.researchgate.net/figure/The-classification-of-desalination-technologies-by-working-principle
_fig1_348425553

treatment processes and provide a sustainable means of addressing the global water crisis [21, 22]. By investing in research, development, and implementation of these advanced technologies, we can take significant strides toward ensuring equitable access to safe and clean water for all.

1.2 Polymeric membranes and their potential in addressing water scarcity

Polymeric membranes are thin, semipermeable barriers made from synthetic polymers. They possess a network of microscopic pores that allow selective passage of certain substances, while blocking others. These membranes are designed to separate and purify liquids or gases, based on differences in molecular size, charge, or solubility [8]. In the context of water scarcity, polymeric membranes have shown tremendous potential in addressing the challenge of limited freshwater resources. Their applications are particularly significant in two key areas:

–   **Desalination:** With a large portion of the Earth covered by seawater, desalination – the process of removing salt and impurities from seawater to produce fresh water – holds immense promise in alleviating water scarcity. Polymeric membranes, especially those used in reverse osmosis, play a critical role in desalination technologies. When seawater is forced through these membranes under high pressure, their tiny pores effectively block salt ions and other impurities, allowing only pure water molecules to pass through. The result is a significant increase in the availability of fresh water, which can be used for drinking, agriculture, and other essential purposes.

–   **Water treatment and purification:** Beyond desalination, polymeric membranes find applications in various water treatment processes as shown in Figure 2. They can effectively remove contaminants, bacteria, viruses, and other harmful substances from different water sources, such as rivers, lakes, and groundwater. By employing polymeric membranes in filtration systems, water can be purified for safe consumption and industrial use, thus mitigating the impacts of water scarcity caused by pollution and limited access to clean water.

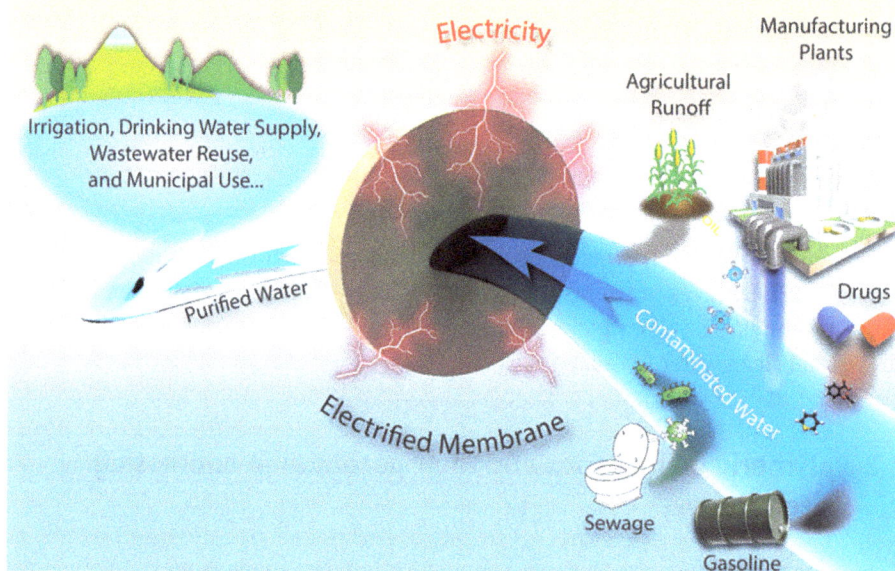

Figure 2: Electrified membranes for water treatment applications.
*https://www.google.com/url?sa=i&url=https%3A%2F%2Fpubs.acs.org%2Fdoi%2F10.1021%2Facses
tengg.1c00015&psig=AOvVaw2BXZDOZ2dw8sEKzHsRwdv1&ust=1690615771279000&source=images&cd=v
fe&opi=89978449&ved=0CBQQ3YkBahcKEwiAuN_a8LCAAxUAAAAAHQAAAAQSA

The versatility of polymeric membranes, coupled with their tunable properties and cost-effectiveness, makes them a favorable option in addressing water scarcity. Researchers and engineers continually explore ways to optimize membrane materials, pore size, and membrane configurations to enhance efficiency, durability, and sustainability [5]. As a result, polymeric membranes are poised to play a crucial role in advancing water treatment technologies and supporting global efforts to ensure a reliable and safe water supply, even in regions facing severe water scarcity challenges.

2 Membranes and its classification

Membranes are thin, selective barriers that play a fundamental role in various processes, including separation, purification, and concentration of substances. They serve as interfaces between two phases, typically liquid or gas, and selectively allow the passage of certain molecules or ions, while blocking others. Membranes are widely utilized in numerous industries, ranging from water treatment and pharmaceuticals to food processing and chemical engineering.

2.1 Classification of membranes

Membranes can be classified based on different criteria, such as their composition, structure, and application. Table 1 provides a general overview of membrane classifications, based on separation mechanisms, pore size ranges, typical applications, advantages, and disadvantages. Membrane performance and characteristics may vary, depending on specific materials, operating conditions, and applications.

Table 1: Classification of membranes.

Membrane classification	Separation mechanism	Pore size range	Typical applications	Advantages	Disadvantages
Microfiltration (MF)	Size exclusion	0.1–10 µm	Clarification of liquids, removal of large particles	High flow rates, low fouling, suitable for large particles	Limited separation of smaller solutes
Ultrafiltration (UF)	Size exclusion	0.001–0.1 µm	Concentration, purification of macromolecules	High rejection of macromolecules, low fouling	Limited separation of smaller solutes
Nanofiltration (NF)	Size exclusion; charge interaction	0.001–0.01 µm	Softening of water, color removal, wastewater treatment	Selective separation of charged molecules, low fouling	Moderate rejection of some small solutes

Table 1 (continued)

Membrane classification	Separation mechanism	Pore size range	Typical applications	Advantages	Disadvantages
Reverse osmosis (RO)	Solution-diffusion	<0.001 μm	Desalination, water purification	High rejection of solutes, versatile applications	High operating pressures, potential fouling
Forward osmosis (FO)	Osmotic pressure	N/A	Water recovery, draw solute concentration	Low energy consumption, environmentally friendly	Limited to specific applications, draw solute regeneration
Organic solvent NF (OSN)	Solution-diffusion	0.001–0.1 μm	Organic solvent purification, solvent recovery	Selective separation of organic solutes, high solvent flux	Limited solvent compatibility, potential fouling
Gas separation	Solution-diffusion, size exclusion	N/A	Air separation, gas purification	High selectivity for specific gases, energy-efficient	Sensitive to impurities, membrane aging

These membrane classifications provide a framework for understanding their diverse applications and functionalities. Membranes play a critical role in various industries, driving advancements in technology and addressing pressing challenges related to water scarcity, environmental protection, and sustainable resource management.

2.2 Structure and properties of polymeric membranes

Polymeric membranes are thin, semipermeable barriers composed of synthetic or natural polymers. They are widely used in various separation processes, such as filtration, desalination, and gas separation, due to their tunable properties and cost-effectiveness. The structure and properties of polymeric membranes are crucial factors that determine their performance in specific applications [23, 24]. Polymeric membranes consist of a dense layer or a porous matrix of polymer chains. The arrangement of these chains defines the membrane's structure and directly influences its separation capabilities. There are two main types of polymeric membrane structures:

- **Dense or nonporous membranes:** In dense membranes, polymer chains are closely packed, leaving little-to-no space between them. These membranes rely on a solution–diffusion mechanism, where small molecules dissolve into the membrane and diffuse through it, while larger molecules are blocked. Dense membranes are commonly used in gas separation applications.

– **Porous membranes:** Porous membranes have a network of interconnected pores, allowing the passage of specific molecules, based on their size and shape. The pore size and distribution can be controlled during membrane fabrication, enabling selectivity for certain solutes or particles. Porous membranes are widely used in liquid separations, such as ultra filtration and reverse osmosis, where they block larger particles and solutes, while allowing smaller ones to pass through.

The properties of polymeric membranes determine their separation efficiency, selectivity, and durability. Key properties include:
– **Permeability:** Permeability refers to the ability of a membrane to allow the passage of certain substances, such as water or specific solutes. The permeability of a polymeric membrane is influenced by its structure, pore size, and polymer material.
– **Selectivity:** Selectivity is the membrane's ability to distinguish between different solutes or particles. Highly selective membranes only allow specific substances to pass through, while rejecting others, based on their size, charge, or chemical properties.
– **Mechanical strength:** Mechanical strength is crucial for membrane integrity and long-term operation. Polymeric membranes should be mechanically robust to withstand pressure and mechanical stress during filtration or separation processes.
– **Chemical compatibility:** The chemical compatibility of polymeric membranes ensures their stability and resistance to chemical degradation when exposed to various solutions or environmental conditions.
– **Fouling resistance:** Membrane fouling is the deposition of unwanted particles or contaminants on the membrane surface, which can decrease performance over time. Polymeric membranes with fouling-resistant properties help maintain their efficiency during continuous operation.
– **Thermal stability:** Some applications, such as gas separation or water treatment at elevated temperatures, require membranes with high thermal stability to withstand harsh conditions.

Polymeric membranes' diverse structure and properties make them versatile tools in addressing separation and purification challenges across industries, offering an array of solutions to improve water quality, recover valuable materials, and promote sustainable practices in various fields.

2.3 Transport mechanisms in polymeric membranes

Polymeric membranes employ various transport mechanisms to facilitate the selective passage of molecules or ions, based on their size, charge, and solubility. The key transport mechanisms in polymeric membranes include:

– **Diffusion:** Diffusion is a fundamental transport mechanism in polymeric membranes. It involves the movement of molecules or solutes from regions of high concentration to regions of low concentration, across the membrane. In dense polymeric membranes, small molecules dissolve into the polymer matrix and diffuse through it. The rate of diffusion depends on the size of the diffusing species and the polymer's permeability to the specific solute.

– **Osmosis:** Osmosis is a special type of diffusion that occurs when a solvent (usually water) moves across a semipermeable membrane to equalize the concentration of solutes on both sides. In osmosis, water molecules move from a region of lower solute concentration to a region of higher solute concentration until equilibrium is reached. This phenomenon is essential in processes like reverse osmosis, where water is separated from dissolved salts.

– **Ultrafiltration:** Ultrafiltration is a size-based separation mechanism in polymeric membranes that selectively allows the passage of molecules below a certain molecular weight cutoff. The membrane's pores act as a physical barrier, retaining larger particles, macromolecules, and colloids, while permitting smaller solutes and solvents to pass through.

– **Reverse osmosis:** Reverse osmosis is a pressure-driven separation process that uses polymeric membranes to remove dissolved salts, ions, and other impurities from water. In this process, external pressure is applied to the solution side, forcing water molecules to move from the high-solute concentration (feed) side to the low-solute concentration (permeate) side, leaving the salts and impurities behind.

– **Donnan exclusion:** Donnan exclusion is a charge-based separation mechanism that occurs in ion-exchange membranes. These membranes have fixed charged groups that attract counterions and repel co-ions. This results in a selective permeation of ions, based on their charge and size, allowing for specific ion separation.

– **Facilitated transport:** In some polymeric membranes, specific functional groups or chemical moieties are introduced to enhance the transport of certain solutes. This facilitated transport mechanism involves interactions between the solute and the functional groups, allowing for higher selectivity and faster transport of targeted species.

These transport mechanisms are crucial for the selective separation of substances in polymeric membranes. Understanding and optimizing these mechanisms play a vital role in tailoring membrane materials and designs for specific applications, such as water purification, gas separation, and other separation processes in various industries.

2.4 Membrane selectivity and rejection mechanisms

Membrane selectivity and rejection mechanisms are fundamental aspects that determine a membrane's ability to selectively separate specific substances from a mixture. These mechanisms play a crucial role in various separation processes, including desalination, filtration, and gas separation. Membrane selectivity refers to a membrane's ability to preferentially allow certain molecules or ions to pass through, while blocking others [25–27]. It is governed by several factors, including pore size, charge, and chemical interactions. The following are common factors influencing membrane selectivity:

a. Pore size: Membranes with specific pore sizes can selectively block or allow molecules or ions, based on their size. Smaller solutes can pass through the membrane pores, while larger solutes are rejected.

b. Charge selectivity: Charged membranes, such as ion-exchange membranes, exhibit selectivity, based on the charge of the solutes. Positively charged membranes attract and pass through negatively charged ions, while repelling similarly charged ions.

c. Chemical interactions: Some membranes have functional groups or chemical moieties that interact selectively with certain solutes. These interactions can enhance or inhibit the passage of specific species through the membrane.

Rejection refers to the percentage of solute or impurity that is blocked or retained by the membrane during a separation process. It is a measure of the membrane's efficiency in removing unwanted substances. The main rejection mechanisms are:

a. Size exclusion: This mechanism occurs in membranes with specific pore sizes. Solutes larger than the pore size are excluded or rejected, while smaller solutes pass through.

b. Charge repulsion: In charged membranes, solutes with the same charge as the membrane are repelled and rejected, while oppositely charged solutes can be attracted and allowed to pass through.

c. Solution–diffusion: In dense polymeric membranes, solution–diffusion is the primary rejection mechanism. Solutes dissolve into the membrane and diffuse through it, based on their solubility in the polymer. Solutes with low solubility in the membrane are more effectively rejected.

d. Steric hindrance: In porous membranes, steric hindrance plays a role in rejection. Larger solutes are unable to pass through the narrow pores due to physical hindrance, leading to rejection.

e. Donnan exclusion: In ion-exchange membranes, Donnan exclusion contributes to rejection by selectively excluding ions, based on charge and size interactions.

Overall, membrane selectivity and rejection mechanisms are critical factors in designing membranes for specific separation processes. Understanding and controlling these mechanisms enable the development of efficient membranes for various applications, ranging from water purification and desalination to gas separation and biotechnology [28].

3 Desalination technologies

Desalination is the process of removing salts and other impurities from seawater or brackish water to produce fresh water suitable for various applications [9, 10]. There are several desalination methods, each with its unique principles and operational characteristics. Below is an overview of some prominent desalination methods:

- **Reverse osmosis (RO):** Reverse osmosis is one of the most widely used desalination methods. It employs a semipermeable membrane that allows water molecules to pass through, while rejecting dissolved salts and other impurities. In this process, seawater is pressurized and forced through the membrane, leaving behind concentrated brine on one side and producing fresh water (permeate) on the other side. RO is energy-efficient and capable of producing high-quality drinking water, making it a popular choice for both large-scale industrial applications and smaller residential systems.
- **Multi-stage flash (MSF) distillation:** MSF distillation is a thermal desalination method that utilizes heat to evaporate seawater. In this process, seawater is heated in multiple stages, and each stage operates at a lower pressure than the previous one. As the water vaporizes, it is condensed to produce fresh water. The method is energy-intensive but can be effective for large-scale desalination plants.
- **Multi-effect distillation (MED):** MED is another thermal desalination method that operates at lower temperatures than MSF. It utilizes multiple evaporation stages, with a decrease in pressure, similar to MSF. The heat released during condensation in each stage is used to heat the seawater in the subsequent stage, making MED more energy-efficient than MSF.
- **Electrodialysis (ED) and electrodialysis reversal (EDR):** ED and EDR are desalination methods based on the principle of ion-exchange membranes and electric fields. In ED, ions are selectively transported through ion-exchange membranes, separating salt and producing fresh water. EDR is a variation where the polarity of the electric field is reversed periodically to reduce membrane fouling.
- **Forward osmosis (FO):** Forward osmosis is an emerging desalination method that utilizes an osmotic gradient to draw water through a semipermeable membrane. A highly concentrated solution (draw solution) with a lower water potential is used to create the osmotic pressure, driving water from the seawater (feed solution) through the membrane. The draw solution is then separated from the water, and fresh water is obtained. FO has the potential to be more energy-efficient than RO, but it is still in the early stages of development and implementation [3, 4].

Each desalination method has its advantages and disadvantages, and the choice of the method depends on factors such as the available feedwater source, energy availability, cost considerations, and environmental impacts. RO remains the most dominant and widely used desalination method due to its efficiency, reliability, and widespread application in both large and small-scale desalination projects [11]. However, ongoing research and advancements in other desalination technologies, such as FO and ED, hold promise for further enhancing the efficiency and sustainability of desalination processes in the future.

3.1 Comparison of different desalination processes in terms of efficiency, energy consumption, and application

Table 2 shows the comparison of different desalination processes, including reverse osmosis (RO), multi-stage flash (MSF) distillation, multi-effect distillation (MED), electrodialysis (ED), and forward osmosis (FO), in terms of efficiency, energy consumption, and typical applications.

The efficiency and energy consumption of each desalination process can vary depending on specific system configurations, feedwater quality, and operational conditions. Additionally, desalination technologies are continually evolving, and ongoing research and innovation may lead to improvements in efficiency and energy consumption in the future.

3.2 Role of polymeric membranes in desalination technologies

Polymeric membranes play a central and indispensable role in various desalination technologies, most notably in the widely used reverse osmosis (RO) process [16, 17]. The unique properties of polymeric membranes make them highly effective in separating salts and other impurities from seawater, brackish water, or other saline sources, producing fresh water suitable for drinking, industrial use, and agricultural applications.

- Reverse osmosis (RO): RO is the dominant desalination method, and polymeric membranes are the core component of RO systems. Polymeric membranes act as semipermeable barriers that selectively allow water molecules to pass through, while rejecting dissolved salts and other contaminants. The thin, dense or porous structure of polymeric membranes ensures efficient separation and high salt rejection rates, typically exceeding 99%. The efficiency and reliability of polymeric membranes in RO make this process the most widely adopted and commercially viable desalination method.
- Forward osmosis (FO): In FO, polymeric membranes are used to create an osmotic gradient between a concentrated draw solution and the feed solution (seawater or brine). The draw solution's low water potential draws water across the polymeric membrane, leaving behind the salts and impurities in the feed solution. Polymeric membranes in FO facilitate the osmotic process, providing a unique

Table 2: Comparison of different desalination processes.

Desalination process	Reverse osmosis (RO)	Electrodialysis (ED)	Multi-effect distillation (MED)	Multi-stage flash (MSF)	Solar desalination	Forward osmosis (FO)
Energy consumption	High	Moderate to high	High	High	Moderate to high	Low to moderate
Feedwater source	Seawater, brackish water	Seawater, brackish water	Seawater, brackish water	Seawater	Seawater, brackish water	Seawater, brackish water
Membrane type	Spiral wound	Ion-exchange membrane	Evaporation	Evaporation	Various	Hollow fiber
Operation pressure	50–1,000 psi	<100 psi	Low (atmospheric)	Low (vacuum)	Varies	Low to moderate
Removal efficiency	95–99% salt removal	90–99% salt removal	90–99% salt removal	90–99% salt removal	Varies	70–90% salt removal
Scale of application	Large scale and municipal	Small to medium scale	Large scale and industrial	Large scale and municipal	Small to large scale	Small to medium scale
Advantages	High salt rejection, widely used, mature technology	Lower energy than RO, selective ion removal	Utilizes waste heat, reliable in harsh environments	High production capacity, thermal efficiency	Utilizes renewable energy, environmentally friendly	Lower energy consumption, gentle on membranes
Disadvantages	High energy requirement, fouling potential	Limited scalability, higher operational costs	High capital cost, complex operation	High energy consumption, potential scaling issues	Weather-dependent, lower production capacity	Limited scalability, draw solute regeneration

and energy-efficient approach to desalination, especially when coupled with other desalination methods like RO.

– Hybrid desalination processes: Polymeric membranes play a crucial role in hybrid desalination processes, where multiple desalination technologies are combined to improve the overall efficiency and cost-effectiveness. For example, RO-FO hybrid systems use the osmotic pressure generated by FO to reduce the energy requirements of the subsequent RO process, leading to energy savings.

– Membrane distillation (MD): Though not primarily a desalination method, MD can be used for desalination, and polymeric membranes are used in membrane distillation applications. MD utilizes polymeric membranes to separate pure water vapor from a heated saline solution, effectively desalinating the water.

– Pretreatment and post-treatment: In desalination plants, polymeric membranes are often employed in pretreatment stages to remove larger particles, colloids, and organisms before the water enters the main desalination process.

Additionally, post-treatment polymeric membranes can be used to ensure the produced water meets the required quality standards before distribution or use. The versatility of polymeric membranes, their controllable pore size, and their compatibility with different feedwater sources make them a preferred choice for desalination technologies [12]. Continuous research and development efforts are dedicated to improving membrane efficiency, reducing fouling, and increasing durability to enhance the overall performance and sustainability of desalination processes using polymeric membranes. As water scarcity continues to be a global concern, the critical role of polymeric membranes in desalination is instrumental in ensuring a reliable supply of fresh water for various sectors and communities worldwide.

4 Membrane fabrication techniques

Polymeric membranes can be fabricated using various techniques, each imparting distinct properties and structures to the membranes. Below are explanations of some common fabrication methods for polymeric membranes:

1. **Phase inversion:** Phase inversion is the most widely used method for fabricating polymeric membranes. It involves the transformation of a homogeneous polymer solution into a membrane structure by inducing phase separation. The process typically consists of three steps: casting, coagulation, and drying. During casting, a polymer solution is spread on a substrate, and a solvent or nonsolvent is added to induce phase separation. The solvent diffuses out of the polymer solution, leading to the formation of a porous structure. The resulting membrane's properties can be tailored by adjusting factors such as polymer concentration, solvent type, and coagulation conditions.

2. **Interfacial polymerization:** Interfacial polymerization is a technique used to fabricate thin-film composite membranes, particularly in reverse osmosis applications. It involves the reaction between two monomers at the interface of two immiscible phases. One phase contains a polymer precursor and the other phase contains a reactant for the monomer. When the two phases come into contact at the membrane surface, polymerization occurs, forming a thin selective layer. This layer enhances the membrane's separation properties, allowing it to be more effective in rejecting salts and impurities.

3. **Sol–Gel method:** The sol–gel method involves the conversion of a liquid precursor (sol) into a gel-like material that solidifies into a membrane structure. In this process, a colloidal suspension (sol) of inorganic or organic materials is cast onto a substrate and undergoes a chemical reaction, leading to the formation of a porous gel. The gel is then dried and sintered to form the final membrane. The sol–gel method is commonly used to fabricate ceramic and hybrid organic–inorganic membranes.

4. **Electrospinning:** Electrospinning is a technique used to produce nanofibrous membranes with high surface area and porosity. In this process, a polymer solution is subjected to a high electric field, causing the solution to form fine jets that elongate and solidify into nanofibers as they travel to a grounded collector. The nanofibers intertwine to create a porous structure with a high aspect ratio. Electrospinning allows the fabrication of membranes with precise control over fiber diameter and porosity, making it suitable for various applications, including filtration and tissue engineering.

5. **Breath figure method:** The breath figure method is a template-assisted technique that utilizes condensation of water droplets to create ordered patterns on a polymer film. This patterned film serves as a template, and upon polymer casting, the pores align with the water droplet pattern, forming a porous membrane. The breath figure method offers control over pore size and distribution and finds application in gas separation and filtration.

Each of these fabrication methods offers unique advantages and is suited for specific membrane applications. Researchers and engineers choose the most appropriate method, based on desired membrane properties, pore size, selectivity, and the intended application of the membrane in desalination, filtration, gas separation, or other separation processes [29].

4.1 Factors influencing membrane morphology and performance

The morphology and performance of polymeric membranes are influenced by a variety of factors, which can be broadly categorized into material-related, fabrication-related, and operational factors [18]. Understanding these factors is crucial for tailor-

Table 3: Factors influencing membrane morphology and performance.

Factor	Description
Polymer type	Choice of polymer material affects membrane properties
Membrane fabrication method	Fabrication technique influences membrane structure and morphology
Polymer concentration in casting solution	Higher concentration can lead to denser membranes
Solvent type and composition	Solvent affects polymer–solvent interactions and phase separation
Additives and cross-linking agents	Addition of modifiers can alter membrane properties
Casting temperature and time	Temperature and time influence membrane crystallinity and porosity
Post-treatment processes (annealing, cross-linking)	Post-treatment impacts membrane structure and performance
Surface modification and functionalization	Surface treatments can enhance selectivity and fouling resistance
Membrane thickness	Thickness affects permeability and solute transport
Porosity and pore size distribution	Pore structure influences solute rejection and flux
Mechanical strength and stability	Membrane integrity affects long-term performance and robustness
Operating parameters (pressure, temperature, flow)	Operating conditions impact membrane efficiency and fouling
Feedwater characteristics (pH, ionic strength)	Feedwater properties affect solute–membrane interactions
Fouling and cleaning behavior	Fouling propensity influences membrane maintenance and lifespan

ing membrane properties to meet specific separation requirements. Table 3 depicts the key factors that influence membrane morphology and performance.

5 Membrane characterization and performance evaluation

Membrane characterization involves various techniques that provide valuable information about the morphology, structure, and performance of polymeric membranes. Some common techniques for membrane characterization include:

- **Scanning electron microscopy (SEM):** SEM is used to visualize the surface morphology and cross-sectional structure of membranes at a high resolution. It provides detailed images that can reveal pore sizes, distribution, and overall membrane structure.
- **Atomic force microscopy (AFM):** AFM is another high-resolution imaging technique that provides topographical information about the membrane surface at the nanoscale. It can be used to study surface roughness, pore sizes, and surface modifications.
- **Porosity measurement:** Various methods can be employed to determine the porosity of membranes. Mercury intrusion porosimetry is a common technique that measures the intrusion of mercury into the membrane pores at different pressures, allowing for the calculation of pore size distribution and porosity.
- **Bubble point test:** The bubble point test determines the minimum pressure required to force a wetting liquid out of the largest pores in the membrane. It provides information about the largest pore size and pore size distribution.
- **Permeability measurement:** Permeability testing assesses the rate at which a specific solvent or gas passes through the membrane under defined conditions. It is an essential parameter for understanding the membrane's transport properties.
- **Contact angle measurement:** Contact angle measurement provides insights into the wettability of the membrane surface, which can influence its fouling behavior and interaction with the feed solution.
- **Fourier transform infrared spectroscopy (FTIR):** FTIR is a technique used to study the chemical composition of the membrane material. It can identify functional groups, verify polymer composition, and detect potential surface modifications.
- **X-ray diffraction (XRD):** XRD is employed to analyze the crystallinity of membranes and identify the presence of any crystalline phases in the polymer material.
- **Water flux measurement:** Water flux measurement determines the rate of water permeation through the membrane under specific pressure conditions, providing insight into the membrane's transport properties.
- **Fouling and scaling analysis:** Techniques like fouling index measurements, fouling potential assessment, and antifouling tests help evaluate the membrane's susceptibility to fouling and scaling in real-world applications.

By employing a combination of these techniques, researchers and engineers can gain a comprehensive understanding of the structure, morphology, and performance of polymeric membranes [30]. This knowledge is crucial for optimizing membrane materials and designs to enhance their efficiency and applicability in various desalination, filtration, and separation processes.

5.1 Performance evaluation parameters

Table 4 provides performance evaluation parameters that are critical in assessing the efficiency and effectiveness of polymeric membranes in desalination, filtration, and other separation processes. By monitoring and optimizing these parameters, membrane manufacturers and users can improve the performance and reliability of their membrane systems.

Table 4: Performance evaluation parameters of polymeric membranes.

Performance evaluation parameters	Description
Flux	The rate of permeation of solvent (usually water) through the membrane. It is expressed as volumetric flow rate per unit area (e.g., $L.m^{-2} \cdot h$ or $m^3.m^{-2} \cdot day$). High flux indicates faster water production.
Salt rejection	The percentage of salts and solutes rejected by the membrane during desalination. It represents the efficiency of the membrane in removing dissolved solids. High salt rejection indicates better desalination performance.
Selectivity	The ability of the membrane to selectively allow certain solutes to pass through while rejecting others. It is often related to salt rejection and other separation parameters. High selectivity indicates better separation performance.
Permeability	A measure of how easily solvent molecules pass through the membrane. It is related to the membrane's porosity and pore size. High permeability indicates efficient transport of water molecules.
Fouling resistance	The ability of the membrane to resist fouling, which is the accumulation of impurities or particles on the membrane surface over time. Higher fouling resistance leads to longer membrane lifespan and stable performance.
Mechanical strength	The ability of the membrane to withstand mechanical stress, pressure, and handling during operation. Adequate mechanical strength is crucial for reliable performance and durability.
Chemical stability	The resistance of the membrane to chemical degradation when exposed to different solvents, solutions, or operating conditions. Chemical stability ensures long-term performance in various environments.
Thermal stability	The ability of the membrane to maintain its structural integrity and performance at elevated temperatures. It is particularly important for thermal desalination processes.
Flux decline rate (FDR)	The rate at which the membrane's flux decreases over time during continuous operation. Low FDR indicates better fouling resistance and longer membrane life.
Fouling reversibility	The extent to which fouling can be reversed through cleaning or regeneration procedures. High fouling reversibility indicates easier maintenance and longer membrane lifespan.

5.2 Standard testing protocols for polymeric membranes

Standard testing protocols for polymeric membranes are essential for ensuring consistent and reliable performance evaluation across different membrane materials and manufacturers [19, 20]. These protocols provide guidelines and procedures for conducting various tests to assess the membrane's properties and performance. Some of the standard testing protocols for polymeric membranes include as depicted in Figure 3:

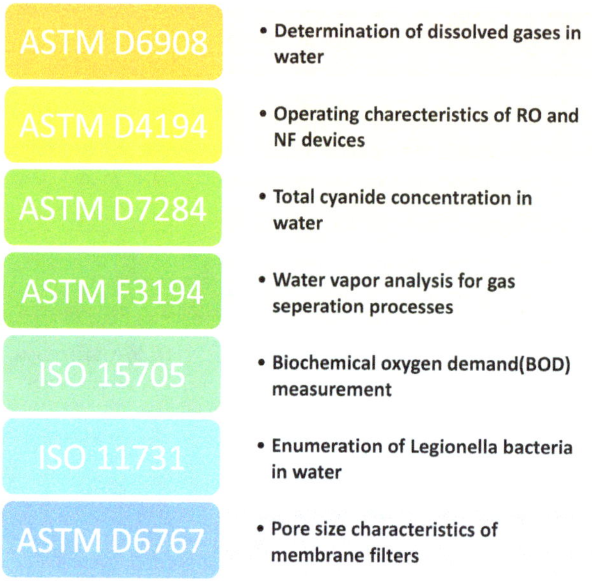

ASTM D6908	• Determination of dissolved gases in water
ASTM D4194	• Operating charecteristics of RO and NF devices
ASTM D7284	• Total cyanide concentration in water
ASTM F3194	• Water vapor analysis for gas seperation processes
ISO 15705	• Biochemical oxygen demand(BOD) measurement
ISO 11731	• Enumeration of Legionella bacteria in water
ASTM D6767	• Pore size characteristics of membrane filters

Figure 3: Standard testing protocols.

These standard testing protocols are essential tools for evaluating the performance and properties of polymeric membranes used in various applications, including desalination, water treatment, gas separation, and environmental monitoring [31, 32]. Adhering to these protocols ensures consistent and reliable assessment of membrane performance across different laboratories and membrane manufacturers.

6 Membrane fouling and scaling

Membrane fouling and scaling are common challenges that can significantly impact the performance and efficiency of polymeric membranes in various separation processes. Understanding the causes and mechanisms of fouling and scaling is crucial for developing strategies to mitigate these issues.

6.1 Membrane fouling

Membrane fouling is a significant challenge encountered in various membrane-based separation processes, including Organic Solvent Nanofiltration (OSN). It refers to the accumulation and deposition of undesired substances, such as particulates, colloids, microorganisms, organic matter, and inorganic salts, on the surface or within the pores of the membrane [6, 47]. This fouling phenomenon leads to a decline in membrane performance, reduced permeability, and increased energy consumption, impacting the efficiency and sustainability of the separation process [13].

– **Particulate fouling:** Fine particles, suspended solids, and colloids in the feedwater can accumulate on the membrane surface, blocking the pores and reducing permeability. Particulate fouling is often influenced by the concentration and size of particles in the feedwater.

– **Biological fouling (biofouling):** Microorganisms, such as bacteria, algae, and fungi, can attach to the membrane surface and form biofilms. These biofilms can grow and spread, leading to reduced membrane performance and increased resistance to solvent flow.

– **Organic fouling:** Organic matter, such as humic substances, proteins, and polysaccharides, can adsorb onto the membrane surface. This leads to pore blockage and reduced membrane permeability.

– **Scaling or inorganic fouling:** High concentrations of inorganic salts, such as calcium carbonate, calcium sulfate, and silica, can precipitate on the membrane surface and within the membrane pores, leading to scaling. Scaling reduces membrane flux and efficiency.

– **Colloidal fouling:** Colloidal particles in the feedwater can interact with the membrane surface through van der Waals forces or electrostatic interactions, causing fouling and reduced membrane performance.

6.2 Scaling

Scaling is a common form of membrane fouling that occurs when certain solutes in the feed water become supersaturated and form solid deposits on the membrane surface. This phenomenon is particularly prominent in desalination and water treatment processes, where high concentrations of dissolved salts, such as calcium carbonate, calcium sulfate, and silica, can precipitate and adhere to the membrane.

– **Temperature and concentration polarization:** In high-temperature processes, the concentration of salts in the feedwater may exceed their solubility limits upon cooling near the membrane surface, leading to scaling.

– **Supersaturation:** Supersaturation occurs when the concentration of certain salts in the feedwater exceeds their solubility limits. As a result, these salts precipitate and form scale deposits on the membrane surface.

– **pH changes:** Changes in pH can alter the solubility of certain salts, leading to their precipitation and scaling on the membrane.
– **Saturation index:** Saturation index indicates the propensity of a solution to scale. A positive saturation index indicates a potential for scaling.

Mitigation of fouling and scaling requires appropriate pretreatment of the feedwater, optimization of operating conditions, and periodic cleaning or maintenance of the membranes. Strategies such as chemical cleaning, backwashing, air scouring, and incorporating antifouling additives can help improve membrane performance and extend membrane lifespan [33]. Additionally, advancements in membrane materials and surface modifications can enhance the fouling and scaling resistance of polymeric membranes.

6.3 Strategies for fouling prevention and mitigation

Fouling is a common issue in membrane-based separation processes, but several strategies can be employed for its prevention and mitigation. By implementing these strategies, the performance and lifespan of polymeric membranes can be significantly improved. Some effective fouling prevention and mitigation strategies include:
    Implementing these fouling prevention and mitigation strategies can enhance the performance and longevity of polymeric membranes, ensuring consistent and reliable operation in various desalination, water treatment, and separation processes.

6.4 Role of membrane surface modification in reducing fouling tendencies

Membrane surface modification plays a crucial role in reducing fouling tendencies and improving the fouling resistance of polymeric membranes as shown below in Figure 4. By altering the properties of the membrane surface, surface modification strategies aim to create a more favorable and antifouling environment. Some of the key roles of membrane surface modification in reducing fouling tendencies include:
– **Enhanced hydrophilicity:** Many fouling materials, such as organic compounds and biofilms, have hydrophobic properties, leading to their strong attachment to hydrophobic membrane surfaces. Surface modification can introduce hydrophilic functional groups, increasing the membrane's affinity to water and reducing fouling adhesion.
– **Reduced surface roughness:** Microscopic roughness on the membrane surface can trap and promote the accumulation of foulants. Surface modification can smooth the membrane surface, minimizing sites for fouling materials to adhere and facilitating fouling removal during cleaning.

Proper Pretreatment	– Filtration – Coagulation/Flocculation
Backwashing	– Reverses flow direction to dislodge and remove accumulated fouling material
Air Scrouring	– Introduces air bubbles to create turbulane and dislodge fouling particles
Periodic Cleaning	– Uses appropriate cleaning agents to remove fouling deposits and restore membrane performance
Surface Modifications	– Coating or functionalization alter the membrane surface properties to reduce fouling
Addition of Antifouling Agents	– Incorporate antifouling additives to prevent or reduce fouling
Crossflow Filteration	– Tangential flow continuously removes fouling particles
Control of Operating Parameters	– Optimize pressure, flow rate, and temperature to influence fouling behaviour
Use of Cleaning-in-Place (CIP) - system	– Automated cleaning without disassembling the membrane module
Feedwater Source Management	– Monitoring and managing water quality to prevent fouling precursors

Figure 4: Strategies for fouling prevention and mitigation.

– **Zwitterionic coatings:** Zwitterionic materials have both positively and negatively charged functional groups, which create a neutral charge on the surface. This charge neutrality repels both positively and negatively charged foulants, leading to reduced fouling adhesion.
– **Antifouling polymers:** Surface modification with antifouling polymers introduces chemical moieties that discourage fouling materials from adhering to the membrane. These polymers act as barriers and prevent fouling substances from reaching the membrane surface.

- **Incorporation of antimicrobial agents:** In applications where biofouling is a concern, surface modification can introduce antimicrobial agents to inhibit the growth and adhesion of microorganisms, reducing the formation of biofilms.
- **Biocidal surfaces:** For severe biofouling challenges, biocidal surface modifications may be applied. These coatings release antimicrobial agents to prevent and control biofouling on the membrane surface.
- **Charge repulsion:** By introducing charged functional groups on the membrane surface, charge repulsion can be utilized to prevent foulants with similar charges from approaching and attaching to the membrane.
- **Pore structure modification:** In nanofiltration and ultrafiltration membranes, surface modification can alter the pore size or distribution, improving selectivity and reducing the passage of foulants.
- **Membrane surface grafting:** Grafting of functional polymers onto the membrane surface can create tailored surface properties, improving fouling resistance based on the specific needs of the application.

Overall, membrane surface modification offers a versatile and effective approach to minimize fouling tendencies and enhance the long-term performance of polymeric membranes. Depending on the application and the type of fouling, specific surface modification strategies can be selected and tailored to optimize the membrane's fouling resistance, leading to more efficient and sustainable membrane-based separation processes.

7 Applications of polymeric membranes in desalination

Polymeric membranes have found widespread applications in desalination processes due to their versatility, cost-effectiveness, and ability to address various water treatment challenges. One of the primary applications of polymeric membranes in desalination is through reverse osmosis (RO) technology. RO membranes, typically made of thin-film composite polyamide layers, effectively remove salts and impurities from seawater and brackish water sources, producing fresh and potable water [14]. This process has become a crucial solution to address global water scarcity, especially in regions with limited access to freshwater resources. The use of polymeric membranes in desalination has several advantages, including high salt rejection rates, low energy consumption, and scalability for large-scale water production. Additionally, the continuous development of advanced polymeric materials and membrane surface modifications has further improved the efficiency and longevity of RO membranes. As desalination becomes an increasingly vital part of sustainable water management, the application of polymeric membranes continues to play a pivotal role in providing clean and safe drinking water to communities worldwide. Polymeric membranes have demonstrated significant success and viability in various desalination applications [34, 35]. Their cost-effectiveness, energy efficiency, and scalability make them an attractive choice for large-scale desalination projects worldwide. Additionally, ongoing research and advancements in membrane technology continue to enhance the economic and environmental feasibility of polymeric membrane-based desalination.

8 Membrane distillation using polymeric membranes

Membrane distillation (MD) is a promising separation process that utilizes a semipermeable membrane to separate volatile components from a liquid mixture. It is based on the principle of vapor pressure difference, where heat is applied to one side of the membrane to vaporize the volatile component, and the vapor passes through the membrane, leaving behind the nonvolatile components in the liquid phase. The vapor is then condensed to recover the purified component. Membrane distillation offers several advantages over traditional separation techniques. It operates at low temperatures, making it energy-efficient and suitable for treating heat-sensitive or concentrated solutions [36]. Additionally, MD can handle high salinity or complex feed streams, making it particularly valuable in desalination and wastewater treatment applications. The process can be driven by various heat sources, including waste heat, solar energy, or geothermal energy, further contributing to its environmental sustainability.

Polymeric membranes are commonly used in membrane distillation due to their versatility, ease of fabrication, and cost-effectiveness. These membranes are typically made from hydrophobic polymers, which allow them to selectively transport water vapor, while rejecting liquid water and nonvolatile solutes. The hydrophobic nature of polymeric membranes prevents wetting by the liquid phase, facilitating vapor transport and minimizing fouling issues [37]. Polymeric membranes used in membrane distillation can be classified into two main categories:
1. **Flat sheet membranes**
2. **Hollow fiber membranes.**

Flat sheet membranes consist of a single, continuous layer of polymeric material, while hollow fiber membranes have a tubular structure with a porous inner membrane surface [15, 16]. The selection of the appropriate membrane type depends on factors such as the specific application, required flux, and fouling tendencies of the feed solution. Polymeric membranes offer various advantages in membrane distillation applications:
– **Hydrophobicity:** The inherent hydrophobic nature of polymeric membranes prevents the penetration of liquid water, ensuring efficient vapor transport and reducing the risk of wetting.

– **Pore size control:** Polymeric membranes can be tailored to have specific pore sizes and distributions, allowing for customization based on the separation requirements of the feed solution.
– **Scalability:** Polymeric membranes are highly scalable and can be manufactured in various sizes and configurations, making them suitable for both small-scale and large-scale applications.
– **Cost-effectiveness:** Compared to other membrane materials, polymeric membranes are generally more cost-effective, making membrane distillation an economically viable option.
– **Chemical compatibility:** Polymeric membranes exhibit good chemical compatibility with a wide range of feed solutions, making them suitable for diverse applications, including desalination, wastewater treatment, and concentration processes.

However, polymeric membranes may have limitations in terms of thermal stability at high temperatures, and susceptibility to degradation by certain chemicals. Nonetheless, ongoing research and material advancements continue to enhance the performance and durability of polymeric membranes, making them increasingly attractive for membrane distillation applications in various industrial and environmental settings.

8.1 Comparative analysis with other distillation techniques

Table 5 shows a comparative analysis of various distillation techniques, including membrane distillation, multi-effect distillation (MED), and vapor compression distillation (VC), presented in a tabular format:

Table 5: Comparative analysis with other distillation techniques.

Distillation technique	Operating principle	Energy consumption	Scalability	Suitability for seawater desalination
Membrane distillation	Utilizes polymeric membranes to facilitate vapor transport, while rejecting nonvolatile contaminants	Low to moderate	Highly scalable	Well-suited due to energy efficiency and ability to handle saline water
Multi-effect distillation (MED)	Utilizes multiple stages (effects) with decreasing pressures to evaporate and condense water	High	Limited scalability	Commonly used in large-scale plants and for brackish water desalination

Table 5 (continued)

Distillation technique	Operating principle	Energy consumption	Scalability	Suitability for seawater desalination
Vapor compression	Utilizes a compressor to increase	Moderate to high	Moderately scalable	Energy-efficient for seawater
Distillation (VC)	The vapor pressure of water vapor for evaporation and condensation			Desalination; requires suitable waste heat source

Operating principle describes the main operating principle of each distillation technique. Membrane distillation relies on polymeric membranes for vapor transport and nonvolatile contaminant rejection. MED involves multiple stages with decreasing pressures to facilitate evaporation and condensation. VC uses a compressor to increase vapor pressure for the distillation process. Energy consumption indicates the relative energy consumption of each technique. Membrane distillation generally has low-to-moderate energy requirements compared to MED and VC. The scalability of each technique is mentioned in the scalability column. Membrane distillation is highly scalable and can be adapted for various capacities [38, 39]. MED has limited scalability due to the need for multiple stages, while VC is moderately scalable. Suitability for seawater desalination assesses the suitability of each distillation technique for desalination, specifically for seawater. Membrane distillation is well suited for seawater desalination due to its energy efficiency and ability to handle saline water. MED is commonly used in large-scale plants and is suitable for brackish water desalination. VC is energy-efficient for seawater desalination but requires a suitable waste heat source for optimal performance [40].

8.2 Emerging trends and future prospects

Current research and developments in polymeric membranes for desalination and distillation are driving significant advancements in membrane technology. Researchers are continuously exploring new materials, fabrication techniques, and surface modifications to enhance the performance and efficiency of polymeric membranes in these applications [41, 42]. One area of focus is the development of novel membrane materials with improved selectivity and fouling resistance, aiming to achieve higher salt rejection and longer membrane lifespan. Additionally, efforts are being made to optimize membrane fabrication processes, such as nanofabrication and thin-film composite methods, to control membrane morphology and tailor properties for specific separation needs. Potential areas for improvement and innovation in polymeric membranes for desalination and distillation include:

- High-temperature stability: Developing polymeric membranes with enhanced thermal stability to withstand higher temperatures in thermal-driven processes like membrane distillation
- Fouling resistance: Designing antifouling surfaces and introducing additives to mitigate fouling tendencies and extend membrane life
- Energy efficiency: Innovating energy-efficient membrane processes, reducing the energy consumption required for separation
- Hybrid membrane systems: Exploring the integration of different membrane technologies and hybrid processes to enhance overall efficiency and reduce operational costs

Future prospects and challenges in the field of polymeric membranes for desalination and distillation are promising, yet complex. The ongoing research in material science, nanotechnology, and membrane surface engineering is expected to lead to breakthroughs in membrane performance, selectivity, and cost-effectiveness. This may enable wider adoption of membrane-based desalination and distillation processes, especially in regions facing water scarcity [43, 44]. However, challenges persist, such as the development of cost-competitive and durable membranes capable of handling highly concentrated or aggressive feed solutions. Additionally, balancing the trade-offs between permeability, selectivity, and fouling resistance remains a key challenge in membrane design. Overall, continued research, collaboration between academia and industry, and advancements in materials science will play a pivotal role in shaping the future of polymeric membranes for desalination and distillation [45]. As technology progresses, membrane-based processes have the potential to contribute significantly to addressing global water challenges, providing sustainable solutions for clean water supply and resource recovery [46].

9 Conclusion

In conclusion, this document explored the significance of polymeric membranes in desalination and distillation processes. Polymeric membranes have proven to be valuable tools in addressing the global water crisis by providing efficient and sustainable solutions for water purification and resource recovery. They offer advantages such as cost-effectiveness, scalability, and energy efficiency, making them suitable for large-scale desalination projects and various water treatment applications. The document highlighted the role of polymeric membranes in key desalination methods, including reverse osmosis, forward osmosis, and membrane distillation. Membrane distillation, in particular, was discussed as a promising separation process that utilizes polymeric membranes to separate volatile components from liquid mixtures, demonstrating its potential in various industrial and environmental settings. Surface modification emerged as a crucial strategy to reduce fouling tendencies and enhance the performance of polymeric membranes. By introducing hydrophilic coatings, zwitterionic functionalities, and antifouling agents, researchers are continually improving the fouling resistance and longevity of these membranes. Moving forward, potential areas for improvement and innovation in polymeric membranes include increasing their thermal stability for high-temperature applications, optimizing energy efficiency, and exploring hybrid membrane systems to maximize overall process efficiency. In conclusion, polymeric membranes play a vital role in the pursuit of sustainable water treatment and desalination technologies. Their continuous development and integration into large-scale projects hold the promise of alleviating water scarcity and supporting global water management efforts. As research progresses and new advancements arise, polymeric membranes are expected to remain at the forefront of innovative solutions for clean water supply and resource recovery. Collaborative efforts between academia, industry, and policymakers will be key to furthering research and pushing the boundaries of polymeric membrane technology, ultimately shaping a more water-secure and sustainable future for generations to come.

References

[1]   Baker, R. W. Membrane Technology and Applications, (3rd ed.), John Wiley & Sons, 2004.
[2]   Warsinger, D. M., Tow, E. W., Nayar, K. G., Maswadeh, L. A., and Lienhard, V. J. H. High-performance flat–sheet polytetrafluoroethylene membranes for membrane distillation. Environ. Sci. Technol., 2015, **49**(11): 6875–6883.
[3]   Achilli, A., Cath, T. Y., and Marchand, E. A. The forward osmosis membrane bioreactor: A low fouling alternative to MBR processes. Desalination, 2009, **239**(1–3): 10–21.
[4]   Zhao, S., Zhang, Y., Vrijenhoek, E. M., and Tang, C. Y. Recent advancements in forward osmosis: Opportunities and challenges. J. Membr. Sci., 2018, **550**: 270–287.
[5]   Vatanpour, V., Madaeni, S. S., Moradian, R., Zinadini, S., and Astinchap, B. Novel antibifouling nanofiltration polyethersulfone membrane fabricated from embedding TiO2 coated multiwalled carbon nanotubes. J. Membr. Sci., 2011, **375**(1–2): 284–294.
[6]   Lee, K. P., and Arnot, T. C. Comparison of fouling properties of polyvinylidene fluoride, polyethersulfone, m-polyphenylene isopthalamide, and cellulose acetate membranes during cross-flow microfiltration of activated sludge. J. Membr. Sci., 2011, **367**(1–2): 340–349.
[7]   Qadir, M., Sharma, B. R., Bruggeman, A., Choukr-Allah, R., and Karajeh, F. Non–conventional water resources and opportunities for water augmentation to achieve food security in water scarce countries. Agric. Water Manag., 2007, **87**(1): 2–22.
[8]   Zhao, C., Sankir, M., and Li, K. Recent advances in membrane distillation processes: Membrane development, configuration design and application exploring. Sep. Purif. Technol., 2019, **211**: 885–909.
[9]   Rautenbach, R., Laubscher, R. F., and Johnson, D. J. Development of a spiral-wound membrane module for membrane distillation. J. Membr. Sci., 2010, **349**(1–2): 41–52.
[10] Alghamdi, A., and Li, K. Membrane distillation of high salinity solutions: A comprehensive review. Desalination, 2014, **342**: 15–30.

[11]   Lai, R. J., and Shih, N. Y. Fabrication and optimization of polyvinylidene fluoride-co-hexafluoropropylene membranes for membrane distillation. J. Membr. Sci., 2012, **394–395**: 115–124.

[12]   Baker, R. W., Lokhandwala, K., and Lonsdale, H. K. Natural gas dehydration by membrane: A comprehensive review. Sep. Sci. Technol., 1998, **33**(5): 591–609.

[13]   Huang, R. Y. M., Hsu, S. C., and Zydney, A. L. Fouling of polymeric membranes during ultrafiltration of calcium alginate solutions. J. Membr. Sci., 2009, **330**(1–2): 127–137.

[14]   Wang, K.Y., Chung, T.-S., and Rajagopalan, R. Dehydration of tetrafluoropropanol (TFP) by pervaporation via novel PBI/BTDA-TDI/MDI co-polyimide (P84) dual-layer hollow fiber membranes. J. Membr. Sci., 2007, **287**: 60–66.

[15]   Ma, G. Q., Li, M., Shi, L. E., and Liu, L. M. Hollow fiber membrane distillation: Modifications, module development, and applications. Ind. Eng. Chem. Res., 2009, **48**(7): 3269–3286.

[16]   Liu, L., Wang, H., and Yang, F. Investigation of polyetherimide hollow fiber membrane distillation for desalination. J. Membr. Sci., 2016, **508**: 1–8.

[17]   McCutcheon, J. R., Elimelech, M., and Cohen, Y. Global challenges in energy and water supply: The promise of engineered osmosis. Environ. Sci. Technol., 2006, **40**(17): 1–8.

[18]   Yip, N. Y., Tiraferri, A., Phillip, W. A., Schiffman, J. D., and Elimelech, M. High performance thin-film composite forward osmosis membranes based on interfacial polymerization. Environ. Sci. Technol., 2010, **44**(10): 3812–3818.

[19]   Su, J. H., Lai, R. J., and Shih, N. Y. Polyacrylonitrile–Grafted Polyvinylidene fluoride membranes for membrane distillation. J. Membr. Sci., 2015, **474**: 179–188.

[20]   Misdan, N., Lau, W. J., Ismail, A. F., and Matsuura, T. A review of fouling in reverse osmosis membranes: Strategies for fouling control. J. Environ. Sci., 2013, **28**: 121–131.

[21]   Warsinger, D. M., Swaminathan, J., Guillen-Burrieza, E., Arafat, H. A., Lienhard, V. J. H., and Zubair, S. M. A review of polymeric membranes and processes for potable water reuse. Prog. Polym. Sci., 2018, **81**: 209–237.

[22]   Sanchez, S., Garcia, A., and Meneses, J. Assessment of membrane distillation configurations for the treatment of desalination brines. Desalination, 2017, **423**: 87–94.

[23]   Zhao, C., Qu, X., and Li, K. Membrane Distillation. In: Sarp, S., Hilal, N. (Eds.) Membrane–Based Salinity Gradient Processes for Water Treatment and Power Generation, Elsevier, 2018, pp. 63–90.

[24]   Cheng, Y. L., and Teoh, M. M. Forward osmosis: Principles, applications, and recent developments. J. Membr. Sci., 2017, **523**: 189–210.

[25]   Ha, M., Jang, J., and Park, J. Development of polyacrylonitrile electrospun nanofiber membranes for direct contact membrane distillation. J. Membr. Sci., 2012, **415–416**: 443–453.

[26]   Kim, S., Lee, S. Y., and Lee, Y. M. A comprehensive review of the synthesis and development of polymeric membranes for forward osmosis. J. Membr. Sci., 2018, **550**: 131–144.

[27]   Al-Mohtar, A. A., and Brule, M. R. Fouling mechanism in membrane distillation using the direct contact membrane distillation technique: A review. Water Res., 2013, **47**(6): 2013–2029.

[28]   Hou, D., Chen, V., and Fane, A. G. Tailoring of hydrophobic–hydrophilic hybrid PVDF composite hollow fiber membranes for MD applications. J. Membr. Sci., 2013, **442**: 219–227.

[29]   Ren, J., Wang, Z., and Lin, L. A comprehensive review of surface modifications for forward osmosis membranes. J. Membr. Sci., 2018, **550**: 269–279.

[30]   Zhao, C., Sun, D., and Li, K. Fouling–resistant composite nanofiltration hollow fiber membranes for membrane distillation. J. Membr. Sci., 2016, **520**: 657–666.

[31]   He, Y., and Ghaffour, N. Membrane distillation for desalination and water treatment – A critical review. J. Membr. Sci., 2016, **498**: 261–281.

[32]   Zuo, J., Ye, W., Zhang, Y., Liu, F., and Chen, C. Role of membrane in membrane distillation and membrane crystallization processes: A comprehensive review. J. Membr. Sci., 2018, **552**: 285–311.

[33]   Huang, R. Y. M., Su, J. H., and Shih, N. Y. Polyvinylidene fluoride/graphene oxide nanocomposite membranes for direct contact membrane distillation. J. Membr. Sci., 2015, **489**: 242–254.

[34] Warsinger, D. M., Swaminathan, J., Guillen–Burrieza, E., Arafat, H. A., Lienhard, V. J. H., and Zubair, S. M. A review of polymeric membranes and processes for potable water reuse. Prog. Polym. Sci., 2018, **81**: 209–237.

[35] Yip, N. Y., Tiraferri, A., Phillip, W. A., Schiffman, J. D., and Elimelech, M. High performance thin-film composite forward osmosis membrane. Environ. Sci. Technol., 2011, **45**(23): 10273–10279.

[36] Park, J., Hong, S., and Lee, Y. M. A comprehensive review of the synthesis and development of polymeric membranes for forward osmosis. J. Membr. Sci., 2018, **550**: 131–144.

[37] Warsinger, D. M., Tow, E. W., Nayar, K. G., Maswadeh, L. A., and Lienhard, V. J. H. High-performance flat–sheet polytetrafluoroethylene membranes for membrane distillation. Environ. Sci. Technol., 2015, **49**(11): 6875–6883.

[38] Liu, Y., and Wang, Z. The hydrophobic modification of PVDF membranes for fouling control in MD process: A review. J. Membr. Sci., 2016, **514**: 595–612.

[39] Chen, Z., and Fane, A. G. Study of PES–g–PEGDA hollow fiber composite membranes with polyvinylpyrrolidone additive for direct contact membrane distillation. J. Membr. Sci., 2013, **431**: 103–115.

[40] Rautenbach, R., Laubscher, R. F., and Johnson, D. J. Development of a spiral-wound membrane module for membrane distillation. J. Membr. Sci., 2010, **349**(1–2): 41–52.

[41] Cheng, Y. L., and Teoh, M. M. Forward osmosis: Principles, applications, and recent developments. J. Membr. Sci., 2017, **523**: 189–210.

[42] Yeo, J., Lee, S., and Im, S. Membrane-based desalination technologies: A review. Desalination, 2019, **454**: 35–60.

[43] Alghamdi, A., and Li, K. Membrane distillation of high salinity solutions: A comprehensive review. Desalination, 2014, **342**: 15–30.

[44] Warsinger, D. M., Swaminathan, J., Guillen-Burrieza, E., Arafat, H. A., Lienhard, V. J. H., and Zubair, S. M. A review of polymeric membranes and processes for potable water reuse. Prog. Polym. Sci., 2018, **81**: 209–237.

[45] Sanchez, S., Garcia, A., and Meneses, J. Assessment of membrane distillation configurations for the treatment of desalination brines. Desalination, 2017, **423**: 87–94.

[46] Kim, S., Lee, S. Y., and Lee, Y. M. A comprehensive review of the synthesis and development of polymeric membranes for forward osmosis. J. Membr. Sci., 2018, **550**: 131–144.

[47] Al–Mohtar, A. A., and Brule, M. R. Fouling mechanism in membrane distillation using the direct contact membrane distillation technique: A review. Water Res., 2013, **47**(6): 2013–2029.

Hind Yaacoubi*, Mahmoud A. Abdulhamid

Chapter 10
Oil fractionation and water/oil emulsion separation using polymer membranes

Abstract: Oil separation using membrane technology is a highly effective and environmentally friendly method that can provide energy savings, environmental protection, and long-term industrial expansion, competing with traditional separation techniques like extraction, adsorption, or distillation. The employment of membranes in virtually every stage of processing is possible. This chapter studied the production of oils that benefit significantly from membrane separation techniques as they are employed for separation, recovery, purification, and dehydration. Various attempts for degumming and deacidification using membrane technology have also been covered. At the same time, several additional applications for membrane separation techniques fall outside this chapter's scope.

Keywords: polymer membranes, water/oil separation, recovery, oil fractionation, anti-fouling

1 Introduction

The oil and gas sector relies on energy-intensive distillation procedures for separation and purification, which together account for 10–15% of energy consumption in the world [1]. Daily processing of about 100 million barrels of crude oil typically involves thermal distillation to separate compounds depending on boiling points [2]. The separation of molecules in the 100–400 g/mol range is accomplished by distillation and extraction procedures. By separating molecules according to their size, shape, or polarity, membrane-based separations become a sustainable option with the potential to lower the energy and water required for oil refining as industries attempt to decrease energy and resource consumption [3]. Membrane separation has drawn much interest because of its advantages: simplicity, high permeation, high efficiency, good stability, lack of secondary contamination, and broad application. Factors including shape, structure, surface chemistry,

---

*Corresponding author: Hind Yaacoubi, Sustainable and Resilient Materials Lab, Center for Integrative Petroleum Research (CIPR), College of Petroleum Engineering & Geosciences (CPG), King Fahd University of Petroleum and Minerals, 31261, Dhahran, Saudi Arabia, e-mail: Hindyaacoubi@outlook.fr
Mahmoud A. Abdulhamid, Sustainable and Resilient Materials Lab, Center for Integrative Petroleum Research (CIPR), College of Petroleum Engineering & Geosciences (CPG), King Fahd University of Petroleum and Minerals, 31261, Dhahran, Saudi Arabia

The original version of this chapter was revised. Unfortunately, the institutional address of the author Hind Yaacoubi was incorrect in the original publication. This has been corrected. We apologize for the mistake.
https://doi.org/10.1515/9783110796032-010

endurance in challenging conditions, anticorrosion, and antifouling performances are commonly considered while preparing membranes. Polymer materials are preferred in the membrane design because they are simple to work with and can be scaled up. Polyether sulfone (PES), polyvinylidene fluoride (PVDF), polyvinyl alcohol (PVA), polyacrylonitrile (PAN), polyethylene (PE), polyamide, and chitosan are among the polymers being researched and used in membranes. Polymer membranes are more effective in separating oil-water emulsions such as PVDF and PES because of their excellent mechanical performance, improved durability (including chemical corrosion and heat resistance), and simple production procedures [4, 5]. The study on the demulsification-separation process needs to be more in-depth, and membrane stability and fouling are still significant issues. These constraints prevent the development of membrane separation technology for practical commercial applications in the future [6]. For this reason, academic researchers must concentrate on closing the most critical knowledge gaps in membrane fabrication, which include the repeatability of materials and short- and long-term process dynamics.

With an emphasis on using of polymeric membranes as a separation technique, this chapter addresses their application in different steps for oil fractionations, especially oil/water separation. First, it begins by outlining the fractionation of crude oil and bio-oil, mainly showing their composition. Then, it describes the application of polymer membranes in refining and recovery, citing the influence of surface chemistry and wettability on the separation efficiency.

2 Fractionation products of oil

2.1 Crude oil

Crude oil is a dark-brown oily fluid composed of thousands of groups of liquid hydrocarbon molecules in varied proportions called "fractions," each of which is the main source of specific commodities such as gasoline, diesel, kerosene, waxes, and asphalt (Figure 1) [7]. Saturates, aromatics, resins, and asphaltenes (SARA fractions) are four chemical group classes it includes [8]. Also called SARA analysis, which divides all types of crude oils depending on their solubility and polarity [9]. The saturate fraction, mainly consisting of nonpolar alkanes with linear or branched chains and alicyclic paraffin, is the light component in many crude oils. The aromatic components often have heteroatoms and heavy metals and are composed of one or more aromatic rings. Resins include a lot of heavy metals and heteroatoms and are made up of aromatic rings and aliphatic side chains. The asphaltenes fraction of crude oil, which is the most polar and the heaviest, comprises complex components and has a more significant concentration of heavy metals and heteroatoms than the resins fraction, even though their chemical components are equal [10]. How asphaltenes and resins interact determines the structure and stability of crude oil [8]. The types of hydrocarbons and potential contaminant species present determine the oil quality [11]. Unconventional oils differ from light (conventional) crudes with a higher viscosity (range from <20 to >1,000,000 cP), higher density, and low API gravity because they contain more asphaltenes and resins as well as more heavy metals (vanadium and nickel) and heteroatoms (sulfur and nitrogen) [12]. It is commonly known that heavy crude oil has an API gravity lower than 22.3, whereas medium and light crude oils have API gravities between 22.3 and 31.1 and equal to or greater than 31.1 [13]. There are now 2 trillion barrels of conventional crude oil reserves globally, 63% are located in Arab nations. And 6 trillion barrels of unconventional crude oil in reserves, which accounts for almost 70% of global fossil fuel reserves [12].

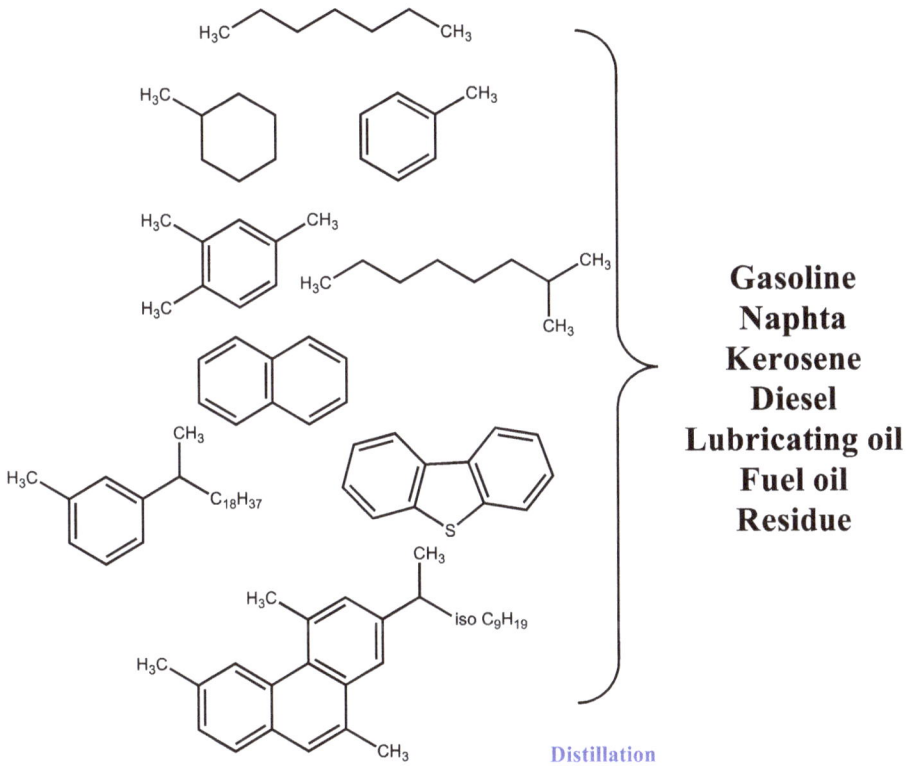

Figure 1: Fractional distillation of the crude oil.

The key bridging process between crude oil and refined products is refining. Light crude oil predominantly produces gasoline, diesel, and aviation fuels. Feedstock for plastics, petrochemicals, other fuels, and road paving is also provided by heavy crude oil. In addition, fuels for automobiles may be made from heavy oil. Thermal procedures like distillation, which are energy-intensive and frequently have low thermodynamic energy efficiency, dominate crude oil separations [3]. Concerns about greenhouse gas emissions have risen steadily over the past few decades due to environmental problems brought on by our excessive usage of fossil fuels. Different approaches, such as bio-based fuels made from bio-resources like biomass, have gradually been suggested to replace conventional fossil fuels.

2.2 Bio-crude

The idea of a bio-refinery is comparable to the petroleum refinery, where biomasses and wastes are converted into bio-based products such as fuels and chemical products through a variety of different biological, chemical, or thermal processes but are more ecologically friendly [14]. It has gained interest as an abundant source of energy due to its qualities as a cheap, accessible, and clean renewable energy source [15]. Primarily, there are two types of bio-refineries: one based on biochemical feedstock production, such as fermentation systems, and another on thermochemical conversion of biomass to bio-crude (usually via pyrolysis or an alternative low-temperature chemical process) [16].

Pyrolysis oil or bio-crude, other names for bio-oil, is a dark liquid composed of a complex of hydrocarbon ketones, aldehydes, phenols, furans, sugars, acids, alcohols, and esters mixture used as an alternative to fossil-based petroleum [17, 18]. This bio-oil has an excellent potential for environmentally friendly thermo-chemical and catalytic processes for manufacturing chemicals, H_2, and automobile fuels [19]. Several thermochemical techniques, including gasification, pyrolysis, fermentation, and liquefaction, can convert biomass into bio-oil as the main products and chemicals [20]. Pyrolysis has been developed to produce bio-oil from biomass (450–500 °C) without oxygen or by liquefaction with a catalyst [21–23]. Depending on the pyrolysis conditions, biomass quality, phenol derivatives, alkanes, and minute amounts of ketone, amine, alcohol, ester, and ether are present and differ in bio-oil produced (Figure 2) [22, 24]. Due to its high viscosity, high water content, high oxygen content, high acidity, high instability, and poor heating value, the quality of bio-oil deteriorates with time and requires extensive further processing before it can be used as a viable alternative to petroleum [23, 25–27]. Furthermore, the bio-oil acidity is a concern, causing corrosion and safety difficulties. Consequently, to improve the stability and quality of bio-oil (physical and chemical upgrading), the complex combination of bio-oil molecules may be divided into different fine chemicals and fractions that are richer in particular classes of chemical compounds by various techniques such as hydrotreating, deoxygenation, esterification, aldol condensation, ketonization, and catalytic cracking [19, 23, 28]. Using the difference in densities, bio-oil can be divided into light and heavy fractions. Typically, the light fraction comprehends low-molecular-weight acids and ketones with high reactivity to catalytic reforming and hydro-deoxygenation reactions [29]. Nevertheless, the heavy fraction has water-insoluble components with different chemical reactivity.

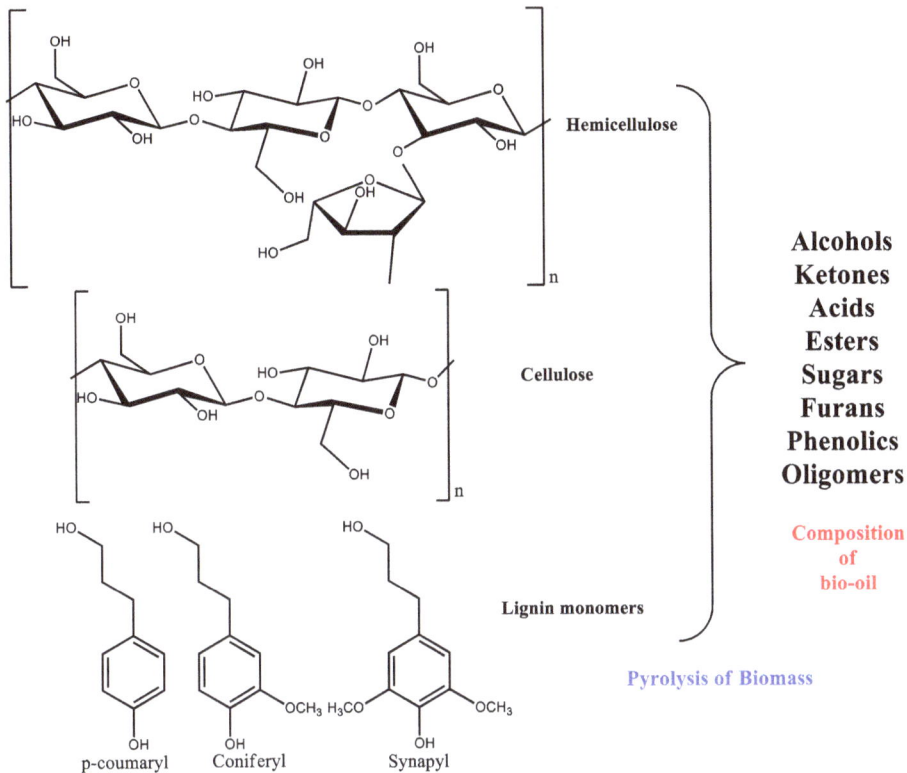

Figure 2: Example of biomass pyrolysis and bio-oil composition.

2.2.1 Lignin

As an abundant and sustainable resource, lignin is becoming increasingly popular as a viable substitute for crude oil in synthesizing bio-based compounds. After cellulose, it is considered the most abundant biopolymer, with 20% of the weight of hardwood and 15–30% of the weight of softwood [30], found in lignocellulosic biomass, made of a complex and heterogeneous polymer with strong chemically bound aromatic units that must be broken to utilize the aromatic compounds in industries [31]. Lignin is considered a desirable feedstock for the synthesis of biofuels and chemicals due to its higher carbon content and lower oxygen concentration despite its more complicated structure [32]. Additionally, by valorizing this waste stream, lignocellulosic bio-refineries may produce a variety of chemicals and materials without using petroleum. The only issue for valorizing lignin is recovering required compounds with high purity and yield. These typical lignin oils have a high-average molecular weight (MW) and include a mixture of many phenolic compounds, making it difficult to isolate pure monomer fractions [33]. Although there are other fractionation methods, including sequential precipitation, solvent extraction, and distillation, fractionation using membranes has some advantages as it requires no phase transitions or additives, causes no thermal damage, and consumes little energy. It divides solutes according to size and affinity.

2.3 Biofuels

Liquid biofuels are alternatives to petroleum due to their high oxygen content, such as biodiesel and bioethanol, whose raw materials are environmentally friendly, biodegradable, widely available with low sulfur concentrations, and safe to use in current engines. Bioethanol and biodiesel are common biofuels that are made from a range of biomass sources, including agricultural and forestry waste, such as sugar cane, pineapple, oil palm (trunk, empty fruit bunch, stalk, and frond), bamboo, corn, and rice husk (Figure 3) [34, 35]. Primarily used in the automotive industry, bioethanol purified is produced through the digestion and fermentation of lignocellulosic biomass by specific microorganisms and purification. Distillation is now the most often used operation to perform this sort of purification procedure, which divides diverse compounds based on their varying boiling points. However, distillation has a high energy need and is unable to separate azeotropic mixtures altogether [36]. Therefore, membrane-based technologies are the most promising for dehydrating organic solvents, separating organic/organic mixtures, and recovering volatile organic compounds from water. Also, water reuse in biodiesel operations has grown with membranes, particularly in low-water locations.

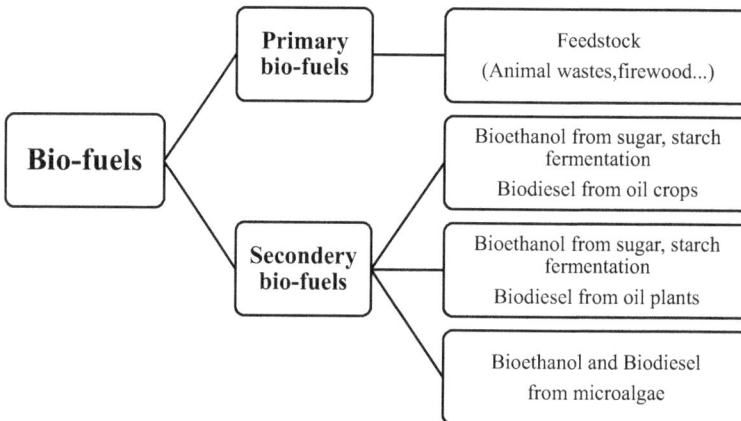

Figure 3: An illustration of raw material-based biofuel classification.

3 Membrane separation process for oil fractionation

Crude oil distillation consumes more than 1,100 TW-hours per year and can extract compounds in separate boiling-point ranges. Combining low-energy membrane separations based on molecular variations in size, shape, and membrane-penetrant interactions with distillation could increase the energy and carbon efficiency of the process by a factor of 10 compared to conventional thermal techniques [3]. In theory, membrane separations using membrane cascades with various separation modalities (such as organic solvent nanofiltration (OSN) and organic solvent reverse osmosis (OSRO)) can be used to convert thermal fractionation of liquid hydrocarbons [37]. Several polymers have been used for the OSRO-based separation of organics with wildly divergent polarity [38], although those for liquid hydrocarbon applications are uncommon. Scalable materials and membranes must be developed for hydrocarbon OSRO separation and testing. Standardization of these must be expanded to include all complicated feeds with more than 1,000 components, such as crude oil fractions. Recently, we showed that light crude oil fractionation utilizing polymer membranes is feasible without the need for phase change. In these trials with a low-stage cut (4%), the membrane successfully rejected molecules over 170 Da and selectively allowed aromatic molecules under this MW to pass through. Conceptually, distinct product streams might be produced from the original crude oil using a series of membranes with various separation methods (such as sorption, diffusion, or size-sieving selectivity) [39]. In the case of bio-fuels, membrane-based technologies are the most used since the demand for bioethanol is increasing, as it can reduce the cost and energy required to produce and purify ethanol.

3.1 Refining

A family of membranes was prepared using spirocyclic polytriazoles with noninterconnected microporosity, rapid reaction rates, and high MWs and solubility in common organic solvents. As demonstrated by the fractionation of whole Arabian light crude oil and atmospheric tower bottom feeds by enriching the low-boiling components and removing trace heteroatom and metal impurities, those membranes present an opportunity to reduce the energy cost of crude oil distillation [2]. Another type of membrane for the hydrocarbon processing industry that separates similarly sized molecules is Torlon® membranes, which continue to function and keep their shape for more than 400 h under harsh OSRO separation conditions. According to the capacity to reject hydrocarbon solvents (185 g/mol) and separate multiple class and size-based a combination using commercial polymers and industrially scalable spinning techniques, these membranes may be able to handle challenging solvent-solute or even solvent-solvent separations found in the hydrocarbon processing industry [40]. More recently, a membrane was fabricated using an N-aryl-linked spirocyclic polymer (SBAD) with higher selectivity than PIM-1, which generally has an MWCO of around

600 Da. It fractionated molecules lighter than 170 Da from crude oil mixtures [39]. A structural change has been done to the same polymer SBAD-1. According to the results, polymer backbone rigidity, rather than aromatic interactions between the solute and polymer, significantly impacts membrane performance in OSRO separations [41]. When employed for ultrafiltration (UF), the hydrophobic membrane (PP) may efficiently remove ash content (90–99%), water (78–82.5%), carbon residue (52.6–65.9%), acidity (44.9–73.3%), calcium (50.1–58.9%), and zinc (28.9–43.8%) from old engine oils. The oil-treated quality improves viscosity, density, color, and wettability toward oil [42]. Multiblock oligomer amine was synthesized, which consists of a core amine segment with two hydrophobic oligomer blocks. Compared to 35 traditional hydrophilic equivalents, these polyamide nanofilms successfully separate molecules based on variations in size and class, enabling >100 times quicker transport of hydrophobic liquids [1]. The polytriazole membrane with 10-nm-thin-selective layers containing subnanometer channels fabricated through film casting, nonsolvent-induced phase separation and thermal cross-linking may enhance 80–95% of the hydrocarbon content (140 g/mol) with less than 10 carbon atoms. These membranes show excellent selectivity and permeability and may be used in hybrid distillation systems for crude oil fractionation since they preferentially separate paraffin over aromatic components [43].

Membrane separation technology has been widely used to recover and fractionate lignin from wasted liquors. There have been relatively few studies published to date that address the fractionation and purification of lignin oils. Technical lignin generated by soda pulping wheat straw was successfully separated from lignin oil derived by catalytic cleavage in supercritical ethanol using polymeric NP030 membrane [31]. Thus, in diafiltration tests on two ethanol-based lignin oil mixtures, using ethanol as the substituting liquid, the NP030 membrane can selectively retain higher MW lignin derivatives while permeating the MW fraction is <1,000 Da. The polyethersulfone nanofiltration membranes fabricated by adding AC nanoparticles (0.1%) to separate lignin from black liquor show a lower membrane fouling and performed best at around 40 °C, 20 bar, and 11.5 pH. The inclusion of AC nanoparticles enhanced the membrane proprieties, and the BL filtration had the highest flux of 9.04 ± 0.29 L/m^2 h at 24 bar, good lignin rejection (88–91% at 12–24 bar), and the highest flux recovery of 98.02% [44].

3.2 Recovery

Due to their affordability and adaptability, hydrophobic polymers are the most often utilized membrane materials for ethanol extraction. Membranes made of polydimethylsiloxane (PDMS) are the most typical and extensively researched membranes for this kind of application. They were used to recover the bioethanol generated from the fermentation of lignocellulosic hydrolyzate [36]. At 40 °C and 2 bar TMP, the superhydrophobic hollow fiber PP membrane used for clarification and dehydration can produce

CPO with 59% nondissolved solid removal, > 99% water removal, and 47% clarity enhancement. The cake filtration dominates the fouling process, and the air scouring could be successfully used to recover the membrane flux with a flux recovery ratio of 90% (backflush duration = 5 min, filtration cycle = 2) [45]. Fiber-reinforced fluorinated polyimide foam membranes with adjustable multi-stage pore structure by freeze-drying and surface modifying were fabricated to increase the wettability and roughness. With good oil resistance, such membranes can separate oil-water emulsions and mixtures more than 10 times and recover heavy oil (30 L/m^2/h) [46]. To create hierarchical pore structures with porosity, a porous hydrophobic PVDF membrane was fabricated through electrospinning can adsorb lubricant oil due to its underwater lyophilic effect and then transfer to hexane solution, which can be quickly recovered through hexane evaporation both in low (17 °C) and relatively high temperature (67 °C). The oil recovered had water content nearly identical to the original samples (0.11, 0.02%; 0.10, 0.02%). This strategy offers an effective way of treating oil-in-water emulsion with high viscosity and concentration [47].

3.3 Degumming

One of the critical steps in the oil refining process is to eliminate the impurities from crude edible oils, especially free fatty acids (FFA), waxes, colored pigments, and particularly phosphatides or so-called gums [48]. Two types of phospholipids are available in the oil: hydratable phosphatides and nonhydratable phosphatides. Removing phospholipids is crucial since their presence affects the quality and storage of the final oil [49]. Due to the similarity of their MWs, which vary from 700 to 900 Da, phospholipids and triacylglycerols are challenging to separate using the molecular sieving separation technique [8]. Currently, phospholipids are typically converted into phospholipid micelles by the effect of organic solvents, hydration, and acidification and then removed by centrifugation [49]. Compared to membranes, centrifugation requires a lot of energy and is expensive. As a result, membrane equipment will inevitably be used more frequently in the oil degumming process.

3.4 Deacidification

Organic acids like naphthenic acid may be present in crude oil and distilled fractions in small concentrations. As an impurity, it can lead to corrosion in processes where the temperature is 200 °C or higher [50]. Also, the acidity of the bio-oil (pH in the range of 2–4) by the presence of low MW organic acids, such as acetic and formic acid, leads to corrosion and safety issues. To improve the quality of bio-oil, these acids and other light organic components should be removed and even will generate useful feedstock for industries [26]. Usually, the acidic components are eliminated by extraction using a polar solvent, such as methanol, which is recovered by distillation and may be recycled after being separated from the processed crude oil system and extract phase, respectively. As an example, polyamide, polyethersulfone, and polytetrafluoroethylene (PTFE) polymeric membranes were used to deacidify sardine oil, with properties influenced by feed pressure and concentration. PTFE membranes showed maximum efficiency regarding FFA reduction, solvent removal, and reduced oil loss. Future applications of hydrophobic membranes could be explored [51]. In another study, PVA cross-linked PVDF membrane for FFA removal from crude palm oil was investigated and showed that the hydrophobicity of PVDF decreased upon PVA cross-linking, and average roughness increased with increasing PVA concentration as the membrane coated with 100 ppm PVA exhibited the highest FFA rejection [52].

3.5 Water/oil separation

The main step of oil refining, which is the separation of water-oil mixtures, is always connected to oil exploration, production, transportation, and refining. For example, the transportation problem; the production of fluids becomes more viscous when there is water. The amount of water in crude oil increases by accelerating pumping and aging of the reservoir. Therefore, since environmental issues are a global problem, the petrochemical industry has recently made investments in the development of methods that enable the maximum amount of oil removal in the quickest amount of time, in line with the goals of the United Nations 2030 Agenda for Sustainable Development and considering their accessibility in developing countries [53]. Conventional oil removal procedures, such as gravity separation, flotation, skimming, absorption, coagulation, and centrifugation, are frequently employed to handle oily wastewater extract [54–56]. However, they have some limitations including low separation efficiency, high cost, poor recovery, excessive energy use, and easy secondary pollutant production [56, 57]. The incomplete separation of the oil/water mixtures is another significant issue with older techniques. While separating the oil-water mixtures, the water may stay in the oil, or the oil may remain in the water. Moreover, when a system has at least two immiscible liquid phases, emulsions also develop due to the employment of agents with surface-active activities during oil production and recovery [58]. So, those techniques can work well for separating immiscible oil and water combinations, but they struggle a lot to separate emulsions. More effective approaches are needed to separate the oil from the water, especially in treating emulsified oil/water combinations, particularly those with oil droplet sizes smaller than 20 μm. Surfactant-stabilized emulsions provide a substantial problem since the droplet size is so minute that it is challenging to separate using traditional techniques. Due to the ineffectiveness of conventional methodologies, researchers are looking into innovative procedures and strategies for effectively separating oil/water emulsions [59]. Those procedures are either ineffectual or cause secondary pollution and energy consumption by creating an electric field or adding chemicals to the emulsions to demulsify them [5]. Compared to those techniques, membrane filtration offers various advantages, including higher-quality permeate, the absence of excessive chemicals during filtering, which produce less waste, and a reduction in energy input [60]. UF membranes may more effectively separate well-emulsified oil droplets from water than MF membranes due to their suitable pore size of 2–50, high porosity, and high-specific surface area; nevertheless, these membranes have poor rejection and demand high operating pressure [61, 62]. Based on oil droplet size, the three most common states of an oil–water combination are free (>150 μm), dispersed (20–150 μm), and emulsified (<20 μm) [61, 63–65]. Meanwhile, the formation of water-in-oil emulsions, oil-in-water emulsions, or more complex ones (multiple emulsions), which become more stable due to the presence of different components in the oil or because water acts as emulsifiers, complicates the complete cleaning of water (Figure 4) [64, 66, 67]. It was also known that interfacial elasticity with both asphaltene aggregation state and asphaltene content is one of the factors affecting emulsion stability [8]. Moreover, the most common form of crude oil emulsions is water-in-oil emulsions, which use natural emulsifiers such as asphaltenes, waxes, resins, solids, naphthenic, aliphatic, and aromatic acids [67].

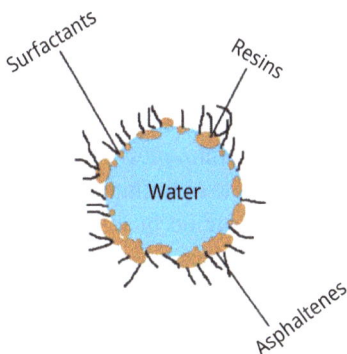

Figure 4: Example of stabilization of water droplet by emulsifiers.

Consequently, as the oil concentration in the emulsion increases, the oil droplet size can increase, resulting in larger average drop sizes [68]. Furthermore, the membrane pore size, which influences the oil-water emulsions filtration, may also vary related to the oil drop size, which depends on the oil concentration in the emulsion [68, 69]. Surface wettability (degree that a liquid spreads on a surface) and morphology are directly connected to oil/water separation effectiveness using membrane technology [70]. It is known that membranes can be either hydrophilic/underwater oleophobic or hydrophobic/oleophilic depending on the wettability (how liquids behave on a surface) of the surface or the materials used in membrane production [61, 71]. Furthermore, super-amphiphilic, super-omniphobic, and Janus membranes can be produced. Therefore, two groups of membranes can be explored related to wettability, super-hydrophobic, and super-oleophilic membranes (a surface with an oil contact angle close to 0° and a water contact angle higher than 150°), which are effective at separating water-in-oil emulsions (heavy crude oil, viscous crude oil, and diesel fuel) because of their selectivity for oil and rejection of water. Contrary to super-hydrophilic and underwater super-oleophobic membranes (an oil contact angle above 150° and a water contact angle of around 0°) that are water-selective, reject oil, and are appropriate for separating oil-in-water (gasoline, kerosene, petroleum ether) [72]. Due to their simplicity of fabrication and low cost, polymeric membranes (both hydrophilic and hydrophobic) garner the most significant interest among membranes, particularly in the oil/water separation process [54]. Compared to hydrophobic membranes, hydrophilic membranes exhibit better antifouling capabilities, allowing water droplets to pass from side to side of the membrane and not block oil droplets. In contrast, the surfaces of hydrophobic materials repel water and enable oil droplets to spread quickly, which mostly exhibit fouling problems [73]. As an example, to separate the oil-in-water emulsions without or with surfactant and extract water under slick oil, Jin et al. [74] prepared a porous super-hydrophilic/underwater super-oleophobic cotton/PA6/PAN membrane via spraying and water vapor-induced phase inversion method. The result shows the efficient separation of 99.9% for petroleum ether-in-water at a pressure of 1.6 kPa, which predicts the use in the oil-water separation industry at a large scale. Thus, prewetting permits water to quickly diffuse on the surface of the hydrophilic membrane and enter into the membrane to produce a hydration layer that can effectively prevent the adhesion of oil to the membrane surface and reduce the membrane fouling. PVDF membranes are known for their low cost and excellent chemical resistance. However, the super-hydrophobic/oleophilic nature of PVDF membrane material makes it challenging to deal with oil contamination during the separation process, and it is crucial to modify membrane surfaces to impart super-hydrophilicity and effective filtration properties. As an example, a hydrophilic PVDF@TP membrane prepared using green tea generates an excellent antifouling ability and stability, good mechanical durability, and underwater super-oleophilicity with separation efficiency (above 98%) even after 10 times of use and separation fluxes (955–1672 L/(m^2 h bar)) for various oil-in-water emulsions [75]. In another research, the adhesive-free super-hydrophobic SiO$_2$-decorated PVDF membranes fabricated by delayed phase inversion exhibit a high separation efficiency (99.95%) for oil/water mixtures and surfactant stabilized water-in-oil emulsions as well as excellent recycling performance which is over 20 cycles [76]. As a result, limited recyclability is one of the major issues with super-hydrophobic and super-oleophilic membranes. So, more efforts are needed to offer more robust and ecologically friendly surfaces to treat the oil/water mixtures [72]. Moreover, hydrophilic separation membranes can be better used to treat this oil-in-water emulsion in the oil industry. Recently, HKUST-1/PDA@SM membrane with super-hydrophilicity and underwater super-oleophobicity fabricated shows excellent mechanical stability, chemical stability, thermal stability, antifouling property, and recyclability [6].

Oil concentration and chemical composition are additional factors affecting the efficacy and choice of oil-water separation systems [68, 77]. Thus, to effectively separate oil/water mixtures, the membrane's pore diameter, porosity, and wettability should all be considered [74]. The most common techniques for fabricating oil-water separation membranes include phase separation, porous material modification, vacuum filtering, and electrospinning [57]. An asymmetric super-hydrophobic PVDF-*co*-PDMS nanofiber membrane with selective layers was prepared using coaxial electrostatic spinning to achieve rapid separation of water-oil emulsions. However, separating crude oil from water is still challenging due to its high viscosity and limited mobility [78]. Another super-hydrophobic and super-oleophilic PAN/s-kaolin composite membrane prepared via electrospinning technology which exhibits micro-nanopore channels shows a separation efficiency above 95% and high separation flux under gravity-driven separation of oil/water mixture [79]. In another study, nanofiber electrospun membranes feature excellent mechanical qualities, a porous structure, a large surface area, and a high surface-to-volume ratio and can be used under low pressure. They exhibit strong oil rejection as a result as well as high flux with excellent antifouling properties and low expenses for operation [80–82]. It was shown that electrospun membranes with hydrophilic-oleophobic surfaces have many advantages over hydrophobic-oleophilic ones, including their anticlogging abilities and capacity to repel oil from an oil/water emulsion under the force of gravity [83, 84]. The sandwich-structured PVDF/rGO/PDMS composite membrane fabricated through a two-step electrospinning technology with unprecedented electrothermal/photothermal effects shows a stable hydrophobicity and lipophilicity, which was beneficial to separate water/viscous crude oil mixtures on sunny/rainy days or at night and exhibited outstanding recyclability [85]. The impact of membrane surface charges on crude oil demulsification and fouling resistance was examined by Wu et al. [86]. For the demulsification procedure, PP-*g*-pDMAEMA and PP-*g*-pOEGMA membranes were used. In water, the membranes showed positive and negative surface charges. The usage of PP-*g*-pDMAEMA led to a 15% improvement in demulsification efficiency compared to the use of PP-*g*-pOEGMA. The authors concluded that positive membrane surface charges facilitated crude oil demulsification. However, the membrane damage that followed the demulsification was made worse.

Shakiba et al. [61] found that PAN/40%PANI@40 °C flux and oil rejection are still high even after 15 cycles of separation of surfactant-stabilized emulsions and have excellent antifouling properties. Moreover, a self-assembling super-hydrophobic mesh covered with PTFE nanofibrous can separate oil/water combinations more than 30 times without decreasing separation efficiency [87]. Even though the hydrophobicity of a membrane might contribute to the membrane fouling phenomena and reduce its life, decreasing permeate flux, increasing energy consumption and operating costs, and increasing cleaning frequency necessitate membrane replacement [65, 68, 88, 89]. Furthermore, the contact angle can define the membrane hydrophobicity and has an impact on the surface roughness, as confirmed by Mahmoudnia et al. [90] in their research experiments based on how multiwall carbon nanotube (MWVNT) and graphene oxide (GO) affected the separation of electrospun PS nanofiber membrane utilizing two crude oils. It was found that MWCNT is more effective than GO as an ideal membrane with 20 wt.% PS, 1.5 wt.% of MWCNT, and 15 mL of either heavy or light oils and may achieve a separation efficiency of nearly 90%. In another study, the comparison of PAN cast membranes, and electrospun PAN nanofiber membranes demonstrated that the concentration of the precursor decreased the hydrophilicity of the cast membranes while increasing the surface roughness and porosity of the electrospun nanofiber membranes. The electrospun PAN nanofiber membranes outperformed the cast PAN membranes in terms of flow at 250 LMH and oil rejection at 97%. Up to 10 cycles of the electrospun PAN membrane's reusability were investigated [91].

The membranes fabricated by applying fluoropolymer coatings to metal meshes using the hot wire chemical vapor deposition technique show separation efficiency over 99% of water and commercial crude oil emulsions and can be maintained by washing and reusing [92]. The best efficiency is attained with a super-hydrophobic coating (WCA = 170) and a minimum pore size (40 m), which shows that the separation efficiency rises when the membrane pore size decreases. Both the surface composition and the microstructure significantly influence the oil dewetting ability. The desorption process depends on the surface composition. To convert the solid-oil interface into the solid-water-oil interface and allow oil droplets to desorb from the surface, the surface has to be sufficiently hydrophilic. The structural characteristics have an impact on this desorption capability [93]. So, high separation efficiency, great permeation flux, and low fouling propensity may all be achieved with membranes because of the particular wettability and regulated pore size. The characteristics of the surface and how they react to water and oil are used to identify the unique wettable membranes. Thus, by adjusting the surface chemistry and texture, the unique wettable membranes can have super-hydrophilic, super-hydrophobic, super-oleophilic, and super-oleophobic characteristics [94]. Another type of membrane that was used to separate oil-water with specific underwater oleophobicity can be provided by bio-polymeric nanofibrous membranes that were inspired by fish skin. With a maximum flow of 2,086 L/m^2/h, it can separate suspended and emulsified oil with a separation efficiency of more than 99% [95]. Designing unique wettable surfaces is greatly influenced by the surface roughness at the micro/nanoscale level and the chemical composition surface. However, due to the quick oil fouling of the membrane, which results in low separation effectiveness, high costs, and the requirement for the disposal of oil-contaminated membranes, membrane-based separation approaches are to be less successful [71].

3.5.1 Antifouling

Membrane fouling is a serious obstacle to the practical application of membrane filtration caused by oil emulsions which lower the membrane stability, raise the operating expenses, and substantially impede the use of membranes [6, 77, 89, 96]. The fouling can be brought on by smaller particles depositing or adhering inside the pore wall, decreasing the pore volume and increasing membrane resistance, or forming of a cake on the membrane surface by a particle that did not enter the pores of the membrane. In this instance, the resistance provided by the membrane is added to the resistance provided by the cake [97]. Moreover, using backwashing, ultrasonic vibration, or periodic flow to remove the fouling from the membrane surface is an option. By doing this, the lifetime of the membrane can be increased [91]. For example, in the case of hydrophobic/oleophilic membranes, fouling results from water density increasing as the water barrier layer forms between light oils and the membrane. The use of membranes in industries still needs to be improved due to this problem which is affected by droplets' deformability, coalescence, membrane wetting by droplets and films, pore blockage, and intrusion by oil [98]. Recently, it has been more common to handle oil-water emulsions using separating membranes, as the oil-water emulsion is still tricky because oil can clog membrane surfaces. Filtration membranes' ability to resist oil fouling depends on the hydration layer that covers their surface and protects them from oil contact and adherence, as the oil repellency can affect membrane permeability [99]. Furthermore, as the fouling develops, the membrane begins to disintegrate and needs to be replaced, increasing the expense of treatment [60]. Many initiatives have been made to understand the causes of membrane fouling and establish strategies to evaluate and control fouling processes. Recent investigations have proven the performance of super-hydrophilic/underwater super-oleophobic membranes over oil fouling [100]. Nevertheless, most hydrophilic membranes are effective at separating wastewater containing low-viscosity oil-containing because the crude oil with a high viscosity fouls the membrane and reduces the separation's effectiveness [101, 102]. So, hydrophilic/underwater oleophobic membranes are better used for oil and water separation due to their clogging decrease. Super-hydrophilic polymeric membranes have greater antifouling performance, a higher permeate flow (4,020–24,828 L/m^2/h), and a contact angle of less than 4 up to close to zero. This encouraged the usage of traditional MSPs in various settings including large-scale institutions [60]. Hence, increasing membrane hydrophilicity is a tried-and-true method of reducing membrane fouling during oily wastewater treatment because the oil interaction with the membrane is restricted by the water adsorbed (entropic barrier) onto the membrane surface. So, incorporating hydrophilic materials or modifying the membrane surface can enhance hydrophilicity [60]. Usually, oil-water emulsion separation can be achieved by changing the wettability of the membrane and its permeability, and membranes created through surface modification with super-hydrophilic materials exhibit a unique wettability. Recently, Jiang et al. [103] reported a new method for preparing hydrogel-coated porous materials for oil/water separation. PAAm-TA coated a copper mesh substrate, PVDF membrane, and PES membrane to give it a super-hydrophilic and underwater super-oleophobic performance. The results show exceptional crude-oil antifouling and self-cleaning capabilities compared to pristine PVDF membrane, pristine PES membrane, and pristine copper mesh in which the crude oil adhered tightly on the surface of substrates.

PEEK is one of the polymers which offer excellent resistance to membranes. Under ideal testing conditions, the PANI/PEEK membrane showed remarkable anti-fouling behavior and flux recovery rate with increased pure water permeation flux to 302.5 LMH [104]. In a membrane made of polydopamine (PDA), Zhan et al. [105] inter-calated GO with halloysite nanotubes. In gravity-driven filtration, the membrane could reject more than 99% of oil and provide a high permeate flow (1,130 L/m^2 h).

Zwitterionic membranes possess a neutral charge due to both cationic and anionic groups in their structure, which helps maintain stability in various water conditions and forms a compact hydration layer [106, 107]. Using those polymers can give membrane resistance to oil adhesion underwater [108]. Recently, the zwitterionic nanogel-modified polyacrylonitrile (ZPAN) nanofibrous membrane show an increase in surface roughness with antifouling and super-wetting performance and underwater super-oleophobic properties under a wide range of salt concentrations and pH values while separating surfactant oil-in-water emulsions as the rejection reaches 99.6% [82]. In another study, thin-film composite UF membranes were prepared with selective layers consisting of zwitterion-containing hydrogels by adding hydrophilic comonomers using the interfacial initiated free radical polymerization method to boost its water permeance and fouling resistance while retaining a high-protein rejection rate [109]. A new study was done using dual-biomimetic techniques to prepare super-wetting membranes by introducing N-oxide zwitterionic (PTMAO) into the membrane surface. The results show exceptional super-hydrophilicity and underwater super-oleophobicity, excellent efficiency of oil-water separation performance, which reached 99.2% under ultra-low pressure, and outstanding stability [110].

The super-wetting gel membrane (PVDF/CD-SA) prepared with a porous and rough coral stone structure shows chemical stability while separating different oil (hexane, toluene, petroleum ether, and tri-chloromethane) water emulsions with a high separation efficiency (98%) [111]. As a result, the coral structure gave the gel-modified membrane a large emulsion flux and prevented the adhesion of the oil layer by a hydration layer formed on the surface of the PVDF/CD-SA, and then it can be restored and used. The temperature polarization effect, which can lessen permeate flow by lowering driving power, is another significant MD-related factor. By using turbulence promoters or spacers close to or on the membrane surface, it is possible to get around this problem and minimize the thickness of the thermal boundary layer. Another tactic is to add metals and nanomaterials to the polymeric matrix of the membrane to boost heat conductivity.

3.5.2 Coalescence

To destabilize emulsified oil droplets, a process known as coalescence is used. During coalescence, small, unstable oil droplets combine to generate bigger droplets that eventually float on the water surface [112]. Even though chemical demulsification is the most widely used and well-established industrial technique, determining the chemical demulsifier dosage can be difficult. For example, if the dosage is insufficient, demulsification will be ineffective, while an excessive amount of demulsifier will further stabilize the emulsions and produce secondary wastewater. Membrane demulsification is a competitive method to destabilize and coalesce the emulsified oil droplets. Oil/water emulsion may coalesce by passing through the pores of a hydrophobic membrane. Because of the contact between droplets and the membrane substance, coalescence also occurs on the membrane surface. However, the interactions between droplets and the membrane surface and the pore wall cause membrane fouling, resulting in reduced flow, poor oil-water phase separation, and shortened membrane lifespan.

4 Conclusions and prospects

Polymeric membranes have become increasingly popular in recent years for oil separation due to their operational simplicity, lack of a phase change requirement, ease of scaling up, and higher energy efficiency than distillation and adsorption. In order to improve polymeric membrane performance, intensive research is being conducted on the biodegradability of polymers to be environmentally friendly. Thus, special attention was given to studying parameters influencing membrane performance, such as fouling and wettability, to avoid the inherent drawbacks of membrane processes during oil separation. So, the conversion of present fossil-based refineries to renewable feed-stocks indicates that the "hybrid" separation system paradigm is likely to prevail for years to come, even fact that bio- and e-refineries are commonly seen as a "greenfield" for advanced separations technologies.

References

[1]   Li, S., Dong, R., Musteata, V.-E., Kim, J., Rangnekar, N. D., Johnson, J. R., Marshall, B. D., Chisca, S., Xu, J., Hoy, S., McCool, B. A., Nunes, S. P., Jiang, Z., and Livingston, A. G. Hydrophobic polyamide nanofilms provide rapid transport for crude oil separation. Science, New York, NY 2022, **377**: 1555–1561.

[2]   Bruno, N. C., Mathias, R., Lee, Y. J., Zhu, G., Ahn, Y., Rangnekar, N. D., Johnson, J. R., Hoy, S., Bechis, I., Tarzia, A., Jelfs, K. E., McCool, B. A., Lively, R., and Finn, M. G. Solution-processable polytriazoles from spirocyclic monomers for membrane-based hydrocarbon separations. Nat. Mater., 2023, **22**: 1540–1547.

[3]   Sholl, D. S., and Lively, R. P. Seven chemical separations to change the world. Nature, 2016, **532**: 435–437.

[4]   Li, B., Qi, B., Guo, Z., Wang, D., and Jiao, T. Recent developments in the application of membrane separation technology and its challenges in oil-water separation: A review. Chemosphere, 2023, **327**: 138528.

[5]   Bongiovanni, R., Nettis, E., and Vitale, A. Fluoropolymers for Oil/water Membrane Separation. In: Ameduri, B., Fomin, S. (Eds.). Opportunities for Fluoropolymers, Elsevier (Progress in Fluorine Science), 2020, pp. 209–246.

[6]   Deng, Y., Bian, H., Dai, M., Liu, X., and Peng, C. Underwater superoleophobic HKUST-1/PDA@SM membrane with excellent stability and anti-fouling performance for oil-in-water emulsion separation. J. Membr. Sci., 2023, **678**: 121655.

[7]   Altawell, N. Crude Oil. In: Najib, J. M. A., Seowou, P., Sykes, L. (Eds.). Rural Electrification, Academic Press, 2021, pp. 39–80.

[8]   Demirbas, A., and Taylan, O. Removing of resins from crude oils. Pet. Sci. Technol., 2016, **34**: 771–777.

[9]   Joshi, S. J., Al-Wahaibi, Y., and Al-Bahry, S. Biotransformation of Heavy Crude Oil and Biodegradation of Oil Pollution by Arid Zone Bacterial Strains. In: Arora, P. K. (Ed.). Microbial Metabolism of Xenobiotic Compounds, Springer Singapore, Singapore, 2019, pp. 103–122.

[10]   Gudiña, E. J., and Teixeira, J. A. Biological Treatments to Improve the Quality of Heavy Crude Oils. In: Heimann, K., Karthikeyan, O. P., Muthu, S. S. (Eds.). Biodegradation and Bioconversion of Hydrocarbons, Springer Singapore, Singapore, 2017, pp. 337–351.

[11]   Gascon, G., Vargas, V., Feo, L., Castellano, O., Castillo, J., Giusti, P., Acavedo, S., Lienemann, C.-P., and Bouyssiere, B. Size distributions of sulfur, vanadium, and nickel compounds in crude oils, residues, and their saturate, aromatic, resin, and asphaltene fractions determined by gel permeation chromatography inductively coupled plasma high-resolution mass spectrometry. Energy Fuels, 2017, **31**: 7783–7788.

[12]   Ismail, W. A., Abdul Raheem, A. S., and Bahzad, D. Biotechnological Approaches for Upgrading of Unconventional Crude Oil. In: Singh, R. P., Manchanda, G., Bhattacharjee, K., Panosyan, H. (Eds.). Microbial Syntrophy-Mediated Eco-enterprising, Academic Press, 2022, pp. 125–175.

[13]   Niyonsaba, E., Wehde, K. E., Yerabolu, R., Kilaz, G., and Kenttämaa, H. I. Determination of the chemical compositions of heavy, medium, and light crude oils by using the distillation, precipitation, fractionation mass spectrometry (DPF MS) method. Fuel, 2019, **255**: 115852.

[14]   Liao, B., Bokhary, A., Cui, L., and Lin, H. A review of membrane technology for integrated forest biorefinery. J. Membr. Sci. Res., 2017, **3**: 120–141.

[15]   Tang, X.-D., Zhou, T.-D., Li, -J.-J., Deng, C.-L., and Qin, G.-F. Experimental study on a biomass-based catalyst for catalytic upgrading and viscosity reduction of heavy oil. J. Anal. Appl. Pyrolysis, 2019, **143**: 104684.

[16]   Lively, R. P. The refinery of today, tomorrow, and the future: A separations perspective. AIChE J., 2021, **67**: e17286.

[17]   Tao, J., Li, C., Li, J., Yan, B., Chen, G., Cheng, Z., Li, W., Lin, F., and Hou, L. Multi-step separation of different chemical groups from the heavy fraction in biomass fast pyrolysis oil. Fuel Process Technol., 2020, **202**: 106366.

[18]   Caroca, E., Serrano, A., Borja, R., Jiménez, A., Carvajal, A., Braga, A. F. M., Rodriguez-Gutierrez, G., and Fermoso, F. G. Influence of phenols and furans released during thermal pretreatment of olive mill solid waste on its anaerobic digestion. Waste Manag., 2021, **120**: 202–208.

[19]   Valle, B., García-Gómez, N., Remiro, A., Gayubo, A. G., and Bilbao, J. Cost-effective upgrading of biomass pyrolysis oil using activated dolomite as a basic catalyst. Fuel Process. Technol., 2019, **195**: 106142.

[20] Biswas, B., Kumar, J., and Bhaskar, T. Advanced Hydrothermal Liquefaction of Biomass for Bio-Oil Production. In: Pandey, A., Larroche, C., Dussap, C.-G., Gnansounou, E., Khanal, S. K., Ricke, S. (Eds.). Biofuels: Alternative Feedstocks and Conversion Processes for the Production of Liquid and Gaseous Biofuels, (2nd ed.), Academic Press, 2019, pp. 245–266.

[21] Agutaya, J. K. C. N., Quitain, A. T., Kam, Y. L., Zullaikah, S., Auresenia, J., Tan, R. R., Assabumrungrat, S., and Kida, T. Chapter 7 – Hydrothermal Liquefaction of Algal Biomass to Bio-oil. In: Yusup, S., Rashidi, N. A. (Eds.). Value-Chain of Biofuels, Elsevier, 2022, pp. 159–180.

[22] Ali, H., Kansal, S. K., and Saravanamurugan, S. Upgradation of Bio-oil Derived from Various Biomass Feedstocks via Hydrodeoxygenation. In: Li, H., Saravanamurugan, S., Pandey, A., Elumalai, S. (Eds.). Biomass, Biofuels, Biochemicals, Elsevier, 2022, pp. 287–308.

[23] Chan, Y. H., Loh, S. K., Chin, B. L. F., Yiin, C. L., How, B. S., Cheah, K. W., Wong, M. K., Loy, A. C. M., Gwee, Y. L., Lo, S. L. Y., Yusup, S., and Lam, S. S. Fractionation and extraction of bio-oil for production of greener fuel and value-added chemicals: Recent advances and future prospects. Chem. Eng. J., 2020, **397**: 125406.

[24] Abdulkhani, A., Zadeh, Z. E., Bawa, S. G., Sun, F., Madadi, M., Zhang, X., and Saha, B. Comparative production of bio-oil from in situ catalytic upgrading of fast pyrolysis of lignocellulosic biomass. Energies, 2023, **16**: 2715.

[25] Chang, S. H. Bio-oil derived from palm empty fruit bunches: Fast pyrolysis, liquefaction and future prospects. Biomass and Bioenergy, 2018, **119**: 263–276.

[26] Rahman, S., Helleur, R., MacQuarrie, S., Papari, S., and Hawboldt, K. Upgrading and isolation of low molecular weight compounds from bark and softwood bio-oils through vacuum distillation. Sep. Purif. Technol., 2018, **194**: 123–129.

[27] Mostafazadeh, A. K., Solomatnikova, O., Drogui, P., and Tyagi, R. D. A review of recent research and developments in fast pyrolysis and bio-oil upgrading. Biomass Convers. Biorefin., 2018, **8**: 739–773.

[28] Brueckner, T. M., Pickup, P. G., and Hawboldt, K. A. Improvement of bark pyrolysis oil and value added chemical recovery by pervaporation. Fuel Process. Technol., 2020, **199**: 106292.

[29] Ribeiro, L. A. B., Martins, R. C., Mesa-Pérez, J. M., and Bizzo, W. A. Study of bio-oil properties and ageing through fractionation and ternary mixtures with the heavy fraction as the main component. Energy, 2019, **169**: 344–355.

[30] Aminzadeh, S., Lauberts, M., Dobele, G., Ponomarenko, J., Mattsson, T., Lindström, M. E., and Sevastyanova, O. Membrane filtration of kraft lignin: Structural characteristics and antioxidant activity of the low-molecular-weight fraction. Ind. Crops. Prod., 2018, **112**: 200–209.

[31] Polizzi, V., Servaes, K., Vandezande, P., Kouris, P. D., Panaite, A. M., Jacobs, G., Hensen, E. J. M., Boot, M. D., and Vanbroekhoven, K. Molecular weight-based fractionation of lignin oils by membrane separation technology. Holzforschung, 2020, **74**: 166–174.

[32] Rinaldi, R., Jastrzebski, R., Clough, M. T., Ralph, J., Kennema, M., Bruijnincx, P. C. A., and Weckhuysen, B. M. Paving the way for lignin valorisation: Recent advances in bioengineering, biorefining and catalysis. Angew. Chem. Int. Ed., 2016, **55**: 8164–8215.

[33] Croes, T., Dutta, A., De Bie, R., Van Aelst, K., Sels, B., and Van der Bruggen, B. Extraction of monophenols and fractionation of depolymerized lignin oil with nanofiltration membranes. Chem. Eng. J., 2023, **452**: 139418.

[34] Mostapha, M., Mohamed, M., Ameen, M., Lam, M. K., and Yusup, S. Upgrading biocrudes derived from agricultural biomass into advanced biofuels: Perspective from Malaysia. Fuel, 2022, **323**: 124300.

[35] Hajilary, N., Rezakazemi, M., and Shirazian, S. Biofuel types and membrane separation. Environ. Chem. Lett., 2019, **17**: 1–18.

[36] Galiano, F., Russo, F., Ursino, C., Castro-Muñoz, R., Criscuoli, A., and Figoli, A. Pervaporation and Membrane Distillation Technology in Biorefinery. In: Iulianelli, A., Cassano, A., Conidi, C., Petrotos, K. (Eds.). Membrane Engineering in the Circular Economy, Elsevier, 2022, pp. 251–280.

[37] Mathias, R., Weber, D. J., Thompson, K. A., Marshall, B. D., Finn, M. G., Scott, J. K., and Lively, R. P. Framework for predicting the fractionation of complex liquid feeds via polymer membranes. J. Membr. Sci., 2021, **640**: 119767.

[38] Liu, C., Dong, G., Tsuru, T., and Matsuyama, H. Organic solvent reverse osmosis membranes for organic liquid mixture separation: A review. J. Membr. Sci., 2021, **620**: 118882.

[39] Thompson, K. A., Mathias, R., Kim, D., Kim, J., Rangnekar, N., Johnson, J. R., Hoy, S. J., Bechis, I., Tarzia, A., Jelfs, K. E., McCool, B. A., Livingston, A. G., Lively, R. P., and Finn, M. G. N-Aryl Linked Spirocyclic Polymers for Membrane Separations of Complex Hydrocarbon Mixtures. Science (New York, N.Y.) 2020, **369**: 310–315.

[40] Jang, H.-Y., Johnson, J. R., Ma, Y., Mathias, R., Bhandari, D. A., and Lively, R. P. Torlon® hollow fiber membranes for organic solvent reverse osmosis separation of complex aromatic hydrocarbon mixtures. AIChE. J., 2019, **65**: e16757.

[41] Thompson, K. A., Mathias, R., Lively, R. P., and Finn, M. G. Structure–function relationships in membrane-based hydrocarbon separations using *N*-aryl-linked spirocyclic polymers. Chem. Mater., 2023, **35**(9): 3464–3469.

[42] Widodo, S., Khoiruddin, K., Ariono, D., Subagjo, S., and Wenten, I. G. Re-refining of waste engine oil using ultrafiltration membrane. J. Environ. Chem. Eng., 2020, **8**: 103789.

[43] Chisca, S., Musteata, V.-E., Zhang, W., Vasylevskyi, S., Falca, G., Abou-Hamad, E., Emwas, A.-H., Altunkaya, M., and Nunes, S. P. Polytriazole membranes with ultrathin tunable selective layer for crude oil fractionation. Science (New York, N.Y.) 2022, **376**: 1105–1110.

[44] Sharma, M., Alves, P., and Gando-Ferreira, L. M. Effect of activated carbon nanoparticles on the performance of PES nanofiltration membranes to separate kraft lignin from black liquor. J. Water Process. Eng., 2023, **52**: 103487.

[45] Wenten, I. G., Victoria, A. V., Tanukusuma, G., Khoiruddin, K., and Zunita, M. Simultaneous clarification and dehydration of crude palm oil using superhydrophobic polypropylene membrane. J. Food Eng., 2019, **248**: 23–27.

[46] Yao, X., Hou, X., and Zhang, R. Flexible and mechanically robust polyimide foam membranes with adjustable structure for separation and recovery of oil-water emulsions and heavy oils. Polymer, 2022, **250**: 124889.

[47] Liu, M., Shen, L., Wang, J., Ding, Y., Zhou, Y., and Liu, F. Continuous separation and recovery of high viscosity oil from oil-in-water emulsion through nondispersive solvent extraction using hydrophobic nanofibrous poly(vinylidene fluoride) membrane. J. Membr. Sci., 2022, **660**: 120876.

[48] Doshi, K., and Shah, S. R. Removal of phospholipids from crude edible oil by PVDF membrane. Int. J. Adv. Res. Ideas Innovations Technol., 2018, **4**(2): 2524–2529.

[49] Hou, Z., Cao, X., Cao, L., Ling, G., Yu, Z., Pang, M., Yang, P., and Jiang, S. The removal of phospholipid from crude rapeseed oil by enzyme-membrane binding. J. Food Eng., 2020, **280**: 109910.

[50] Brinkmann, T., and Filiz, V. Organic solvent nanofiltration. In: Schäefer, A. I., Fane, A. G. (Eds.) Nanofiltration. Wiley, 2021, pp. 889–932.

[51] Charanyaa, S., Vaisali, C., Belur, P. D., and Regupathi, I. Screening of polymeric membranes for membrane assisted deacidification of sardine oil. Resource-Efficient Technologies, 2016, **2**: S119–S123.

[52] Azmi, R. A., Goh, P. S., Ismail, A. F., Lau, W. J., Ng, B. C., Othman, N. H., Noor, A. M., and Yusoff, M. S. A. Deacidification of crude palm oil using PVA-crosslinked PVDF membrane. J. Food Eng., 2015, **166**: 165–173.

[53] Da Costa Cunha, G., Pinho, N. C., Alves Silva, I. A., Silva, L. S., Santana Costa, J. A., Da Silva, C. M. P., and Romão, L. P. C. Removal of heavy crude oil from water surfaces using a magnetic inorganic-organic hybrid powder and membrane system. J. Environ. Manage., 2019, **247**: 9–18.

[54] Obaid, M., Mohamed, H. O., Yasin, A. S., Yassin, M. A., Fadali, O. A., Kim, H., and Barakat, N. A. M. Under-oil superhydrophilic wetted PVDF electrospun modified membrane for continuous gravitational oil/water separation with outstanding flux. Water Res., 2017, **123**: 524–535.

[55] Al-Maas, M., Hussain, A., Minier Matar, J., Ponnamma, D., Hassan, M. K., Al Ali Al-Maadeed, M., Alamgir, K., and Adham, S. Validation and application of a membrane filtration evaluation protocol for oil-water separation. Journal of Water Process Engineering, 2021, **43**: 102185.

[56] Baig, N., and Kammakakam, I. Special wettable *Azadirachta indica* leaves like microarchitecture mesh filtration membrane produced by galvanic replacement reaction for layered oil/water separation. Chemosphere, 2023, **313**: 137544.

[57] Ren, W., Pan, J., Gai, W., Pan, X., Chen, H., Li, J., and Huang, L. Fabrication and characterization of PVDF-CTFE/SiO$_2$ electrospun nanofibrous membranes with micro and nano-rough structures for efficient oil-water separation. Sep. Purif. Technol., 2023, **311**: 123228.

[58] Faisal, W., and Almomani, F. A critical review of the development and demulsification processes applied for oil recovery from oil in water emulsions. Chemosphere, 2022, **291**: 133099.

[59] Baig, N., Abdulazeez, I., and Aljundi, I. H. Low-pressure-driven special wettable graphene oxide-based membrane for efficient separation of water-in-oil emulsions. NPJ Clean Water, 2023, **6**: 40.

[60] Amaral, M. C. S., Lebron, Y., and Moreira, V. Oily Wastewater Treatment by Membrane-assisted Technologies. In: Basile, A., Cassano, A., Rahimpour, M. R., Makarem, M. A. (Eds.). Advanced Technologies in Wastewater Treatment, Elsevier, 2023, pp. 213–255.

[61] Shakiba, M., Nabavi, S. R., Emadi, H., and Faraji, M. Development of a superhydrophilic nanofiber membrane for oil/water emulsion separation via modification of polyacrylonitrile/polyaniline composite. Polym. Adv. Technol., 2021, **32**: 1301–1316.

[62] Ge, J., Zong, D., Jin, Q., Yu, J., and Ding, B. Biomimetic and superwettable nanofibrous skins for highly efficient separation of oil-in-water emulsions. Adv. Funct. Mater., 2018, **28**: 1705051.

[63] Ahmad, T., Guria, C., and Mandal, A. A review of oily wastewater treatment using ultrafiltration membrane: A parametric study to enhance the membrane performance. J. Water Process. Eng., 2020, **36**: 101289.

[64] Dmitrieva, E. S., Anokhina, T. S., Novitsky, E. G., Volkov, V. V., Borisov, I. L., and Volkov, A. V. Polymeric membranes for oil-water separation: A review. Polymers, 2022, **14**, 980–1005.

[65] Junaidi, N. F. D., Othman, N. H., Fuzil, N. S., Mat Shayuti, M. S., Alias, N. H., Shahruddin, M. Z., Marpani, F., Lau, W. J., Ismail, A. F., and Aba, N. D. Recent development of graphene oxide-based membranes for oil–water separation: A review. Sep. Purif. Technol., 2021, **258**: 118000.

[66] Raya, S. A., Mohd Saaid, I., Abbas Ahmed, A., and Umar, A. A. A critical review of development and demulsification mechanisms of crude oil emulsion in the petroleum industry. J. Pet. Explor. Prod. Technol., 2020, **10**: 1711–1728.

[67] Faizullayev, S., Adilbekova, A., Kujawski, W., and Mirzaeian, M. Recent demulsification methods of crude oil emulsions – Brief review. J. Pet. Sci. Eng., 2022, **215**: 110643.

[68] Almojjly, A., Johnson, D., and Hilal, N. Investigations of the effect of pore size of ceramic membranes on the pilot-scale removal of oil from oil-water emulsion. J. Water Process. Eng., 2019, **31**: 100868.

[69] Kumar, S., Nandi, B., Guria, C., and Mandal, A. Oil removal from produced water by ultrafiltration using polysulfone membrane. Braz. J. Chem. Eng., 2017, **34**: 583–596.

[70] Dansawad, P., Yang, Y., Li, X., Shang, X., Li, Y., Guo, Z., Qing, Y., Zhao, S., You, S., and Li, W. Smart membranes for oil/water emulsions separation: A review. Adv. Membranes, 2022, **2**: 100039.

[71] Baig, U., Gondal, M. A., Dastageer, M. A., and Falath, W. S. Rapid fabrication of textured membrane with super-wettability using simple spray-coating of Pd-doped WO3 nanoparticles for efficient oil-water separation, Colloids and Surfaces A. Physicochem. Eng. Aspect., 2021, **609**: 125643.

[72]   Baig, N., Sajid, M., Salhi, B., and Abdulazeez, I. Special Wettable Membranes for Oil/Water
        Separations: A Brief Overview of Properties, Types, and Recent Progress. Colloids and Interfaces,
        2023, **7**: 11.
[73]   Jana, K. K., Bhowal, A., and Das, P. Nanocoated Membranes for Oil/water Separation. In: Das, P.,
        Manna, S., Pandey, J. K. (Eds.). Advances in Oil-Water Separation, Elsevier, 2022, pp. 207–230.
[74]   Jin, K., Zhao, Y., Fan, Z., Wang, H., Zhao, H., Huang, X., Hou, K., Yao, C., Xie, K., and Cai, Z. A facile and
        green route to fabricate fiber-reinforced membrane for removing oil from water and extracting
        water under slick oil. J. Hazard. Mater., 2021, **416**: 125697.
[75]   Xu, J., Cui, J., Sun, H., Wu, Y., Xue, C., Xie, A., and Li, C. Facile preparation of hydrophilic PVDF
        membrane via tea polyphenols modification for efficient oil-water emulsion separation. Colloids
        Surf. A Physicochem. Eng. Asp., 2023, **657**: 130639.
[76]   Wei, C., Dai, F., Lin, L., An, Z., He, Y., Chen, X., Chen, L., and Zhao, Y. Simplified and robust adhesive-
        free superhydrophobic SiO$_2$-decorated PVDF membranes for efficient oil/water separation.
        J. Membr. Sci., 2018, **555**: 220–228.
[77]   Tanudjaja, H. J., Hejase, C. A., Tarabara, V. V., Fane, A. G., and Chew, J. W. Membrane-based
        separation for oily wastewater: A practical perspective. Water Res., 2019, **156**: 347–365.
[78]   Cheng, X. Q., Jiao, Y., Sun, Z., Yang, X., Cheng, Z., Bai, Q., Zhang, Y., Wang, K., and Shao, L.
        constructing scalable superhydrophobic membranes for ultrafast water–oil separation. ACS
        Nano, 2021, **15**: 3500–3508.
[79]   Zhang, T., Zhang, C., Zhao, G., Li, C., Liu, L., Yu, J., and Jiao, F. Electrospun composite membrane with
        superhydrophobic-superoleophilic for efficient water-in-oil emulsion separation and oil adsorption.
        Colloids Surf. A Physicochem. Eng. Asp., 2020, **602**: 125158.
[80]   Yin, H., Zhao, J., Li, Y., Liao, Y., Huang, L., Zhang, H., and Chen, L. Electrospun SiNPs/ZnNPs-SiO$_2$/TiO$_2$
        nanofiber membrane with asymmetric wetting: Ultra-efficient separation of oil-in-water and water-
        in-oil emulsions in multiple extreme environments. Sep. Purif. Technol., 2021, **255**: 117687.
[81]   Ge, J., Zhang, J., Wang, F., Li, Z., Yu, J., and Ding, B. Superhydrophilic and underwater
        superoleophobic nanofibrous membrane with hierarchical structured skin for effective oil-in-water
        emulsion separation. J. Mater. Chem. A, 2017, **5**: 497–502.
[82]   Zang, L., Zheng, S., Wang, L., Ma, J., and Sun, L. Zwitterionic nanogels modified nanofibrous
        membrane for efficient oil/water separation. J. Membr. Sci., 2020, **612**: 118379.
[83]   Shakiba, M., Abdouss, M., Mazinani, S., and Reza Kalaee, M. Super-hydrophilic electrospun PAN
        nanofibrous membrane modified with alkaline treatment and ultrasonic-assisted PANI in-situ
        polymerization for highly efficient gravity-driven oil/water separation. Sep. Purif. Technol., 2023,
        309: 123032.
[84]   Justin Koh, J., Pang, P., Chakraborty, S., Kong, J., Sng, A., Anukunwithaya, P., Huang, S., Koh, X. Q.,
        Thenarianto, C., Thitsartan, W., Daniel, D., and He, C. Presence, origins and effect of stable surface
        hydration on regenerated cellulose for underwater oil-repellent membranes. J. Colloid. Interface.
        Sci., 2023, **635**: 197–207.
[85]   Zhou, J., Li, K., Chen, Z., Su, X., Xie, H., Yao, L., Wu, Y., Zhang, X., Chen, L., Wu, X., and
        Wu, W. Hydrophobic sandwich-like PVDF/rGO/PDMS composite membrane with fast
        electrothermal/photothermal response for crude oil removal. J. Environ. Chem. Eng., 2023, **11**:
        109840.
[86]   Wu, J., Wei, W., Li, S., Zhong, Q., Liu, F., Zheng, J., and Wang, J. The effect of membrane surface
        charges on demulsification and fouling resistance during emulsion separation. J. Membr. Sci., 2018,
        563: 126–133.
[87]   Chen, C., Du, C., Weng, D., Mahmood, A., Feng, D., and Wang, J. Robust superhydrophobic
        polytetrafluoroethylene nanofibrous coating fabricated by self-assembly and its application for
        oil/water separation. ACS Appl. Nano Mater., 2018, **1**: 2632–2639.

[88]   Moser, P. B., Bretas, C., Paula, E. C., Faria, C., Ricci, B. C., Cerqueira, A. C. F. P., and
       Amaral, M. C. S. Comparison of hybrid ultrafiltration-osmotic membrane bioreactor and conventional
       membrane bioreactor for oil refinery effluent treatment. Chem. Eng. J., 2019, **378**: 121952.
[89]   Tanudjaja, H. J., and Chew, J. W. Assessment of oil fouling by oil-membrane interaction energy
       analysis. J. Membr. Sci., 2018, **560**: 21–29.
[90]   Mahmoudnia, S., Gharehaghaji, A. A., Bahrami, S. H., and Razbin, M. Investigating the effect of
       carbon nanotubes and graphene oxide on water/crude oil separation efficiency in polystyrene
       nanofiber membrane. J. Text. Inst., 2023, 1–9.
[91]   Diwan, T., Abudi, Z. N., Al-Furaiji, M. H., and Nijmeijer, A. A competitive study using electrospinning
       and phase inversion to prepare polymeric membranes for oil removal. Membranes, 2023, **13**: 474.
[92]   Melnik, A., Bogoslovtseva, A., Petrova, A., Safonov, A., and Markides, C. N. Oil-water separation on
       hydrophobic and superhydrophobic membranes made of stainless steel meshes with
       fluoropolymer coatings. Water, 2023, **15**: 1346.
[93]   Chen, Q., Wang, T., Tang, L., Zeng, Z., and Zhu, B. Study on the structure-activity relationship
       between oil dewetting self-cleaning and surface morphology for crude oil pollution treatment and
       crude oil/water separation. J. Environ. Chem. Eng., 2023, **11**: 109092.
[94]   Chen, C., Weng, D., Mahmood, A., Chen, S., and Wang, J. Separation mechanism and construction of
       surfaces with special wettability for oil/water separation. ACS Appl. Mater. Interfaces, 2019, **11**:
       11006–11027.
[95]   Oh, S., Bang, J., Jin, H.-J., and Kwak, H. W. Green fabrication of underwater superoleophobic
       biopolymeric nanofibrous membranes for effective oil–water separation. Adv. Fiber Mater., 2023, **5**:
       603–616.
[96]   Wang, Y., He, Y., Yu, J., Li, H., Li, S., and Tian, S. A freestanding dual-cross-linked membrane with
       robust anti-crude oil-fouling performance for highly efficient crude oil-in-water emulsion
       separation. Colloids Surf. A Physicochem. Eng. Asp., 2022, **654**: 130117.
[97]   Bardhan, A., Akhtar, A., and Subbiah, S. Microfiltration and Ultrafiltration Membrane Technologies.
       In: Nayak, S. K., Dutta, K., Gohil, J. M. (Eds.). Advancement in Polymer-Based Membranes for Water
       Remediation, Elsevier, 2022, pp. 3–42.
[98]   Tummons, E., Han, Q., Tanudjaja, H. J., Hejase, C. A., Chew, J. W., and Tarabara, V. V. Membrane
       fouling by emulsified oil: A review. Sep. Purif. Technol., 2020, **248**: 116919.
[99]   Li, S., Zhang, L., Tian, S., He, Y., and Guo, X. Mineralized cupric phosphate/alginate gel alternately
       multilayer-wrapped nanofibrous membrane with robust anti-crude oil pollution for oily wastewater
       purification. J. Membr. Sci., 2023, **669**: 121280.
[100] Qu, M., He, D., Luo, Z., Wang, R., Shi, F., Pang, Y., Sun, W., Peng, L., and He, J. Facile preparation of a
       multifunctional superhydrophilic PVDF membrane for highly efficient organic dyes and heavy metal
       ions adsorption and oil/water emulsions separation. Colloids Surf. A Physicochem. Eng. Asp., 2022,
       637: 128231.
[101] Zhou, W., Fang, Y., Li, P., Yan, L., Fan, X., Wang, Z., Zhang, W., and Liu, H. Ampholytic chitosan/
       alginate composite nanofibrous membranes with super anti-crude oil-fouling behavior and
       multifunctional oil/water separation properties. ACS Sustain. Chem. Eng., 2019, **7**: 15463–15470.
[102] Peng, Y., Wen, G., Gou, X., and Guo, Z. Bioinspired fish-scale-like stainless steel surfaces with robust
       underwater anti-crude-oil-fouling and self-cleaning properties. Sep. Purif. Technol., 2018, **202**:
       111–118.
[103] Jiang, Y., Xian, C., Xu, X., Zheng, W., Zhu, T., Cai, W., Huang, J., and Lai, Y. Robust PAAm-TA hydrogel
       coated PVDF membranes with excellent crude-oil antifouling ability for sustainable emulsion
       separation. J. Membr. Sci., 2023, **667**: 121166.
[104] Lin, Z., Cao, N., Li, C., Sun, R., Li, W., Chen, L., Sun, Y., Zhang, H., Pang, J., and Jiang, Z. Micro-
       nanostructure tuning of PEEK porous membrane surface based on PANI in-situ growth for
       antifouling ultrafiltration membranes. J. Membr. Sci., 2022, **663**: 121058.

[105] Zhan, Y., He, S., Wan, X., Zhao, S., and Bai, Y. Thermally and chemically stable poly(arylene ether nitrile)/halloysite nanotubes intercalated graphene oxide nanofibrous composite membranes for highly efficient oil/water emulsion separation in harsh environment. J. Membr. Sci., 2018, **567**: 76–88.

[106] Zhang, J., Zhang, F., Wang, A., Lu, Y., Li, J., Zhu, Y., and Jin, J. Zwitterionic nanofibrous membranes with a superior antifouling property for gravity-driven crude oil-in-water emulsion separation. Langmuir, 2019, **35**: 1682–1689.

[107] Han, L., Tan, Y. Z., Xu, C., Xiao, T., Trinh, T. A., and Chew, J. W. Zwitterionic grafting of sulfobetaine methacrylate (SBMA) on hydrophobic PVDF membranes for enhanced anti-fouling and anti-wetting in the membrane distillation of oil emulsions. J. Membr. Sci., 2019, **588**: 117196.

[108] Zhu, Y., Wang, J., Zhang, F., Gao, S., Wang, A., Fang, W., and Jin, J. Zwitterionic nanohydrogel grafted PVDF membranes with comprehensive antifouling property and superior cycle stability for oil-in-water emulsion separation. Adv. Funct. Mater., 2018, **28**: 1804121.

[109] Aguiar, A. O., Yi, H., and Asatekin, A. Fouling-resistant membranes with zwitterion-containing ultra-thin hydrogel selective layers. J. Membr. Sci., 2023, **669**: 121253.

[110] Yang, J., Lin, L., Tang, F., and Zhao, J. Superwetting membrane by co-deposition technique using a novel N-oxide zwitterionic polymer assisted by bioinspired dopamine for efficient oil–water separation. Sep. Purif. Technol., 2023, **318**: 123965.

[111] Ma, W., Lin, L., Yang, J., Liu, Z., Li, X., Xu, M., Li, X., Wang, C., Xin, Q., and Zhao, K. Coral stone-inspired superwetting membranes with anti-fouling and self-cleaning properties for highly efficient oil–water separation. J Ind Eng Chem, 2023, **120**: 231–243.

[112] Chiam, C.-K., and Sarbatly, R. Mechanisms of oil coalescing through water-filled and air-filled membrane pores. Mater. Today Proc., 2023.

Mahmoud A. Abdulhamid

Chapter 11
Advancement and recent development of polymer membranes for organic solvent nanofiltration

Abstract: Organic solvent nanofiltration (OSN) membranes require stable polymers with good solubility to facilitate membrane fabrication. The resultant membrane should also exhibit excellent solvent resistance. Therefore, development of stable polymer membranes for OSN applications has seen dramatic progress in the last decade. Various approaches and membrane fabrication techniques have been employed to develop thin film composite (TFC) membranes, integrally skinned asymmetric membranes (ISA), hollow fiber membranes (HF), and carbon molecular sieve membranes. This chapter will discuss the different types of polymers used to develop OSN membranes and their performance. We will also discuss the various preparation methods and characterization techniques.

Keywords: polymer membranes, separation, organic solvent nanofiltration

1 Introduction

Filtration has been the keystone of various manufacturing processes, from water treatment to the pharmaceutical and chemical sectors [1, 2]. In its basic form, filtration utilizes a porous medium to separate particles or substances from a fluid. While rudimentary filtration techniques serve their purpose in many applications, the drive for efficiency, selectivity, and environmental sustainability has pushed the boundaries of filtration technology. At the forefront of this innovation is nanofiltration, a specialized filtration process that operates at the molecular level to achieve high levels of separation [3]. The beginning of nanofiltration has unlocked novel potentials, allowing for high-selectivity separation of molecules based on size and chemistry. As ecological concerns rise and rigorous quality controls are imposed on manufacturing,

Acknowledgment: This work was supported by the College of Petroleum Engineering and Geoscience (CPG), King Fahd University of Petroleum and Minerals (KFUPM).

Mahmoud A. Abdulhamid, Sustainable and Resilient Materials Lab, Center for Integrative Petroleum Research (CIPR), College of Petroleum Engineering and Geosciences (CPG), King Fahd University of Petroleum and Minerals (KFUPM), Dhahran, 31261, Saudi Arabia,
e-mail: Mahmoud.abdulhamid@kfupm.edu.sa

https://doi.org/10.1515/9783110796032-011

there is an increasing need for efficient and reliable separation processes. Organic Solvent Nanofiltration (OSN) has gained significant attention due to its ability to operate in nonaqueous systems. OSN stands out as an essential subset of membrane-based technologies because of its versatility in solvent selection, process controllability, and lower energy requirements compared to traditional separation methods like distillation or crystallization. The essence of OSN lies in the membrane's ability to selectively permeate target solutes or solvents through a semipermeable barrier, effectively separating them from their mixtures [3].

Nanofiltration is a pressure-driven membrane separation process that fills the gap between ultrafiltration and reverse osmosis. Whereas ultrafiltration is proficient at removing macromolecules and particles ranging from 0.01 to 0.1 microns, and reverse osmosis excels at desalting solutions, nanofiltration operates in the middle ground. It is effective for separating small organic molecules and ions, usually in the range of 0.001–0.01 microns. This makes nanofiltration an incredibly versatile tool, capable of separating solutes based on size, charge, and even chemical affinity to the membrane. Nanofiltration showed various benefits over other separation techniques in terms of: i) Energy efficiency – while the traditional separation techniques like distillation and crystallization often require significant amounts of energy for phase changes or heat input [4], nanofiltration operates at lower pressures compared to reverse osmosis and generally requires less energy, making it an economically viable option; ii) Selectivity: the high selectivity of nanofiltration membranes allows for precise separations, be it in organic solvent nanofiltration, desalting of dairy products, or the selective removal of contaminants from water [5]; iii) Operational flexibility: unlike processes that require specific operating conditions such as temperature or pH, nanofiltration can often be adapted easily to existing process streams, making it highly versatile [3]; iv) Environmental Impact: nanofiltration allows for the recycling of solvents and other materials, thereby reducing waste and lowering the environmental impact of industrial operations [1]; v) Scalability: with advancements in membrane technology, nanofiltration systems have become increasingly scalable, allowing for easier adoption in both small-scale specialized applications and large-scale industrial processes [6]; vi) Integration with other processes: due to its moderate operating conditions, nanofiltration can be easily integrated into existing process chains, often serving as a pretreatment or posttreatment step, enhancing the efficiency of the overall system.

Polymer membranes have attracted attention due to their excellent stability, flexibility, and tunability [7, 8]. The membrane separation performance can be tailored and fine-tuned by varying polymer structures or functional groups [9, 10]. This flexibility credits polymer membranes over ceramic and inorganic membranes, which suffer from rigidity and scalability issues. Polymer membranes have been fabricated either as flat-sheet or hollow fiber. The flat-sheet membranes are usually fabricated as: i) dense membranes by solvent evaporation [11], ii) thin-film composite (TFC) membranes on top of nonwoven support by interfacial polymerization [12], iii) integrally skinned asymmetric membranes by phase inversion in various coagulation baths [13], or iv) carbon molecular sieve membranes by thermal annealing at elevated temperature [14]. Solution-processable polymers tend to dissolve in organic solvents based on the chemical structure of the polymer. For instance, the polymer contains polar groups such as -OH, -COOH, -NR$_2$, and -SO$_3$H, demonstrating good solubility in polar aprotic solvents such as *N*-methyl-2-pyrrolidone (NMP), dimethylformamide (DMF) and tetrahydrofuran (THF) [10, 11, 15, 16]. However, for OSN membranes, the polymer needs to be soluble in organic solvents and the membrane needs to be solvent-resistant simultaneously, making it challenging to develop this type of material. Therefore, different approaches were considered to enhance the stability of the membrane after fabrication, including thermal annealing and crosslinking [14, 17].

Poly(ether-ether-ketone) (PEEK) and polybenzimidazole (PBI) were investigated for OSN due to their high thermal and chemical stability [18–20]. For instance, PEEK demonstrates excellent solvent resistance ability; however, it only dissolves in harsh acids such as sulfuric acid, which is considered as toxic and harmful to the environment. Abdulhamid et al. has developed a new family of PEEK named intrinsically microporous PEEK (iPEEK) [21] to overcome the solubility issue of PEEK. iPEEK contains kinked units like triptycene, Tröger's base, and spirobisindane, which help increase the polymer porosity and enhance the solubility. The iPEEK-based membranes showed excellent performance, two- to sixfold higher, compared to sulfonated PEEK (SPEEK). Furthermore, the performance of iPEEK was found to be tuned by controlling polymer molecular weight [22], in which the lower molecular weight iPEEK displayed honey-like morphology, and controlling crosslinker type and size [17]. Further development on PEEK was achieved by controlling the functional group existing in the polymer chain from polar to nonpolar groups. For instance, Alqadhi et al. explored the effect of SO$_2$ and CH$_3$ functionalities on the solubility, stability, and performance of the resulting membranes (Figure 1) [23]. Interestingly, the presence of the SO$_2$ group enhanced the permeance of the polar solvents due to polar–polar interactions. This chapter aims to comprehensively understand polymer membranes used in OSN, discussing their types, fabrication techniques, characterization methods, and applications. We will also delve into the current research trends and future prospects of polymer membranes in OSN, providing the reader with an in-depth overview of this dynamic field.

2 Basic principles of nanofiltration

Nanofiltration is a membrane-based separation technique that bridges the gap between ultrafiltration and reverse osmosis, offering a suite of unique capabilities in terms of molecular separation and treatment efficiency. Understanding its basic principles is crucial for optimizing its performance and extending its range of applications. This section will explore the key principles underlying nanofiltration, including the mechanisms of separation, driving forces, and challenges specific to organic sol-

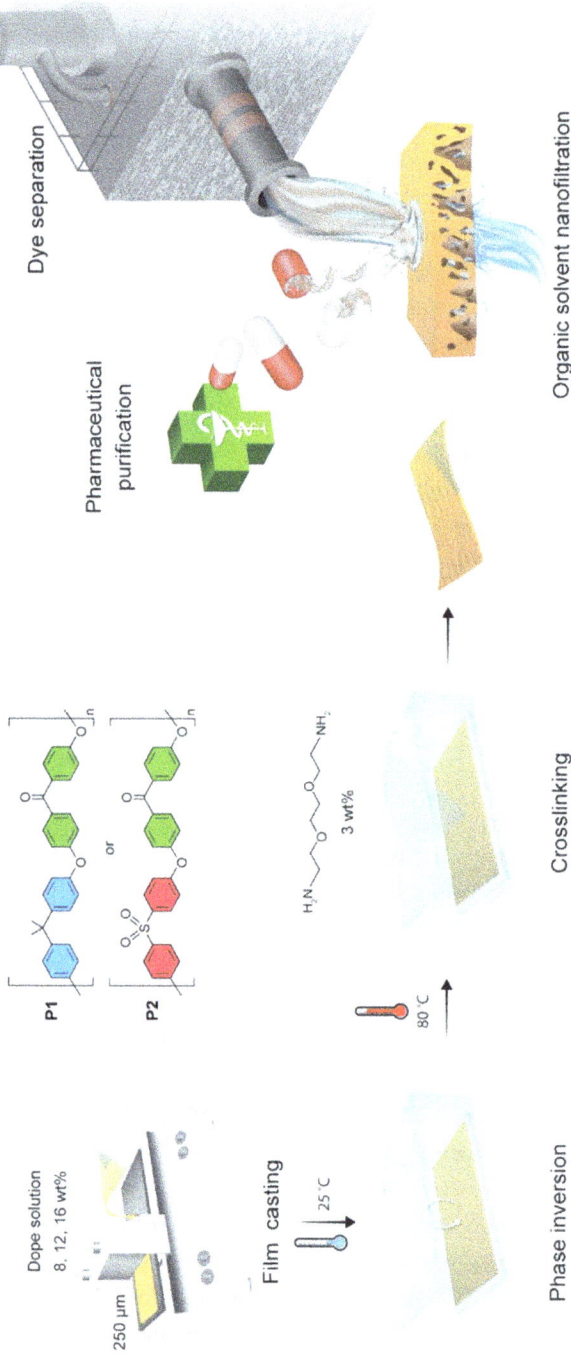

Figure 1: Schematic illustration for the preparation, crosslinking and application of PEEK-based membranes [23].

vents. Mechanisms of separation: In OSN separation, various crucial factors play important roles in determining the permeability and selectivity; these factors are as follows:

1. **Size exclusion**: Sieving or size exclusion is a critical factor in the primary mechanisms of nanofiltration. The membrane acts as a physical barrier that allows smaller molecules or ions to pass through while retaining larger molecules. This is particularly useful for separating solutes of close molecular weights. A molecular weight cutoff (MWCO) will be determined for each membrane and can vary from >100 g/mol to <2,000 g/mol.

2. **Charge exclusion**: Some polymer membranes are developed with charges species, allowing them to selectively allow or reject ions based on their charge. This is essential in applications such as water softening or selective ion removal. Furthermore, incorporating cations or anions into the membrane allows them to be utilized for anion-exchange membranes or cation-exchange membranes.

3. **Solvent interaction**: It is found that different solvents can have different binding energies with the membranes based on the polymer structure. Simulations have proven that the presence of different functionalities in the polymer structure can directly affect the performance. Additionally, some membranes are designed to have affinity to specific solvents, providing an additional level of selectivity. This is crucial in applications like organic solvent nanofiltration.

4. **Adsorption**: In some instances, the membrane may also exhibit adsorptive properties, where molecules are temporarily held on the surface of the membrane, affecting their transmission rates, permeance, and selectivity. The adsorption properties could also be affected by polymer and membrane porosity.

It is worth mentioning that various driving forces should be considered for OSN applications:

1. **Pressure**: Nanofiltration is primarily a pressure-driven process. The transmembrane pressure propels the solvent and certain solutes through the membrane while retaining others. Operating pressures generally range from 5 to 30 bar, depending on the application and membrane type. Pressure-independent studies are usually reported in publications to demonstrate the effect of pressure on permeability and selectivity.

2. **Concentration gradient**: Osmotic pressure due to concentration differences across the membrane can also serve as a driving force, although this is generally less significant than the applied mechanical pressure.

3. **Temperature**: Higher temperatures can improve solute diffusivity and membrane permeability, although this may also bring about challenges such as membrane fouling and degradation. Additionally, membrane thermal stability plays a vital role as the membrane should be thermally stable to resist the high temperature during the experiments.

However, with the current development in the OSN membranes, several challenges remain in focus of researchers to address. Below are the main challenges that we are continuously trying to overcome:

1. **Chemical stability**: Dealing with organic solvents requires excellent membrane chemical stability, which becomes a critical factor. Many polymer membranes may swell, dissolve, or degrade in the presence of certain organic solvents, requiring careful membrane selection. Therefore, methods like crosslinking and thermal annealing have shown good improvement in membrane stability but significantly affect the performance [14, 17]. Therefore, further development is needed.

2. **Membrane fouling**: Organic solvents often contain complex mixtures of solutes that can foul the membrane, reducing efficiency over time and affecting the membrane's performance. Therefore, developing membranes with antifouling properties is essential for OSN. Moreover, several approaches like coating were used to introduce an extra thin layer with antifouling properties to protect the membranes without affecting its separation performance.

3. **Viscosity**: The permeance through the membranes depends also on the viscosity of the solution. Organic solvents may have different viscosities compared to water, affecting flow rates and filtration performance. Specialized pump systems and membranes may be required to handle higher-viscosity fluids for efficient separation.

4. **Solvent recovery**: Some applications are required to recover the solvent due to its high. Therefore, membranes must be chosen for their separation capabilities and how effectively they can recover solvents for reuse.

Understanding the basic principles of nanofiltration is pivotal for harnessing its full capabilities. It operates through a combination of size exclusion, charge exclusion, solvent interaction, and, at times, adsorption. Being a pressure-driven process, its operating conditions can be fine-tuned to achieve desired separation outcomes. However, additional challenges such as chemical compatibility, fouling, and solvent recovery must be considered when organic solvents are involved. By comprehending these fundamental aspects, researchers and engineers can make informed decisions on membrane selection, system design, and operating conditions, thereby broadening the applicability and improving the efficiency of nanofiltration processes.

3 Polymer science fundamentals in nanofiltration

The performance of the nanofiltration membranes is deeply affected by polymer science, as a majority of the membranes used are polymer-based. The use of polymers in nanofiltration offers several advantages, such as ease of fabrication, tunable properties, and cost-effectiveness. In this section, we delve into the fundamental aspects of polymer science that are critical to understanding and optimizing polymer membranes for nanofiltration applications.

Polymer structure and properties: as the performance of the OSN membranes remained affected by the chemical structure of the final polymer, several key factors need to be taken into consideration for polymer development:

1. **Monomers and polymers**: Polymers are made from small units called monomers; the final polymer structure, properties, stability, and performance will totally depend on the nature of the monomers. For instance, Abdulhamid et al. demonstrated the importance of the monomer unit on the performance and stability of the final polymer membrane product. Intrinsically microporous PEEK (iPEEK) was prepared by utilizing unique monomers known as kinked monomers [21]. Varying the monomers from triptycene to spirobisindane and Tröger's base has resulted in significantly different performance (Figure 2). Furthermore, Alqadhi et al. reported the development of two novel PEEK polymers by fixing the difluoro ketone unit and varying the diol monomer. The two diol monomers used exhibited different functional groups (SO_2 and CH_3) that affect the membrane performance [23]. Therefore, a careful selection of the monomer units is required.

2. **Molecular weight**: The length of a polymer chain is usually expressed in terms of molecular weight. It is known that higher molecular weight polymers often lead to membranes with excellent mechanical strength and increased stability. Recently, Abdulhamid et al. reported the effect of polymer molecular weight on the membrane morphology for iPEEK-based membranes. For instance, it was found that the lower molecular weight polymer demonstrates honey-like surface morphology, which exceeded the performance of the higher molecular weight polymer by 26% (Figure 3) [22].

3. **Crosslinking**: Crosslinking was used due to the need for chemically stable membranes. The crosslinked membranes showed higher thermal and chemical stability relative to the pristine polymer. Moreover, it was found that crosslinker type and chemical structure can affect the crosslinked membranes' performance. For instance, Abdulhamid et al. demonstrated that the crosslinker size and functional group position (*ortho* vs. *para*) could improve the flux by 120% (Figure 4) [17]. Nonetheless, excessive crosslinking can make the membranes too rigid and decrease permeability.

4. **Functional groups**: The presence of specific functional groups on a polymer can impart certain selective interactions with solutes, enhancing the membrane's ability to discriminate between different molecules. Additionally, the presence of functional groups can improve the polymer's solubility in various organic solvents. For instance, having -OH or -COOH groups can make the polymer better soluble in polar aprotic solvents, while alkyl groups can improve the solubility of chlorinated solvents. Additionally, excess polar groups can lead to enhanced solubility on water, making the fabrication of OSN membranes more difficult.

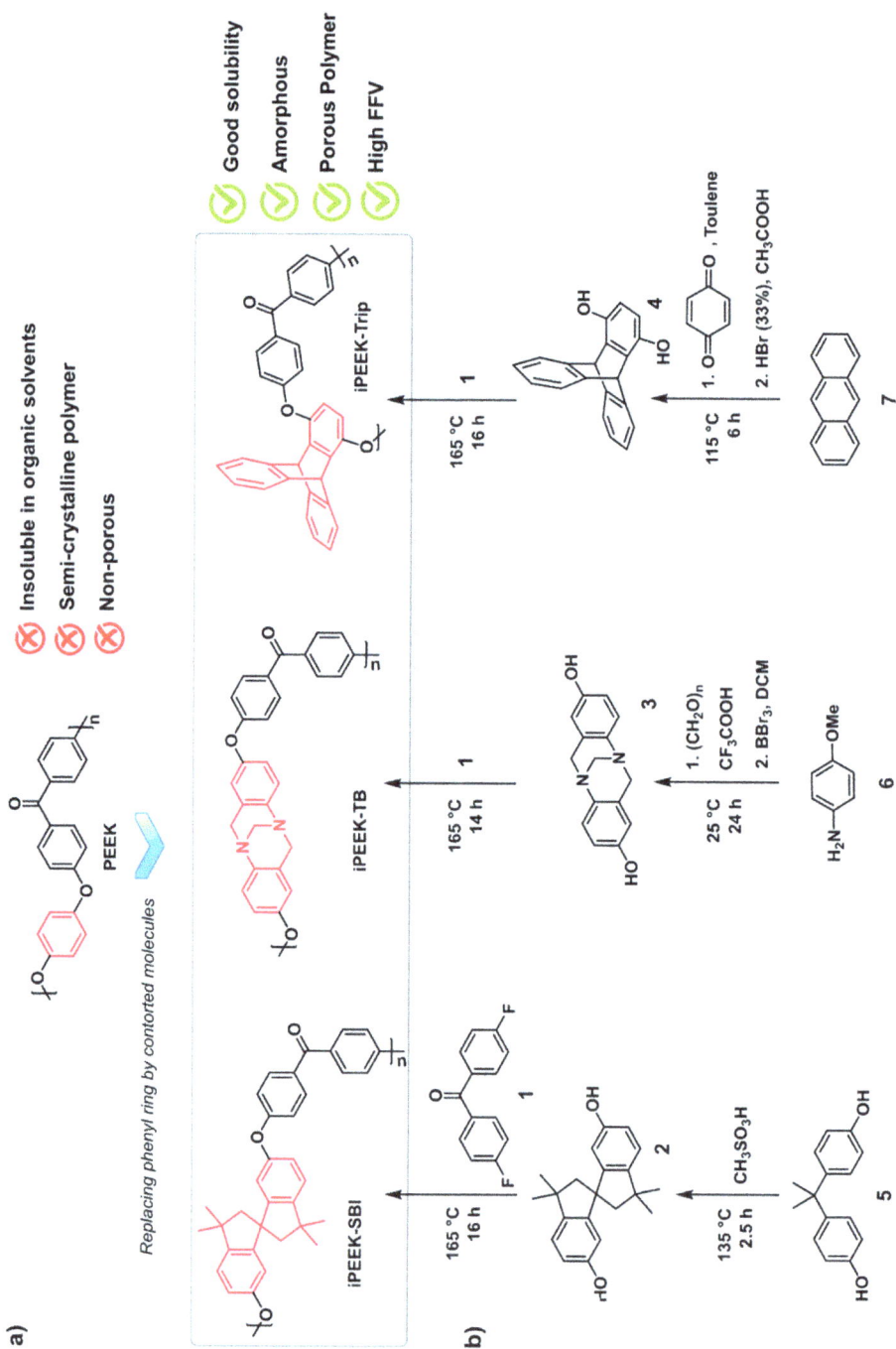

Figure 2: Synthesis of iPEEK polymers with three different kinked structures and comparison of their properties to commercial PEEK [21].

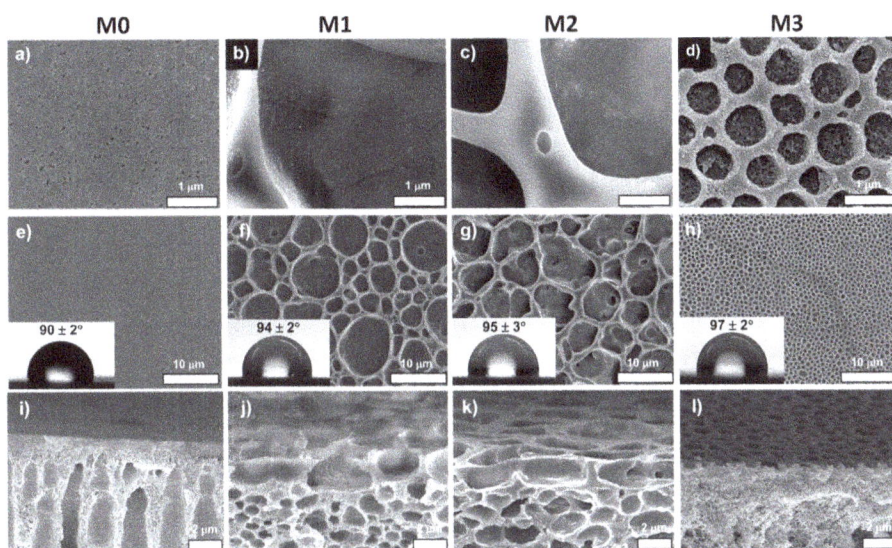

Figure 3: SEM images of IPEEK-SBI membranes. The images show the effect of polymer molecular weight on the surface and cross-section morphology; a–h represent surface morphology, while i–l represent cross-section morphology. M0 is a membrane made from high molecular weight polymer and M1–M3 are made from low molecular weight polymer at different dope solution concentration [22].

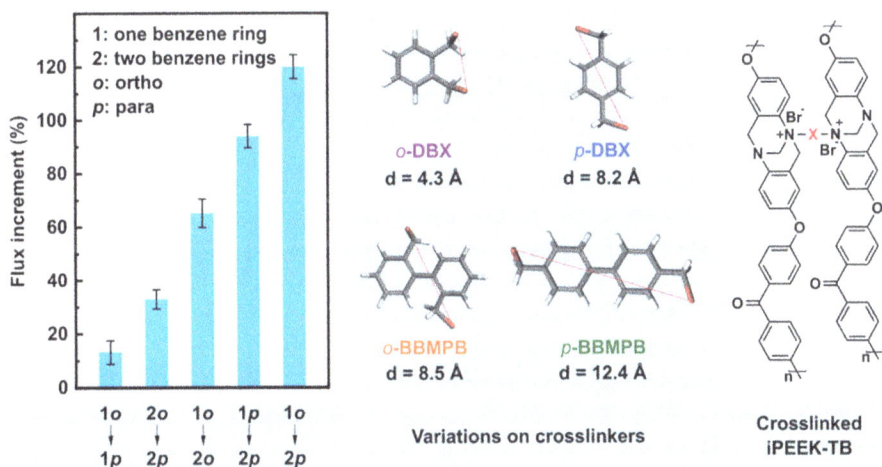

Figure 4: Schematic illustration of the effect of crosslinker on the OSN membranes performance [17].

5. **Tunability**: An important feature that should be also considered for polymer development is its potential for tunability to fine-tune its performance towards OSN applications. Membranes can be modified by posttreatment through techniques like grafting, introducing new functional groups that can tailor the membrane's selectivity and per-

meability, crosslinking or thermal annealing in which the polymer can be converted to carbon molecular sieve membranes. Polymers with high carbon content are suitable for carbonization. Still, the molecular weight and mechanical strength of the initial polymer play an essential role in fabricating the carbonized membranes.

6. **Cost-effectiveness**: Compared to ceramic or metallic membranes, polymer membranes are generally more cost-effective to manufacture and scale, making them a popular choice for industrial applications. Nonetheless, some monomers are quite expensive and challenging to scale up; thus, researchers should consider developing polymers from cheap starting materials.

It is worth mentioning that polymer membranes were fabricated in different forms including: i) flat-sheet membranes using various polymers such as polyimides [24], polysulfone [25], cellulose acetate [26], Polyvinylidene Fluoride (PVDF) [27] and PEEK [18]; ii) hollow-fiber membranes using polymers like polyacrylonitrile (PAN) [28] and polypropylene (PP) [29]; iii) TFC membranes using polymers such as polyamides [30] and polymers of intrinsic microporosity [31]; and iv) carbon molecular sieve membranes using polyimide of intrinsic microporosity [14]. The intersection of polymer science and nanofiltration technology represents a dynamic and evolving field. Understanding the fundamental aspects of polymers, from their molecular structure to their mechanical and chemical properties, is crucial for designing and optimizing membranes. The choice of polymer can dramatically affect not only the effectiveness of the separation but also the long-term stability and operational cost of the nanofiltration system. By considering these principles and criteria, one can make more informed decisions in selecting and designing polymer membranes, enabling more efficient and targeted separations in nanofiltration applications.

4 Fabrication techniques of polymer membranes for nanofiltration

The performance and efficacy of polymer membranes in nanofiltration are strongly influenced by their fabrication methods. Various techniques exist to produce membranes with specific structures, porosities, and surface functionalities to meet the demands of a wide array of applications. This section will delve into the prevalent fabrication techniques, their underlying principles, and their impact on membrane properties.

Phase inversion
1. **Non-solvent-induced phase separation (NIPS)**: This is one of the most common techniques for fabricating integrally skinned asymmetric membranes. A polymer solution is cast onto a support and then immersed in a non-solvent bath (water or

alcohol generally). The interaction between the solvent and non-solvent leads to phase separation, creating a porous or finger-like structure [32].

2. **Thermally induced phase separation (TIPS)**: In this method, a polymer is dissolved in a solvent that is miscible at high temperatures but immiscible at low temperatures. Cooling the solution results in phase separation and membrane formation [33].

3. **Vapor-induced phase separation (VIPS)**: Similar to NIPS, but instead of immersing the polymer solution in a non-solvent, it is exposed to non-solvent vapor. This allows for more control over the rate of phase separation [34].

Interfacial polymerization

1. **TFC membranes**: Often used to fabricate high-performance composite membranes, this technique involves forming a thin, selective polymer layer on porous support by interfacial polymerization of two reactive monomers, mainly diamines and acyl chloride-based monomers [35].

2. **Dual-layer composite membranes**: In this variation, two different types of monomers are sequentially polymerized at the interface, resulting in a dual-layer membrane with custom properties [36].

Electrospinning

1. **Nanofibrous membranes**: A high-voltage electric field is used to draw ultrafine fibers from a polymer solution; the fibers are then collected to form a porous membrane. This technique allows for the fabrication of membranes with high surface area-to-volume ratios [37].

Solvent casting

1. **Homogeneous membranes**: A polymer solution is poured into a flat petri dish and evaporated, leaving a thin and homogeneous polymer film behind. This method is straightforward but offers less control over membrane structure compared to phase inversion or electrospinning. However, the membranes resulting from this method offer low permeability for liquids but high gas permeability with good selectivity [38].

Advanced techniques

1. **3D printing**: Additive manufacturing allows for the precise design of membrane architectures, offering unparalleled control over pore size and distribution [39].

Considerations in fabrication

1. **Material selection**: The choice of polymer and solvent/non-solvent systems is crucial for achieving desired membrane properties.

2. **Process parameters**: Factors like temperature, concentration, and casting speed can significantly influence membrane structure and performance.

3. **Scale-up**: The transition from laboratory-scale to industrial-scale production often presents challenges in maintaining consistent membrane quality.

The fabrication techniques employed in making polymer membranes for nanofiltration directly impact their structural and functional properties. Whether one opts for phase inversion methods for their versatility or specialized techniques like electrospinning for high-performance applications, understanding the fundamentals of these fabrication methods is essential for both the design and selection of membranes. Advances in fabrication technologies continue to expand the boundaries of what is possible in membrane performance, heralding new opportunities for innovation in nanofiltration applications.

5 Membrane characterization techniques

Understanding a polymer membrane's performance, reliability, and limitations in nanofiltration applications requires thorough characterization. Several techniques are available to provide quantitative and qualitative insights into polymer membrane properties, such as morphology, chemical composition, mechanical strength, and transport phenomena. This section will delve into the key characterization methods and discuss how they contribute to a comprehensive understanding of membrane performance.

Morphological characterization
1. **Scanning electron microscopy (SEM)**: One of the most widely used techniques, SEM provides high-resolution images of the membrane surface and cross-section, offering insights into pore size, distribution, and surface roughness.
2. **Atomic force microscopy (AFM)**: For even higher resolution and three-dimensional surface profiling, AFM can be employed. It can be beneficial for characterizing the nano-scale features of the membrane. The membrane roughness and high-resolution surface morphology can be seen.

Chemical characterization
1. **Fourier-transform infrared spectroscopy (FTIR)**: This technique identifies specific functional groups and chemical bonds in the polymer, which can be crucial for understanding chemical resistance, compatibility, and potential leaching issues. It is also a valuable tool to characterize the membranes after crosslinking or any other post-modification treatment.
2. **X-ray photoelectron spectroscopy (XPS)**: Provides detailed information on the elemental composition of the membrane surface and can be essential for assessing surface modifications or coatings.

Mechanical properties
1. **Tensile testing**: Offers quantitative data on the mechanical strength of the membrane, including tensile strength and elongation at break, which are important for applications under high-pressure conditions.
2. **Dynamic mechanical analysis (DMA)**: Used to assess the membrane's mechanical properties as a function of temperature, which can be crucial for high-temperature applications.

Transport properties
1. **Permeability testing**: Evaluates how effectively the membrane allows the passage of solvents and solutes, often assessed using a dead-end or cross-flow filtration setup.
2. **Salt rejection or molecular sieving tests**: Measures the membrane's selectivity for different solutes, providing valuable data for applications like desalination or organic solvent nanofiltration.

Advanced characterization techniques
1. **Contact angle measurements**: Assessing the wettability of the membrane can be crucial for applications that involve hydrophilic or hydrophobic interactions. It is also affected by the membrane roughness, which may increase as the roughness increases the contact angle.
2. **Thermal analysis**: Techniques like thermogravimetric analysis (TGA) or Differential Scanning Calorimetry (DSC) can offer insights into the thermal stability and transitions of the polymer membrane.

Characterizing a polymer membrane for nanofiltration applications is a multifaceted process that requires a synergistic application of various techniques. Each characterization method offers a unique set of information that collectively contributes to holistic understanding of the membrane's performance under different operational conditions. From basic tests that provide morphological data to advanced in situ analysis for real-time insights, membrane characterization remains a cornerstone in nanofiltration system design, selection, and optimization processes.

6 Other applications of polymer membranes in nanofiltration

The diverse portfolio of polymer membranes developed for nanofiltration is reflected in their equally diverse range of applications. Whether the aim is drinking water purification or separating complex organic mixtures, polymer membranes play a crucial role. This section focuses on the key sectors where nanofiltration has proven invaluable, detailing the particular challenges and benefits associated with each.

Water treatment
1. **Contaminant removal**: Specialized membranes can selectively remove trace organic compounds, heavy metals, and other potential pollutants, improving water quality for drinking and industrial purposes.

Pharmaceutical industry
1. **Solvent recovery**: Organic solvent nanofiltration (OSN) can recover and purify valuable pharmaceutical manufacturing solvents, offering economic and environmental benefits.
2. **Biomolecule separation**: Specially designed membranes can separate complex biomolecules such as proteins and enzymes, aiding drug discovery and manufacturing processes.

Food and beverage
1. **Concentration processes**: Nanofiltration membranes can concentrate fruit juices, dairy products, and other beverages without affecting their nutritional quality owing to low operating temperatures.
2. **Sugar refinement**: The technology aids in the purification and decolorization of liquid sugar solutions, producing high-quality syrups.

Chemical industry
1. **Petrochemical separations**: High-performance membranes can be used to separate hydrocarbons, contributing to more efficient and environmentally friendly petrochemical processes.
2. **Acid/base recovery**: Membranes with high chemical resistance can recover acids and bases from industrial waste streams, enabling their reuse and reducing disposal costs.

Textile industry
1. **Dye recovery**: Membranes can recover and concentrate dyes from wastewater streams, reducing environmental impact and enabling reuse.
2. **Water recycling**: High-flux membranes can purify and recycle water used in textile manufacturing, reducing water consumption and discharge of polluted water.

Environmental applications
1. **Wastewater treatment**: Beyond the traditional use of removing contaminants, nanofiltration membranes can also recover valuable materials from wastewater, such as nutrients in agricultural runoff.

2. **Landfill leachate**: Specialized membranes can treat landfill leachate, reducing its environmental impact.

Cutting-edge applications
1. **Battery recycling**: Innovations in membrane technology contribute to recycling materials from lithium-ion batteries.
2. **Carbon capture**: Emerging research suggests that nanofiltration may play a role in capturing and separating carbon dioxide from industrial emissions.
3. **Oil-fractionation**: Fractioning oil using polymer membranes is a new promising technology that shows great potential for separating different fractions of oil, which could be an alternative technology for distillation.

The applications of polymer membranes in nanofiltration are as vast as they are impactful, serving industries from water treatment to pharmaceuticals to environmental protection. With ongoing advances in membrane fabrication and characterization techniques, the range of applications continues to expand, offering new avenues for efficiency, sustainability, and innovation. As the understanding of membrane science grows, so does the potential to address some of the most pressing challenges of our time through nanofiltration technologies. Continued exploration and adoption of polymer membranes in these applications present opportunities for technological advancement and solutions to critical global challenges such as water scarcity and environmental pollution.

7 State-of-the-art advancements in polymer membranes for OSN nanofiltration

The exploration of advanced materials has been pivotal in the evolution of membrane technologies. One such remarkable material is poly(ether-ether-ketone), commonly known as PEEK. This case study aims to illuminate the trajectory of PEEK in the domain of polymer membranes for organic solvent nanofiltration, underlining its state-of-the-art developments and real-world applications. PEEK is a high-performance thermoplastic polymer known for its exceptional mechanical properties, thermal stability, chemical resistance, semicrystalline morphology and insolubility in organic solvents [40]. Therefore, due to its high solvent resistance, PEEK was utilized to develop membranes for OSN. The early method for fabricating PEEK-based membranes was dissolving them in harsh acidic solutions, mainly sulfuric acid and methanesulfonic acid, to reduce their sulfonation rate [18]. Thus, PEEK membranes with a low degree of sulfonation (<4%) were prepared and tested for OSN. The sulfonated PEEK (SPEEK) demonstrated acetonitrile permeance of 1.95 L m^{-2} h^{-1} bar^{-1} and styrene dimer rejection of 56%. Further development on PEEK was reported by Sandra et al. where a modified PEEK precursor was prepared, followed by membrane casting, and then

PEEK regeneration by removing the prior modification [41]. This approach allowed preparation of PEEK without sulfonating it. Later, Abdulhamid et al. developed a new strategy to synthesize solution-processable PEEK by replacing the simple aromatic ring with kinked monomers that improve solubility and significantly enhance porosity. The obtained intrinsically microporous PEEK (iPEEK) demonstrated a microporosity feature with excellent solubility in polar aprotic solvents and high performance, exceeding the SPEEK two- to sixfold [21]. The development of PEEK has continued to improve the developed membranes in thermal and chemical stability to mechanical stability and long-term performance. Thus, since membranes tend to foul, swell, and lose their mechanical properties, several methods were reported to improve the stability of PEEK-based membranes for OSN. For instance, the long-term stability of iPEEK was investigated for six months of aging and the membranes remained stable with no notable reduction in performance [22]. Furthermore, crosslinking of PEEK membranes shows great potential in improving the solvent resistance in polar aprotic solvent [17]. Interestingly, the crosslinking was found to play a critical role in improving the chemical and mechanical stability and controlling the flux and rejection. Surprisingly, selecting crosslinkers with specific sizes and group distribution at *meta* or *para* positions can significantly affect the performance. Abdulhamid et al. enhanced the iPEEK-based membrane by 120% by using a large crosslinker with crosslinking groups at *para* positions.

Furthermore, varying the functional groups on the PEEK chains can also improve the solubility and processability of the polymers. Recently, Alqadhi et al. showed the effect of incorporating SO_2 group into the PEEK chain on the polymer solubility in polar aprotic and green solvents [38]. Additionally, it was found that the presence of a polar group (SO_2) can directly affect the performance of polar solvents due to polar-polar interactions between the polymer chains and the solvents. The journey of PEEK from a high-performance polymer to a transformative material in nanofiltration demonstrates the immense potential in material science advancements for membrane technologies. The state-of-the-art developments in PEEK membranes are not merely academic exercises but are increasingly being translated into real-world solutions. As the science behind PEEK evolves, it offers a promising pathway for addressing some of the most critical filtration challenges across multiple industries. As the field of nanofiltration continues to advance, materials like PEEK will likely remain at the forefront of research and development, pushing the boundaries of what is achievable in terms of efficiency, selectivity, and sustainability.

8 Future prospects in polymer membranes for nanofiltration

The evolving landscape of nanofiltration presents a fertile ground for innovation, especially with the advent of advanced polymer materials, computational tools, and engineering strategies. While current technologies have made substantial progress in addressing the numerous challenges in various industries, there is still sufficient room for improvement and innovation. This section outlines the future prospects and avenues for research and development in the field of polymer membranes for nanofiltration.

Material advancements

1. **Bio-derived polymers**: Developing and incorporating bio-based and biodegradable polymers could offer a more sustainable path for membrane technologies. It is worth mentioning that the utilization of bio-based materials for nanofiltration is growing as it provides cheap and biodegradable membranes.
2. **Multilayered structures**: Future designs could incorporate multilayered or gradient structures that offer optimized performance across different parameters, such as selectivity and flux. It can also be afforded by blending polymers that have high permeability with polymers that have high selectivity to fine-tune the performance and achieve a good combination of permeability and selectivity.

Computational modeling

1. **Predictive analytics**: Machine learning and AI-based models could significantly streamline the process of membrane design by predicting performance metrics and offering data-driven insights.
2. **Molecular dynamics simulations (MDS)**: Advanced simulations can offer microscopic insights into membrane behavior, enabling more targeted engineering approaches. Some papers report the interaction between solvents and polymers and their functional group to understand the structure/property relationship better. Nonetheless, more studies are needed and more understanding of using MDS will be helpful to achieve better membranes.

Fabrication techniques

1. **3D Printing**: This could allow for rapid prototyping and customization of membrane structures, even at the industrial scale.
2. **Green manufacturing**: Environmentally-friendly solvents and processes can lower the ecological footprint of membrane fabrication.

Internet of Things (IoT)

1. **Smart membranes**: Integrating sensors and connectivity could make membranes part of the IoT ecosystem, enabling remote monitoring and control.

2. **Automated systems**: AI-driven automation could manage and optimize large-scale filtration systems in real time, minimizing human error and operational costs.

Regulatory changes

1. **Standardization**: As technologies advance, new standards and regulations will likely be introduced to ensure safety, efficiency, and environmental compliance.
2. **Policy-driven adoption**: Governmental incentives and mandates could accelerate the adoption of advanced nanofiltration technologies in various sectors.

Global challenges

1. **Water scarcity**: Increasing focus will likely be placed on developing membranes that can efficiently handle seawater desalination and wastewater treatment.
2. **Environmental remediation**: The escalating global pollution crisis will require advanced air and water purification filtration solutions.

The future of polymer membranes in nanofiltration is brimming with possibilities. With ongoing research focusing on material science, computational modeling, and engineering innovation, the next generation of membranes could offer unprecedented performance and versatility. Collaborative efforts between academia, industry, and governmental bodies will likely catalyze these advancements, shaping a future where nanofiltration technologies are indispensable in addressing some of the most pressing global challenges. As we move forward, it is clear that the evolution of polymer membranes for nanofiltration is a critical facet of technological progress, promising transformative solutions that have the potential to impact humanity on a global scale.

9 Conclusions and recommendations

The narrative of polymer membranes in nanofiltration is a merger of scientific curiosity, engineering innovation, and practical utility. Throughout this chapter, we have traversed the complexities of polymer science, the intricacies of membrane fabrication, and the expansive scope of applications that range from industrial to environmental. As we conclude, it is imperative to encapsulate key findings and extend recommendations for future progress in this interdisciplinary field.

9.1 Conclusions

1. **Material science**: Polymers like poly (ether-ether-ketone) (PEEK) and others have established themselves as versatile and robust materials for constructing mem-

branes with high performance in various scenarios. The chemistry has offered a few potential solutions to tailor polymer membranes and enhance their stability and performance.

2. **Fabrication technologies**: Advances in fabrication techniques like phase inversion, electrospinning, and 3D printing have paved the way for membranes with complex architectures and enhanced functionalities. Additionally, developing new fabrication techniques will improve the fabrication scalability, reduce the cost, and increase energy efficiency.

3. **Characterization**: Modern analytical techniques have become instrumental in understanding the physicochemical properties of membranes, thereby guiding iterative design improvements. Techniques like SEM, AFM, ToF-SIMs, TEM, and XPS have showed great potential in characterizing the membranes and giving insights into the surface and cross-section morphology and chemistry.

4. **Applications**: The widespread adoption of polymer membranes in sectors like water treatment, chemical industries, and environmental protection testifies to their efficacy and adaptability. The OSN membranes are nanofiltration membranes and can be applied for water treatment applications like water/oil separation, dye and heavy metal removal.

5. **Future prospects**: With a shift towards sustainable materials, computational modeling, and IoT integration, the future is ripe with opportunities for transformational impacts. Developing sustainable polymers or utilizing the current polymers such as chitosan and cellulose to make stable membranes is ongoing. The sustainable membranes will offer sustainable solutions for membrane development and utilization, which lead to a decrease in the CO_2 emission and energy used. Furthermore, utilizing computational modeling allows better understanding of the structure/property relationships and development of specific criteria for polymers to achieve the best performance.

9.2 Recommendations

1. **Interdisciplinary collaboration**: A multidisciplinary approach involving chemists, engineers, and computational scientists can accelerate the pace of innovation.

2. **Focus on sustainability**: Introducing bio-derived and recyclable materials should be a priority to address environmental concerns.

3. **Invest in research and development**: Both private and public sectors should allocate more resources for the R&D of advanced membrane materials and technologies.

4. **Adopt standardization**: An industry-wide adoption of standardized testing and evaluation methods can help objectively assess and compare various membrane technologies.

5. **Policy framework**: Regulatory bodies should formulate policies that encourage the adoption of advanced filtration technologies, especially in critical areas like water scarcity and pollution control.
6. **Public awareness**: Initiatives should be taken to educate the public and industry stakeholders about the advantages and potential of nanofiltration technologies to drive demand and adoption.
7. **Scalability**: Research should focus on lab-scale efficacy and the scalability of the technology for real-world industrial applications.

References

[1] Abdulhamid, M. A., and Muzamil, K. Recent progress on electrospun nanofibrous polymer membranes for water and air purification: A review. Chemosphere, 2023, **310**. https://doi.org/10.1016/j.chemosphere.2022.136886.

[2] Buonomenna, M. G., and Bae, J. Organic solvent nanofiltration in pharmaceutical industry. Sep. Purif. Rev., 2015, **44**(2): 157–182. https://doi.org/10.1080/15422119.2014.918884.

[3] Marchetti, P., Jimenez Solomon, M. F., Szekely, G., and Livingston, A. G. Molecular separation with organic solvent nanofiltration: A critical review. Cheme. Rev., 2014, **114**(21): 10735–10806. https://doi.org/10.1021/cr500006j.

[4] Diawara, C. K. Nanofiltration process efficiency in water desalination. Sep. Purif. Rev., 2008, **37**(3): 302–324. https://doi.org/10.1080/15422110802228770.

[5] Zhang, H., He, Q., Luo, J., Wan, Y., and Darling, S. B. Sharpening nanofiltration: Strategies for enhanced membrane selectivity. ACS Appl. Mater. Interfaces, 2020, **12**(36): 39948–39966. https://doi.org/10.1021/acsami.0c11136.

[6] See Toh, Y. H., Lim, F. W., and Livingston, A. G. Polymeric membranes for nanofiltration in polar aprotic solvents. J. Memb. Sci., 2007, **301**(1–2): 3–10. https://doi.org/10.1016/j.memsci.2007.06.034.

[7] Ulbricht, M. Advanced functional polymer membranes. Polymer (Guildf)., 2006, **47**(7): 2217–2262. https://doi.org/10.1016/j.polymer.2006.01.084.

[8] Topuz, F., Abdellah, M. H., Budd, P. M., and Abdulhamid, M. A. Advances in Polymers of Intrinsic Microporosity (PIMs)-based materials for membrane, environmental, catalysis, sensing and energy applications. Polym. Rev., 2023. https://doi.org/10.1080/15583724.2023.2236677.

[9] Ma, X., Abdulhamid, M. A., and Pinnau, I. Design and synthesis of polyimides based on carbocyclic pseudo-Tröger's base-derived dianhydrides for membrane gas separation applications. Macromolecules, 2017, **50**(15): 5850–5857. https://doi.org/10.1021/acs.macromol.7b01054.

[10] Ma, X., Abdulhamid, M., Miao, X., and Pinnau, I. Facile synthesis of a hydroxyl-functionalized Tröger's base diamine: A new building block for high-performance polyimide gas separation membranes. Macromolecules, 2017, **50**(24): 9569–9576. https://doi.org/10.1021/acs.macromol.7b02301.

[11] Abdulhamid, M. A., Ma, X., Miao, X., and Pinnau, I. Synthesis and characterization of a microporous 6FDA-polyimide made from a novel carbocyclic pseudo Tröger's base diamine: Effect of bicyclic bridge on gas transport properties. Polymer (Guildf)., 2017, **130**: 182–190. https://doi.org/10.1016/j.polymer.2017.10.017.

[12] Park, S. H., Yang, C., Ayaril, N., and Szekely, G. Solvent-resistant thin-film composite membranes from biomass-derived building blocks: Chitosan and 2,5-furandicarboxaldehyde. ACS Sustain. Chem. Eng., 2022, **10**(2): 998–1007. https://doi.org/10.1021/acssuschemeng.1c07047.

[13]   Obidara, T. O., Azeem, M. A., Lawal, D. U., Alfaraj, M. A., Abdulhamid, M. A., and Baroud, T. N. Novel hexa-fluorinated intrinsically porous polyimide membranes for the desalination of high saline water by air-gap membrane distillation. Desalination, 2023, **566**. https://doi.org/10.1016/j.desal.2023. 116948.

[14]   Abdulhamid, M. A., Hardian, R., and Szekely, G. Carbon molecular sieve membranes with integrally skinned asymmetric structure for organic solvent nanofiltration (OSN) and organic solvent reverse osmosis (OSRO). Appl. Mater. Today, 2022, **28**. https://doi.org/10.1016/j.apmt.2022.101541.

[15]   Abdulhamid, M. A., Genduso, G., Ma, X., and Pinnau, I. Synthesis and characterization of 6FDA/3,5-diamino-2,4,6-trimethylbenzenesulfonic acid-derived polyimide for gas separation applications. Sep. Purif. Technol., 2021, **257**. https://doi.org/10.1016/j.seppur.2020.117910.

[16]   Abdulhamid, M. A., Genduso, G., Wang, Y., Ma, X., and Pinnau, I. Plasticization-resistant carboxyl-functionalized 6FDA-polyimide of intrinsic microporosity (PIM-PI) for membrane-based gas separation. Ind. Eng. Chem. Res., 2020, **59**(12): 5247–5256. https://doi.org/10.1021/acs.iecr.9b04994.

[17]   Abdulhamid, M. A., Hardian, R., and Szekely, G. Waltzing around the stereochemistry of membrane crosslinkers for precise molecular sieving in organic solvents. J. Memb. Sci., 2021, **638**. https://doi.org/10.1016/j.memsci.2021.119724.

[18]   da Silva Burgal, J., Peeva, L. G., Kumbharkar, S., and Livingston, A. Organic solvent resistant poly (ether-ether-ketone) nanofiltration membranes. J. Memb. Sci., 2015, **479**: 105–116. https://doi.org/ 10.1016/j.memsci.2014.12.035.

[19]   da Silva Burgal, J., Peeva, L., and Livingston, A. Towards improved membrane production: Using low-toxicity solvents for the preparation of PEEK nanofiltration membranes. Green Chem., 2016, **18**(8): 2374–2384. https://doi.org/10.1039/c5gc02546j.

[20]   Fei, F., Cseri, L., Szekely, G., and Blanford, C. F. Robust covalently crosslinked polybenzimidazole/ graphene oxide membranes for high-flux organic solvent nanofiltration. ACS Appl. Mater. Interfaces, 2018, **10**(18): 16140–16147. https://doi.org/10.1021/acsami.8b03591.

[21]   Abdulhamid, M. A., Park, S. H., Vovusha, H., Akhtar, F. H., Ng, K. C., Schwingenschlögl, U., and Szekely, G. Molecular engineering of high-performance nanofiltration membranes from intrinsically microporous poly(ether-ether-ketone). J. Mater. Chem. A, 2020, **8**(46): 24445–24454. https://doi.org/ 10.1039/d0ta08194a.

[22]   Abdulhamid, M. A., Park, S. H., Zhou, Z., Ladner, D. A., and Szekely, G. Surface engineering of intrinsically microporous poly(ether-ether-ketone) membranes: From flat to honeycomb structures. J. Memb. Sci., 2021, **621**. https://doi.org/10.1016/j.memsci.2020.118997.

[23]   Alqadhi, N., Abdellah, M. H., Nematulloev, S., Mohammed, O. F., Abdulhamid, M. A., and Szekely, G. Solution-processable poly(ether-ether-ketone) membranes for organic solvent nanofiltration: From dye separation to pharmaceutical purification. Sep. Purif. Technol., 2024, **328**: 125072. https://doi.org/10.1016/j.seppur.2023.125072.

[24]   Han, R., Xie, Y., and Ma, X. Crosslinked P84 copolyimide/MXene mixed matrix membrane with excellent solvent resistance and permselectivity. Chin. J. Chem. Eng., 2019, **27**(4): 877–883. https://doi.org/10.1016/j.cjche.2018.10.005.

[25]   Liu, Q., Wu, X., and Zhang, K. Polysulfone/polyamide-SiO2 composite membrane with high permeance for organic solvent nanofiltration. Membranes (Basel), 2018, **8**: 4. https://doi.org/ 10.3390/membranes8040089.

[26]   Abdellah, M. H., Oviedo, C., and Szekely, G. Controlling the degree of acetylation in cellulose-based nanofiltration membranes for enhanced solvent resistance. J. Memb. Sci., 2023, **687**: 122040. https://doi.org/10.1016/j.memsci.2023.122040.

[27]   Kang, G. D., and Cao, Y. M. Application and modification of poly(vinylidene fluoride) (PVDF) membranes – A review. J. Memb. Sci., 2014, **463**: 145–165. https://doi.org/10.1016/j.memsci.2014. 03.055.

[28] Qin, J. J., Cao, Y. M., Li, Y. Q., Li, Y., Oo, M. H., and Lee, H. Hollow fiber ultrafiltration membranes made from blends of PAN and PVP. Sep. Purif. Technol., 2004, **36**(2): 149–155. https://doi.org/10.1016/S1383-5866(03)00210-7.

[29] Lv, Y., Yu, X., Jia, J., Tu, S. T., Yan, J., and Dahlquist, E. Fabrication and characterization of superhydrophobic polypropylene hollow fiber membranes for carbon dioxide absorption. Appl. Energy, 2012, **90**(1): 167–174. https://doi.org/10.1016/j.apenergy.2010.12.038.

[30] Jimenez Solomon, M. F., Bhole, Y., and Livingston, A. G. High flux hydrophobic membranes for organic solvent nanofiltration (OSN)-Interfacial polymerization, surface modification and solvent activation. J. Memb. Sci., 2013, **434**: 193–203. https://doi.org/10.1016/j.memsci.2013.01.055.

[31] Li, S., Zhang, R., Yao, Q., Su, B., Han, L., and Gao, C. High flux thin film composite (TFC) membrane with non-planar rigid twisted structures for organic solvent nanofiltration (OSN). Sep. Purif. Technol., 2022, **286**. https://doi.org/10.1016/j.seppur.2022.120496.

[32] Garcia, J. U., Iwama, T., Chan, E. Y., Tree, D. R., Delaney, K. T., and Fredrickson, G. H. Mechanisms of asymmetric membrane formation in nonsolvent-induced phase separation. ACS Macro Lett., 2020, **9**(11): 1617–1624. https://doi.org/10.1021/acsmacrolett.0c00609.

[33] Li, D., Krantz, W. B., Greenberg, A. R., and Sani, R. L. Membrane formation via thermally induced phase separation (TIPS): Model development and validation. J. Memb. Sci., 2006, **279**(1–2): 50–60. https://doi.org/10.1016/j.memsci.2005.11.036.

[34] Ismail, N., Venault, A., Mikkola, J. P., Bouyer, D., Drioli, E., and Tavajohi Hassan Kiadeh, N. Investigating the potential of membranes formed by the vapor induced phase separation process. J. Memb. Sci., 2020, **597**: 117601. https://doi.org/10.1016/j.memsci.2019.117601.

[35] Lau, W. J., Ismail, A. F., Misdan, N., and Kassim, M. A. A recent progress in thin film composite membrane: A review. Desalination, 2012, **287**: 190–199. https://doi.org/10.1016/j.desal.2011.04.004.

[36] Sun, S. P., Chan, S. Y., and Chung, T. S. A slow-fast phase separation (SFPS) process to fabricate dual-layer hollow fiber substrates for thin-film composite (TFC) organic solvent nanofiltration (OSN) membranes. Chem. Eng. Sci., 2015, **129**: 232–242. https://doi.org/10.1016/j.ces.2015.02.043.

[37] Feng, C., Khulbe, K. C., and Matsuura, T. Recent progress in the preparation, characterization, and applications of nanofibers and nanofiber membranes via electrospinning/interfacial polymerization. J. Appl. Polym. Sci., 2010, **115**(2): 756–776. https://doi.org/10.1002/app.31059.

[38] Alqadhi, N., Abdellah, M. H., Nematulloev, S., Mohammed, O. F., Abdulhamid, M. A., and Szekely, G. Solution-processable poly(ether-ether-ketone) membranes for organic solvent nanofiltration: From dye separation to pharmaceutical purification. Sep. Purif. Technol., 2024, **328**: 125072. https://doi.org/10.1016/j.seppur.2023.125072.

[39] Tijing, L. D., Dizon, J. R. C., Ibrahim, I., Nisay, A. R. N., Shon, H. K., and Advincula, R. C. 3D printing for membrane separation, desalination and water treatment. Appl. Mater. Today, 2020, **18**: 100486. https://doi.org/10.1016/j.apmt.2019.100486.

[40] Hendrix, K., Koeckelberghs, G., and Vankelecom, I. F. J. Study of phase inversion parameters for PEEK-based nanofiltration membranes. J. Memb. Sci., 2014, **452**: 241–252. https://doi.org/10.1016/j.memsci.2013.10.048.

[41] Aristizábal, S. L., Chisca, S., Pulido, B. A., and Nunes, S. P. Preparation of PEEK membranes with excellent stability using common organic solvents. Ind. Eng. Chem. Res., 2020, **59**(12): 5218–5226. https://doi.org/10.1021/acs.iecr.9b04281.

Rifan Hardian, Diana G. Oldal, Zulfida Mohamad Hafis Mohd Shafie,
Mahmoud A. Abdulhamid, Gyorgy Szekely

Chapter 12
Advanced membranes from interpenetrating polymer networks

Abstract: The entanglement of two or more polymer networks can be conceptualized in such a way that they are concatenated without the presence of chemical bonds binding these two polymers to each other. This concept has been achieved in interpenetrating polymer networks (IPNs), which exhibit various unique properties, such as mechanical robustness, chemical stability, and tunable property. Accordingly, owing to these distinctive properties, IPNs have emerged as suitable candidates for various applications. To control the properties of IPNs, the use of two polymer networks has proven to be an effective method. In this chapter, IPNs are defined, and their types and fabrication processes are described. Further, the utilization of IPNs in the fabrication of advanced membrane and its applications in gas separation, water purification, nonaqueous separation, and batteries are elaborated. Last, the summary of the current state of IPNs-based membranes along with future outlooks are discussed.

Keywords: interpenetrating polymer networks (IPNs), membrane, gas separation, water purification, nanofiltration, battery

1 Introduction

Interpenetrating polymer networks (IPNs) are systems consisting of two crosslinked polymers that are physically intertwined, but not chemically linked [1]. IPNs differ from polymer complex and graft copolymers, which consist of chemical bonds and/or a low degree of crosslinking. IPNs can be defined based on the architectures/crosslinking of the macromolecules and their fabrication process. Based on the fabrication processes,

Rifan Hardian, Diana G. Oldal, Gyorgy Szekely, Advanced Membranes and Porous Materials Center, Physical Science and Engineering Division, King Abdullah University of Science and Technology (KAUST), Thuwal, 23955-6900, Saudi Arabia
Zulfida Mohamad Hafis Mohd Shafie, School of Chemistry, Chemical Engineering and Biotechnology, Nanyang Technological University, 62, Nanyang Drive, 637459 Singapore; School of Chemical Engineering, Universiti Sains Malaysia, Penang, Malaysia; Laboratoire Réactions & Génie des Procédés, Université de Lorraine, Nancy, France
Mahmoud A. Abdulhamid, Sustainable and Resilient Materials Lab, Center for Integrative Petroleum Research (CIPR), College of Petroleum Engineering and Geosciences (CPG), King Fahd University of Petroleum and Minerals (KFUPM), Dhahran, 31261, Saudi Arabia

https://doi.org/10.1515/9783110796032-012

IPNs can be classified as simultaneous IPNs, in which two monomers are simultaneously mixed and crosslinked, and sequential IPNs, in which the two monomers are sequentially mixed and crosslinked [2]. Based on the type of crosslinking, IPNs can be classified as semi- and full IPNs, depending on whether one or both of the monomers are crosslinked [2]. To achieve phase separation, IPNs may undergo nucleation-growth or decomposition kinetics. Nucleation-growth kinetics involves the generation of spheres of the second phase within the matrix of the first phase, whereas decomposition kinetics involves the production of interconnected cylinders of the second phase within the matrix of the first phase. The IPNs generated by these processes comprises a linear polymer entangled with a crosslinked polymer network through a combination of phase separation and entrapped molecular-level mixing.

Although the constituent polymers in IPNs are incompatible and there is negligible bonding between the two polymers, phase segregation is restrained by the formation of networks between the two polymers. Hence, the mixing of these two polymers results in the homogenization of their properties; thus, only one glass transition temperature is generally expected for IPNs. Commonly, the glass transition temperature range of IPNs is broader than that of the neat constituent polymers, and it occurs at temperatures intermediate to the glass transition temperature of the pure components.

Generally, IPNs are efficient and effective for developing porous templates for various applications. For example, a previous study employed semi-IPNs (sIPNs) as porogen, which can be easily extracted by selective degradation after generating the required porous template [1]. Porous polymers can be developed by removing the linear non-crosslinked polymer from the polymer matrix. Grand et al. [3, 4] reported the synthesis of sIPNs porous polymer using a linear polylactic acid (PLA) and crosslinked polymethylmethacrylate (PMMA). Consequently, porous templates were fabricated by the selective removal of PLA from the media, while preventing the collapse of the crosslinked network by controlling crosslinking and polymer miscibility.

During porogen removal process, it is essential to ensure that the crosslinked polymer maintains its rigidity to prevent the collapse of the porous template. To maintain the porosity of the template, Widmaier and Sperling [5] developed sequential-IPNs, which utilize two polymers networks: a polymer network crosslinked with divinylbenzene (DVB) and another network crosslinked with acrylic anhydride (AAn). The porous structure was achieved by removing the AAn-containing network via hydrolysis in a basic medium, after which selective extraction was performed [2, 5]. Furthermore, studies have investigated a combination of the simultaneous- and semi-IPNs synthesis processes. For example, a previous study synthesized sim-semi-IPNs by incorporating ring-opening polymerization via the free-radical copolymerization of ε-caprolactone and styrene with DVB [6]. Subsequently, a porous network polymer was generated using hydrolysis and removing the linear poly(ε-caprolactone) [6].

IPNs have been employed in numerous applications such as dielectric materials, automotive parts, medical applications, aerospace applications, and thermosetting foams [7]. Further, the polymerization and crosslinking of monomers absorbed in a stretched network have been introduced as a method to fabricate IPNs that can be potentially applied in mechanical actuators, such as artificial muscles [8]. Particularly, owing to its robust mechanical property and chemical stability, the potential application of IPNs as advanced membranes for various applications has attracted attention. In addition to the inherent robustness of IPNs, their use of two polymer networks enables the control of their morphology to simultaneously improve their membrane selectivity and mechanical properties, which will enable their application as membrane separators.

Advanced membranes have attracted widespread attention for several industrial applications. Owing to its low cost and environmental friendliness, membrane technology has emerged as an effective alternative to conventional thermal separation for phase separation (either liquid or gas). Particularly, as phase separation requires materials that exhibit high mechanical stability and chemical resistance and organic solvents, IPNs, which exhibits these properties, have attracted attention for phase separation.

In the field of renewable energy production and storage, electrochemical systems – such as batteries – have emerged as an attractive solution, considering the expanding market for electric vehicles, portable electronics, smart grid, and off-grid energy storage. Membrane separator is an essential component of redox flow batteries (RFBs) as it facilitates the transport of supporting ions, while preventing the permeation of reactive species. Particularly, IPNs-based membranes have attracted attention for application in batteries owing to their extraordinary mechanical, thermal, and chemical stability, which enable them withstand harsh alkaline environment.

Although IPNs have been thoroughly investigated for decades, fundamental research and/or applications of IPNs in the field of advanced membranes are yet to be explored. In this section, the fabrication of IPNs and their applications for advanced membranes will be discussed in details. The scope of this section includes the fabrication of IPNs-based membranes and their applications for gas separation, aqueous and nonaqueous separation, and battery applications. In addition, by tailoring the chemistry of the constituent polymers in IPNs, a synergy can be achieved whereby the performance exceeds that of the constituent neat polymer.

2 Fabrication of interpenetrating polymer network membranes

Typically, IPNs can be defined as an intimate entanglement of two crosslinked networks, and the stability of bulk IPNs and their surface morphologies can be attributed to the interlocked structure of crosslinked components. Various fabrication routes have been reported for the synthesis of IPNs, such as in situ crosslinking and selective and nonselective sequential crosslinking (Figure 1(a)). Typically, IPN systems are used in membrane fabrication to ensure the superior stability of membranes compared to other multicomponent systems. Moreover, s-IPNs are well known for improving the

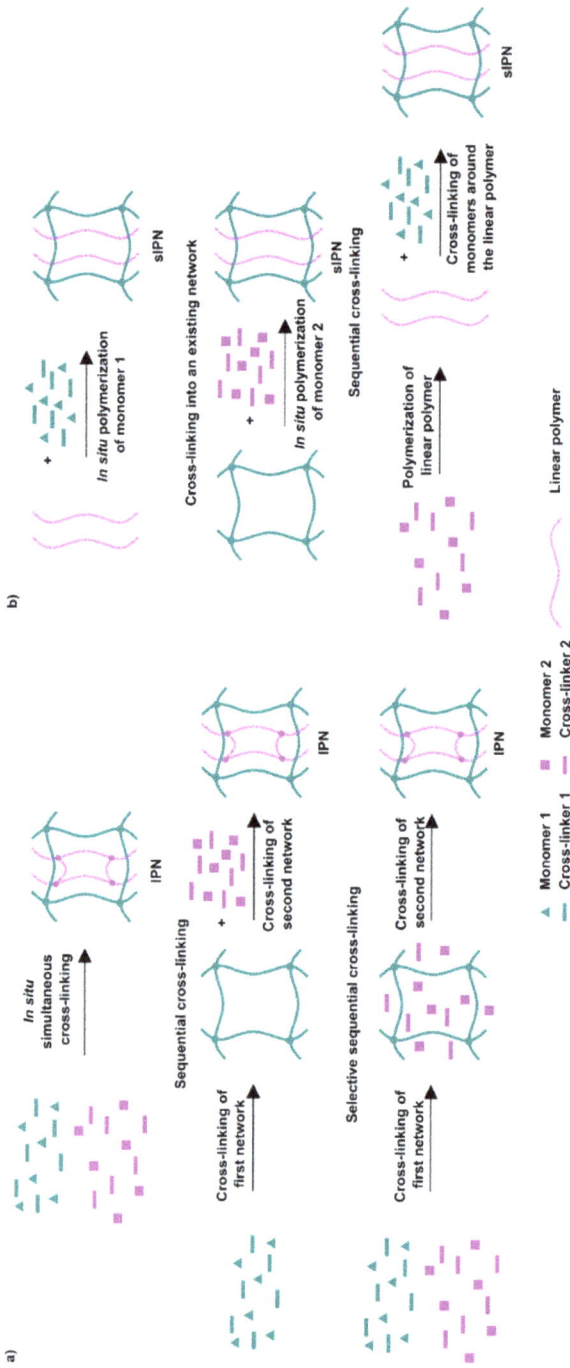

Figure 1: Synthesis process of (a) interpenetrating polymer networks (IPNs) and (b) a semi-interpenetrating polymer networks (sIPNs) showing the in situ and sequential synthesis routes.

chemical stability of membranes. Generally, sIPNs are fabricated by entrapping a linear polymer within the network of another polymer, thus achieving a synergistic effect [9]. Figure 1(b) shows the synthetic routes of sIPNs. The main difference between IPNs and sIPNs is that two or more networks are interlocked in IPNs, whereas only one of the polymer components is present as a crosslinked network in sIPNs (Figure 2).

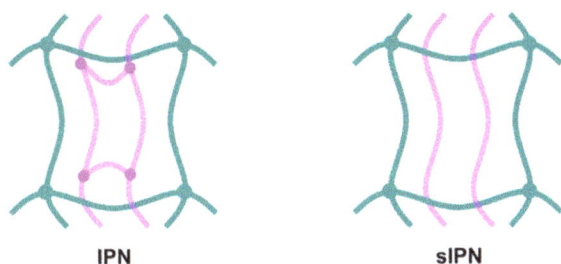

IPN   sIPN

Figure 2: Structural difference between IPNs and sIPNs. The circles indicate the crosslinking points.

Various methods have been developed for the fabrication of IPNs-based membranes including monomer immersion, solution-casting, sequential crosslinking, in situ crosslinking, radiation crosslinking, and thermal crosslinking. Each of this method will be further elaborated in the following section.

Monomer immersion is a sequential technique for obtaining IPNs and is illustrated in Figure 1(a). This method consists of two steps: obtaining the first polymer network, followed by the crosslinking of the second polymer network. Turner and Cheng [10] employed the monomer immersion sequential method to fabricate IPNs with bicontinuous morphologies using polydimethylsiloxane (PDMS)–poly(methacrylic acid) composition. First, they fabricated a crosslinked PDMS film, after which the film was immersed into a methacrylic acid monomer solution. The monomer immersion method reduced the monomer concentration gradient in the pre-IPNs film, which was caused by evaporation or surface thermodynamics, resulting in the formation of a uniform, bicontinuous membrane morphology.

Solution-casting is a versatile and straightforward method for membrane fabrication. In addition, it can be applied for the fabrication of sIPN membranes, where a linear polymer and monomers are dissolved in an appropriate solution, which results in the dispersion of the monomers within the polymer matrix phase. Subsequently, the resulting solution is cast into a thin film, followed by the crosslinking of the monomer, which is affected by diverse factors, such as temperature or the introduction of self-crosslinkable materials. Kim et al. [11] reported the fabrication of an sIPN membrane based on a self-crosslinkable comb copolymer for CO_2 separation. To this end, they introduced a self-crosslinkable poly(glycidyl methacrylate-*g*-polypropylene glycol)-*co*-poly(oxyethylene methacrylate) comb copolymer into a Pebax matrix (Figure 3), and the membranes

(a)

Pebax

(b)

Self-cross-linkable PGP-POEM copolymer

Blending →

(c)

Pebax/PGP-POEM semi-IPN membrane

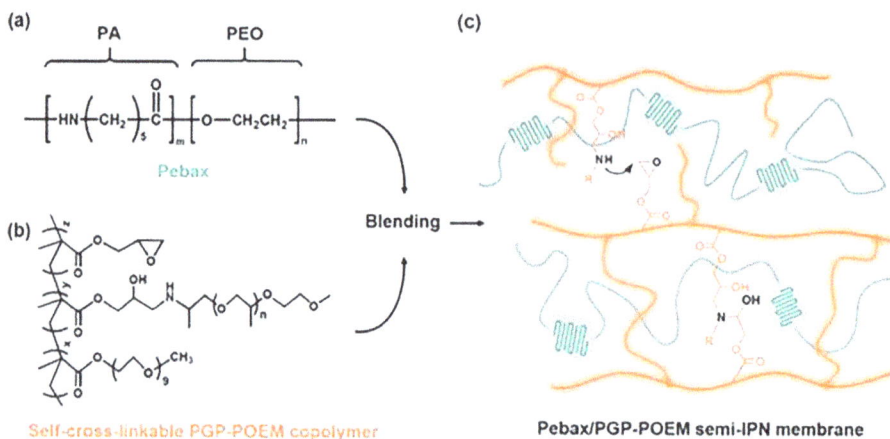

Figure 3: Illustration of the fabrication process of Pebax/PGP-POEM sIPNs. PA: polyamide, PEO: poly(ethylene oxide) PGP-POEM: poly(glycidyl methacrylate-*g*-polypropylene glycol)-*co*-poly(oxyethylene methacrylate) (reprinted from literature [11]).

were prepared using solution casting. The introduction of a self-crosslinkable polymer is a novel approach for the fabrication of IPNs and/or sIPNs as it does not require additional crosslinking agents, catalysts, or heat treatment to achieve the crosslinking reaction.

Sequential crosslinking is a step-wise method for fabricating either IPNs or sIPNs. This process can be performed competitively or selectively. In this method, a crosslinked network or linear polymer is synthesized in the first step, after which a complete network is fabricated in the second step. A previous study successfully fabricated sIPNs anion exchange membranes (AEMs) using a quaternized poly(styrene) copolymer and crosslinked poly(phenylene oxide) network [12]. To obtain sIPN membranes, they achieved crosslinking through the azide-alkyne cycloaddition (click reaction) of functionalized crosslinkable components, which is a selective and easy method. Another study fabricated sIPN AEMs using quaternized chitosan and polystyrene [13]. The sIPN membranes were fabricated by the polymerization of styrene monomers in an emulsion of quaternized chitosan in acetic acid aqueous solution, followed by the crosslinking of quaternized chitosan to form IPNs with polystyrene. The fabricated sIPN membranes possessed superior chemical and mechanical stabilities. Figure 1(b) shows an illustration of a typical example of sequential crosslinking, wherein styrene monomers are simultaneously present with quaternized chitosan molecules, and the styrene monomers were polymerized, after which their mixture was cast into a membrane. Subsequently, a crosslinking agent was added to the solution to cross link the quaternized chitosan molecules, thus obtaining a sIPNs membrane containing polystyrene as a linear polymer.

In situ crosslinking is one of the smoothest and most straightforward IPNs and sIPNs fabrication methods. Figure 1(a) and b) shows the fabrication process of IPNs using in situ crosslinking, wherein the monomers are in situ simultaneously polymerized and cross-linked, resulting in IPNs, or monomers are in situ polymerized and crosslinked around an existing linear polymer chain. In this process, nature-inspired polydopamine (PDA) can be utilized to fabricate high-performance organic solvent nanofiltration (OSN) membranes by forming IPNs. Zhao et al. [14] developed a solvent-stable IPNs by the in situ polymerization of dopamine within a polybenzimidazole (PBI) support (Figure 4). They prevented the chemical crosslinking of the PBI backbone and achieved exceptional thermal and chemical stability, wherein the PBI functioned as the linear chain. In addition, they revealed that dopamine monomers were entrapped within the PBI support, where the in situ polymerization occurred; thus, the PDA branches surrounded the linear PBI chain, resulting in a physically crosslinked sIPN. The remarkable stability of the resulting membrane was attributed to the strong adhesion property of PDA, which was inspired by mussels, and the sIPNs.

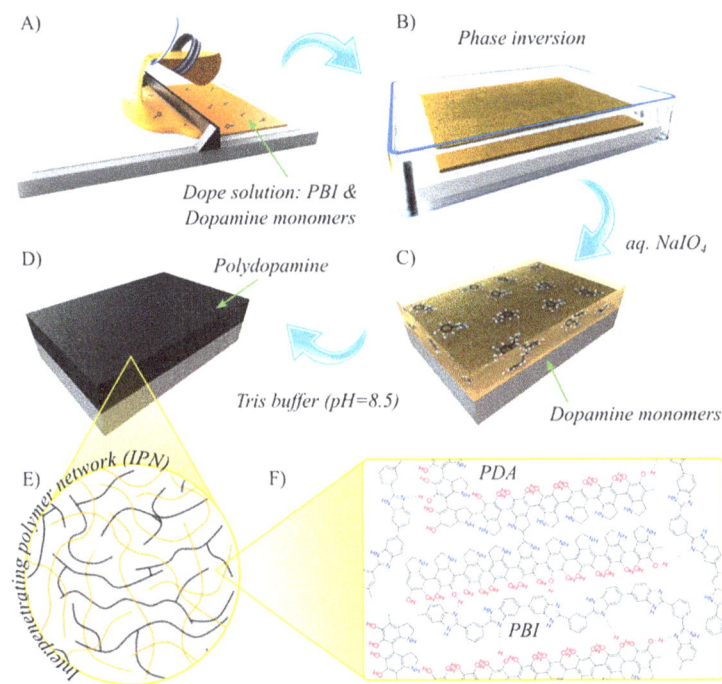

Figure 4: Schematic illustration of the membrane fabrication process of polydopamine (PDA)/ polybenzimidazole (PBI) sIPNs using in situ polymerization (reprinted from literature [14]).

Another study performed the in situ synthesis of crosslinking networks to synthesize sIPNs for fuel cell membranes [15]. To this end, diamine crosslinkers and bromomethylated poly(phenylene oxide) were subjected to an alkylation reaction for the reinforcement of poly (ether–ether–ketone) membrane. In addition, a study developed

IPNs-modified ultrafiltration membranes with boron retention property by the in situ polymerization of an N-methyl-ᴅ-glucamine-based boron-selective monomer [16]. A cellulose ultrafiltration membrane was employed to function as a primary network for the in situ polymerization of the boron-selective monomer inside the membrane pores, resulting in the formation of IPNs membrane. The in situ polymerization of the vinyl monomers enhanced the retention, permeability, and surface characteristics of the obtained ultrafiltration membranes.

Radiation crosslinking involves the use of a crosslinking inducing source, such as ultraviolet, electron beam to avoid the use of catalysts or crosslinking materials. The use of radiation crosslinking or grafting for the fabrication of sIPNs membranes has been proposed. For example, a previous study fabricated sIPNs membranes by the high-energy irradiation of Nafion®, poly(vinylidene fluoride) (PVDF), and vinyl monomers to achieve the crosslinking of PVDF and the vinyl monomers within the Nafion® membrane [17]. Consequently, the prepared membranes exhibited significantly enhanced mechanical and dimensional stability compared to the pristine Nafion® membrane. Furthermore, another study fabricated sIPNs as a selective layer of a thin-film composite membrane to enhance its permselectivity and antifouling properties of the membrane using radiation-induced polymerization [18]. The sIPNs were obtained by the polymerization of N-vinyl-2-pyrrolidone monomers into linear poly(N-vinyl-2-pyrrolidone) chains within the polyamide (PA) network using UV irradiation.

Thermal crosslinking involves the use of temperature as a crosslinking inducer for the fabrication of IPNs membranes. Proton-conducting sIPNs membranes based on poly (vinyl alcohol) (PVA) crosslinked by functionalized nanocrystals in the presence of sulfonated poly(ether–ether–ketone) (PEEK) were fabricated using solution-casting followed by thermal-crosslinking [19]. The resulting sIPNs membranes exhibited excellent thermal and mechanical properties. In addition, another study performed a thermal esterification between poly(acrylic acid) (PAA) and PVA as a crosslinking path within Nafion® to fabricate a proton-exchange membrane [20]. The chemical crosslinking of sIPNs was achieved via the esterification reaction of the hydroxyl groups of PVA and carboxylic acid groups in PAA, thus inducing a three-dimensional crosslinked network. The fabricated sIPNs membrane exhibited enhanced mechanical toughness and thermal and dimensional stability, while preserving the advantageous properties of the Nafion® membranes as a proton-exchange membrane. Temperature-tailorable porous and rigid sIPNs structures for gas separation have been fabricated by the interpenetration of linear thermally rearranged polymers into network thermally rearranged polymers (Figure 5) [21]. The sIPNs was obtained by incorporating the crosslinked network into the polymer matrices, which enhanced the gas transport property of the sIPNs membrane. In addition, the study revealed the importance of applied temperature for IPNs membrane fabrication.

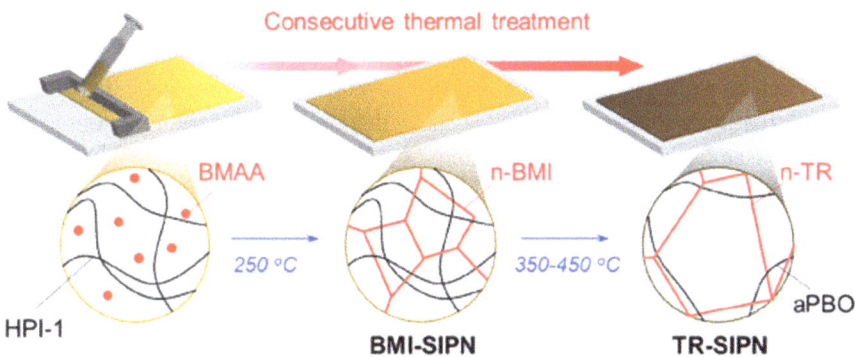

Figure 5: Schematic illustration of the fabrication process of thermally rearranged semi-interpenetrating polymer network showing consecutive thermal treatments and potential chain arrangements of every step. HPI-1: hydroxyl polyimide; BMAA: bismaleamic acid; n-BMI: bismaleimide-based polymer network; BMI-SIPNs: bismaleimide semi-interpenetrating network; n-TR: thermally rearranged polymer network; aPBO: thermally rearranged polybenzoxazole; TR-SIPNs: thermally rearranged semi-interpenetrating polymer network (obtained from literature [21]).

3 Membranes based on interpenetrating polymer networks for gas separation

Over the past few decades, membrane-based gas separation technology has attracted attention owing to its high potential for providing excellent solutions to solve environmental and energy crisis [22]. Membranes fabricated using high-performance polymers, such as polysulfone (PSF) and polyimides (PIs), have been fabricated and investigated for gas separation applications. However, the trade-off between the permeability and selectivity of these membranes has limited their further application [22]. Accordingly, tremendous molecular design and engineering efforts have been devoted to fine-tune the separation performance of these membranes to simultaneously achieve a good permeability and selectivity [23]. For example, polymers with intrinsic microporosity exhibit significantly enhanced membrane separation performance over other commercial polymers, such as PSF and cellulose acetate. Moreover, studies have introduced polar functional groups, such as hydroxyl [24], carboxyl [25], sulfonyl [26], and bromine [27], to polymers to control the permeability and selectivity of membranes. However, compared to functionalized polymers, pristine polymers exhibit notable aging and plasticization [25]. Therefore, to overcome this limitation, mixed matrix membranes (MMM) have been developed by the combination of high-performance polymers with organic and inorganic porous materials, such as activated carbon [28], metal-organic frameworks (MOFs) [29], and covalent-organic frameworks [30]. The performance of MMM has been significantly improved by incorporating fillers, but the interfacial defects and phase separation of these membranes affect their mechanical properties, thus affecting its overall performance. To address this, IPNs have emerged as an excellent solution to address interfacial defects and phase separation issues and enhance membrane stability.

Among the three different IPNs, sIPNs have demonstrated excellent potential for membrane-based gas separation applications. A study investigated the effect of incorporating semi-IPNs into membranes and found that the resulting composite membranes exhibited enhanced gas transport and permeance compared to the pristine membranes [31]. For example, Kurdi et al. fabricated a bismaleimide (BMI)/PSF semi-IPN, and found that the oxygen permeance of the IPN increased by 12 folds relative to the PSF without reducing the O_2/N_2 selectivity [31]. Particularly, PSF exhibited an O_2 permeance of 29 GPU, whereas BMI/PSF exhibited an O_2 permeance of 366 GPU (both membranes exhibited an O_2/N_2 selectivity of 1.9) [31]. It is worth mentioning that the BMI was gradually polymerized inside the PSF solution at room temperature and in the presence of proton donor molecules [32]. In addition, the lack of catalysts or initiators facilitated the in situ polymerization. The IBM polymerization was visually confirmed by an enhancement in the viscosity of the polymer solution and a notable change in the color of the solution [32]. The color of the BMI/PSF polymer solution changed after 40 days of the reaction and turned bright after 80 days. Similarly, a study incorporated polyetherimide (PEI) into BMI and utilized the polymer to fabricate composite membranes for gas separation. The resulting BMI/PEI membrane demonstrated an oxygen permeance of 341 GPU, which is 14 times higher than that of pristine PEI with the same O_2/N_2 selectivity (2.3) [32]. This indicates that the incorporation of semi-IPNs can enhance the permeance of high-performance polymers without affecting the gas selectivity [32]. In addition, the trade-off behavior has been eliminated by combining a porous polymer, which provides high permeability and internal crosslinking connections that result in the enhancement of the selectivity.

A study fabricated sIPNs membranes for gas separation from linear PI entwined by crosslinking poly(ethylene oxide) (PEO) [33]. The crosslinking degree was fine-tuned by controlling the loading amount of PEO into the PI (Figure 6). The sIPNs (PI/PEO) exhibited significantly enhanced membrane mechanical properties compared to the pristine PEO counterparts and seven times higher tensile strength compared to the crosslinked PEO. Furthermore, the CO_2 permeability of PI/PEO was notably improved by 3.5-fold relative to that of the pristine PI membranes [33]. Particularly, pristine PI exhibited a CO_2 permeability of 19.7 barrer with a CO_2/N_2 selectivity of 20 [34], whereas PI/PEO exhibited a CO_2 permeability of 90 barrer with a CO_2/N_2 selectivity of 46 [33]. Moreover, they found that PI/PEO membranes fabricated using short amine-terminated PEO600 exhibited lower gas permeabilities and selectivities relative to those fabricated using long amine-terminated PEO2000. For example, the CO_2 permeability of S-24-2000 film (90.3 barrer) was 120% higher than that of S-24-600 film (41.3 barrer) at a PEO content loading of 70 wt.%. The higher permeability of the 2,000-series film was attributed to the enhancement of the diffusion coefficient with an increase in the membrane porosity [33]. Furthermore, nanohybrid sIPN has been developed containing high PEO loading for CO_2 separation, exhibiting CO_2/N_2 selectivity of 79.6, therefore surpassing the 2008 Robeson upper bound [35].

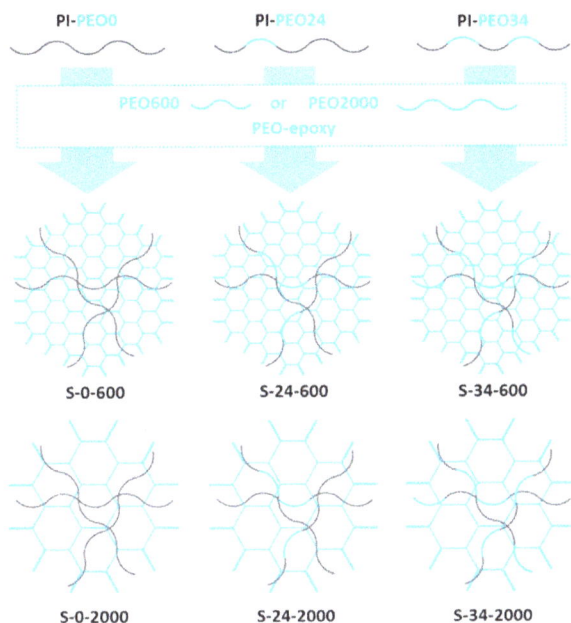

Figure 6: Proposed schematic of the IPNs prepared from PEO/PI at varying poly(ethylene oxide) (PEO) content and crosslinking degree (reprinted from literature [33]).

Integrally skinned asymmetric membranes fabricated using sIPNs of PEI (ULTEM®) and poly(ethylene glycol) diacrylate (PEGDa) in 1-methyl-2-pyrrolidinone (NMP) have been investigated for the separation of pure and mixed gases [36]. The membranes were investigated at a constant pressure and variable volume at 22 °C and 1.35 MPa. The results revealed that the selectivity of the sIPN membranes for the separation of pure and mixed gas at different feed concentrations was very similar, indicating the absence of plasticization [36]. The selectivity of the membrane for CO_2/N_2 mixed gas was as high as 50 at a PEGDa content of 6%, which was comparable to its selectivity for pure gas. Further, the selectivity of the EI/PEGDa membrane for CO_2/N_2 was notably higher than that of PEI. Additionally, the CO_2/CH_4 selectivity (43) of semi-IPNs-based membranes was comparable to the dense film selectivity of PEI (39) and 100% higher than that of dense PEGDa membrane. These results suggest that the PEGDa/PEI semi-IPNs membranes exhibited synergistic properties, wherein the selectivity of the obtained composite membrane was higher than that of the two pristine materials (i.e., PEI and PEGDa) [36].

Semi-IPNs have been employed to fine-tune the fractional free volume (FFV) cavities of PI membranes to control their gas separation performance [37]. To this end, sIPNs were fabricated using azido-containing units within a rigid PI scaffold using in situ polymerization. The gas permeability performance was controlled by varying the PI backbone and functional groups [37]. The FFV of 6FDA-NDA PI decreased from 1.88 to 0.43% after the incorporation of 50% azide (6FDA-NDA/azide (50:50)). Similarly, the FFV decreased from 1.86 to 1.01% after the incorporation of 30% azide into 6FDA-TMPDA. The reduction in the FFV resulted in a decrease in the gas permeability and a significant increase in the gas selectivity. For example, the H_2 permeability of pure 6FDA-NDA decreased from 77.6 to 21.1 barrer after the incorporation of 50% azide. This reduction in permeability resulted in a prominent enhancement in the H_2/N_2 selectivity, which increased from 44.6 in 6FDA-NDA to 105 in 6FDA-NDA/azide (50:50) [37].

Furthermore, sIPNs have been employed for the preparation of carbon molecular seize membranes for gas separation using a 6FDA-TMPDA/azide precursor [38]. The incorporation of IPNs enables the preservation of the porosity after pyrolysis and results in the generation of cavities within the carbon membranes. The azide concentration did not exhibit a linear correlation with the gas permeation. For example, 10 wt.% azide loading increased the CO_2 permeability of pyrolyzed membranes (at 550 °C) by 36% (from 6,810 to 9,290 barrer), whereas 30 wt.% azide reduced the CO_2 permeability by 46% (from 6,810 to 3,640 barrer) compared to that of 6FDA-TMPDA, with no significant changes in the CO_2/N_2 selectivity [38]. However, at 800 °C, pyrolyzed 6FDA-TMPDA/azide sIPNs membranes exhibited a linear correlation with the azide loading concentration and the CO_2 permeability, in which the gas permeability decreases with an increase in the azide concentration. The highest CO_2/CH_4 selectivity was achieved by membranes containing 30 wt.% azide, indicating that the pyrolysis temperature and azide concentration are important factors affecting the fine-tuning of the separation performance [38].

Recently, thermally rearranged semi-PNIs (TR-SPINs) membranes were developed and investigated for olefin/paraffin separation [21]. The TR-SPINs exhibited enhanced porosity and gas permeability without substantial loss in the pair-gas selectivity owing to their unique micropores and structural morphology. For example, TR-SPINs-30 exhibited a Brunauer–Emmet–Teller (BET) surface area of 900 m^2/g, which was 68% higher than that of pristine polybenzoxazoles (PBO) (535 m^2/g) [21]. In addition, the TR-SPINs-30 exhibited six-fold enhanced CO_2 permeability than PBO under the same experimental conditions. Moreover, the TR-SPINs membranes demonstrated plasticization resistance under condensable CO_2 and hydrocarbons, indicating the benefit of using IPNs for improving membrane separation performance [21].

Ionic liquid-based semi-IPNs membranes have been fabricated and investigated for gas separation [39]. Three ionic liquids (ILs), namely [C_2mim][C(CN)$_3$], [C_2mim][NTf$_2$], and [C_2mim][FSI], with maximum loading of 66 wt.% have been employed. IPN-membranes prepared by postswollen with 66 wt.% [C_2mim][NTf$_2$] IL exhibited enhanced CO_2 permeability (609 barrer) compared to the membranes prepared using 40 wt.% of [C_2mim][NTf$_2$] (341 barrer). Particularly, sIPNs/66 wt.% [C_2mim][C(CN)$_3$] exhibited the highest CO_2/N_2 selectivity (52.7) relative to other IL-based membranes [39].

Moreover, a highly CO_2-permeable ion gel membrane was developed that contains a large amount of [Emim][B(CN)$_4$] and is composed of an IPN from poly(vinylidene fluoride-*co*-hexafluoropropylene) and crosslinkable poly(ethyl acrylate-*co*-N-succinimide acrylate) [40]. The IPN membrane exhibited high mechanical strength and selective CO_2 permeability and CO_2/N_2 permselectivity.

Further, IPNs membranes have been fabricated using self-crosslinkable comb copolymers and employed for CO_2 capture. The obtained membranes demonstrated a notable increase in both solubility and diffusivity of CO_2 with an increase in the poly (glycidyl methacrylate-*g*-polypropylene glycol)-*co*-poly(oxyethylene methacrylate) (PGP–POEM) loading percentage. For example, the CO_2 permeability of sIPN membranes containing 40 wt.% PGP–POEM increased by approximately 2.5-fold (up to 236.6 barrer) compared to that of Pebax with comparable CO_2/N_2 selectivity (38.8) [11].

Polyethylene glycol (PEG)-based IPN membranes have been fabricated using solvent-free method for CO_2 separation. The dual network was constructed using Aza-Michael addition and UV-polymerization. The CO_2 permeability of the resulting membranes increased by up to 48-fold compared to that of the pristine membranes [41]. Moreover, PEG in situ polymerization has been employed to seal the internal voids of nanofibrous IPNs membranes prepared from PEO/PAN for CO_2 separation. Compared to pure PEO membranes, the PEO/PAN membranes exhibited a slight reduction in CO_2 permeability owing to the barriers of PAN nanofibers toward CO_2 molecules. Nevertheless, the CO_2/N_2 selectivity of each nanofibrous membrane improved by 13–15 compared to that of the PEO membrane [42]. Liu et al. [43] reported three 2D metalloporphyrin nanosheets crosslinked with PEO sIPN that inducing a novel "rubber-band" straightening effect, and they achieved CO_2 permeability of 398 barrer and high CO_2/N_2 separation selectivity of 81.

IPNs membranes exhibit excellent performance for gas separation with suitable membrane mechanical properties. The separation performance of IPNs exhibits a strong dependence on the structural and composition characteristics of the linear and crosslinked polymer chains. Thus, the performance can be fine-tuned by tailoring the molecular structure of both polymers. Additionally, it is essential to evaluate the long-term stability of IPNs membranes for gas separation application. Therefore, it is essential to conduct further studies on the aging separation and performance of IPNs in addition to its plasticization resistance. The reported gas permeability and selectivity of IPN-based membranes compared to those of pristine membranes are tabulated in Table 1.

Table 1: Gas permeability and selectivity of IPN-based membranes compared to those of pristine membranes.

Polymer designation	Permeability (barrer)			Selectivity		Ref.
	CO_2	N_2	CH_4	CO_2/N_2	CO_2/CH_4	
6FDA-TMPDA[a]	315	19.4	15.5	16.2	20.3	[37]
6FDA-TMPDA[b]	6,810 ± 125	277 ± 0.6	115 ± 2.1	24.6 ± 0.4	59.4 ± 0.1	[38]
6FDA-TMPDA[c]	1,460 ± 17	46.6 ± 0.1	23.5 ± 0.8	31.3 ± 0.3	62.2 ± 1.3	[38]
6FDA-TMPDA/azide (90-10)[a]	133	7.12	5.29	18.6	25	[37]
6FDA-TMPDA/azide (90-10)[b]	9,290 ± 170	358 ± 17	236 ± 19	26.0 ± 0.8	39.6 ± 2.5	[38]
6FDA-TMPDA/azide (90-10)[c]	851 ± 9.1	25.6 ± 0.2	7.73 ± 0.2	33.2 ± 0.1	110 ± 1.3	[38]
6FDA-TMPDA/azide (70-30)[a]	38.6	1.97	1.22	19.6	31.6	[37]
6FDA-TMPDA/azide (70-30)[b]	3,640 ± 17	151 ± 0.2	54.9 ± 1.1	24.2 ± 0.1	66.4 ± 1.0	[38]
6FDA-TMPDA/azide (70-30)[c]	280 ± 7.0	8.85 ± 0.2	1.71 ± 0.1	31.7 ± 0.1	164 ± 6.0	[38]
6FDA-NDA	38	1.74	1.02	22	37	[37]
6FDA-NDA/azide (90-10)[a]	12.9	0.54	0.26	24	49.6	[37]
6FDA-NDA/azide (70-30)[a]	6.61	0.28	0.14	23.6	47	[37]
6FDA-NDA/azide (50-50)[a]	4.99	0.2	0.11	25	45.4	[37]
PEI	1.4	–	–	23 ± 6	–	[36]
PEGDa/PEI-2%	1.2	–	–	28 ± 5	–	[36]
PEGDa/PEI-4%	1.5	–	–	24 ± 2	–	[36]
PEGDa/PEI-6%	1.2	–	–	48 ± 12	–	[36]
PEGDa/PEI-8%	1.0	–	–	31 ± 6	–	[36]
S-0-600	40.0 ± 0.9	0.8 ± 0.02	–	51.6 ± 2.7	–	[33]
S-24-600	41.3 ± 0.9	0.8 ± 0.02	–	49.2 ± 2.5	–	[33]
S-34-600	54.3 ± 1.6	1.0 ± 0.04	–	52.1 ± 3.5	–	[33]
S-0-2000	73.9 ± 1.8	1.5 ± 0.04	–	49.0 ± 2.5	–	[33]
S-24-2000	90.3 ± 1.9	1.9 ± 0.04	–	46.2 ± 2.0	–	[33]
S-34-2000	86.3 ± 2.0	1.9 ± 0.05	–	46.2 ± 2.3	–	[33]
PEO/NBR sIPN	116 ± 0.7	3.4 ± 0.01	–	34 ± 0.1	–	[39]
sIPN/50 wt.% [C_2mim][NTf_2]	341 ± 2.6	11.6 ± 0.12	–	29.3 ± 0.1	–	[39]
sIPN/66 wt.% [C_2mim][NTf_2]	609 ± 2.7	21.9 ± 0.35	–	22.7 ± 0.45	–	[39]
Pebax/PGP-POEM (60 wt.%)	271.8	9.3	–	29.2		[11]

[a]Pristine.
[b]pyrolyzed at 550 °C.
[c]pyrolyzed at 800 °C.

4 Membranes based on interpenetrating polymer networks for water purification

Since the seminal breakthrough by Loeb and Sourirajan in 1962, research on synthetic membrane for water purification has flourished rapidly over the past 60 years. To further elevate its potential, various state-of-the-art modifications for synthetic membranes have been proposed, from minor modification of currently available membranes to major revamp of the entire separation process design. For example, membrane capacitive deionization has continuously attracted attention as an alternative desalination membrane technology for a more energy efficient desalination process [44, 45]. For example, 3D-printed membranes have been predicted to be the future of membrane fabrication for improved control of the microstructures of the formed membranes [46, 47]. Meanwhile, water filtration across inorganic membranes, such as nanoporous single-layer or laminated multilayer graphene, enable improved water flux owing to its nanoscale thickness, while enabling molecular separation through its nanopores [48].

Furthermore, MMMs that incorporates the benefit of inorganic fillers and the flexibility and low cost of polymeric membrane matrix have attracted significant attention, particularly in antifouling membrane research [49]. Particularly, the development of nanotechnology has further propelled the MMM research, which could not be previously achieved. Other studies have attempted to develop a more sustainable membrane fabrication approach by utilizing new green solvents [50] or eluding the limitation of the current membranes through the synthesis of new polymers. For example, polymer of intrinsic microporosity (PIM) has emerged as a promising solution. Although PIM has been extensively applied for gas separation, the promising potential of PIM for water desalination, biofuel dehydration, and antibiotics removal from wastewater has been recently reported [51, 52].

IPN-based membranes have attracted attention for the fabrication of various types of water purification membrane. Particularly, IPN-based membranes have attracted attention owing to their ability to synergically blend already developed polymers into a single membrane matrix. Nevertheless, compared to simple blending, entanglement and interlocking of the polymer chains in IPNs at the molecular level enable enhanced membrane stability and peculiar properties, which are not commonly observed in base polymers, without the need for extensive chemical modifications [1].

For pressure-driven membrane separation, augmented antifouling properties have attracted attention. Pressure-driven membrane separation comprises an increase in the hydrophilicity of the membrane to minimize the adsorption and deposition of hydrophobic contaminants on the surface of the membrane. Hence, blending the base polymer matrix of membranes with hydrophilic polymers during the preparation of casting solutions has emerged as a method for modifying the characteristics of membranes. However, simple blending tends to result in the leaching out of these materials during the coagulation period or during the filtration process, resulting in the gradual loss of the antifouling potential of the membrane [53–55]. Hence, polymer blending is more commonly used as for the formation of pores, while polymer grafting is performed if the hydrophilicity is to be retained [56, 57]. IPNs can be an alternative solution for improved retention of the polymers, as IPNs enable the retaining of normally leached polymers through the formation of 3D-intercalated polymer chains. This simple concept can be observed even for the simple polymer blending of high-molecular weight hydrophilic polymers, such as polyvinyl pyrrolidone (PVP), which are better retained compared to its low-molecular weight counterpart owing to the improved intermingling of longer PVP chains with the base PSF polymer chains [58]. Nevertheless, as linear PVP chains exhibit limited physical restriction, retention of polymer blend remains low or moderate with only 40–50 wt.% of the polymer being retained after 5 min in a water bath. This retention can be improved if IPNs are designed instead, as the formed PVP would have a more homogenous interpenetration or will be physically locked around the nonlinear base polymer chains. Figure 7 shows the integration of hydrophilic polymers, such as PVP, using IPNs with a PA layer fabricated using the conventional interfacial polymerization (IP) process. Moreover, the recently observed polymer–polymer stability of IPNs enables a larger range of miscibility of the polymers, enabling for possible unique phase separation pathways and subsequent membrane morphologies not exhibited by their blended counterparts.

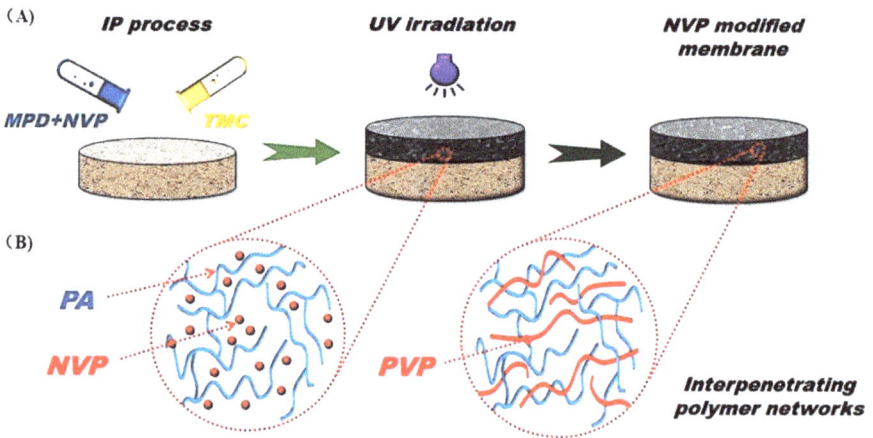

Figure 7: Schematic of the (A) fabrication process of N-vinylpyrrolidone (NVP) modified reverse osmosis (RO) membrane and (B) structure of PA/PVP sIPNs (reprinted from literature [18]).

A semi-IPN membrane fabricated from PVDF and PAA for oil–water emulsion separation exhibited enhanced hydrophilicity than pure PVDF membrane [59]. Similar improvement in hydrophilicity was also observed for sIPNs of PA and PVP for RO membrane, resulting in higher water flux and enhanced fouling mitigation over time compared to pristine PA thin-film composite membrane [18]. In addition, 1,3-diamino-2-propanol (DAPL)-modified PA RO membrane has been reported for improving antifouling properties [60]. Meanwhile, sIPNs of polystyrene sodium sulfonate (PSS) and polypiperazinetrimesamide have been reported for nanofiltration membrane, which exhibit an enhanced flux recovery ratio improvement of approximately 6.9% higher than that of neat polypiperazinetrimesamide membrane, suggesting enhanced antifouling properties [61]. In addition, the incorporation of sIPNs polymeric nanoparticles into the base casting solution through blending has been reported. For example, PES blended with PES/PVP sIPN nanoparticles exhibited significant reduction in adsorbed protein fouling

[62]. A novel UV-initiated modification process has been applied by Miao et al. [63] to fabricate a sIPN TFC RO membrane showing excellent antifouling property and high water flux of 77.5 L/m^2/h and 99.44% NaCl rejection, demonstrating scalability and easy operation making the proposed modification of great significance for industrial application. MMM IPN membranes have also been reported for water remediation purposes. Gupta et al. [64] developed a copper-substituted polyoxometalate-soldered IPN membrane via in situ electrostatic attachment. The membrane showed enhanced surface roughness, pore structure and hydrophilic nature, and excellent antifouling and mechanical properties with high salt rejection (~98%) and high dye removal (>97%).

Compared to blended membranes, IPNs membranes are advantageous in that they enable the integration of otherwise incompatible or immiscible polymers without undergoing phase separation [1]. This opens various new possibilities that could not be previously achieved through simple polymer blending. Previous studies have reported that a PA layer in a TFC RO membrane could be modified with polyisobutylene (PIB), an immiscible and hydrophobic moiety, to alter the pore size of the membrane for enhanced boron removal. In addition, another study achieved a boron rejection of 93.12% when 0.3% PIB was added compared to when just a pristine PA layer was used (81.36%) at a pH of 5.8 [65]. IPNs could also provide the necessary anchoring point for subsequent surface grafting process. It has been reported that grafting PVA on the membrane surface of DAPL-PA IPNs membrane results in the formation of a more stable and uniform PVA layer, resulting in long-term stability [60]. Numerous opportunities are also available beyond those directly reported for water purification. For example, PDMS-polyzwitterion IPNs [66] and silicone elastomer-hydrogel IPNs [67] have been recently reported as antifouling coatings, which can be modified and adapted for water purification application. In addition, the adhesion between multicomposite membrane layers can be improved through the formation of interlayer IPNs. Figure 8 shows the prevention of delamination through the incorporation of IPNs between immiscible polymer layers. The grafting of 4-fluoro-2-(trifluoromethyl)benzylamine (FTB) monomer to P84® PI as a selective layer polymer resulted in enhanced intermolecular interactions between the fluorine atoms from FTB and the carboxyl and ether groups of the Ultem® substrate during the coextrusion of the membrane [68]. This results in the formation of IPNs at the interface and was noted as the source of the adhesion between the poorly miscible polymers. However, this resulted in decreased water permeance with slightly increased salt rejection, which may be attributed to the slightly different properties of the formed sIPNs interlayer compared to that of P84® PI.

IPNs, as in polymer blending, can affect the properties of casting solutions, thus affecting the morphology and permeation performances of the resulting membranes. However, the molecular interlocking of polymer chains in IPNs may result in differences in their kinetics and thermodynamic properties compared to those of their blending counterpart [1]. This could enable researchers modify the characteristics of the resulting membrane by preforming IPNs prior to phase inversion. A previous study reported a significant increase in the solution viscosity by crosslinking PDMS

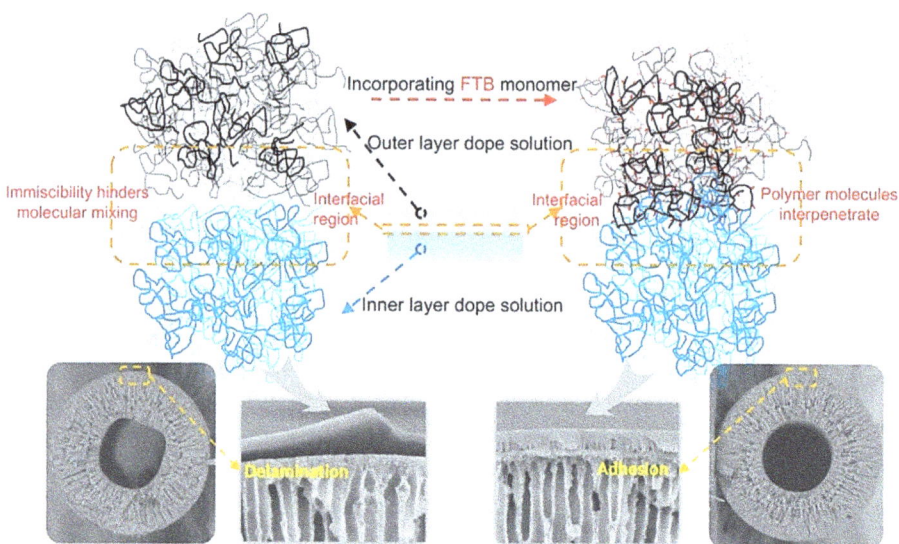

Figure 8: Illustration of the incorporation of 4-fluoro-2-(trifluoromethyl)benzylamine (FTB), which enhanced the adhesion at the membrane interface through IPN formation (reprinted from literature [68]).

after blending with PVDF, thus forming semi-IPNs structure (as compared to simply blending the polymers). This results in different membrane structures and characteristics after the phase inversion process despite the similar PDMS/PVDF mass ratio of the casting solution [69, 70]. Moreover, the minimization of the compaction effect of the layers in TFC RO membranes by IPNs has been reported [18]. The increased mechanical strength enables their application as hydrogel-based water purification membranes, which normally lack the necessary self-supporting properties. For example, the self-standing properties of IPNs gel fabricated using Ca^{2+} crosslinked alginate network and covalently crosslinked poly(acrylamide) network and its high antifouling and water flux performance for dye removal have been reported [71]. Moreover, sIPN structure has been constructed by intertwining MXene laminates by the long chain of PAA macromolecules showing high permeability and high rejection to dyes, together with long-term stability enabling them for effective wastewater treatment [72]. Nature-inspired electrostatically soldered green tea extract of epigallocatechin gallate was used for a sequential triple IPN membrane augmenting its hydrophilicity, antifouling, and antimicrobial properties [73]. The IPN membrane was able to achieve ~99% dye rejection and ~98% desalination efficiency, and it demonstrated noncytotoxic properties against mammalian cell lines making it biocompatible. Solvent-resistant nanofiltration sIPN membrane has been fabricated by Nozad et al. [74] consisting of hydroxyl-terminated polybutadiene (HTPB) and multifunctional isocyanate (MFI) embedded into the polyphenylsulfone (PPSU) membrane matrix. The membrane showed high resistance against organic solvents such as DMF and successfully separated dyes from water and organic solvents up to 98%. In addition, commercially already existing nanofitration membranes could be surface-modified via IPNs resulting in increasing seawater permeability by 300% without compromising the high rejections of original membranes [75].

Further, the application of IPNs for ionic separation in aqueous solution has been reported. In addition, the modification of ultrafiltration cellulose membranes into IPNs via an in situ polymerization with N-methyl-D-glucamine to fabricate membranes with high-boron retention capacity between 14.8% (at pH 9.0) and 20.5% (at pH 5.0) has been reported [16]. In addition, the use of water-soluble polymers as metal chelators enables the removal of heavy metals, which cannot be efficiently removed by ultrafiltration membranes. This can be achieved via the formation of a sufficiently large cation-polymer complex. The use of IPNs could enable the immobilization of the complexing polymers in the base membrane matrix. The ability of the sIPNs of PVA networks with PEI to enable the removal of various heavy metal ions from aqueous solutions through this method has been reported [76]. Similarly, another study attempted the fabrication of pH-responsive and antifouling ultrafiltration membranes synthesized using the sIPNs of PES and in situ polymerized poly(methyl methacrylate-co-acrylic acid) (P(MMA-AA)) and poly(methyl methylacrylate-co-4-vinyl pyridine) (P(MMA-4VPy)) copolymer [77]. The resulting membrane exhibited good Cu^{2+} adsorption capacity. In addition to IPNs immobilization, polymer enhanced ultrafiltration (PEUF) could also function with the water-soluble polymers introduced directly into the feed. Here, hydrogels have attracted significant attention owing to their ability to utilize various functional groups in a three-dimensional structure for the removal of heavy metal ions from wastewater. The relatively denser hydrogel matrices of IPNs enable a faster and higher sorption of ionic species compared to single network hydrogels [78]. Figure 9 shows the mechanism of the utilization of the structure of IPNs hydrogel for increasing its adsorption capacity of heavy metal ion [79]. Combined with stimuli-responsive modification for controlled regeneration, this opens for various possibility of IPNs hydrogel application for water filtration membrane through PEUF.

Despite the potential benefits of IPNs membrane compared to that of the blended membrane, it is important to note that without a full 3D interlocking network between the polymer chains, sIPNs configuration are susceptible to the leaching of the interpenetrated hydrophilic species. Although no research on the comparison of the leaching properties of blended versus IPNs membrane has been conducted, studies on chitosan-PVA semi-IPNs hydrogel have reported the potential leaching properties of PVA [80]. Nevertheless, the leaching process can be enhanced using a high PVA content for hydrogel formation compared to the possible leaching in membranes where PVA would be introduced in relatively smaller amount. In contrast, PVA/PEI semi-IPNs were reported to partially dissolve over the course of 70 days with up to 5.3% of PEI and 6.5% of PVA [76]. However, PSS leaching from PSS/polypiperazine trimesamide semi-IPNs nanofiltration membrane was reported to be negligible with a rela-

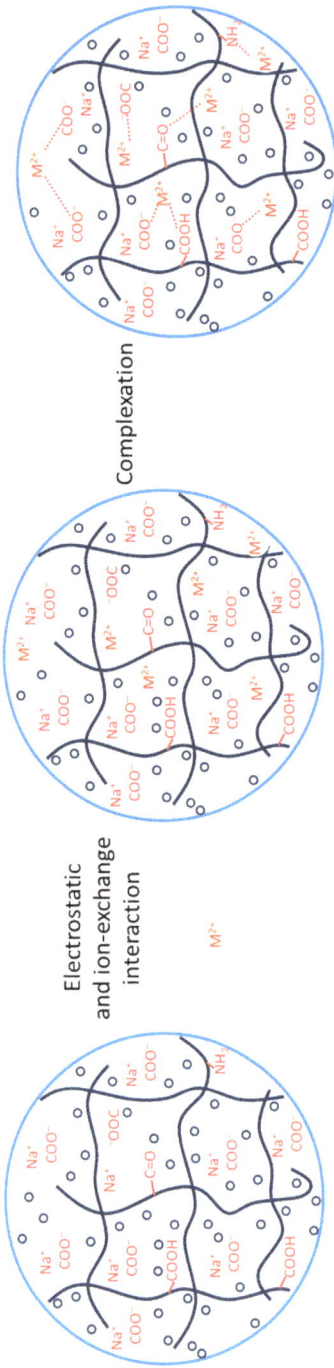

Figure 9: Mechanism of heavy metal ion adsorption on IPNs hydrogel (adapted from literature [79]).

tively stable Na_2SO_4 rejection and flux over the course of 36 days [61]. In summary, achieving a good polymer–polymer interlocking balance for stability, while optimizing the separation performances might be a promising method to enhance the properties of IPN membranes compared to polymer blends.

5 Membranes based on interpenetrating polymer networks for nonaqueous separation

Nonaqueous separations are an important aspect of chemical industries as majority of industrial chemical synthesis occurs in organic solutions [81]. The use of molecularly selective membranes through the pervaporation of nonaqueous mixtures has emerged as an alternative separation method, which may be more energy efficient than the conventional distillation method. Moreover, OSN and organic solvent reverse osmosis enables the occurrence of these separations without phase changes, which would further reduce the energy cost and facilitate the separation of thermal-sensitive products. However, the limited stability of polymeric membranes in nonaqueous conditions has limited its further application [82].

Efforts have been devoted to augment the solvent resistance of polymers through chemical crosslinking, such as in PI and PBI membranes [83, 84]. However, the increase in the membrane rigidness and loss in flux performance have limited the further application of this method. To overcome this problem, the use of inorganic membranes, such as membranes fabricated from solvent-stable porous silica-zirconia ceramics and stacked graphene oxides, has been proposed [85, 86]. Ceramics and carbon-based membranes are advantageous not only because of their superior solvent resistancy but also because of their superior pore size that can be controlled down to a few nanometers for precise molecular sieving separation. Nevertheless, the high manufacturing cost and difficulty for defect-free large scale fabrication have limited the practical application of inorganic membranes. Hence, the use of novel microporous structures, such as PIM or MMM, has emerged as an effective method for enhancing the stability and performance of polymeric membranes, while maintaining the economic viability of membranes for industrial usage [87]. For example, the microporosity of PIM is a significant advantage for both pervaporation and OSN, as the molecular sieving properties of inorganic membranes can be utilized in polymeric structures [88, 89]. In contrast, MMM, which utilizes MOFs, could enhance the selectiveness of conventional solvent-resistant polymers toward similarly sized molecules [90]. The incorporation of hydrophobic structures for nonaqueous separations to increase the apolar solvents permeance of the membrane is promising, whereas the incorporation of fillers could minimize the swelling and compaction of the base polymer matrix [82, 91].

In addition to the previously mentioned membranes, IPN-based membranes exhibit promising results compared to conventional polymeric membranes for addressing the swelling or dissolution of polymers in organic solvents. For example, the use of IPNs can increase the stability of membranes in polar aprotic solvent (PAS) without the need for extensive covalent crosslinking, which may degrade the mechanical flexibility of the material. This solvent-resistant property of IPNs enables the formation of high chemical stability, yet highly permeable membranes for OSN and pervaporation: two of the main methods used for nonaqueous separations using membranes. Some examples of the use of IPNs-based membranes in nonaqueous conditions are summarized in Table 2.

Table 2: Reported use of IPN-based membranes for nonaqueous separations.

Polymers	Fabrication method	Tested solvent	Tested application	Ref.
SBR–PMMA	In situ polymerization of MMA in crosslinked SBR matrix	Benzene, toluene, xylene, CCl_4, ethyl methyl ketone, ethyl acetate	Solvent diffusion experiment	[92]
NR–PVA	Simultaneous crosslinking of both NR and PVA blends casting solution using GA	Benzene	Solvent diffusion experiment	[93]
PDA–PBI	In situ polymerization of PDA monomer dispersed in PBI matrix	HMPT, DMF, DMAc, DMSO, NMP, PC, cyrene	OSN (polystyrene markers)	[14]
HTPB–PPSU	Crosslinking of HTPB using MFI in PPSU casting solution	Acetone, n-heptane, DMF, NMP, DMSO, THF	OSN (dye markers)	[74]
Polyacrylate–PSU	Physical blending of PSU, dipentaerythritol penta-acrylate and 2,4,6-trimethylbenzoyl-diphenyl-phosphine oxide, followed by UV-curing	IPA, toluene, xylene, butyl acetate, acetone, DMF, THF, NMP	OSN (dye markers)	[94]
HEC–PP (IPNs at interface)	Cross-flow deposition of HEC on PP, followed by crosslinking using GA	EtOH, THF, ethyl acetate	OSN (dye markers)	[95]
FTB-grafted P84® PI–Ultem® PEI (IPNs at interface)	Cocasting or coextrusion of P84® PI grafted with FTB and Ultem® PEI	EtOH	OSN (tetracycline and vitamin B12)	[68]

Table 2 (continued)

Polymers	Fabrication method	Tested solvent	Tested application	Ref.
TMSC–PDMS	Physical blending of TMSC/PDMS, followed by PDMS crosslinking and TMSC hydrolysis	MeOH, hexane, EtOH, IPA, THF, toluene, chloroform, acetone, ACN	OSN (polystyrene and dye markers) and pervaporation (EtOH/H_2O)	[96]
PVA–PSSH	Physical blending of PVA/PSSH, followed by the crosslinking of PVA	Propanol, propanoic acid, ester (propyl propanoate)	Pervaporation assisted esterification (propanol-propanoic/H_2O)	[97]
PDMS–PAA	Simultaneous polymerization and crosslinking of PAA and PDMS blend	MeOH, toluene	Pervaporation (MeOH/toluene)	[98]

SBR: styrene butadiene rubber, PMMA: poly methyl methacrylate, CCl_4: carbon tetrachloride, NR: natural rubber, PVA: polyvinyl alcohol, GA: glutaraldehyde, PDA: polydopamine, PBI: polybenzimidazole, HMPT: hexamethylphosphoric triamide, DMF: dimethyl formamide, DMAc: dimethyl acetamide, DMSO: dimethyl sulfoxide, NMP: N-methyl-2-pyrrolidone, PC: propylene carbonate, OSN: organic solvent nanofiltration, HTPB: hydroxyl-terminated polybutadiene, PPSU: polyphenylsulfone, MFI: multifunctional isocyanate, THF: tetrahydrofuran, PSF: polysulfone, UV: ultraviolet, IPA: isopropanol, HEC: hydroxyethyl cellulose, PP: polypropylene, EtOH: ethanol, FTB: 4-fluoro-2-(trifluoromethyl) benzylamine, PI: polyimide, PEI: polyetherimide, TMSC: trimethylsilyl cellulose, PDMS: polydimethyl siloxane, MeOH: methanol, ACN: acetonitrile, H_2O: water, PSSH: poly(styrene sulfonic acid), PAA: poly(amic acid).

The high chemical stability of PBI and PDA-based IPN membranes fabricated using in situ polymerization for OSN application has been reported [14]. This method enables the maintenance of the integrity of PBI, which is normally affected by PAS, in various conventional and green PAS at a PDA incorporation loading of as low as 1–3%, while maintaining a good permeation performance and solvent stability in a wide temperature range from –10 to 100 °C.

Another study reported that sIPN membranes fabricated from PPSU achieve a high solvent resistivity when HTPB and MFI were incorporated into the casting solution [74]. Although the resulting IPN membranes only exhibited partial stability in PAS, it exhibited lower swelling percentage in ethanol (up to 84% less) and isopropanol (up to 72% less) with increasing HTPB to PPSU wt.% (at 40 wt.% maximum). The stability of styrene butadiene rubber (SBR) and PMMA) was also evaluated in both sIPNs (MMA not crosslinked with DVB) and full-IPNs (MMA crosslinked with DVB) configuration [92]. The formed IPN membranes exhibited enhanced swelling resistance in nonpolar solvents, similar resistance in acid, bases, and salt solution, and higher swelling in polar solvent. Nevertheless, the comparison of the results of the sIPNs and full-IPNs revealed that the formation of crosslinked PMMA resulted in enhanced solvent resistancy and up to 60% reduction in solvent uptake; however, this also resulted in stiffer membranes and lower solvent permeability. This suggested that it is essential to control the degree of interpenetration of the polymer chains in IPNs to maintain a good balance between stability and performance. Recently, Cavalcante et al. reported the fabrication of biobased IPN membranes from agarose and natural rubber latex, that successfully purged a carcinogenic impurity at a concentration of 5 ppm with a negligible 0.56% pharmaceutical loss [133].

The use of IPNs enables the casting of immiscible polymers as a homogeneous membrane structure. However, nonaqueous permeance can be used as various solvents with different properties (i.e., being polar or nonpolar). Thus, the appropriate membrane should be determined based on the material properties, where a polar solvent functions better in hydrophilic membranes, while nonpolar solvents function better in hydrophobic membranes [91]. In addition, IPNs enable the application of the fabricated membrane as a universal membrane for the separation of both solvents. For example, the fabrication of cellulose/PDMS interpenetrating network has been reported [96], wherein a polymer blend of trimethylsilyl cellulose (TMSC) and PDMS was fabricated, followed by a hydrolysis reaction to convert the TMSC back into cellulose. The fabricated membrane exhibited high solvent stability, suggesting the formation of an IPN membrane structure between the hydrophilic cellulose and hydrophobic PDMS, which are otherwise incompatible. A good solvent permeance across both polar and nonpolar solvents and pervaporation of ethanol/water mixture was observed through this preparation method. Another study reported the application of IPNs, such as in PVA/poly(styrene sulfonic acid) (PSSH), for pervaporation-assisted esterification [97]. Although their blends dissolve rapidly in water, IPNs structure enables the trapping of PSSH in the polymer matrix by crosslinked PVA, resulting in a minimal loss of PSSH. In addition, the layering of PVA/PSSH IPNs on another PVA sublayer enables the extraction of water by the hydrophilic PVA, while enabling the catalyzation of the propanol-propanoic esterification reaction by the sulfonic acid groups in PSSH during pervaporation process.

The concept of multilayer composite membrane has attracted increasing attention. The adhesion properties between multilayer composite membranes can also be improved through IPNs formation for nonaqueous separation. For example, FTB-grafted P84® polyimide and Ultem® interfacial IPNs formation previously reported for water purification has also been tested for OSN application [68]. Meanwhile, hydroxyethyl cellulose (HEC)/ polypropylene (PP) IPNs was formed at the interface upon the crosslinking of HEC, and the layer–layer bonding was observed to be exceptionally strong despite their difference in hydrophilicity [95]. In addition, studies have reported the development of solvent-resistant support for subsequent multilayer composite formation using IPNs [94]. A previous study reported the formation of PSF/polyacrylate IPNs by the UV curing of the fabricated layer to form crosslinked polyacrylate (Figure 10). Although crosslinked polyacrylate was initially reported for OSN application and tested for isopropanol (IPA) and dimethylformamide (DMF) permeation with Rose Bengal as the marker, studies have found that the loss of rejection after soaking in DMF suggested that the membrane would swell in the tested PAS. Subsequent work to optimize the solvent-resistant support of multilayer composite membranes for various applications have been reported [99, 100].

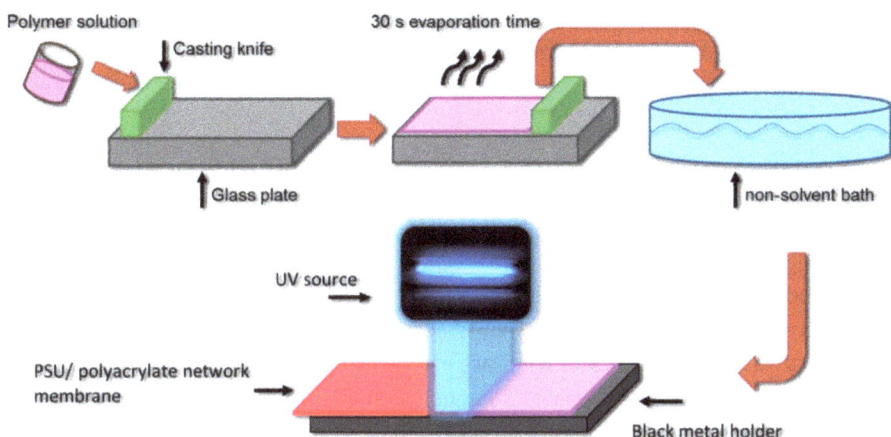

Figure 10: Nonsolvent-induced phase separation (NIPS), followed by UV-LED curing for production of PSU/polyacrylate IPNs membrane support (reprinted from literature [100]).

Although the application of IPN membranes in nonaqueous conditions is still relatively new and studies have focused on the application of the improved chemical and mechanical stability of the the membrane, its advantages beyond the well-known benefits have attracted attention. For example, membranes from molecularly imprinted polymers (MIP) known as molecularly imprinted membranes have emerged as a promising alternative for separation requiring molecular recognition. Particularly, studies on MIP in nanoparticle have attracted increasing attention owing to its increased surface-to-volume ratio and its specific binding capacity [101, 102]. The immobilization of nano-MIP in membrane matrix is a promising approach for easier particle handling and regeneration. The use of IPNs may enable the retention of nano-MIP and prevent interfacial defects between the nano-MIP and the bulk polymer matrix. Direct imprinting on PBI as both functional polymer and OSN membrane has been reported for the selective removal of active pharmaceutical ingredients (API), such as 2-aminopyrimidine from similarly-sized catalyst 4-dimethylaminopyridine and the larger roxithromycin API [103], suggesting the promising application of hybrid MIP-OSN for multicomponent separations. The ability to separately optimize MIP recognition capability and membrane transport properties using IPNs may be considered as the bridge for maintaining proper MIP immobilization. To date, several attempts to immobilize imprinted nanoparticles, such as by sandwiching in between membranes for enantiomer separation of chiral amino acid derivative [104, 105] or simply blending the nano-MIP in the casting membrane solution [106, 107], have been reported, but no nano-MIP dispersed membranes have been reported for separation in nonaqueous condition, and studies on the possible leaching of the dispersed MIP after extended usage are rare. Following the trend of its aqueous separation counterpart, it is expected that studies on the integration of IPNs-based nano-MIP in OSN membrane would be seen in the next few years owing to its niche potential, particularly in pharmaceutical industries or racemic separation.

6 Membranes based on interpenetrating polymer networks for battery applications

Environmental and materials sustainability issues have emerged as essential considerations in current and future technologies, which focus on renewable energy production and storage. Currently, solar and wind are examples of renewable energy sources that are intermittently available. Thus, it is important to utilize energy storage materials in the generation of renewable energy to ensure the collection and storage of renewable energy for optimal use. Electrochemical energy storage, such as batteries, has emerged as an attractive solution owing to it potential application in electric vehicles, portable electronics, smart grid, and off-grid energy storage. Particularly, RFBs have attracted attention for large-scale energy storage owing to their relatively low production costs, long service lifetimes, and independent scaling of energy and power density [108].

RFBs are fabricated using various arrays and design elements that can impact its battery performance. Generally, RFBs consist of cathode and anode chambers, flowable electrolytes, and a membrane separator (Figure 11). Electrolyte solutions for the two half cells, which are stored in separate tanks and pumped, pass through a semipermeable membrane that enables the transfer of charge-balancing ions and solvent, while preventing the interchange of redox-active species. The energy density of RFBs is controlled by the number of electrons involved in redox reactions, the concentration of redox-active species, and cell voltage. In contrast, the power density is affected by the electrocatalytic properties of the materials, active surface area of the electrodes, ionic and electronic conductivity of the electrodes, and ohmic resistance of the batteries [109, 110].

In RFBs, membranes are utilized to separate the two electrode tanks to prevent the mixing of anolyte and catholyte. The membranes should be permeable for supporting ions, but impermeable for the reactive species. In addition, the membranes should be chemically stable against oxidation and acid-catalyzed reactions, such as hydrolysis [109]. Generally, membranes for battery application can be classified into porous membranes, cation exchange membranes (CEMs), AEMs, and acid-doped membranes.

6.1 Porous membranes

Ultrafiltration membranes can be used as porous membranes in RFBs with polymeric redox couples. For instance, a commercial Celgard 2400 membrane (39% porosity and 28 nm average pore diameter) rejected 86.3% of a redox-active polymer at a polymer

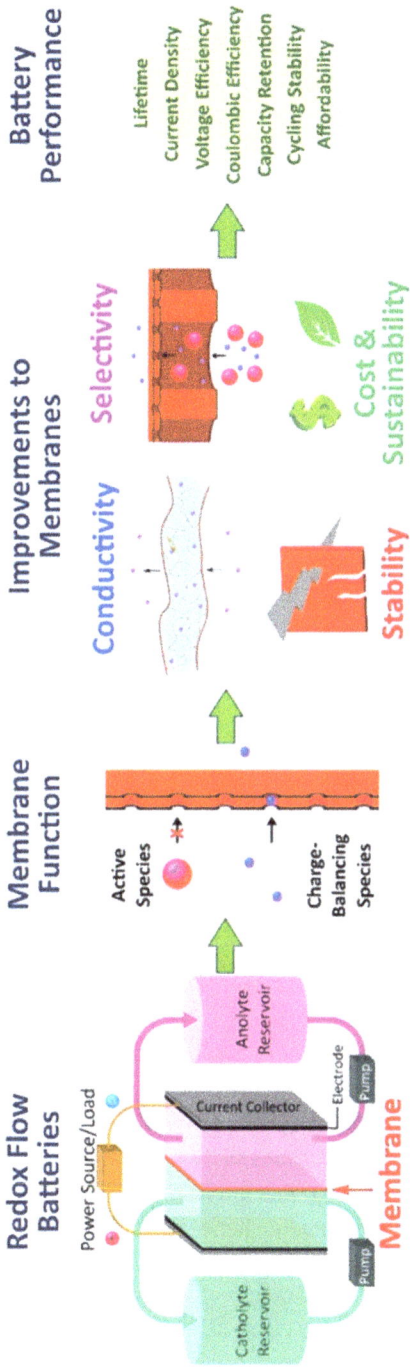

Figure 11: Schematic of RFB membranes allowing transport of selected ions while keeping redox-active species separate. Improvements to membranes will yield enhanced battery performance (reprinted with permission from literature [111]. Copyright (2021) American Chemical Society).

molecular weight of 318 kDa [112]. Further, in nanofiltration membranes, the size exclusion effect can be used to separate protons and vanadium ions [113, 114]. To this end, a poly(acrylonitrile) membrane has been prepared using a phase inversion process. Recently, ultrafiltration membranes have been considered for RFBs application. For example Wessling and coworkers [115] utilized a PIM-1 layer to coat a porous polyacrylonitrile membrane. The PIM-1 was designed to exhibit a high free volume with a rigid, kinked polymer backbone. The nanosized distance between the polymer chains is hydrophobic, which limits its water absorption and blocks large vanadium ions, while enabling the entry of protons into the polymer and their interaction with the water molecules in the nanopores of PIM-1.

6.2 Cation exchange membranes (CEMs)

CEMs consist of polymers substituted with acidic groups, such as sulfonic or phosphoric acids, which easily dissociate into a mobile cation and immobilized counter ion. Variation in the hydrophilicities of the backbone (hydrophobic) and the functional groups (hydrophilic) results in phase separation into hydrophobic domains (which provide mechanical integrity) and hydrophilic domains (which provide conductivity) [109]. CEMs exhibit several advantages such as commercial availability, high conductivity, high chemical stability, and structural variety. However, CEMs are susceptible to the crossover of metals (e.g., vanadium ions). The most prominent examples of CEMs membrane are Nafion and sulfonated poly(ether–ether–ketone) (sPEEK). sPEEK-based composite proton exchange membranes regulated by local sIPN have been developed for vanadium flow battery [116]. The membrane demonstrated outstanding proton conductivity and ion selectivity as well as superior energy efficiency against Nafion membranes. Nanofibrous sPEEK IPN membranes have also been employed as efficient materials for vanadium RFB [117].

6.3 Anion exchange membranes (AEMs)

The positively charged functional groups of AEMs electrostatically repel cations to inhibit crossover. The suitability of AEMs for vanadium/cerium RFBs systems has been demonstrated by Ramani and coworkers [118]. They found that employing trimethylammonium-functionalized poly(ether ketone)-based generated a chemically and mechanically stable AEMs membrane and prevented the mixing of the active species. Compared to Nafion system, which exhibits a capacity retention of 55% after 20 cycles, the AEMs-based membrane exhibited a capacity loss of <5%. However, the alkaline degradation of the polymer backbone and low alkaline stability of the cationic groups have limited the effective application of AEMs-based membranes. Jung et al. [119] developed polydiallylammonium interpenetrating cationic network ion-solvating membranes for anion exchange membrane water electrolyzers, which demonstrated excellent performance and high alkaline stability compared to a commercial anion exchange membrane. Moreover, an sIPN anion exchange membrane based on quaternized polyepichlorohydrin and PVA demonstrated excellent acid stability, mechanical, and thermal properties making it a suitable candidate for acid recovery by diffusion dialysis [120].

6.4 Interpenetrating polymer networks (IPNs)

The application of IPN-based membranes in batteries has attracted attention owing to their extraordinary mechanical, thermal, and chemical stability to withstand harsh alkaline environment. Particularly, the application of IPN-based membranes in RFBs has attracted increased attention owing to their beneficial operation conditions and rapid development in the past decades. To ensure the superior stability of vanadium RFBs, sIPN membranes have been developed using PSF and PVP as a hydrophobic and hydrophilic polymer, respectively [121]. In addition, PSF have been commonly applied as sIPNs in vanadium RFB systems in the fabrication of amphoteric membranes [122, 123].

In addition to polymeric IPN membranes, mixed-matrix IPNs have been developed for application in RFB systems. Shin et al. [124] reported the use of PVDF polymeric support and the introduction of silica nanoparticles into the membranes to achieve the low crossover of active species. The fabrication of sIPNs through the addition of three different monomers enables the control of the membrane structure from dense to porous and from aliphatic or aromatic.

IPN membranes are broadly applied for improving the performance of lithium-ion batteries. For example, in situ UV-assisted reactions have been employed to fabricate IPN polymer electrolyte membranes using modified poly(ethylene glycol), plasticizers, and a lithium salt to form MMMs with improved ionic conductivity [125, 126]. Also, IL-supported IPNs have been developed using UV-induced polymerization [127]. However, in addition to UV-cured IPNs, thermally induced free radical polymerized sIPNs have been designed by the in situ crosslinking of a dimethacrylate oligomer with a long thermoplastic linear PEO chain and a lithium salt [128]. Outstanding chemical and oxidation stability of sIPNs have been reported for lithium metal, indicating a safe and long-lasting polymer electrolyte owing to the incorporation of the advantageous properties of both precursors.

In situ polymerization has been widely employed for the fabrication of sIPN membranes for lithium batteries as a single-ion conducting membrane, which commonly involves the addition of a metal ion into the precursors [129, 130]. Nevertheless, in situ sol-gel crosslinking has been employed to develop single-ion conductors for lithium batteries [131]. Therefore, the development of IPNs membranes for lithium batteries predominantly focuses on MMMs.

In summary, several types of membrane have been explored as a separator for battery systems. Among the types of membranes discussed in this section, IPN-based membranes exhibit interesting properties as a battery separator owing to its intrinsic chemical and mechanical stability that makes it withstand harsh alkaline environment. One of the notable examples is reported by Wang et al. [132] where they developed high-performance PVDF separators based on crystallization template for lithium-ion batteries. Nevertheless, further research still needs to be conducted, particularly to address the need for a smart membrane separator that can respond to external stimuli, such as temperature or voltage, which will enable the development of advanced controllable batteries.

7 Conclusion and outlooks

7.1 Conclusion

The use of two polymer networks enables the control of the properties of IPN-based membranes with mechanical robustness, chemical stability, and tunable separation performance. By tailoring the chemistry of the constituent polymers in IPNs, a synergy can be achieved, whereby the performance exceeds that of the constituent neat polymers. Numerous methods have been developed to fabricate IPN-based membranes that can offer process flexibility depending on the objectives. These methods include monomer immersion, solution-casting, sequential crosslinking, in situ crosslinking, radiation crosslinking, and thermal crosslinking. IPN membranes have demonstrated excellent performance for gas separation, water purification, and nonaqueous separation with suitable membrane mechanical properties. The separation performance of IPNs exhibits a strong dependence on the structural and composition characteristics of the constituent linear and crosslinked polymer chains. Thus, the performance can be fine-tuned by tailoring the molecular structure of both polymers. Particularly, IPN-based membranes exhibit interesting properties for battery separators owing to its intrinsic chemical and mechanical stability that enables its resistance toward harsh alkaline environment. Moreover, the presence of two distinct polymers that constitute the IPNs enable the separate functionalization of each polymer, such that the mechanical and chemical strength of one polymer is fine-tuned, and the conductivity of the other is fine-tuned. In summary, achieving a good polymer–polymer interlocking balance for stability, while optimizing the separation performances has emerged as a key aspect to enable the superiority of IPN membranes to polymer blend membranes.

7.2 Outlooks

One of the important aspects of membrane applications is the long-term stability upon continuous operation. This data will ensure the feasibility of the membranes for industrial applications and commercialization. Therefore, for future research on IPNs-based membranes, the aging and long-term performance, as well as plasticization resistance, should be investigated. Moreover, despite the potential benefits of IPN membranes owing to its interlocking network between the polymer chains, the interpenetrated species of sIPNs configuration are susceptible to leaching. Thus, investigating the leaching potential of semi-IPNs, particularly after long-term operation, is essential. Further research should be conducted, particularly to address the need for smart IPN-based membrane separators that can respond to external stimuli, such as temperature, pH, solvent, or voltage, which will enable the development of advanced controllable membranes for molecular separations or battery applications. The need for precise molecular sieving in membrane technology has also directed research toward fine-tuning the pore size of the membranes. This area of IPN-based membranes is yet to be explored, for example, by fabricating MMM IPNs and MOFs. The pore regularity in MOFs can facilitate precise molecular sieving and the entangled polymers in IPNs can potentially interact with the MOFs particles to enable the formation of defect-free membrane and to avoid the formation of interfacial hole between the MOFs and the polymer. This will be expected to enhance the selectivity of the filtration membranes. Further, the use of machine learning and artificial intelligence to systematically explore the potential polymer candidates to achieve particular characteristics of IPNs is a promising research direction. The advancement in computer science along with a more available and accessible database will emerge as valuable resources to virtually and effectively design potential IPNs compared to conventional trial and error laboratory experiment. Another research area that is worth investigating is molecular simulations to understand the interaction between the two polymers governing the IPNs from molecular level. Molecular simulations can also be applied to study the transport phenomena in IPNs, which is very relevant for their applications in membranes for liquid and gas separations, as well as in battery applications. Last, in line with the sustainability principles, the needs for alternative polymer and crosslinker sources from greener or naturally derived materials are highly demanded. Moreover, IPN-based membranes fabrication can be further explored by employing green solvents.

References

[1]   Silverstein, M. S. Interpenetrating polymer networks: So happy together? Polymer, 2020, **207**: 122929. https://doi.org/10.1016/j.polymer.2020.122929.
[2]   Roland, C. M. Interpenetrating polymer networks (IPN): Structure and mechanical behavior. Encycl. Polym. Nanomater., 2013, 1–9. https://doi.org/10.1007/978-3-642-36199-9_91-1.

[3]   Rohman, G., Lauprêtre, F., Boileau, S., Guérin, P., and Grande, D. Poly(D,L-lactide)/poly(methyl methacrylate) interpenetrating polymer networks: Synthesis, characterization, and use as precursors to porous polymeric materials. Polymer, 2007, **48**(24): 7017–7028. https://doi.org/10. 1016/j.polymer.2007.09.044.

[4]   Rohman, G., Grande, D., Lauprêtre, F., Boileau, S., and Guérin, P. Design of porous polymeric materials from interpenetrating polymer networks (IPNs): Poly(DL-lactide)/poly(methyl methacrylate)-based semi-IPN systems. Macromolecules, 2005, **38**(17): 7274–7285. https://doi.org/10.1021/ma0501390.

[5]   Widmaier, J.-M., and Sperling, L. H. The use of labile crosslinker for morphological studies in interpenetrating polymer networks. Br. Polym. J., 1984, **16**(1): 46–48. https://doi.org/10.1002/pi. 4980160109.

[6]   Utroša, P., Žagar, E., Kovačič, S., and Pahovnik, D. Porous polystyrene monoliths prepared from in situ simultaneous interpenetrating polymer networks: Modulation of morphology by polymerization kinetics. Macromolecules, 2019, **52**(3): 819–826. https://doi.org/10.1021/acs.macro mol.8b01923.

[7]   Sperling, L. H. Interpenetrating polymer networks: An overview. Adv. Chem.: Am. Chem. Soc., 1994, **239**: 3–38. https://doi.org/10.1021/ba-1994-0239.ch001.

[8]   Ha, S. M., Yuan, W., Pei, Q., Pelrine, R., and Stanford, S. Interpenetrating polymer networks for high-performance electroelastomer artificial muscles. Adv. Mater., 2006, **18**(7): 887–891. https://doi.org/ 10.1002/adma.200502437.

[9]   Chikh, L., Delhorbe, V., and Fichet, O. (Semi-)interpenetrating polymer networks as fuel cell membranes. J. Membr. Sci., 2011, **368**(1): 1–17. https://doi.org/10.1016/j.memsci.2010.11.020.

[10]   Turner, J. S., and Cheng, Y.-L. Preparation of PDMS–PMAA interpenetrating polymer network membranes using the monomer immersion method. Macromolecules, 2000, **33**(10): 3714–3718. https://doi.org/10.1021/ma991873k.

[11]   Kim, N. U., Park, B. J., Park, M. S., Park, J. T., and Kim, J. H. Semi-interpenetrating polymer network membranes based on a self-crosslinkable comb copolymer for CO2 capture. Chem. Eng. J., 2019, **360**: 1468–1476. https://doi.org/10.1016/j.cej.2018.10.152.

[12]   Xue, J., Liu, L., Liao, J., Shen, Y., and Li, N. Semi-interpenetrating polymer networks by azide–alkyne cycloaddition as novel anion exchange membranes. J. Mater. Chem. A, 2018, **6**(24): 11317–11326. https://doi.org/10.1039/C8TA02177E.

[13]   Wang, J., He, R., and Che, Q. Anion exchange membranes based on semi-interpenetrating polymer network of quaternized chitosan and polystyrene. J. Colloid Interface Sci., 2011, **361**(1): 219–225. https://doi.org/10.1016/j.jcis.2011.05.039.

[14]   Zhao, D., Kim, J. F., Ignacz, G., Pogany, P., Lee, Y. M., and Szekely, G. Bio-inspired robust membranes nanoengineered from interpenetrating polymer networks of polybenzimidazole/polydopamine. ACS Nano, 2019, **13**(1): 125–133. https://doi.org/10.1021/acsnano.8b04123.

[15]   Liu, X., Zhang, Y., Deng, S., Li, C., Dong, J., Wang, J., Yang, Z., Wang, D., and Cheng, H. Semi-interpenetrating polymer network membranes from SPEEK and BPPO for high concentration DMFC. ACS Appl. Energy Mater., **2018**, acsaem.8b01041. https://doi.org/10.1021/acsaem.8b01041.

[16]   Palencia, M., Vera, M., and Rivas, B. L. Modification of ultrafiltration membranes via interpenetrating polymer networks for removal of boron from aqueous solution. J. Membr. Sci., 2014, **466**: 192–199. https://doi.org/10.1016/j.memsci.2014.05.003.

[17]   Zhou, B., Pu, H., Pan, H., and Wan, D. Proton exchange membranes based on semi-interpenetrating polymer networks of Nafion® and poly(vinylidene fluoride) via radiation crosslinking. Int. J. Hydrogen Energy, 2011, **36**(11): 6809–6816. https://doi.org/10.1016/j.ijhydene.2011.02.115.

[18]   Wang, C., Wang, Z., Yang, F., and Wang, J. Improving the permselectivity and antifouling performance of reverse osmosis membrane based on a semi-interpenetrating polymer network. Desalination, 2021, **502**: 114910. https://doi.org/10.1016/j.desal.2020.114910.

[19] Malik, R. S., Soni, U., Chauhan, S. S., Kumar, D., and Choudhary, V. Semi-interpenetrating polymer networks of poly (vinyl alcohol)-functionalized nanocrystals/sulfonated poly (ether ether ketone) (PVA-FNCs/SPEEK) as fuel cell membrane. Mater. Today Commun., 2021, **29**: 102897. https://doi.org/10.1016/j.mtcomm.2021.102897.

[20] Al Munsur, A. Z., Goo, B.-H., Kim, Y., Kwon, O. J., Paek, S. Y., Lee, S. Y., Kim, H.-J., and Kim, T.-H. Nafion-based proton-exchange membranes built on cross-linked semi-interpenetrating polymer networks between poly(acrylic acid) and poly(vinyl alcohol). ACS Appl. Mater. Interfaces, 2021, **13**(24): 28188–28200. https://doi.org/10.1021/acsami.1c05662.

[21] Lee, W. H., Seong, J. G., Bae, J. Y., Wang, H. H., Moon, S. J., Jung, J. T., Do, Y. S., Kang, H., Park, C. H., and Lee, Y. M. Thermally rearranged semi-interpenetrating polymer network (TR-SIPN) membranes for gas and olefin/paraffin separation. J. Membr. Sci., 2021, **625**: 119157. https://doi.org/10.1016/j.memsci.2021.119157.

[22] Iulianelli, A., and Drioli, E. Membrane engineering: Latest advancements in gas separation and pre-treatment processes, petrochemical industry and refinery, and future perspectives in emerging applications. Fuel Process. Technol., 2020, **206**: 106464. https://doi.org/10.1016/j.fuproc.2020.106464.

[23] Sanaeepur, H., Ebadi Amooghin, A., Bandehali, S., Moghadassi, A., Matsuura, T., and Van der Bruggen, B. Polyimides in membrane gas separation: Monomer's molecular design and structural engineering. Prog. Polym. Sci., 2019, **91**: 80–125. https://doi.org/10.1016/j.progpolymsci.2019.02.001.

[24] Ma, X., Abdulhamid, M., Miao, X., and Pinnau, I. Facile synthesis of a hydroxyl-functionalized Tröger's base diamine: A new building block for high-performance polyimide gas separation membranes. Macromolecules, 2017, **50**(24): 9569–9576. https://doi.org/10.1021/acs.macromol.7b02301.

[25] Abdulhamid, M. A., Genduso, G., Wang, Y., Ma, X., and Pinnau, I. Plasticization-resistant carboxyl-functionalized 6FDA-polyimide of intrinsic microporosity (PIM-PI) for membrane-based gas separation. Ind. Eng. Chem. Res., 2020, **59**(12): 5247–5256. https://doi.org/10.1021/acs.iecr.9b04994.

[26] Abdulhamid, M. A., Genduso, G., Ma, X., and Pinnau, I. Synthesis and characterization of 6FDA/3,5-diamino-2,4,6-trimethylbenzenesulfonic acid-derived polyimide for gas separation applications. Sep. Purif. Technol., 2021, **257**: 117910. https://doi.org/10.1016/j.seppur.2020.117910.

[27] Abdulhamid, M. A., Ma, X., Ghanem, B. S., and Pinnau, I. Synthesis and characterization of organo-soluble polyimides derived from alicyclic dianhydrides and a dihydroxyl-functionalized spirobisindane diamine. ACS Appl. Polym. Mater., 2019, **1**(1): 63–69. https://doi.org/10.1021/acsapm.8b00036.

[28] Weigelt, F., Georgopanos, P., Shishatskiy, S., Filiz, V., Brinkmann, T., and Abetz, V. Development and characterization of defect-free Matrimid® mixed-matrix membranes containing activated carbon particles for gas separation. Polymer, 2018, **10**(1): 51. https://doi.org/10.3390/polym10010051.

[29] Liu, G., Chernikova, V., Liu, Y., Zhang, K., Belmabkhout, Y., Shekhah, O., Zhang, C., Yi, S., Eddaoudi, M., and Koros, W. J. Mixed matrix formulations with MOF molecular sieving for key energy-intensive separations. Nat. Mater., 2018, **17**(3): 283–289. https://doi.org/10.1038/s41563-017-0013-1.

[30] Cheng, Y., Ying, Y., Zhai, L., Liu, G., Dong, J., Wang, Y., Christopher, M. P., Long, S., Wang, Y., and Zhao, D. Mixed matrix membranes containing MOF@COF hybrid fillers for efficient CO_2/CH_4 separation. J. Membr. Sci., 2019, **573**: 97–106. https://doi.org/10.1016/j.memsci.2018.11.060.

[31] Kurdi, J., and Kumar, A. Synthesis and characterization of modified bismaleimide/polysulfone semi-interpenetrating polymer networks. J. Appl. Polym. Sci., 2006, **102**(1): 369–379. https://doi.org/10.1002/app.23946.

[32] Kurdi, J., and Kumar, A. Formation and thermal stability of BMI-based interpenetrating polymers for gas separation membranes. J. Membr. Sci., 2006, **280**(1–2): 234–244. https://doi.org/10.1016/j.memsci.2006.01.024.

[33] Kline, G. K., Zhang, Q., Weidman, J. R., and Guo, R. PEO-rich semi-interpenetrating polymer network (s-IPN) membranes for CO2 separation. J. Membr. Sci., 2017, **544**: 143–150. https://doi.org/10.1016/j. memsci.2017.09.005.

[34] Wiegand, J. R., Smith, Z. P., Liu, Q., Patterson, C. T., Freeman, B. D., and Guo, R. Synthesis and characterization of triptycene-based polyimides with tunable high fractional free volume for gas separation membranes. J. Mater. Chem. A, 2014, **2**(33): 13309–13320. https://doi.org/10.1039/ c4ta02303j.

[35] Li, S., Zhang, K., Liu, C., Feng, X., Wang, P., and Wang, S. Nanohybrid Pebax/PEGDA-GPTMS membrane with semi-interpenetrating network structure for enhanced CO_2 separations. J. Membr. Sci., 2023, **674**: 121516. https://doi.org/10.1016/j.memsci.2023.121516.

[36] Saimani, S., Dal-Cin, M. M., Kumar, A., and Kingston, D. M. Separation performance of asymmetric membranes based on PEGDa/PEI semi-interpenetrating polymer network in pure and binary gas mixtures of CO_2, N_2 and CH_4. J. Membr. Sci., 2010, **362**(1–2): 353–359. https://doi.org/10.1016/j.mem sci.2010.06.045.

[37] Low, B. T., Chung, T. S., Chen, H., Jean, Y. C., and Pramoda, K. P. Tuning the free volume cavities of polyimide membranes via the construction of pseudo-interpenetrating networks for enhanced gas separation performance. Macromolecules, 2009, **42**(18): 7042–7054. https://doi.org/10.1021/ ma901251y.

[38] Low, B. T., and Chung, T. S. Carbon molecular sieve membranes derived from pseudo-interpenetrating polymer networks for gas separation and carbon capture. Carbon, 2011, **49**(6): 2104–2112. https://doi.org/10.1016/j.carbon.2011.01.045.

[39] Gouveia, A. S. L., Bumenn, E., Rohtlaid, K., Michaud, A., Vieira, T. M., Alves, V. D., Tomé, L. C., Plesse, C., and Marrucho, I. M. Ionic liquid-based semi-interpenetrating polymer network (sIPN) membranes for CO2 separation. Sep. Purif. Technol., 2021, **274**: 118437. https://doi.org/10.1016/j.sep pur.2021.118437.

[40] He, S., Kamio, E., Zhang, J., Matsuoka, A., Nakagawa, K., Yoshioka, T., and Matsuyama, H. Development of an ion gel membrane containing a CO2-philic ionic liquid in interpenetrating semi-crystalline and crosslinkable polymer networks. J. Membr. Sci., 2023, **685**: 121912. https://doi.org/ 10.1016/j.memsci.2023.121912.

[41] Deng, J., Dai, Z., Yan, J., Sandru, M., Sandru, E., Spontak, R. J., and Deng, L. Facile and solvent-free fabrication of PEG-based membranes with interpenetrating networks for CO_2 separation. J. Membr. Sci., 2019, **570–571**: 455–463. https://doi.org/10.1016/j.memsci.2018.10.031.

[42] Sun, T., Zheng, W., Chen, J., Dai, Y., Li, X., Ruan, X., Yan, X., and He, G. Nanofibers interpenetrating network mimicking "reinforced-concrete" to construct mechanically robust composite membrane for enhanced CO2 separation. J. Membr. Sci., 2021, **639**: 119749. https://doi.org/10.1016/j.memsci. 2021.119749.

[43] Liu, N., Cheng, J., Liu, L., Gao, S., Hou, W., Luo, M., Zhang, H., Ye, B., and Zhou, J. Crosslinking two-dimensional metalloporphyrin (Me-TCPP) nanosheet with poly(ethylene) glycol semi-interpenetrating polymer network for ultrahigh CO2/N2 separation selectivity via "rubber-band". Straightening Effect. J. Membr. Sci., 2023, **676**: 121537. https://doi.org/10.1016/j.memsci.2023.121537.

[44] McNair, R., Szekely, G., and Dryfe, R. A. W. Ion-exchange materials for membrane capacitive deionization. ACS EST Water, 2021, **1**(2): 217–239. https://doi.org/10.1021/acsestwater.0c00123.

[45] Suss, M. E., Porada, S., Sun, X., Biesheuvel, P. M., Yoon, J., and Presser, V. Water desalination via capacitive deionization: What is it and what can we expect from it? Energy Environ. Sci., 2015, **8**(8): 2296–2319. https://doi.org/10.1039/C5EE00519A.

[46] Mazinani, S., Al-Shimmery, A., Chew, Y. M. J., and Mattia, D. 3D printed fouling-resistant composite membranes. ACS Appl. Mater. Interfaces, 2019, **11**(29): 26373–26383. https://doi.org/10.1021/ac sami.9b07764.

[47] Chowdhury, M. R., Steffes, J., Huey, B. D., and McCutcheon, J. R. 3D printed polyamide membranes for desalination. Science, 2018, **361**(6403): 682–686. https://doi.org/10.1126/science.aar2122.

[48] Boretti, A., Al-Zubaidy, S., Vaclavikova, M., Al-Abri, M., Castelletto, S., and Mikhalovsky, S. Outlook for graphene-based desalination membranes. NPJ Clean Water, 2018, **1**(1): 5. https://doi.org/10.1038/s41545-018-0004-z.

[49] Qadir, D., Mukhtar, H., and Keong, L. K. Mixed matrix membranes for water purification applications. Sep. Purif. Rev., 2017, **46**(1): 62–80. https://doi.org/10.1080/15422119.2016.1196460.

[50] Figoli, A., Marino, T., Simone, S., Nicolò, E. D., and Drioli, E. Towards non-toxic solvents for membrane preparation: A review. Green Chem., 2014, **26**: 4034–4059. https://doi.org/10.1039/C4GC00613E.

[51] Shi, Q., Zhang, K., Lu, R., and Jiang, J. Water desalination and biofuel dehydration through a thin membrane of polymer of intrinsic microporosity: Atomistic simulation study. J. Membr. Sci., 2018, **545**: 49–56. https://doi.org/10.1016/j.memsci.2017.09.057.

[52] Alnajrani, M. N., and Alsager, O. A. Removal of antibiotics from water by polymer of intrinsic microporosity: Isotherms, kinetics, thermodynamics, and adsorption mechanism. Sci. Rep., 2020, **10**(1): 794. https://doi.org/10.1038/s41598-020-57616-4.

[53] Wan, L.-S., Xu, Z.-K., and Wang, Z.-G. Leaching of PVP from polyacrylonitrile/PVP blending membranes: A comparative study of asymmetric and dense membranes. J. Polym. Sci. B Polym. Phys., 2006, **44**(10): 1490–1498. https://doi.org/10.1002/polb.20804.

[54] Marbelia, L., Bilad, M. R., and Vankelecom, I. F. J. Gradual PVP leaching from PVDF/PVP blend membranes and its effects on membrane fouling in membrane bioreactors. Sep. Purif. Technol., 2019, **213**: 276–282. https://doi.org/10.1016/j.seppur.2018.12.045.

[55] Mavukkandy, M. O., Bilad, M. R., Giwa, A., Hasan, S. W., and Arafat, H. A. Leaching of PVP from PVDF/PVP blend membranes: Impacts on membrane structure and fouling in membrane bioreactors. J. Mater. Sci., 2016, **51**(9): 4328–4341. https://doi.org/10.1007/s10853-016-9744-7.

[56] Lou, D., Hou, Z., Yang, H., Liu, Y., and Wang, T. Antifouling membranes prepared from polyethersulfone grafted with poly(ethylene glycol) methacrylate by radiation-induced copolymerization in homogeneous solution. ACS Omega, 2020, **5**(42): 27094–27102. https://doi.org/10.1021/acsomega.0c02439.

[57] Folgado, E., Ladmiral, V., and Semsarilar, M. Towards permanent hydrophilic PVDF membranes. Amphiphilic PVDF-*b*-PEG-*b*-PVDF triblock copolymer as membrane additive. Eur. Polym. J., 2020, **131**: 109708. https://doi.org/10.1016/j.eurpolymj.2020.109708.

[58] Matsuyama, H., Maki, T., Teramoto, M., and Kobayashi, K. Effect of PVP additive on porous polysulfone membrane formation by immersion precipitation method. Sep. Sci. Technol., 2003, **38**(14): 3449–3458. https://doi.org/10.1081/SS-120023408.

[59] Ma, Y., Chen, X., Wang, S., Dong, H., Zhai, X., Shi, X., Wang, J., Ma, R., and Zhang, W. Significantly enhanced antifouling and separation capabilities of PVDF membrane by synergy of semi-interpenetrating polymer and TiO_2 gel nanoparticles. J. Ind. Eng. Chem., **2021**: S1226086X21006420. https://doi.org/10.1016/j.jiec.2021.11.038.

[60] Chen, F.-F., Su, T., Zhao, X.-T., Pan, J.-F., and Liu, L.-F. A rigid-flexible interpenetrating polyamide reverse osmosis membrane with improved antifouling property fabricated via two step modifications. J. Membr. Sci., 2021, **637**: 119625. https://doi.org/10.1016/j.memsci.2021.119625.

[61] Polisetti, V., and Ray, P. Thin film composite nanofiltration membranes with polystyrene sodium sulfonate–polypiperazinetrimesamide semi-interpenetrating polymer network active layer. J. Appl. Polym. Sci., 2020, **137**(44): 49351. https://doi.org/10.1002/app.49351.

[62] Zhao, W., Huang, J., Fang, B., Nie, S., Yi, N., Su, B., Li, H., and Zhao, C. Modification of polyethersulfone membrane by blending semi-interpenetrating network polymeric nanoparticles. J. Membr. Sci., 2011, **369**(1–2): 258–266. https://doi.org/10.1016/j.memsci.2010.11.065.

[63] Miao, Y., Wang, C., Wang, J., and Wang, Z. A novel UV-initiated modification process for fabricating high-performance TFC RO membrane. J. Membr. Sci., 2023, **666**: 121158. https://doi.org/10.1016/j.memsci.2022.121158.

[64] Sen Gupta, R., Mandal, S., Arya, S., Dutta, S., Manna, K., Safikul Islam, S., Pathan, S., and Bose, S. Copper-substituted polyoxometalate-soldered interpenetrating polymeric networks membranes for water remediation. Chem. Eng. J., 2023, **461**: 141949. https://doi.org/10.1016/j.cej.2023.141949.

[65] Wang, S., Zhou, Y., and Gao, C. Novel high boron removal polyamide reverse osmosis membranes. J. Membr. Sci., 2018, **554**: 244–252. https://doi.org/10.1016/j.memsci.2018.03.014.

[66] Dundua, A., Franzka, S., and Ulbricht, M. Improved antifouling properties of polydimethylsiloxane films via formation of polysiloxane/polyzwitterion interpenetrating networks. Macromol. Rapid Commun., 2016, **37**(24): 2030–2036. https://doi.org/10.1002/marc.201600473.

[67] Tian, S., Jiang, D., Pu, J., Sun, X., Li, Z., Wu, B., Zheng, W., Liu, W., and Liu, Z. A new hybrid silicone-based antifouling coating with nanocomposite hydrogel for durable antifouling properties. Chem. Eng. J., 2019, **370**: 1–9. https://doi.org/10.1016/j.cej.2019.03.185.

[68] Wang, Z.-Y., Fu, Z.-J., Shao, -D.-D., Lu, M.-J., Xia, Q.-C., Xiao, H.-F., Su, B.-W., and Sun, S.-P. Bridging the miscibility gap to fabricate delamination-free dual-layer nanofiltration membranes via incorporating fluoro substituted aromatic amine. J. Membr. Sci., 2020, **610**: 118270. https://doi.org/10.1016/j.memsci.2020.118270.

[69] Qu, Z.-T., Duan, S.-Y., Li, -B.-B., Sun, D., and Gu, Y.-L. PDMS/PVDF microporous membrane with semi-interpenetrating polymer networks for vacuum membrane distillation. J. Appl. Polym. Sci., 2018, **135**(8): 45792. https://doi.org/10.1002/app.45792.

[70] Sun, D., Liu, M.-Q., Guo, J.-H., Zhang, J.-Y., Li, -B.-B., and Li, D.-Y. Preparation and characterization of PDMS-PVDF hydrophobic microporous membrane for membrane distillation. Desalination, 2015, **370**: 63–71. https://doi.org/10.1016/j.desal.2015.05.017.

[71] Hu, J., Chen, Y., Lu, J., Fan, X., Li, J., Li, Z., Zeng, G., and Liu, W. A self-supported gel filter membrane for dye removal with high anti-fouling and water flux performance. Polymer, 2020, **201**: 122531. https://doi.org/10.1016/j.polymer.2020.122531.

[72] Zhang, Y., Li, S., Huang, R., He, J., Sun, Y., Qin, Y., and Shen, L. Stabilizing MXene-based nanofiltration membrane by forming analogous semi-interpenetrating network architecture using flexible poly(acrylic acid) for effective wastewater treatment. J. Membr. Sci., 2022, **648**: 120360. https://doi.org/10.1016/j.memsci.2022.120360.

[73] Dutta, S., Sen Gupta, R., Manna, K., Safikul Islam, S., and Bose, S. 'Green-tea' extract soldered triple interpenetrating polymer network membranes for water remediation. Chem. Eng. J., 2023, **472**: 145008. https://doi.org/10.1016/j.cej.2023.145008.

[74] Nozad, E., Poursattar Marjani, A., and Mahmoudian, M. A novel and facile semi-IPN system in fabrication of solvent resistant nano-filtration membranes for effective separation of dye contamination in water and organic solvents. Sep. Purif. Technol., 2022, **282**: 120121. https://doi.org/10.1016/j.seppur.2021.120121.

[75] Vargas-Figueroa, C., Pino-Soto, L., Beratto-Ramos, A., Rivas, B. L., Tapiero, Y., Palacio, D. A., Melendrez, M. F., and Borquez, R. Surface modification of nanofiltration membranes by interpenetrating polymer networks and their evaluation in water desalination. ACS Appl. Polym. Mater., 2023, **5**(7): 4910–4920. https://doi.org/10.1021/acsapm.3c00501.

[76] Bessbousse, H., Verchère, J.-F., and Lebrun, L. Characterisation of metal-complexing membranes prepared by the semi-interpenetrating polymer networks technique. Application to the removal of heavy metal ions from aqueous solutions. Chem. Eng. J., 2012, **187**: 16–28. https://doi.org/10.1016/j.cej.2011.12.079.

[77] Han, Z., Cheng, C., Zhang, L., Luo, C., Nie, C., Deng, J., Xiang, T., and Zhao, C. Toward robust pH-responsive and anti-fouling composite membranes via one-pot in-situ cross-linked copolymerization. Desalination, 2014, **349**: 80–93. https://doi.org/10.1016/j.desal.2014.06.025.

[78] Dragan, E. S. Design and applications of interpenetrating polymer network hydrogels. A review. Chem. Eng. J., 2014, **243**: 572–590. https://doi.org/10.1016/j.cej.2014.01.065.

[79] Wang, W., Kang, Y., and Wang, A. One-step fabrication in aqueous solution of a granular alginate-based hydrogel for fast and efficient removal of heavy metal ions. J. Polym. Res., 2013, **20**(3): 101. https://doi.org/10.1007/s10965-013-0101-0.

[80] Wang, T., Turhan, M., and Gunasekaran, S. Selected properties of pH-sensitive, biodegradable chitosan–poly(vinyl alcohol) hydrogel. Polym. Int., 2004, **53**(7): 911–918. https://doi.org/10.1002/pi.1461.

[81] Galizia, M., and Bye, K. P. Advances in organic solvent nanofiltration rely on physical chemistry and polymer chemistry. Front. Chem., 2018, **6**: 511. https://doi.org/10.3389/fchem.2018.00511.

[82] Hermans, S., Mariën, H., Van Goethem, C., and Vankelecom, I. F. Recent developments in thin film (nano)composite membranes for solvent resistant nanofiltration. Curr. Opin. Chem. Eng., 2015, **8**: 45–54. https://doi.org/10.1016/j.coche.2015.01.009.

[83] Valtcheva, I. B., Kumbharkar, S. C., Kim, J. F., Bhole, Y., and Livingston, A. G. Beyond polyimide: Crosslinked polybenzimidazole membranes for organic solvent nanofiltration (OSN) in harsh environments. J. Membr. Sci., 2014, **457**: 62–72. https://doi.org/10.1016/j.memsci.2013.12.069.

[84] Li, C., Li, S., Lv, L., Su, B., and Hu, M. Z. High solvent-resistant and integrally crosslinked polyimide-based composite membranes for organic solvent nanofiltration. J. Membr. Sci., 2018, **564**: 10–21. https://doi.org/10.1016/j.memsci.2018.06.048.

[85] Tsuru, T., Miyawaki, M., Kondo, H., Yoshioka, T., and Asaeda, M. Inorganic porous membranes for nanofiltration of nonaqueous solutions. Sep. Purif. Technol., 2003, **32**(1–3): 105–109. https://doi.org/10.1016/S1383-5866(03)00074-1.

[86] Akbari, A., Meragawi, S. E., Martin, S. T., Corry, B., Shamsaei, E., Easton, C. D., Bhattacharyya, D., and Majumder, M. Solvent transport behavior of shear aligned graphene oxide membranes and implications in organic solvent nanofiltration. ACS Appl. Mater. Interfaces, 2018, **10**(2): 2067–2074. https://doi.org/10.1021/acsami.7b11777.

[87] Jue, M. L., Koh, D.-Y., McCool, B. A., and Lively, R. P. Enabling widespread use of microporous materials for challenging organic solvent separations. Chem. Mater., 2017, **29**(23): 9863–9876. https://doi.org/10.1021/acs.chemmater.7b03456.

[88] Cook, M., Gaffney, P. R. J., Peeva, L. G., and Livingston, A. G. Roll-to-roll dip coating of three different PIMs for organic solvent nanofiltration. J. Membr. Sci., 2018, **558**: 52–63. https://doi.org/10.1016/j.memsci.2018.04.046.

[89] Wu, X. M., Guo, H., Soyekwo, F., Zhang, Q. G., Lin, C. X., Liu, Q. L., and Zhu, A. M. Pervaporation purification of ethylene glycol using the highly permeable PIM-1 membrane. J. Chem. Eng. Data, 2016, **61**(1): 579–586. https://doi.org/10.1021/acs.jced.5b00731.

[90] Li, B., Japip, S., and Chung, T.-S. Molecularly tunable thin-film nanocomposite membranes with enhanced molecular sieving for organic solvent forward osmosis. Nat. Commun., 2020, **11**(1): 1198. https://doi.org/10.1038/s41467-020-15070-w.

[91] Marchetti, P., Solomon, M. F. J., Szekely, G., and Livingston, A. G. Molecular separation with organic solvent nanofiltration: A critical review. Chem. Rev., **2014**: 72.

[92] James, J., Thomas, G. V., Pramoda, K. P., and Thomas, S. Transport behaviour of aromatic solvents through styrene butadiene rubber/poly [methyl methacrylate] (SBR/PMMMA) interpenetrating polymer network (IPN) membranes. Polymer, 2017, **116**: 76–88. https://doi.org/10.1016/j.polymer.2017.03.063.

[93] Johns, J., and Nakason, C. Novel interpenetrating polymer networks based on natural rubber/poly (vinyl alcohol). Polym.-Plast. Technol. Eng., 2012, **51**(10): 1046–1053. https://doi.org/10.1080/03602559.2012.689053.

[94] Altun, V., Remigy, J.-C., and Vankelecom, I. F. J. UV-cured polysulfone-based membranes: Effect of co-solvent addition and evaporation process on membrane morphology and SRNF performance. J. Membr. Sci., 2017, **524**: 729–737. https://doi.org/10.1016/j.memsci.2016.11.060.

[95] Liu, M., Wu, Y., Wu, Y., Gao, M., Lü, Z., Yu, S., and Gao, C. Cross-flow deposited hydroxyethyl cellulose (HEC)/polypropylene (PP) thin-film composite membrane for aqueous and non-aqueous nanofiltration. Chem. Eng. Res. Des., 2020, **153**: 572–581. https://doi.org/10.1016/j.cherd.2019.11.003.

[96] Puspasari, T., Chakrabarty, T., Genduso, G., and Peinemann, K.-V. Unique cellulose/polydimethylsiloxane blends as an advanced hybrid material for organic solvent nanofiltration and pervaporation membranes. J. Mater. Chem. A, 2018, **6**(28): 13685–13695. https://doi.org/10.1039/C8TA02697A.

[97] Nguyen, Q. T., M'Bareck, C. O., David, M. O., Métayer, M., and Alexandre, S. Ion-exchange membranes made of semi-interpenetrating polymer networks, used for pervaporation-assisted esterification and ion transport. Mater. Res. Innovations, 2003, **7**(4): 212–219. https://doi.org/10.1007/s10019-003-0253-3.

[98] Garg, P., Singh, R. P., and Choudhary, V. Selective polydimethylsiloxane/polyimide blended IPN pervaporation membrane for methanol/toluene azeotrope separation. Sep. Purif. Technol., 2011, **76**(3): 407–418. https://doi.org/10.1016/j.seppur.2010.11.012.

[99] Van den Mooter, P.-R., Daems, N., and Vankelecom, I. F. J. Preparation of solvent resistant supports through formation of a semi-interpenetrating polysulfone/polyacrylate network using UV cross-linking – Part 1: Selection of optimal UV curing conditions. React. Funct. Polym., 2019, **136**: 189–197. https://doi.org/10.1016/j.reactfunctpolym.2018.12.015.

[100] Van den Mooter, P.-R., Daems, N., and Vankelecom, I. F. J. Preparation of solvent resistant supports through formation of a semi-interpenetrating polysulfone/polyacrylate network using UV cross-linking – Part 2: Optimization of synthesis parameters for UV-LED curing. React. Funct. Polym., 2020, **146**: 104403. https://doi.org/10.1016/j.reactfunctpolym.2019.104403.

[101] Ulbricht, M. Membrane separations using molecularly imprinted polymers. J. Chromatogr. B, 2004, **804**(1): 113–125. https://doi.org/10.1016/j.jchromb.2004.02.007.

[102] Wackerlig, J., and Schirhagl, R. Applications of molecularly imprinted polymer nanoparticles and their advances toward industrial use: A review. Anal. Chem., 2016, **88**(1): 250–261. https://doi.org/10.1021/acs.analchem.5b03804.

[103] Székely, G., Valtcheva, I. B., Kim, J. F., and Livingston, A. G. Molecularly imprinted organic solvent nanofiltration membranes – Revealing molecular recognition and solute rejection behaviour. React. Funct. Polym., 2015, **86**: 215–224. https://doi.org/10.1016/j.reactfunctpolym.2014.03.008.

[104] Lehmann, M., Brunner, H., and Tovar, G. E. M. Selective separations and hydrodynamic studies: A new approach using molecularly imprinted nanosphere composite membranes. Desalination, 2002, **149**(1–3): 315–321. https://doi.org/10.1016/S0011-9164(02)00754-3.

[105] Gkementzoglou, C., Kotrotsiou, O., and Kiparissides, C. Synthesis of novel composite membranes based on molecularly imprinted polymers for removal of triazine herbicides from water. Ind. Eng. Chem. Res., 2013, **52**(39): 14001–14010. https://doi.org/10.1021/ie400479c.

[106] Melvin Ng, H. K., Leo, C. P., and Abdullah, A. Z. Selective removal of dyes by molecular imprinted TiO_2 nanoparticles in polysulfone ultrafiltration membrane. J. Environ. Chem. Eng., 2017, **5**(4): 3991–3998. https://doi.org/10.1016/j.jece.2017.07.075.

[107] Jantarat, C., Tangthong, N., Songkro, S., Martin, G. P., and Suedee, R. S-Propranolol imprinted polymer nanoparticle-on-microsphere composite porous cellulose membrane for the enantioselectively controlled delivery of racemic propranolol. Int. J. Pharm., 2008, **349**(1–2): 212–225. https://doi.org/10.1016/j.ijpharm.2007.07.030.

[108] Yuan, X.-Z., Song, C., Platt, A., Zhao, N., Wang, H., Li, H., Fatih, K., and Jang, D. A review of all-vanadium redox flow battery durability: Degradation mechanisms and mitigation strategies. Int. J. Energy Res., 2019, **43**(13): 6599–6638. https://doi.org/10.1002/er.4607.

[109] Ye, R., Henkensmeier, D., Yoon, S. J., Huang, Z., Kim, D. K., Chang, Z., Kim, S., and Chen, R. Redox flow batteries for energy storage: A technology review. J. Electrochem. Energy Conversion Storage, 2017, **15**: 1. https://doi.org/10.1115/1.4037248.

[110] Wu, Y., and Holze, R. Electrocatalysis at electrodes for vanadium redox flow batteries. Batteries, 2018, **4**(3): 47. https://doi.org/10.3390/batteries4030047.

[111] Machado, C. A., Brown, G. O., Yang, R., Hopkins, T. E., Pribyl, J. G., and Epps, T. H. Redox flow battery membranes: Improving battery performance by leveraging structure–property relationships. ACS Energy Lett., 2021, **6**(1): 158–176. https://doi.org/10.1021/acsenergylett.0c02205.

[112] Nagarjuna, G., Hui, J., Cheng, K. J., Lichtenstein, T., Shen, M., Moore, J. S., and Rodríguez-López, J. Impact of redox-active polymer molecular weight on the electrochemical properties and transport across porous separators in nonaqueous solvents. J. Am. Chem. Soc., 2014, **136**(46): 16309–16316. https://doi.org/10.1021/ja508482e.

[113] Zhang, H., Zhang, H., Li, X., Mai, Z., and Zhang, J. Nanofiltration (NF) membranes: The next generation separators for all vanadium redox flow batteries (VRBs)? Energy Environ. Sci., 2011, **4**(5): 1676–1679. https://doi.org/10.1039/C1EE01117K.

[114] Zhang, H., Zhang, H., and Li, X. Nanofiltration membranes for vanadium flow battery application. ECS Trans., 2013, **53**(7): 65. https://doi.org/10.1149/05307.0065ecst.

[115] Chae, I. S., Luo, T., Moon, G. H., Ogieglo, W., Kang, Y. S., and Wessling, M. Ultra-high proton/vanadium selectivity for hydrophobic polymer membranes with intrinsic nanopores for redox flow battery. Adv. Energy Mater., 2016, **6**(16): 1600517. https://doi.org/10.1002/aenm.201600517.

[116] Qian, P., Li, L., Wang, H., Sheng, J., Zhou, Y., and Shi, H. SPEEK-based composite proton exchange membrane regulated by local semi-interpenetrating network structure for vanadium flow battery. J. Membr. Sci., 2022, **662**: 120973. https://doi.org/10.1016/j.memsci.2022.120973.

[117] Qian, P., Wang, H., Sheng, J., Zhou, Y., and Shi, H. Ultrahigh proton conductive nanofibrous composite membrane with an interpenetrating framework and enhanced acid-base interfacial layers for vanadium redox flow battery. J. Membr. Sci., 2022, **647**: 120327. https://doi.org/10.1016/j.memsci.2022.120327.

[118] Yun, S., Parrondo, J., and Ramani, V. A vanadium–cerium redox flow battery with an anion-exchange membrane separator. ChemPlusChem., 2015, **80**(2): 412–421. https://doi.org/10.1002/cplu.201402096.

[119] Jung, J., Park, Y. S., Hwang, D. J., Choi, G. H., Choi, D. H., Park, H. J., Ahn, C.-H., Hwang, S. S., and Lee, A. S. Polydiallylammonium interpenetrating cationic network ion-solvating membranes for anion exchange membrane water electrolyzers. J. Mater. Chem. A, 2023, **11**(20): 10891–10900. https://doi.org/10.1039/D3TA01511D.

[120] Gong, Y., Chen, W., Shen, H. Y., and Cheng, C. Semi-interpenetrating polymer-network anion exchange membrane based on quaternized polyepichlorohydrin and polyvinyl alcohol for acid recovery by diffusion dialysis. Ind. Eng. Chem. Res., 2023, **62**(13): 5624–5634. https://doi.org/10.1021/acs.iecr.3c00026.

[121] Zeng, L., Zhao, T. S., Wei, L., Zeng, Y. K., and Zhang, Z. H. Polyvinylpyrrolidone-based semi-interpenetrating polymer networks as highly selective and chemically stable membranes for all vanadium redox flow batteries. J. Power Sources, 2016, **327**: 374–383. https://doi.org/10.1016/j.jpowsour.2016.07.081.

[122] Yu, H., Xia, Y., Zhang, H., Gong, X., Geng, P., Gao, Z., and Wang, Y. Improved chemical stability and proton selectivity of semi-interpenetrating polymer network amphoteric membrane for vanadium redox flow battery application. J. Appl. Polym. Sci., 2021, **138**(6): 49803. https://doi.org/10.1002/app.49803.

[123] Gan, R., Ma, Y., Li, S., Zhang, F., and He, G. Facile fabrication of amphoteric semi-interpenetrating network membranes for vanadium flow battery applications. J. Energy Chem., 2018, **27**(4): 1189–1197. https://doi.org/10.1016/j.jechem.2017.09.017.

[124] Shin, S.-H., Kim, Y., Yun, S.-H., Maurya, S., and Moon, S.-H. Influence of membrane structure on the operating current densities of non-aqueous redox flow batteries: Organic–inorganic composite membranes based on a semi-interpenetrating polymer network. J. Power Sources, 2015, **296**: 245–254. https://doi.org/10.1016/j.jpowsour.2015.07.045.

[125] Wang, L., Li, N., He, X., Wan, C., and Jiang, C. Macromolecule plasticized interpenetrating structure solid state polymer electrolyte for lithium ion batteries. Electrochim. Acta, 2012, **68**: 214–219. https://doi.org/10.1016/j.electacta.2012.02.067.

[126] Dam, T., Jena, S. S., and Ghosh, A. Ion dynamics, rheology and electrochemical performance of UV cross-linked gel polymer electrolyte for Li-Ion battery. J. Appl. Phys., 2019, **126**(10): 105104. https://doi.org/10.1063/1.5112149.

[127] More, S. S., Khupse, N. D., Ambekar, J. D., Kulkarni, M. V., and Kale, B. B. Ionic liquid-supported interpenetrating polymer network flexible solid electrolytes for lithium-ion batteries. Energy Fuels, 2022, **36**(9): 4999–5008. https://doi.org/10.1021/acs.energyfuels.2c00551.

[128] Nair, J. R., Destro, M., Bella, F., Appetecchi, G. B., and Gerbaldi, C. Thermally cured semi-interpenetrating electrolyte networks (s-IPN) for safe and aging-resistant secondary lithium polymer batteries. J. Power Sources, 2016, **306**: 258–267. https://doi.org/10.1016/j.jpowsour.2015.12.001.

[129] Huo, S., He, Y., Hu, Z., Bao, W., Chen, W., Wang, Y., Zeng, D., Cheng, H., and Zhang, Y. New insights into designation of single-ion conducting gel polymer electrolyte for high-performance lithium metal batteries. J. Membr. Sci., 2022, **647**: 120287. https://doi.org/10.1016/j.memsci.2022.120287.

[130] Pan, Q., Jiang, S., Li, Z., Liu, Y., Du, Y., Zhao, N., Zhang, Y., and Liu, J.-M. Highly porous single ion conducting membrane via a facile combined "structural self-assembly" and in-situ polymerization process for high performance lithium metal batteries. J. Membr. Sci., 2021, **636**: 119601. https://doi.org/10.1016/j.memsci.2021.119601.

[131] Hu, Z., Ji, F., Zhang, Y., Guo, W., Jing, X., Bao, W., Qin, J., Huo, S., Li, S., Zhang, Y., Fan, W., and Cheng, H. Siloxane-type single-ion conductors enable composite solid polymer electrolyte membranes with fast Li+ transporting networks for dendrite-proof lithium-metal batteries. Chem. Eng. J., 2023, **468**: 143857. https://doi.org/10.1016/j.cej.2023.143857.

[132] Wang, J., Shen, J., Shi, J., Li, Y., You, J., and Bian, F. Crystallization-templated high-performance PVDF separator used in lithium-ion batteries. J. Membr. Sci., 2023, **670**: 121359. https://doi.org/10.1016/j.memsci.2023.121359.

[133] Cavalcante, J., Oldal, D. G., Peskov, M. V., Beke, A. K., Hardian, R., Schwingenschlögl, U., and Szekely, G. Biobased interpenetrating polymer network membranes for sustainable molecular sieving. ACS Nano, 2024. https://doi.org/10.1021/acsnano.3c10827.

Index

https://doi.org/10.1515/9783110796032-013

www.ingramcontent.com/pod-product-compliance
Lightning Source LLC
Chambersburg PA
CBHW080703220326
41598CB00033B/5292